The
Biochemistry
of Plants

A COMPREHENSIVE TREATISE

Volume 10

P. K. Stumpf and E. E. Conn

EDITORS-IN-CHIEF

*Department of Biochemistry
and Biophysics
University of California
Davis, California*

Volume 1 **The Plant Cell**   *N. E. Tolbert, Editor*

Volume 2 **Metabolism and Respiration**   *David D. Davies, Editor*

Volume 3 **Carbohydrates: Structure and Function**   *Jack Preiss, Editor*

Volume 4 **Lipids: Structure and Function**   *P. K. Stumpf, Editor*

Volume 5 **Amino Acids and Derivatives**   *B. J. Miflin, Editor*

Volume 6 **Proteins and Nucleic Acids**   *Abraham Marcus, Editor*

Volume 7 **Secondary Plant Products**   *E. E. Conn, Editor*

Volume 8 **Photosynthesis**   *M. D. Hatch and N. K. Boardman, Editors*

Volume 9 **Lipids: Structure and Function**   *P. K. Stumpf, Editor*

Volume 10 **Photosynthesis**   *M. D. Hatch and N. K. Boardman, Editors*

Volume 11 **Biochemistry of Metabolism**   *David D. Davies, Editor*

Volume 12 **Physiology of Metabolism**   *David D. Davies, Editor*

Volume 13 **Methodology**   *David D. Davies, Editor*

Volume 14 **Carbohydrates**   *Jack Preiss, Editor*

Volume 15 **Molecular Biology**   *Abraham Marcus, Editor*

Volume 16 **Intermediary Nitrogen Metabolism**   *B. J. Miflin, Editor*

THE BIOCHEMISTRY OF PLANTS

A COMPREHENSIVE TREATISE

Volume 10
Photosynthesis

M. D. Hatch
and N. K. Boardman, editors

Division of Plant Industry
Commonwealth Scientific and Industrial Research Organization
Canberra City, Australia

ACADEMIC PRESS, INC.
Harcourt Brace Jovanovich, Publishers
San Diego New York Berkeley Boston
London Sydney Tokyo Toronto

COPYRIGHT © 1987 BY ACADEMIC PRESS, INC.
ALL RIGHTS RESERVED.
NO PART OF THIS PUBLICATION MAY BE REPRODUCED OR
TRANSMITTED IN ANY FORM OR BY ANY MEANS, ELECTRONIC
OR MECHANICAL, INCLUDING PHOTOCOPY, RECORDING, OR
ANY INFORMATION STORAGE AND RETRIEVAL SYSTEM, WITHOUT
PERMISSION IN WRITING FROM THE PUBLISHER.

ACADEMIC PRESS, INC.
1250 Sixth Avenue, San Diego, California 92101

United Kingdom Edition published by
ACADEMIC PRESS INC. (LONDON) LTD.
24–28 Oval Road, London NW1 7DX

Library of Congress Cataloging in Publication Data
(Revised for vol. 10)

The Biochemistry of plants.

    Includes bibliographies and indexes.
    Contents: v. 1. The plant cell.—v. 2. Metabolism
and respiration.—[etc.]—v. 10. Photosynthesis.
    1. Botanical chemistry. I. Stumpf, Paul K.
(Paul Karl), Date . II. Conn, Eric E.
QK861.B48   581.19'2   80-13168
ISBN 0—12—675410—1 (v. 10) (alk. paper)

PRINTED IN THE UNITED STATES OF AMERICA

87 88 89 90   9 8 7 6 5 4 3 2 1

Contents

List of Contributors   ix

General Preface   xi

Preface to Volume 10   xiii

1   The Molecular Basis of Chloroplast Development
    J. KENNETH HOOBER
       I.   Introduction   2
      II.   The Chloroplast Genome   4
    III.   Replication of the Chloroplast Genome   14
    IV.   Regulation of Plastid Gene Expression   15
     V.   Nuclear Genes that Encode Plastid Proteins   21
    VI.   Phytochrome   31
   VII.   Advances in Chlorophyll Synthesis   34
  VIII.   The Chloroplast Envelope   39
    IX.   Processing of Cytoplasmically Made Precursors of
           Plastid Proteins   44
     X.   Assembly of Complexes in Thylakoid Membranes   48
    XI.   Conclusions   61
          References   62

                                                              v

2   Composition, Organization, and Dynamics of the Thylakoid Membrane in Relation to Its Function
J. BARBER
I.	Introduction	76
II.	Thylakoid Lipids	78
III.	Thylakoid Proteins	84
IV.	Thylakoid Membrane Organization	105
V.	Thylakoid Membrane Dynamics	110
VI.	Final Comment	119
	References	121

3   Rubisco: Structure, Mechanisms, and Prospects for Improvement
T. JOHN ANDREWS AND GEORGE H. LORIMER
I.	Introduction	132
II.	The Catalytic Bifunctionality of Rubisco: A Fixed Constraint in the Architecture of Life?	133
III.	Structure and Synthesis	139
IV.	Mechanisms of Carbamylation and Catalysis	153
V.	Mechanisms for Modulating Activity	189
VI.	Subunit Interactions	195
VII.	Evolution of Rubisco	205
VIII.	Prospects for a Better Rubisco	209
	References	211

4   The CO_2-Concentrating Mechanism in Aquatic Phototrophs
MURRAY R. BADGER
I.	Introduction	220
II.	Historical Perspectives	221
III.	The Significance of the CO_2-Concentrating Mechanism	223
IV.	The Essential Components of the CO_2-Concentrating Mechanism	224
V.	The CO_2-Concentrating Mechanism in Cyanobacteria and Green Microalgae	226
VI.	Induction of the CO_2-Concentrating Mechanism	247
VII.	Evidence for the CO_2-Concentrating Mechanism in Other Aquatic Phototrophs	254
VIII.	The Ecological Importance of the CO_2-Concentrating Mechanism	264
	References	270

5 Biochemistry of C$_3$–C$_4$ Intermediates
GERALD E. EDWARDS AND MAURICE S. B. KU

       I.   Introduction   276

I.	Introduction	276
II.	Occurrence: Families, Genera, and Species	279
III.	Properties of Intermediates	281
IV.	Carbon Isotope Composition of Intermediates Relative to Mechanism of Photosynthesis	307
V.	Biochemistry of Photosynthesis in Intermediates in Relation to Developmental and Environmental Factors	310
VI.	Genetically Based Differences in the Biochemistry of Photosynthesis within an Intermediate Species	312
VII.	Features of Hybrids between Different Photosynthetic Types	312
VIII.	A Theoretical Scheme of Evolution of C$_4$ Photosynthesis Based on Intermediate Species	316
IX.	Concluding Remarks	320
	References	322

6 Control of Photosynthetic Sucrose Formation
MARK STITT, STEVE HUBER, AND PHIL KERR

I.	Introduction	328
II.	Biosynthesis of Sucrose in Leaves	329
III.	Metabolite Levels during Sucrose Synthesis	338
IV.	Regulatory Properties of the Individual Steps in Sucrose Synthesis	349
V.	Regulation of Fructose 2,6-Bisphosphate Concentration	367
VI.	Coarse Control of SPS	377
VII.	Coordinate Control of Sucrose Formation by SPS and the Cytosolic Fructose 1,6-Bisphosphatase	382
VIII.	Adaptation of the Regulation of Sucrose Synthesis	390
IX.	Limitation of Photosynthesis by Sucrose Synthesis	392
	References	403

Index   411

Contents of Other Volumes   422

List of Contributors

Numbers in parentheses indicate the pages on which the authors' contributions begin.

T. John Andrews (131), Department of Environmental Biology, Research School of Biological Sciences, Australian National University, Canberra City, A.C.T. 2601, Australia

Murray R. Badger (219), Department of Environmental Biology, Research School of Biological Sciences, Australian National University, Canberra City, A.C.T. 2601, Australia

J. Barber (75), AFRC Photosynthesis Research Group, Department of Pure and Applied Biology, Imperial College of Science and Technology, London SW7 2BB, England

Gerald E. Edwards (275), Department of Botany, Washington State University, Pullman, Washington 99164

J. Kenneth Hoober (1), Department of Biochemistry, Temple University School of Medicine, Philadelphia, Pennsylvania 19141

Steve Huber (327), USDA/ARS Departments of Crop Science and Botany, North Carolina State University, Raleigh, North Carolina 27650

Phil Kerr (327), Central Research and Development Department, E.I. Du Pont de Nemours and Company, Wilmington, Delaware 19898

Maurice S. B. Ku (275), Department of Botany, Washington State University, Pullman, Washington 99164

George H. Lorimer (131), Central Research and Development Department, E.I. Du Pont de Nemours and Company, Wilmington, Delaware 19898

Mark Stitt (327), Lehrstuhl für Pflanzenphysiologie, Universität Bayreuth, D-8580 Bayreuth, Federal Republic of Germany

General Preface

In 1950, a new book entitled "Plant Biochemistry" was authored by James Bonner and published by Academic Press. It contained 490 pages, and much of the information described therein referred to animal or bacterial systems. This book had two subsequent editions, in 1965 and 1976.

In 1980, our eight-volume series entitled "The Biochemistry of Plants: A Comprehensive Treatise" was published by Academic Press; this multivolume, multiauthored treatise contained 4670 pages.

Since 1980, the subject of plant biochemistry has expanded into a vigorous discipline that penetrates all aspects of agricultural research. Recently a large number of research-oriented companies have been formed to explore and exploit the discipline of plant biochemistry, and older established chemical companies have also become heavily involved in plant-oriented research. With this in mind, Academic Press and the editors-in-chief of the treatise felt it imperative to update these volumes. Rather than have each chapter completely rewritten, it was decided to employ the approach used so successfully by the editors of *Methods in Enzymology,* in which contributors are invited to update those areas of research that are most rapidly expanding. In this way, the 1980 treatise constitutes a set of eight volumes with much background information, while the new volumes both update subjects that are rapidly developing and discuss some wholly new areas. The editors-in-chief have therefore invited the editors of the 1980 volumes to proceed on the basis of this concept. As a result, new volumes are forthcoming on lipids; general metabolism, including respiration; carbohydrates; amino acids; molecular biology; and photosynthesis. Additional volumes will be added as the need arises.

Once again we thank our editorial colleagues for accepting the important task of selecting authors to update chapters for their volumes and bringing their volumes promptly to completion. And once again we thank Mrs. Billie Gabriel and Academic Press for their assistance in this project.

P. K. Stumpf
E. E. Conn

Preface to Volume 10

Volume 8 of this series was intended to be a reasonably complete and cohesive coverage of the biochemical aspects of photosynthesis. When we were asked by the editors-in-chief to update this volume, our task was to identify areas of photosynthesis where there had been major advances in the past five years. As a result we present here a collection of topics which are often treated in more detail than those appearing in Volume 8. The present volume should not be regarded as a cohesive unit in itself; it is meant to complement Volume 8 and certainly not to replace it.

Probably the most surprising development in the field of photosynthesis during this decade has been the realization that not only certain photosynthetic bacteria and microalgae but apparently also other aquatic phototrophs possess mechanisms for concentrating CO_2 or bicarbonate in their cells. These mechanisms provide a new dimension to our understanding of nature's battle to cope with the need to use ribulose 1,5-bisphosphate carboxylase as the key carbon-assimilating step of photosynthesis. The problem, of course, is to find ways of minimizing the wasteful oxygenase reaction catalyzed by this enzyme. Both the inorganic carbon-concentrating mechanism and the properties of ribulose 1,5-bisphosphate carboxylase are topics considered in this volume.

Some terrestrial angiosperms adopted the C_4 pathway as an alternative means of providing a sufficiently high ratio of CO_2 to O_2 to reduce oxygenase activity. One chapter in this volume deals with the remarkable group of plants with physiological and biochemical features intermediate between C_3 and C_4 plants. Initially it seemed most likely that these so-called C_3–C_4 intermediates originate when C_4 plants lose certain capacities critical to C_4 photosynthesis. It now appears much more likely that we are seeing evolution in progress; that is, at least some intermediates originate from C_3 species and are at various stages in the process of becoming C_4 plants. These intermediates often occur in environments where even partial C_4 operation may provide marginal advantage under some conditions, and where full C_4 operation would be a distinct advantage. Certainly the studies of these species are providing a fascinating insight into the likely paths of evolution of C_4 photosynthesis.

The last few years have seen enormous progress in our understanding of the molecular basis underlying the structure of the chloroplast thylakoid membrane and its biogenesis during the maturation of the chloroplast. The techniques of molecular biology have been used to elucidate the structure of many of the genes, both nuclear and chloroplast, which code for chloroplast proteins. The information for the transfer and processing of those proteins, which are synthesized in the cytoplasm but destined for the chloroplast, and the information for the location and orientation of the protein chains in the thylakoid membranes are contained in the amino acid sequences of the proteins. The chapter on chloroplast biogenesis in this volume summarizes the remarkable advances in our knowledge of the molecular processes governing the development of this organelle.

The volume also includes an authoritative account of the molecular organization of the thylakoid membrane in relation to its functions for photosynthetic electron transport and energy conservation. It emphasizes the dynamics of the thylakoid membrane including the role of protein phosphorylation and the lateral distribution of electrical charge and protein components between the thylakoids in the grana and stroma regions.

We would like to take this opportunity to thank the contributors to this volume for their authoritative accounts of these rapidly moving fields.

M. D. Hatch
N. K. Boardman

The Biochemistry of Plants

A COMPREHENSIVE TREATISE

Volume 10

The Molecular Basis of Chloroplast Development

1

J. KENNETH HOOBER

I. Introduction
II. The Chloroplast Genome
    A. Strategies for Identification of Chloroplast Genes
    B. The Chloroplast Genome in Higher Plants
    C. The Chloroplast Genome in Algae
    D. The Presence of Introns in Chloroplast Genes
    E. Plastome "Mutants"
III. Replication of the Chloroplast Genome
IV. Regulation of Plastid Gene Expression
    A. Chloroplast DNA-Dependent RNA Polymerase
    B. Chloroplast Gene Promoters
    C. Regulation of Plastid Gene Expression by Light
    D. Posttranscriptional Control of Protein Synthesis
V. Nuclear Genes that Encode Plastid Proteins
    A. *rbcS* Genes
    B. *cab* Genes
    C. Other Identified Nuclear Genes
    D. Regulation of Nuclear Gene Expression
VI. Phytochrome
VII. Advances in Chlorophyll Synthesis
    A. Synthesis of 5-Aminolevulinate
    B. Conversion of Protoporphyrin IX to Protochlorophyllide
    C. Synthesis of Chlorophyllide *a*
    D. Chlorophyll RC I
    E. Synthesis of Chlorophyllide *b*
VIII. The Chloroplast Envelope
IX. Processing of Cytoplasmically Made Precursors of Plastid Proteins
X. Assembly of Complexes in Thylakoid Membranes
    A. The Complexes
    B. The Cytochrome b_6/f Complex
    C. Plastocyanin and Other Luminal Proteins

The Biochemistry of Plants, Vol. 10

    D. The Photosystems
    E. Entry of Cab Proteins into Chloroplasts
  XI. Conclusion
       References

I. INTRODUCTION*

Over the last few years, truly remarkable advances have been made in knowledge of the molecular processes of chloroplast development. Building on the extensive foundation of morphological and biochemical data, investigators have proceeded to examine the structure of nuclear and chloroplast genes, the mechanisms that control expression of these genes, and the properties of the gene products. Although it has been known for over 20 years that both nuclear and organellar genomes encode organelle proteins, mechanisms of how this cooperation is achieved have only recently begun to emerge. A major impetus to this expansion of knowledge was the discovery that proteins derived from the nuclear DNA–cytoplasmic ribosome system are synthesized initially in precursor form by translation of poly(A)-containing mRNA (Dobberstein *et al.*, 1977). Intensive effort was then directed toward understanding the process of uptake and processing of these precursors by chloroplasts. Subsequently, genes that encode the most prominent precursor polypeptides were isolated from nuclear DNA through the use of cDNA probes that were synthesized from mRNA with reverse transcriptase (Bedbrook *et al.*, 1980; Coruzzi *et al.*, 1983). Determination of the nucleotide sequence of the DNA filled in information on the amino acid sequences of the N-terminal extensions, which are part of the precursor forms, and of the mature polypeptides.

Perhaps an even greater stimulus to work on chloroplast development was the discovery that physical maps of the circular chloroplast genome could be constructed (Hermann *et al.*, 1976; Bedbrook and Bogorad, 1976; Gray and Hallick, 1977). The fact that all chloroplast DNA molecules from a given organism are, for practical purposes, identical allowed localization of genes to specific restriction fragments and the determination of the sequences of these genes. In a short span of time, the organization of the chloroplast genome of several plant species has been outlined and the sequences of many of the plastid genes have been determined. Currently, the phenomenology of chloroplast development is being explained by the properties of

* Abbreviations: kbp, kilo base pairs; cDNA, DNA made from mRNA templates; LHC-1, LHC-2, light-harvesting complexes of photosystem 1 and photosystem 2, respectively; LHCP, apoprotein of light-harvesting complex-2; *cab*, Cab, the genes and gene products, respectively, of the chlorophyll *a/b*-binding proteins of LHC-2 (synonymous with LHCP); P_r, P_{fr}, the red-absorbing and far-red-absorbing forms of phytochrome, respectively; *rbcL*, RbcL, *rbcS*, RbcS, the genes and gene products of the large and small subunits of ribulose 1,5-bisphosphate carboxylase, respectively. Designations for plastid genes are included in the legend to Fig. 1.

the gene products as deduced from the nucleotide sequences of the genes and by the action of flanking cis elements in the DNA that regulate expression of these genes. Deduction of the properties of the gene products from the codon-derived amino acid sequences has spurred the development of hypotheses regarding the fate of the polypeptides. This has led to the design of incisive experiments that have demonstrated how the polypeptides reach their site of function and how the chloroplast structures, as a result, are built.

Current work is predicated on the assumption that knowledge of the structure of the genes encoding the necessary polypeptides and of the mechanisms that control expression of these genes will lead to an understanding of the assembly and function of the chloroplast. This is not a trivial assumption, for this approach has already paid huge dividends. But, there is much remaining to be learned, including the mechanisms of transport of proteins across the chloroplast envelope, the processing of precursors within the stroma, the sorting of proteins within the organelle to their final destinations, formation of multimolecular complexes, and the assembly of thylakoid membranes.

In this review, I have not included a description of the morphology of chloroplast development or of the accompanying biochemical changes. The discussion evolved as a survey of various developmental processes and thus is not a detailed analysis of each aspect. Unfortunately, because of the scope of this review, it cannot be exhaustive. Many excellent reviews on particular aspects of this subject have been published recently (e.g., Whitfeld and Bottomley, 1983; Tobin and Silverthorne, 1985; Cifferi and Dure, 1983; van Vloten-Doting *et al.*, 1985). A volume describing a wide spectrum of developmental aspects of the chloroplast has recently appeared (Akoyunoglou and Senger, 1986).

This review will focus mainly on development of the chloroplast in higher plants and green algae, the Chlorophyta. Information is just beginning to emerge about the biochemistry of plastid development in the rich variety of eukaryotic algae among the division Chromophyta. These latter organisms promise to be extremely interesting. Kühsel and Kowallik (1987) determined the physical map of the chloroplast genome of the brown alga *Dictyota dichotoma*. Friedman and Alberte (1986) and Fawley and Grossman (1986) found that apoproteins of the major light-harvesting complex in the chloroplast of the diatom *Phaeodactylum tricornutum* are synthesized in the cytoplasm as larger precursors just as in the Chlorophyta. These observations are particularly interesting because they pose the question of how a cytoplasmically derived protein is transferred into a chloroplast, which is surrounded by the two membranes of the chloroplast endoplasmic reticulum in addition to the two membranes of the chloroplast envelope (Gibbs, 1981; Ludwig and Gibbs, 1985).

An indication that these additional membranes around the chloroplast in chromophytic algae indeed provide a significant barrier to transport is suggested by work on ribulose bisphosphate carboxylase. Reith and Cattolico (1985a,b) found that the chloroplasts of *Olisthodiscus* contain the genes for both the large and small subunits of this enzyme, in contrast to the situation in chlorophytes in which the gene for the small subunit is in the cell nucleus. Furthermore, Plumley *et al.* (1986) found that this enzyme in several chlorophyll *c*-containing, chromophytic algae is structurally and antigenically different from the enzyme in chlorophytes. There is a rich diversity among the chromophytic algae, which will continue to provide intriguing questions regarding chloroplast development for years to come.

II. THE CHLOROPLAST GENOME

A. Strategies for Identification of Chloroplast Genes

Since the discovery that a map of restriction endonuclease cleavage sites could be constructed with chloroplast DNA (Bedbrook and Bogorad, 1976; Gray and Hallick, 1978; Whitfeld *et al.,* 1978; Rochaix, 1978), intense effort has been focused on transforming this physical map into a genetic map. Several strategies have been used to determine the location of genes on the chloroplast genome. The simplest procedure is hybridization of purified RNA molecules with restriction fragments, although this procedure is limited to abundant species such as ribosomal and transfer RNAs. Identification of genes that code for proteins, however, is more complex. The first approach toward this end took advantage of the phenomenon of light induction of a large number of chloroplast genes (Bedbrook *et al.,* 1978; Rodermel and Bogorad, 1985). Restriction fragments were identified that hybridized with RNA species that increased in amount in cells exposed to light. Further characterization of a fragment was accomplished by selection of the mRNA complementary to one strand of the fragment, translation of the mRNA after dissociation of the hybrid (*hybrid select–translation*), and then identification of the product by screening for immunoprecipitation with antibodies against known proteins. Alternatively, addition of a denatured restriction fragment to an *in vitro* system translating chloroplast RNA would inhibit synthesis of polypeptides coded by sequences in the fragment (*hybrid arrest–translation*). The exact position of a gene and the amino acid sequence of the gene product then were obtained as "open reading frames" after determination of the nucleotide sequence of the DNA.

A third strategy for gene identification, which is a variation of the second, involved amplification of a restriction fragment through cloning in *Escherichia coli* and analysis of expression of genes in the fragment by *in vitro* transcription–translation experiments. The presence of genes on a fragment can be determined more simply, however, by screening a population of *E.*

coli cells that contain a library of chloroplast DNA fragments. Clones expressing chloroplast genes then are identified by immunological procedures with antibodies against known proteins as probes.

A fourth approach has been afforded by the vast bank of known gene sequences, particularly from bacterial cells. Nucleotide sequencing of chloroplast DNA has revealed a number of open reading frames, which have been identified through a search for nucleotide and amino acid sequence homology with known prokaryotic (usually *E. coli*) genes or gene products. Once a gene is characterized in one plant species, the same gene in others is identified by nucleotide homology. This approach has been highly successful and efficient, but is restricted to those genes that have prokaryotic counterparts with sufficient homology to allow unequivocal identification. Fortunately, this seems to apply to most chloroplast genes.

A fifth means of gene identification has recently been developed (Fish *et al.*, 1985; Steinmetz *et al.*, 1986). Antibodies can now be generated against synthetic peptide epitopes that correspond to a segment of a protein known only by its amino acid sequence as deduced from the nucleotide sequence of an open reading frame. The antibody then becomes the probe for identification of the location and function of the protein. As more such open reading frames are found, as the result of continued sequencing of the chloroplast genome, this procedure should prove to be a powerful tool. In particular, this latter approach should allow identification of minor polypeptides that are difficult to characterize by conventional means.

B.   The Chloroplast Genome in Higher Plants

The size of the chloroplast genome from land plants, with few exceptions, is within the range of 120–160 kbp. Among the algae, however, variation is much wider, with the range extending at least from 86 kbp to 292 kbp (Palmer, 1985). The size of the genome in organisms such as *Acetabularia* may even exceed this range (Schweiger *et al.*, 1986). As the number of species for which data on chloroplast DNA is available continues to expand rapidly, substantial differences may well appear, which will make building tidy schemes for evolution of the chloroplast genome difficult but interesting.

The milestone of determining the entire nucleotide sequence was recently achieved for the 121,024-bp DNA molecule in liverwort chloroplasts (Ohyama *et al.*, 1986) and for the 155,844-bp molecule in tobacco chloroplasts (Shinozaki *et al.*, 1986b). The total potential coding capacity for the chloroplast genome can now be established for these organisms. Contained within the molecule are genes for the four ribosomal RNAs (23 S, 16 S, 5 S and 4.5 S), 37 transfer RNA genes encoding 32 species (liverwort and tobacco differ in the number of such genes), and open reading frames for possibly 55 polypeptides ranging in length from 31 to 2136 amino acids. Of this latter group, 40–45 have been positively identified by their coding sequences, which include those for 19 ribosomal proteins and 18–20 thylakoid membrane proteins.

Fish and Jagendorf (1982) and Nivison and Jagendorf (1984) found that 39 membrane polypeptides and about 60 soluble proteins were radioactively labeled by protein synthesis in isolated pea chloroplasts. The total number of labeled polypeptides thus exceeded the number of genes for polypeptides in the chloroplast genome. Partly responsible for the excessive number of polypeptides may have been processing of precursor forms of some proteins. Mullet *et al.* (1986) provided a further clarification of this situation by discovering that a portion of the translation products in isolated chloroplasts resulted from pausing of ribosomes at discrete sites on plastid mRNA. The cause of the pausing is not known, but their conclusion was that perhaps only eight soluble and 18 membrane proteins are synthesized in chloroplasts. Whether the synthesis of ribosomal proteins was detected in these experiments was not clear. Not all the gene products of chloroplast DNA have been identified, but it appears that there is reasonable agreement between the number of proteins synthesized in the organelle and the number of genes determined to reside in the DNA.

Genetic maps of the circular chloroplast DNA molecule usually are dominated by a pair of inverted repeats of 10–26 kbp, within which are found the ribosomal RNA genes. Figure 1 shows the relative positions of genes for a number of proteins on linearized maps of the large single-copy region of genomes from maize, spinach, wheat, pea, and liverwort. The maps in Fig. 1 are incomplete and are intended to show only relative patterns of gene order in these few species. Most angiosperms have a highly conserved gene order in the single-copy regions, which is typified by the spinach genome (145 kbp). Genes for thylakoid membrane proteins were the first major group to be placed on the map, although the *rbcL* gene, which encodes the large subunit of ribulose bisphosphate carboxylase, a soluble protein, also was among the first to be identified.

Not shown on the maps are 19 known genes for ribosomal proteins, which have been identified by a number of laboratories (Sijben-Müller *et al.*, 1986; Fromm *et al.*, 1986; Torazawa *et al.*, 1986; Shinozaki *et al.*, 1986a,b; Posno *et al.*, 1986; Ohyama *et al.*, 1986). This number agrees with previous results that indicated that about one-third of a total of 50–55 such proteins are synthesized within the organelle (Eneas-Filho *et al.*, 1981; Schmidt *et al.*, 1983). The gene for the protein synthesis elongation factor Tu$_{chl}$ (*tufA*) occurs in chloroplast DNA in *Euglena* (Passavant *et al.*, 1983; Montandon and Stutz, 1984), but not in the organelle DNA of higher plants. Genes for the protein synthesis initiation factor IF-1 (*infA*) and four subunits of RNA polymerase are in chloroplast DNA (Sijben-Müller *et al.*, 1986; Ohyama *et al.*, 1986), but none have been found for DNA polymerase. Whether additional genes for factors involved in protein synthesis and for properties of

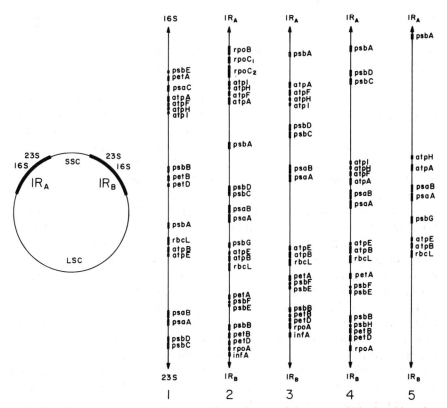

Fig. 1. The order and approximate positions of some of the genes within the chloroplast genome from 1, pea; 2, liverwort; 3, spinach; 4, wheat; and 5, maize. The general pattern of the circular plastid genome from most higher plants, with two inverted repeat regions designated IR_A and IR_B, is illustrated on the left. The ribosomal RNA genes are located within the repeat regions, as indicated. The inverted repeats separate the small single-copy region (SSC) from the large single-copy region (LSC). The linear maps on the right represent the large single-copy regions, between IR_A (top) and IR_B (bottom). The map for pea, which does not contain inverted repeats, is positioned such that the ribosomal RNA genes are oriented analogous to those in IR_A on the other genomes. The maps show mostly genes for thylakoid proteins and do not include the relatively large number of genes for ribosomal proteins and tRNAs. Clusters of genes are generally similar between organisms, but the order and orientation of the clusters differ markedly. Maps were based on data presented in the following: pea (Cozens *et al.*, 1986; Berends *et al.*, 1986; Lehmbeck *et al.*, 1986), liverwort (Ohyama *et al.*, 1986); spinach (Gruissem and Zurawski, 1985; Hennig and Herrmann, 1986; Sijben-Müller *et al.*, 1986), wheat (Hird *et al.*, 1986), and maize (Bogorad *et al.*, 1983; Rodermel and Bogorad, 1985; Steinmetz *et al.*, 1986). Genes are designated by the abbreviated nomenclature and encode the following proteins: *atpA*, *atpB*, *atpE*, subunits α, β, and ε of ATP synthase (CF_1), respectively; *atpF*, *atpH*, *atpI*, subunits I, III, and IV of the membrane component (CF_0) of ATP synthase, respectively; *psaA*, *psaB*, photosystem 1 chlorophyll a-binding apoproteins (P_{700}); *psaC*, possibly another P_{700} apoprotein; *psbA*, M_r 32,000 herbicide-binding protein of photosystem 2; *psbB*, "M_r 51,000" photosystem 2 chlorophyll a-binding apoprotein; *psbC*, "M_r44,000" photosystem 2 chlorophyll a-binding apoprotein; *psbD*, photosystem 2 D2 protein; *psbE*, *psbF*, cytochrome b_{559} and an associated 4000-dalton polypeptide, respectively; *psbG*, unidentified photosystem 2 protein; *petA*, cytochrome f; *petB*, cytochrome b_{563} (b_6); *petD*, subunit IV of the cytochrome b_6/f complex; *rbcL*, large subunit of ribulose bisphosphate carboxylase; *rpoA*, *rpoB*, *rpoC*, subunits α, β, and β' of RNA polymerase, respectively; *infA*, protein synthesis initiation factor IF-1.

plastid derivatives other than the chloroplast, such as the chromoplast (Piechulla *et al.*, 1985), are in the chloroplast genome of any organisms remains to be determined.

Considerable effort has been directed toward identification of tRNA genes (Bergmann *et al.*, 1984; Hallick *et al.*, 1984; Chu *et al.*, 1985; Mubumbila *et al.*, 1985; Gruissem and Zurawski, 1985; Steinmetz and Weil, 1986). For a few amino acids, genes encoding more than one isoaccepting species of tRNA exist. Yet with only a total number of 32 different species of tRNA (Crouse *et al.*, 1986; Ohyama *et al.*, 1986), this number could support protein synthesis only if considerable wobble is allowed, that is, if only the first two positions provide the requisite fit in decoding the mRNA, or if posttranscriptional modifications occur to yield different species. It must be noted that the chloroplast, in contrast to the mitochondria, uses the universal genetic code.

Although chloroplast genome maps for most organisms still are incomplete, it seems that basically the same set of genes is in all. However, it is interesting to compare the positions of several of the genes. The distributions of most of the genes within the circular genomes of maize, spinach, and wheat are similar. Each is characterized by the two large inverted repeats, which contain the genes for ribosomal RNAs and several tRNAs. This inverted repeat structure may tend to stabilize the genome (Palmer and Thompson, 1982; Palmer, 1983), although homologous recombination apparently can occur within the repeats, which reverses the orientation of the small single-copy region.

Several sets of closely linked genes exist, which are transcribed as units (Palmer, 1985). However, a few genes are positioned differently among the various organisms, apparently the result of inversions. In spinach, the *atp* operon A, which includes genes *atpI, atpH, atpF,* and *atpA,* is located near, and transcribed from the same strand as, the *psbA* gene. This operon in maize and wheat, however, is located near the *psaA* gene and in opposite orientation from *psbA*. The orientation of the *psbC* and *psbD* genes is the same in wheat (and perhaps maize) as in spinach, but their positions relative to the *atp* operon A is reversed.

The general patterns exemplified by maize, spinach, and wheat, which have been extensively studied, seem to extend to other angiosperm species in which chloroplast DNA contains two sets of ribosomal RNA genes in inverted repeats. For example, genes in the tobacco genome (Sugita *et al.*, 1984; Torazawa *et al.*, 1986; Shinozaki *et al.*, 1986b) are at positions similar to those on the spinach genome. However, the map for the pea chloroplast genome is quite different from that of the other four shown in Fig. 1. The pea genome is smaller (120 kbp) and contains only one set of ribosomal genes (Palmer and Thompson, 1981; Chu and Tewari, 1982). The general gene order and relative orientations also show little similarity to the others. The pattern for the pea genome extends to two other members of the *Leguminosae* family, broad bean and chickpea (Koller and Delius, 1980; Chu and Tewari, 1982; Shinozaki *et al.*, 1984). Another genome that differs markedly from the spinach type is in geranium chloroplasts. The inverted repeat regions in this latter DNA molecule are three times larger (76 kbp) than in spinach and include many of the genes found in single-copy regions in other species (Palmer *et al.*, 1987).

The evolutionary basis for the diversity is not yet clear. Among the group shown in Fig. 1, spinach, pea, and wheat are C_3 plants whereas maize is a C_4 plant. C_4 metabolism, as an adaptation of plants to harsh environments, seems to have had multiple origins, some of which were quite ancient (Moore, 1982; Thomasson *et al.*, 1986). But although the gene maps for wheat and maize, both monocotyledons, are similar, those for spinach and pea, which are dicotyledons, are different. The pressure for reduction of the ancestral prokaryotic genome after the early endosymbiotic event(s) may have been quite strong, so that the process occurred relatively rapidly until a "limit" was reached. As the gene-sorting process proceeded, it is unlikely that only one of these cells gave rise to the current variety of plant species. Sorting could have occurred in parallel in different cells, to yield similar gene patterns, but rearrangements also may have provided the variety of patterns that currently exist. The basis of the apparent stability of the genome, however, considering the time since divergence of monocotyledons and dicotyledons, is not understood (Palmer, 1985). Possibly, the multiplicity of chloroplast DNA molecules within the organelle favors conservation of sequences through recombination events.

It is worth noting, however, a striking discovery that has come from the elucidation of the *atp* operons. Each of these operons, *atp* operons A and B, are transcribed as a unit to yield polycistronic mRNA (Hennig and Herrmann, 1986; Cozens *et al.*, 1986). The order of the genes within these operons in pea and spinach is the same as that in prokaryotic *E. coli* and the cyanobacterium *Synechococcus* (Fig. 2). Although several genes, those that code for the γ, δ, and CF_0-II subunits of ATP synthase, now reside in nuclear DNA in the plant cells, the order of the remaining genes provides strong evidence for a prokaryotic origin for chloroplasts. Other examples of chloroplast genes being encoded in operons that are organized as in *E. coli*, but with some genes deleted, also have been noted by Sijben-Müller *et al.* (1986), which provides further support for this conclusion. A rather extraordinary mechanism must have precisely dissected out specific genes for transfer to the nucleus.

The genes remaining in the chloroplast from the prokaryotic operons suggest that gene transfer to the nucleus during evolution may have occurred individually and selectively. The question arises as to why a chloroplast genome remained at all. Perhaps the conclusion reached by von Heijne (1986a) regarding the necessity for a mitochondrial genome applies also to the chloroplast. He suggested that the extent of nonpolar amino acid sequences within a protein may determine how the cytoplasmic protein synthetic system handles the protein. If nonpolar sequences are extensive, they may be recognized as signals for cotranslational protein export and be se-

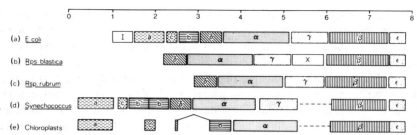

Fig. 2. Comparison of the *atp* operons in the heterotrophic bacterium *Escherichia coli,* the photosynthetic bacterium *Rhodospirillum,* the cyanobacterium *Synechococcus,* and in chloroplasts. In the cyanobacterium and chloroplasts, the operon is separated into two distant portions, with *atp* operon A including the gene for the α subunit (*atpA*) and *atp* operon B containing the gene for the β subunit (*atpB*) (see Fig. 1). In chloroplasts, genes marked a, c, and b encode subunits IV, III, and I of CF_0, respectively; the latter contains an intron. The gene order implies an evolutionary relationship between these organisms, and, in particular, strongly supports a common origin for plant chloroplasts and cyanobacteria. [From Cozens *et al.* (1986), with permission.]

creted from the cell. Or, a portion of such proteins may be exported, so that the proteins span the cell membrane. Any gene that was successfully transferred from the organellar genome to the nucleus during evolution had to code for a protein that could escape such a preexisting "export" mechanism (von Heijne, 1986a). Hydropathy plots of many chloroplast gene products indeed show extensive nonpolar regions within the amino acid sequences (for examples, see Morris and Herrmann, 1984; Steinmetz *et al.,* 1986; Cozens *et al.,* 1986; Hennig and Herrmann, 1986). This hypothesis also provides a rationale for the striking difference between the properties of the N-terminal extensions that determine uptake by organelles as compared to secretion (von Heijne, 1985, 1986b) (see Section IX for more details.)

The final accounting of the total number of genes in chloroplast DNA has been accomplished for liverwort and tobacco, but an understanding of the generality of genome content must await knowledge of the complete nucleotide sequence for other organisms. This is not an unreasonable expectation and should be obtained within a few years at the current rate of progress in this area. It may turn out that all chloroplast genomes, particularly in higher plant species, contain the same set of genes, with perhaps only minor variations.

C. The Chloroplast Genome in Algae

The general features of the chloroplast genome from the alga *Chlamydomonas* are similar to those from higher plants. However, the algal genome, with a size of 195 kbp, is larger, and the inverted repeats are separated by a much larger unique sequence and reside almost opposite from each other on the map (Rochaix, 1978; Rochaix *et al.,* 1984b). Informational content in the spacer region between the ribosomal RNA genes also differs between *Chlamydomonas* and higher plants (Schneider *et al.*, 1985; Schneider and Rochaix, 1986). The ribosomal RNA gene cluster in the 175-kbp chloroplast genome from *Chlorella ellipsoidea* differs yet again from that of *Chlamydomonas*. The spacer region (4.8 kpb) between the 16 S and 23 S rRNA genes is about twice the length of that in other organisms, and it differs in its content of tRNA genes and open reading frames. Also within the spacer region are numerous short direct and inverted repeats that could be involved in specific secondary structures. Most surprising about the *Chlorella* genome, however, is the finding that the 23 S rRNA gene is coded on the opposite strand from the 16 S rRNA gene and in opposite orientation (Yamada and Shimaji, 1986). This arrangement has not been found thus far in any other organism and implies that the ribosomal genes are not co-transcribed as a unit in *Chlorella* as occurs in other species. The *Euglena* chloroplast genome (145 kbp), on the other hand, is strikingly different from the others in containing tandem repeats of ribosomal RNA genes (Gray and Hallick, 1978; Rawson *et al.*, 1978; Jenni and Stutz, 1979; Hollingsworth *et al.*, 1984). The numbers of such genes vary among species of *Euglena*, from one to five complete sets plus an additional 16 S RNA gene (Koller *et al.*, 1984a; Palmer, 1985). The distribution of genes around the genome in *Euglena* also is significantly different from that in other chloroplasts.

Among the chromophyte algae, marked variations in gene content and order exist. In algae that harvest light with phycobilisomes, genes for the phycobiliprotein subunits reside in plastid DNA and the proteins are translated from mRNAs that lack of 3′-polyadenylate extension (Steinmüller *et al.*, 1983; Belford *et al.*, 1983; Lemaux and Grossman, 1985). The genome in the chromophytic alga *Olisthodiscus* contains, within the inverted repeat regions, closely linked sets of *rbcL* and *rbcS* genes, which possibly are cotranscribed (Reith and Cattolico, 1985a, 1986). DNA in the cyanelle within the protozoan *Cyanophora paradoxa* has many similarities to chloroplast DNA, yet also contains adjacent *rbcL* and *rbcS* genes, which are co-transcribed to produce a polycistronic mRNA (Starnes *et al.*, 1985) such as occurs in cyanobacterial cells (Nierzwicki-Bauer *et al.*, 1984; Shinozaki and Sugiura, 1985). The currently held concept is that chloroplasts evolved from cyanobacteria-like organisms, which contained closely linked *rbcL* and *rbcS* genes in their chromosome. In chlorophyte algae and the higher plants, however, thse genes are segregated in the chloroplast and nuclear DNAs, respectively.

D. The Presence of Introns in Chloroplast Genes

Sequences for a large number of chloroplast genes have been determined. In general, the coding sequences of chloroplast genes are continuous, without interruptions. However, interruptions are not rare. Of the 145 kbp of the

Euglena genome, about 35 kbp can be accounted for by at least 50 introns (Koller *et al.*, 1985). The *rbcL* gene, which codes for the large subunit of ribulose bisphosphate carboxylase, in *Euglena* contains nine introns (Koller *et al.*, 1984b). No introns occur in this gene in *Chlamydomonas* and in higher plants (Dron *et al.*, 1982). The *atpF* gene in spinach and pea recently was found to contain an intron (Bird *et al.*, 1985; Hennig and Herrmann, 1986; Cozens *et al.*, 1986).

In *Euglena* four introns are within the *psbA* gene, which codes for the M_r 32,000 herbicide-binding protein (Q_B) of photosystem 2 (Hollingsworth *et al.*, 1984; Koller *et al.*, 1985). A protein designated D1 in the chloroplast membranes of *Chlamydomonas* is the product of the *psbA* gene in this alga (Erickson *et al.*, 1984, 1986). Whereas in higher plants the *psbA* gene is just outside the inverted repeat region (Fig. 1), in *Chlamydomonas* this gene has the unusual feature of being located just within the inverted repeat and therefore present in two copies per genome. The *Chlamydomonas* gene contains four introns (Erickson *et al.*, 1984). Nevertheless, the amino acid sequence of the algal polypeptide is about 95% homologous with the spinach polypeptide and functions in the same manner, as established by the identification of base changes in this gene from herbicide-resistant cells (Erickson *et al.*, 1985; Bennoun *et al.*, 1986). Most of the amino acid changes that do occur between species reside near the C-terminus, which is lost during processing of the initial translated polypeptide to the mature form (Marder *et al.*, 1984). The homology between the chloroplast *psbA* gene and the analogous gene in cyanobacteria is quite high (Mulligan *et al.*, 1984), which indicates that this gene has been highly conserved during evolution.

The D2 protein in *Chlamydomonas* is similar in size and has significant amino acid sequence homology with the D1 protein (Rochaix *et al.*, 1984a; Erickson *et al.*, 1986). However, in contrast to the *psbA* gene, which encodes D1, the *psbD* gene encoding D2 lacks introns. These proteins possibly serve as the apoproteins of the primary and secondary stable electron acceptors of photosytem 2 (Trebst, 1986). Interestingly, mutations in the *psbD* gene results also in the absence of D1 and the lack of accumulation of other photosystem 2 core proteins (Erickson *et al.*, 1986).

Perhaps the most unusual chloroplast gene arrangements have turned up in genes that encode ribosomal proteins. The *rpl16* gene, which encodes protein 16 in the large ribosomal subunit, in *Spirodela* is split close to its 5′ end by an unusually large intron of 1411 bp (Posno *et al.*, 1986). In tobacco chloroplasts, the *rps16* gene is interrupted by an 860-bp intron (Shinozaki *et al.*, 1986a). Most interesting, however, is the arrangement of the *rps12* gene. In tobacco the major portion of the gene, including the 3′ end, is located just upstream from the 16 S rRNA gene, within the inverted repeat, but on the opposite strand (Fromm *et al.*, 1986; Torazawa *et al.*, 1986). In liverwort this segment is just outside the inverted repeat (Fukuzawa *et al.*, 1986). This portion of the *rps12* gene is highly homologous to the *Euglena* gene. In

Euglena, rps12 is linked with *rps7,* which also is true for higher plants, but the algal genes contain no introns (Montandon and Stutz, 1984). In contrast, the 3' segment of the *rps12* gene is interrupted by a 540-bp intron in tobacco and a 500-bp intron in liverwort. But surprisingly, the 5' end of the gene in tobacco has turned up 30 kbp downstream on the same strand from the 3' end, in the nearest inverted repeat, or 90 kbp away on the opposite strand in the other repeat (Torazawa *et al.,* 1986). In liverwort the 5' exon of the gene is located nearly opposite on the circular map (60 kbp away) and on the opposite strand (Fukuzawa *et al.,* 1986). How these fragments are spliced together to make the functional mRNA remains an interesting question.

Several genes for tRNAs in higher plants and algae contain introns that interrupt the coding sequence, usually within the anticodon loop (Bogorad *et al.,* 1983). In *Chlamydomonas,* the gene for the 23 S rRNA also contains a large intron, within which is an open reading frame that may code for an enzyme ("maturase") involved in the resolution of the intron from the primary RNA transcript (Rochaix *et al.,* 1985). Boundaries of the introns contain homologous sequences that apparently are involved in the splicing reactions.

The origins of introns, and their reason for existence, have caused much speculation yet remain puzzles. Senapathy (1986) recently proposed that in primitive cells the upper limit for reading frames was 600 nucleotides (200 codons). Because random nucleotide sequences would generate stop codons at this average frequency, he suggested that introns developed to accomodate such sequences in the genes for larger proteins. In chloroplasts, however, short genes such as those for tRNAs and ribosomal proteins contain numerous introns, whereas genes for large proteins, such as RbcL (Dron *et al.,* 1982) and the photosystem 2 reaction center proteins (Westhoff *et al.,* 1983; Morris and Herrmann, 1984), do not contain introns.

E. Plastome "Mutants"

The cooperation between the chloroplast and nuclear genomes has been extensively studied and is established. The balance that has developed through evolution, however, can be disturbed by mutations. Specific recessive nuclear mutant genes ("plastome mutator" genes), when in a homozygous condition, cause rather frequent plastid mutations. The mechanism of this mutagenesis is unknown. The plastid mutations subsequently are inherited in a non-Mendelian, uniparental pattern typical of organelle genes (Hagemann, 1986). Incompatibility between plastid and nuclear genomes also can occur through variations in chloroplast genomes. A fascinating demonstration of this latter situation was found by Lindenhahn *et al.* (1985) in *Oenothera hookeri.* In the past, the variegated phenotype of some strains of this organism had been attributed to the recessive nuclear "plastome mutator" gene. These investigators found that, in fact, the white leaves contained different chloroplast genomes from the green leaves. Gordon *et al.* (1982) identified five restriction patterns of the plastid DNA of wild-type variants. Lindenhahn *et al.* (1985) suggested that inadvertent pollination resulted in "hybrid" plants, such that the green leaves contain the complementary type I chloroplast genome, whereas the white leaves contain a type III genome, which is not compatible with the nuclear genotype and thus results in bleached plastids (hybrid variegation).

These observations suggest that a minimum exists beyond which the informational content of chloroplast DNA cannot be reduced without loss of function. Such a situation may also limit interspecies exchanges of genetic information. For viable plants, the necessary interplay between nucleus and plastid apparently requires a rather stringent match of compatible genomes.

In summary, the chloroplast genome is a plastic entity. Although there are many similarities in the genetic content of the DNA and in gene order among a variety of organisms, there also are significant differences in these aspects. Nature apparently found several paths toward achieving a minimal set of genes that would maintain the structure and function of the oranelle.

III. REPLICATION OF THE CHLOROPLAST GENOME

A model for semiconservative replication of chloroplast DNA was presented by Tewari *et al.* (1976) in which the origins of replication of the two strands were separated some distance apart on the genome. Replication was suggested to begin by formation of displacement loops (D-loops) at each site of initiation. Evidence for this model has been obtained for several species. In corn and pea chloroplasts, replication begins by formation of D-loops at sites located 7.1 kbp apart (Ravel-Chapuis *et al.*, 1982; Koller and Delius, 1982). Waddell *et al.* (1984) found by electron microscopy that these D-loops form at two positions 6.5 kbp apart on the *Chlamydomonas* genome. Replication proceeds by expansion of the D-loops toward each other and then on around the molecule in both directions.

Vallet *et al.* (1984) identified four segments in *Chlamydomonas* DNA that promoted autonomous replication of DNA vectors in yeast cells and suggested that they were possibly sites of origin. Three of the sequences are clustered within a region of 7 kbp; the fourth is about 50 kbp distant from the others. The sequences have a high AT content and show significant homology. Similar regions have been found in DNA from *Chlorella* (Yamada *et al.*, 1986).

McKown and Tewari (1984; also Tewari, 1986) purified a DNA polymerase from chloroplasts of pea. The enzyme contained only a single subunit of about M_r 87,000. It was optimally active with DNA that was degraded to a limited extent with deoxyribonuclease I. The enzyme had no detectable nuclease activity. In these respects, the enzyme seems to be quite different from the replicative DNA polymerase III from *E. coli*, which has a subunit of M_r 140,000 that is responsible for polymerization and two additional subunits of M_r 25,000 and 10,000 as part of the core enzyme (Kornberg, 1983). The gene for the chloroplast DNA polymerase apparently resides in the nucleus (Ohyama *et al.*, 1986).

IV. REGULATION OF PLASTID GENE EXPRESSION

A. Chloroplast DNA-Dependent RNA Polymerase

RNA polymerases have been partially purified from chloroplasts of maize (Bottomley *et al.*, 1971; Smith and Bogorad, 1974; Kidd and Bogorad, 1979, 1980), pea (Tewari and Goel, 1983), spinach (Briat and Mache, 1980; Lerbs *et al.*, 1983, 1985), and *Euglena gracilis* (Gruissem *et al.*, 1983), and from cells of *Chlamydomonas reinhardtii* (Surzycki and Shellenbarger, 1976). Two types of polymerases seem to be present in the plastid, a soluble enzyme that is active in tRNA and mRNA synthesis and another that is involved in ribosomal RNA synthesis. This latter enzyme is in a tight complex that also includes chloroplast DNA, DNA polymerase, and other proteins. Most, if not all, of the RNA polymerase activity in higher plant chloroplasts is associated with such complexes, which are much more active in transcription as the complex than after reconstitution of the purified enzyme with substrate DNA. The complex contains about 30 different polypeptides (Briat and Mache, 1980), which makes identification of the subunit structure of the RNA polymerase difficult. In *E. coli*, DNA-dependent RNA polymerase has a subunit structure of $\alpha_2\beta\beta'\sigma$ (Burgess, 1976). The subunit molecular weights estimated from electrophoretic mobility for the *E. coli* enzyme are 150,000 (β'), 145,000 (β), 41,000 (α), and 86,000 (σ). Preparations of the polymerase from pea chloroplasts contain subunit polypeptides of M_r 180,000, 140,000, 110,000, 95,000, 65,000, 47,000, and 27,000 (Tewari and Goel, 1983). Maize chloroplast RNA polymerase also contains polypeptides of M_r 180,000 and 140,000, among a total of 14 ranging in size from 180,000 to 27,000 (Kidd and Bogorad, 1979, 1980). The polymerase preparation from spinach was initially reported to consistently have five polypeptides of M_r between 69,000 and 15,000 (Briat and Mache, 1980). More recently, the spinach enzyme was found to contain potential subunits of M_r 145,000, 110,000, 102,000, 80,000, 75,000, and 38,000 (Lerbs *et al.*, 1983, 1985). Whether these patterns indicate similarities in structure between the chloroplast and prokaryotic polymerases is difficult to assess. As Lerbs *et al.* (1985) comment, the possibility is open that several of the polypeptides in the preparations from chloroplasts, in particular those of M_r 80,000 and 75,000, are produced by proteolysis during purification.

The purified chloroplast RNA polymerase is completely dependent on added DNA. The activity is greater with denatured than with native DNA (Tewari and Goel, 1983), although Jolly and Bogorad (1980) showed that the maize chloroplast polymerase actively transcribed chloroplast genes within a supercoiled plasmid. A protein similar to the sigma factor of *E. coli* also is required for activity of the chloroplast polymerase, but no gene homologous to the *E. coli* gene for this subunit was found in chloroplast DNA (Ohyama *et al.*, 1986). An M_r 27,500 polypeptide, designated "S" factor, that activated purified core RNA polymerase was purified from maize chloroplasts (Jolly and Bogorad, 1980). A protein of similar activity, which also activated the core polymerase of *E. coli*, was purified from *Chlamydomonas* cells (Surzycki and Shellenbarger, 1976). The purified chloroplast RNA polymerases studied so far are strongly inhibited by salt solutions but not by either of the inhibitors rifamycin or α-amanitin (Bottomley *et al.*, 1971; Kidd and Bogorad, 1980; Tewari and Goel, 1983).

Open reading frames with partial homology to the genes for the α (*rpoA*), β (*rpoB*), and β′ (*rpoC*) subunits of *E. coli* RNA polymerase were found in chloroplast DNA from higher plants (Sijben-Müller *et al.*, 1986; Ohme *et al.*, 1986; Ohyama *et al.*, 1986). Watson and Surzycki (1983) found that at low stringency, DNA fragments containing the β (*rpoB*) and β′ (*rpoC*) genes of *E. coli* hybridized to chloroplast DNA and also to nuclear DNA of *Chlamydomonas*. However, Lerbs *et al.* (1985) could not detect hybridization of the bacterial *rpoB* and *rpoC* genes with spinach chloroplast DNA. Moreover, the latter investigators found that all subunits of RNA polymerase were translated from poly(A)-rich RNA, which implied that they were encoded in nuclear DNA and translated on cytoplasmic ribosomes. Thus, it seems that additional work is needed to clarify our understanding of the sites of synthesis of the RNA polymerase subunits.

B. Chloroplast Gene Promoters

Chloroplast DNA is transcribed into organellar rRNAs, tRNAs, and mRNAs. The nucleotide sequences of a large number of these chloroplast genes have been determined, including long stretches of their 5′- and 3′-flanking regions. The nucleotide sequences of the flanking regions of the genes have been carefully examined for potential transcriptional control elements. Particular interest has been directed toward identification of promoter and termination sequences. In order to identify the promoter sequence, the start site for transcription must be determined, which is usually accomplished by first hybridizing a cloned restriction fragment that includes the 5′ end of the gene with its complementary mRNA. The hybrid then is treated with S_1 nuclease, which digests only single-stranded DNA (Berk and Sharp, 1977; Weaver and Weissmann, 1979). By comparing the length of the protected fragment with the sequence of the strand from the 3′ end, the start of transcription can be determined to within ±1 base pair. In prokaryotic-type genes, one element of the promoter, the "−10" region or "Pribnow box," is located around the −10 nucleotide from the transcription start site. Another element, the "−35" region, is located further upstream, separated from the "−10" region by 17–18 base pairs (Rosenberg and Court, 1979; Hawley and McClure, 1983; McClure, 1985). In eukaryotic-type genes, the primary promoter element is 20–30 base pairs upstream from the transcription initiation site (Breathnach and Chambon, 1981; McKnight and Kingsbury, 1982).

Link (1984) determined the start site of transcription of the *psbA* gene, which encodes the M_r 32,000, photosystem 2 herbicide-binding protein. By deleting short segments of the region 5' to this start site, he identified a segment 28–35 nucleotides upstream that is required for efficient and selective initiation of transcription at the *in vivo* start site. Within this segment is the sequence TTGACA, which is identical to the consensus sequence of the "−35" element of prokaryotic promoters. Nucleotides −5 to −10 are TATACT, a sequence also found in many prokaryotic promoters (Rosenberg and Court, 1979; Hawley and McClure, 1983). Kung and Lin (1985) and Hanley-Bowdoin and Chua (1987) surveyed most of the known chloroplast gene sequences, many of which have had the transcriptional initiation site determined by S_1 nuclease mapping. In the 5'-flanking region of all these genes are sequences that closely resemble the prokaryotic promoter. The consensus sequence of the "chloroplast promoter" from their analysis is TTGACA at the "−35" region and TATAAT at the "−10" site. These sequences are identical to those of the consensus prokaryotic promoter (Rosenberg and Court, 1979; McClure, 1985). Furthermore, the frequency of the consensus base at each position is similar in chloropalsts and prokaryotic genes. In chloroplast genes, the TTG triplet in the "−35" site is highly conserved, as are the TA—T bases in the "−10" hexamer (Kung and Lin, 1985). These two functional regions in chloroplast promoters are separated by about 17 base pairs, which is also the optimal distance in prokaryotic promoters. Increasing the distance from 18 to 20 base pairs in the *rbcL* promoter drastically reduced the level of *in vitro* transcription (Kung and Lin, 1985).

The similarity in sequence between the chloroplast promoters and bacterial promoters is striking. Furthermore, the strongest bacterial promoters are those that contain the consensus sequence in both the "−10" and "−35" sites, with a separation of 16–17 nucleotides (Amann *et al.*, 1983). Thus, the chloroplast sequences also would be expected to act as strong promoters. Bacterial cells that contain such strong promoters must be protected by repressor proteins from overproduction of the gene product (Amann *et al.*, 1983), which otherwise may become toxic to the cell. It is very likely, therefore, that expression of chloroplast genes also is regulated in some manner, possibly by regulatory proteins. Dairi *et al.* (1985) showed that bacterial promoters containing T in the first position in the "−10" sequence do not require a positive regulator. Because T is present almost exclusively at this position in chloroplast promoters, if control mechanisms similar to those in bacteria also operate in the plastid, they are probably of the negative type, that is, they function to repress expression. Whatever the mechanism, expression of many of the plastid genes is induced by light, a phenomenon that is poorly understood (see Section IV,C).

The sequences TTGACA and TATACT at the "−35" and "−10" sites, respectively, of the *psbA* promoter are the same in wheat, tobacco, maize, spinach, mustard, and soybean (Kung and Lin, 1985; Hanley-Bowdoin and Chua, 1987). This sequence differs by only one base in the "−10" region from the prokaryotic consensus sequence, and thus should act as a relatively strong promoter. Indeed, in isolated chloroplasts, this protein is the most abundantly synthesized membrane protein (Morgenthaler and Mendiola-Morgenthaler, 1976; Ellis, 1977; Fish and Jagendorf, 1982). Moreover, the mRNA for this protein is abundant in chloroplasts (Mattoo *et al.,* 1984; Fromm *et al.,* 1985). The *rbcL* gene promoter is identical in tobacco, maize, spinach, and wheat, with the sequence TTGCGC at the "−35" region and TACAAT at the "−10" site. In pea and broadbean, organisms whose chloroplast DNA lacks an inverted repeat, the promoter for the *rbcL* gene differs only in one position in the "−10" region, with the sequence TAGAAT. The *rbcL* promoter may be the strongest in the chloroplast genome, because mRNA for the RbcL polypeptide is the most abundant in the chloroplast (Fromm *et al.,* 1985; Inamine *et al.,* 1985). The RbcL polypeptide also is the most abundant in the chloroplast, partly because of its stability and partly because it is the predominant product of protein synthesis in isolated chloroplasts (Morgenthaler and Mendiola-Morgenthaler, 1976; Ellis, 1981).

A sequence similar to the eukaryotic promoter was recognized by Link (1984) near −30 in the 5′-flanking region of the *psbA* gene and in the gene for the P_{680} protein of photosystem 2 (*psbB*). When the "−35" region of the prokaryotic promoter was excised from the upstream region of the *psbA* gene, the "eukaryotic-like" promoter sequence still permitted a low level of specific transcription with a plastid-derived RNA polymerase. Whether both promoters are functional in the chloroplast, or are involved in developmentally regulated expression of this gene, as suggested by Link (1984), has not been established.

C. Regulation of Plastid Gene Expression by Light

The *psbA* gene, which encodes the M_r 32,000, photosystem 2 herbicide-binding protein (Q_B), is among a group of plastid genes whose transcription is strongly stimulated by light in etioplasts (Bedbrook *et al.,* 1978; Rodermel and Bogorad, 1985). The mechanism of this induction is not entirely clear, but it seems to involve segments of the DNA in the 5′-flanking region, which in turn increase the rate of transcription. It is not known whether induction by light requires a regulatory protein, although as noted above in Section IV,B, the promoter of this gene is similar to the prokaryotic consensus sequence and thus should be highly active. It is interesting to note that after the gene is induced, it apparently remains active. In fully differentiated chloroplasts, the amount of *psbA* mRNA remains high, even after prolonged periods in the dark (Fromm *et al.*, 1985). However, although the mRNA level remains abundant after transfer to the dark, in *Spirodela* chloroplasts the protein is not made at a significant rate under these conditions (Fromm *et al.*, 1985). An analysis of the effect of light on the level of *rbcL* transcripts in pea chloroplasts has provided a similar result (Inamine *et al.*, 1985). These results suggest that after initial induction of expression of the gene, further synthesis of the protein is controlled at the translational level (see Section IV,D).

In addition to the above well-characterized genes, Zhu *et al.* (1985) found that in maize the levels of transcripts from the genes for the photosystem 1 reaction center apoproteins (*psaA* and *psaB*); the α, β and ε subunits of the coupling factor CF_1 (ATP synthase) (*atpA, atpB,* and *atpE*); subunit III of CF_0, the membrane component of the coupling factor (*atpH*); a photosystem 2 reaction center apoprotein (*psbD*); and an unidentified protein were increased twofold to fourfold by only a 1-min pulse of red (660 nm) light. A 5-min irradiation with far-red (725 nm) light, immediately after the red-light pulse, strongly attenuated the increase in mRNA pools, which is the hallmark of effects of phytochrome. Thus, expression of both nuclear and plastid genes seems to be regulated by phytochrome. The mechanism of this effect, though, remains unclear.

Another level of regulation, at present not understood, occurs during differentiation of cell types in C_4 plants. Mesophyll cells fix CO_2 via phosphoenol pyruvate carboxylase, a cytosolic enzyme; chloroplasts in these cells lack ribulose 1,5-bisphosphate carboxylase activity, and although the *rbcL* gene is present in the chloroplast genome of these cells, it is not transcribed (Link *et al.*, 1978; Jolly *et al.*, 1981). Bundle-sheath cells, in contrast, contain ribulose 1,5-bisphosphate carboxylase but lack phosphoenol pyruvate carboxylase. RbcL mRNA is detectable only in bundle-sheath cells, whereas mRNA for phosphoenol pyruvate carboxylase is detectable only in mesophyll cells (Broglie *et al.*, 1984; Schuster *et al.*, 1985). Bundle-sheath cells also essentially lack photosystem 2 activity and the light-harvesting complex-2 apoproteins (Broglie *et al.*, 1984; Bassi *et al.*, 1985; Schuster *et al.*, 1985). Levels of mRNA for the light-harvesting apoproteins are correspondingly low in bundle sheath cells. Bundle-sheath cells also contain less than 10% of the amount of *psbA* mRNA present in mesophyll cells, and synthesis of the protein cannot be detected after translation of bundle-sheath mRNA (Schuster *et al.*, 1986). Thus a tissue-specific mechanism develops that affects expression of genes in both chloroplast and nucleus as the leaves in C_4 plants differentiate. How this occurs is a major question that is being attacked by current research.

D. Posttranscriptional Control of Protein Synthesis

Evidence is mounting that control at posttranscriptional levels may be very important in regulating synthesis of chloroplast proteins. Chloroplasts isolated from plants that were transferred to darkness maintain high levels of mRNA for a number of proteins, even though the proteins no longer accumulate *in vivo*. Fromm *et al.* (1985) observed that in *Spirodella* the initial increase in the rate of synthesis of the *psbA* gene product, when plants were first exposed to light, closely paralleled the increase in the level of mRNA. However, the extent of synthesis of the polypeptide declined rapidly when plants were transferred to the dark, without a loss of mRNA. The increase in RbcL mRNA in illuminated plants is variable (Inamine *et al.*, 1985; Berry *et al.*, 1986), but in most plants light also has a strong effect on synthesis of this polypeptide. Again, after transfer to the dark, synthesis of the RbcL polypeptide markedly decreases, whereas the mRNA level is stable.

Convincing evidence for both transcriptional and translational control in chloroplasts was obtained by Herrin *et al.* (1986) during the cell cycle of *Chlamydomonas* cells. The *psbA, psbD,* and *rbcL* genes were actively transcribed early in the light period of the cycle, but translatable mRNA for the proteins (D1, D2, and RbcL, respectively) remained at high levels throughout the light–dark cycle. However, synthesis of the proteins, which occurred at high rates in the light, could not be detected in these cells during the dark period.

Although these results could be explained by rapid turnover of the polypeptides in the dark, which may occur in some cases, recent studies indicate that newly synthesized RbcL polypeptides are stable (Berry *et al.*, 1986). Thus, mechanisms exist to regulate synthesis of proteins over the short term at the level of translation. The trivial explanation for a decrease in protein synthesis in the dark is that insufficient ATP is available to drive the process. This may be one factor but it cannot be the full answer. Moreover, translational control may not be confined to the chloroplast, since evidence has been obtained for such control of synthesis of the apoproteins of the light-harvesting complex-2 in the cytoplasm (Slovin and Tobin, 1982). Kirk and Kirk (1985) observed marked effects of light on the synthesis of both nuclear- and chloroplast-DNA encoded proteins in *Volvox*, which they interpreted as regulation at the translational level. These observations have revealed an important type of regulation of gene expression, the magnitude of which may be unique to plants.

Posttranslational processes also appear to be important in keeping the chloroplast tidy. Under conditions in which synthesis of chloroplast proteins is inhibited by addition of antibiotics, transfer of plants to the dark, or exposure to high temperatures that cause depletion of chloroplast ribosomes, the cytoplasmically synthesized counterparts still are imported by the chloroplast but then are rapidly degraded. This scenario has been observed for the major chlorophyll-binding proteins (Bennett, 1981; Bellemare *et al.*, 1982), RbcS, the small subunit of ribulose bisphosphate carboxylase (Schmidt and Mishkind, 1983; Mishkind and Schmidt, 1983), and for the γ and δ subunits of the coupling factor CF_1 (Biekmann and Feierabend, 1985). Leto *et al.* (1985) discovered a converse situation, in which a mutant in a nuclear gene caused an increased rate of degradation of two chloroplast DNA-encoded polypeptides, the M_r 48,000 chlorophyll *a*-binding protein in photosystem 2 reaction centers and the M_r 34,500 "atrazine-binding" protein associated with photosystem 2. In such cases, it seems that inability to assemble a complete functional complex in thylakoid membranes because of the absence of one component causes breakdown of the remaining members of the complex.

A rapid, light-induced breakdown of the M_r 32,000, herbicide-binding protein was discovered recently (Kyle *et al.*, 1984; Ohad *et al.*, 1984, 1985). This protein contains a bound quinone that serves as an electron acceptor for photosystem 2. High light intensities result in damage to the protein, which causes loss of electron transport function. Subsequently, a highly efficient but unidentified protease degrades the protein.

V. NUCLEAR GENES THAT ENCODE PLASTID PROTEINS

Most proteins of the plastid, as also of the mitochondrion, are encoded in nuclear DNA. The reduction of the plastid genome by transfer of genes to the nucleus required a compensatory mechanism for reentry of the gene products into the plastid. According to von Heinje (1986a), the only genes that remained in the organelle may have been those whose products cannot return specifically to the plastid. The consequence of nuclear dominance over the organelle has been amply demonstrated (Ellis, 1981). Intensive research is now directed toward identifying the nuclear gene products, isolation of the corresponding genes, and analysis of the mechanisms involved in reentry of the products into the organelle.

Over 100 different soluble and 20–30 membrane polypeptides that are synthesized *in vitro* from poly(A)-rich RNA are imported by isolated chloroplasts (Grossman *et al.*, 1982). The maximum number of nuclear DNA-encoded, chloroplast proteins is not yet known. The predominant nuclear gene products are the small subunit of ribulose 1,5-bisphosphate carboxylase/oxygenase (RbcS) and the chlorophyll *a/b*-binding proteins (LHCP or Cab) of the light-harvesting complex-2 (LHC-2). Messenger RNA for these two types of polypeptides is abundant in green cells, as assayed by the production of its translation products *in vitro* (Dobberstein *et al.*, 1977; Viro and Kloppstech, 1980; Broglie *et al.*, 1981; Tobin, 1981; Hoober *et al.*, 1982). Because of the abundance of these mRNA species, complementary DNA (cDNA) species synthesized from total mRNA templates by reverse transcriptase likewise were enriched in coding sequences for these proteins. This enrichment allowed efficient selection of clones of *E. coli* that contained plasmids with appropriate cDNA inserts (Bedbrook *et al.*, 1980; Broglie *et al.*, 1981; Tittgen *et al.*, 1986). Subsequently, the cloned cDNA species, which were sequenced (Bedbrook *et al.*, 1980; Coruzzi *et al.*, 1983), became valuable probes for the detection and isolation of the genes in restriction fragments of nuclear DNA.

A. *rbcS* Genes

Ribulose 1,5-bisphosphate carboxylase/oxygenase, which catalyzes fixation of CO_2 in photosynthetic organisms, is composed of eight copies each of two nonidentical subunits. The large subunit, RbcL (M_r 55,000), is encoded by chloroplast DNA, as described in Section II. The small subunit, RbcS (M_r 14,000–16,500, depending on the species), is encoded by nuclear DNA. RbcS is the most prominent polypeptide synthesized *in vitro* by translation of poly(A)-rich RNA (Dobberstein *et al.*, 1977; Viro and Kloppstech, 1980; Tobin, 1981; Hoober *et al.*, 1982). RbcS mRNA is transcribed from a family of 6–10 genes in higher plants, which varies according to the species (Coruzzi *et al.*, 1984; Dean *et al.*, 1985a,b; Tumer *et al.*, 1986). In petunia, eight *rbcS* genes exist, which are arranged in three subgroups based on sequence homology (Dean *et al.*, 1985b). One subgroup contains six genes, five of which are closely linked. The other two subgroups consist of one gene each. Because the three subgroups can be distinguished by hybridization to cloned cDNA species, at least one member of each group must be expressed in the leaf tissue. The alga *Chlamydomonas* contains two complete *rbcS* genes and an additional truncated, homologous sequence, which are closely linked (Goldschmidt-Clermont, 1986; Goldschmidt-Clermont and Rahire, 1986).

Introns have been found in all higher plant *rbcS* genes, but they vary in number and size. In wheat, a monocotyledon, only one intron exists, which is within the region that encodes the N-terminal portion of the mature RbcS (Broglie *et al.*, 1983). Pea and soybean, both dicotyledons, contain *rbcS* genes that have two introns, one of which is in the same position as the one in the wheat gene (Coruzzi *et al.*, 1984). The genes in petunia also contain two introns located at the same positions as in pea and soybean (Dunsmuir *et al.*, 1983), but one *rbcS* gene in petunia contains an additional intron (Dean *et al.*, 1985a; Tumer *et al.*, 1986). A tobacco *rbcS* gene also contains three introns (Mazur and Chui, 1985). Goldschmidt-Clermont and Rahire (1986) found three introns in both of the *Chlamydomonas rbcS* genes, but their positions differ in every case from the introns in *rbcS* genes in other species, which are reviewed in this paper.

Although some *rbcS* coding sequences may be pseudogenes that lack expression elements, usually more than one *rbcS* gene is expressed. However, different genes within a family are expressed at different levels. Tumer *et al.* (1986) determined by hybridization with gene-specific probes that one gene, which has two introns, in petunia accounts for 40% of the total *rbcS* transcripts, whereas a second gene, which contains three introns, accounts for only 4–5% of the transcripts. On the other hand, Dean *et al.* (1985a) used the frequency of cDNA copies to estimate the expression level and found that the three-intron gene accounts for 47% of the total *rbcS* mRNA. The difference between these two reports may lie in the different techniques used to estimate abundance of specific transcripts, but the results illustrate the variation in expression of different genes of this family. Expression of these genes is induced by light (see Section V,D), and it is possible that each member of a family may respond to a different degree. Also, expression seems to be related to the state of development and the function of the chloroplasts (Tobin and Silverthorne, 1985; Fluhr and Chua, 1986; Fluhr *et al.*, 1986).

The *rbcS* gene transcripts in pea are translated into products that contain an N-terminal extension of 57 amino acids as compared to the mature polypeptide (Coruzzi *et al.*, 1984). In the alga *Chlamydomonas*, this extension consists of 45 amino acids (Schmidt *et al.*, 1979; Hurt *et al.*, 1986; Goldschmidt-Clermont and Rahire, 1986). Within the coding region for the mature RbcS, amino acid sequences rarely differ, particularly for polypeptides synthesized from the gene family within a single species. In contrast, homology within the N-terminal extension is less conserved, with at least five amino acid substitutions detected between members of the gene family in petunia (Tumer *et al.*, 1986).

B. *cab* Genes

It is interesting to note that the major gene products encoded by nuclear DNA are chloroplast proteins. Whereas RbcS is the predominant soluble plastid polypeptide encoded by nuclear DNA, the predominant thylakoid membrane polypeptides are the chlorophyll *a/b*-binding proteins, designated Cab proteins or light-harvesting chlorophyll *a/b*–protein complex apoproteins (LHCPs). The abundance of Cab mRNA allowed efficient screening for cDNA clones (Coruzzi *et al.*, 1983), which were then used as probes to isolate the nuclear genes. In a similar manner to the *rbcS* genes, the *cab* genes occur as a multigene family, with the number of coding sequences ranging from possibly as few as four in *Arabidopsis* (Leutwiler *et al.*, 1986) to at least 16 in petunia (Dunsmuir, 1985). In other species, at least seven genes occur in wheat (Lamppa *et al.*, 1985a), eight in pea (Coruzzi *et al.*, 1983; Pollans *et al.*, 1985), and about 12 in *Lemna* (Karlin-Neumann *et al.*, 1985). The observed heterogeneity in the polypeptides extracted from purified LHC-2 in some cases can be attributed to polypeptides made from different coding sequences (Schmidt *et al.*, 1981; Dunsmuir, 1985), but in other cases, multiple polypeptides arise from a single cloned gene (Kohorn *et al.*, 1986) or from several genes that yield completely homologous mature polypeptides as in *Arabidopsis* (Leutwiler *et al.*, 1986). Thus, posttranslational modifications also contribute to heterogeneity of the functional proteins. The *cab* genes in *Chlamydomonas* have not been sequenced, although it is apparent that both gene divergence and posttranslational modifications contribute to the heterogeneity of Cab proteins in this organism (Hoober *et al.*, 1980, 1982; Shepherd *et al.*, 1983; Marks *et al.*, 1985).

Several *cab* genes have been isolated and characterized from pea (Cashmore, 1984), wheat (Lamppa *et al.*, 1985a), petunia (Dunsmuir, 1985), *Lemna* (Karlin-Neumann *et al.*, 1985; Kohorn *et al.*, 1986), and tomato (Pichersky *et al.*, 1985). In tomato, which contains a minimum of 13 *cab* genes, four are arranged in tandem on chromosome 2, whereas another group of three is arranged in tandem, with one gene in opposite orientation, on chromosome 3 (Pichersky *et al.*, 1985). Linkage of *cab* genes also occurs in petunia (Dunsmuir, 1985) and perhaps is common among other species as well. All these genes characterized so far encode polypeptides of 264–269 amino acids, with 266 and 267 the more common numbers. The coding sequence of most *cab* genes that have been sequenced is uninterrupted by introns, but a *cab* gene isolated from *Lemna* includes an 84-bp intron within the sequence that encodes the N-terminal portion of the mature polypeptide (Karlin-Neumann *et al.*, 1985). Possibly other intron-containing *cab* genes also exist. Considerable divergence in the coding sequence exists upstream from the position of the intron of the *Lemna* gene, that is, in the sequence for the N-terminal portion of the mature polypeptide and the N-terminal extension, as compared to another member of the *Lemna cab* gene family (Kohorn *et al.*, 1986). The intron, though, has not disturbed the amino acid sequence at the site of insertion. Nucleotide sequences between *cab* genes of different species show 10–20% divergence, but the amino acid sequences are more highly conserved (Pichersky *et al.*, 1985). Dunsmuir (1985) noted that the amino acid sequences in several portions of the Cab proteins were invariant among the gene family in petunia, which suggests that these segments may be critical for the association of the protein with chlorophyll or other membrane components.

The length of the N-terminal extension, designated the transit sequence, on the Cab protein is uncertain. The mature polypeptide as isolated from thylakoid membranes contains a blocked N-terminus (Hoober *et al.*, 1980; Mullet, 1983), a feature that is common to other chloroplast proteins and has complicated identification of the first amino acid. Mullet (1983) found a hexapeptide that was cleaved from membrane-bound Cab proteins by trypsin. The sequence of this peptide occurs near the N-terminus of the Cab amino acid sequence deduced from the nucleotide sequence for the pea gene. From the assumption that this peptide is near the N-terminus of the mature protein, the N-terminal extension is between 33 and 37 amino acids long, depending on the species (Cashmore, 1984; Dunsmuir, 1985; Lamppa et al., 1985a). This would correspond to a change in M_r of about 4000 as the precursor form of the polypeptide is cleaved after entry into the chloroplast, in agreement with experimental observations (Schmidt et al., 1981; Kohorn et al., 1986). In Chlamydomonas, on the other hand, although one precursor of M_r 30,000 provides a mature protein of M_r 26,000, indicating an extension similar in size to that in higher plants, a second major precursor of M_r 31,500 yields a mature protein of M_r 29,500, a difference of only 2000 (Marks et al., 1985).

Hydropathy plots of the amino acid sequences predicted by the nucleotide sequences of the cab genes revealed three internal, hydrophobic segments (Dunsmuir, 1985; Karlin-Neumann et al., 1985). Assuming that these are potential membrane-spanning segments, Karlin-Neumann et al. (1985) proposed a model for the orientation of the polypeptide in thylakoid membranes (Fig. 3). The model suggests that a large portion of the protein, including the N-terminal region and a highly polar internal segment, is exposed on the stromal surface. Extensive evidence exists for exposure of the N-terminus of the protein on the stromal side of the membrane. Mild treatment of thylakoid membranes with proteolytic enzymes cleaves a short segment (10–20 amino acids long) from this end of the protein (Steinback et al., 1979; Chua and Blomberg, 1979; Mullet, 1983; Delepelaire and Wollman, 1985). Proteolytic cleavage in the presumably exposed, internal segment surprisingly is not observed. And, although this internal segment, proposed to be exposed on the stromal surface, is enriched in acidic amino acids, Ryrie and Fuad (1982) found that only two or three carboxyl groups reacted with carbodiimides, a number that could be accomodated by the acidic amino acids in the exposed N-terminal segment. It is not likely that bound chlorophyll protects the Cab proteins from digestion, because the same pattern of proteolysis was obtained when these proteins were integrated into membranes in vivo with or without chlorophyll synthesis (J. K. Hoober, unpublished results). It is not clear, therefore, whether the model proposed by Karlin-Neumann et al. (1985) explains all the structural features of the membrane-associated Cab proteins.

C. Other Identified Nuclear Genes

Isolation of nuclear genes that encode plastid proteins has depended on the availability of a cDNA probe. Yet plastid proteins are considered as products of nuclear genes if translated from poly(A)-rich RNA in an in vitro system, which implies synthesis on cytoplasmic ribosomes. In the past, persistent synthesis of a polypeptide in the presence of inhibitors of chloroplast ribosomes also was used as evidence for this site of synthesis.

Recently two cDNA sequences were prepared and characterized to less

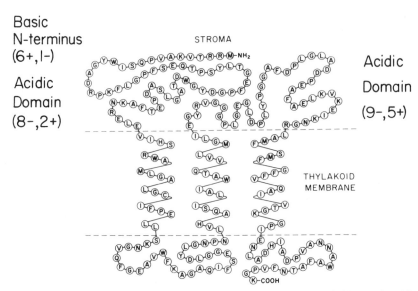

Fig. 3. A model for the association of a chlorophyll *a/b*-binding (Cab) protein with the thylakoid membrane. The model was based on evidence for residence of the N-terminus of the protein on the stromal surface of the membrane and on the existence of three extended hydrophobic sequences within the primary sequence, which could serve as membrane-spanning segments. According to the model, about half of the protein is outside the lipid bilayer on the stromal side. The notations on either side of the figure indicate that the N-terminus of the protein is basic and approximately balances the charges of a contiguous acidic domain, both of which would be expected to extend from the stromal surface if the first hydrophobic segment acted as a stop-transfer sequence. Removal of the basic N-terminus by proteolysis apparently results in charge repulsion between acidic domains on the protein within juxtaposed membranes and consequently the unstacking of grana (Mullet, 1983). The model suggests that a central acidic domain also is exposed on the stromal surface. See text for further discussion. [From Karlin-Neumann *et al.* (1985), with permission.]

abundant transcripts than the two discussed above. A cDNA derived from ferredoxin mRNA was cloned and found to contain the complete coding sequence for the polypeptide (Smeekens *et al.,* 1985a). The gene codes for a polypeptide of 15,300, which is 5600 larger than the mature ferredoxin molecule. This N-terminal extension in the ferredoxin precursor contains 48 amino acids. Hybridization of restriction endonuclease digests of nuclear DNA with the cDNA probe revealed complementarity with a single fragment, which suggests that only a single copy of the ferredoxin gene exists (Smeekens *et al.,* 1985a).

Smeekens *et al.* (1985b) also isolated a cDNA clone that contained the complete coding sequence for the precursor of plastocyanin. The precursor, as deduced from the nucleotide sequence, consists of 165 amino acids, of which 66 comprise the N-terminal extension. Inspection of this somewhat-longer-than-usual N-terminal extension reveals two portions. An N-terminal segment of 46 amino acids has the typical characteristics of a transit sequence, in that it is rich in basic and hydroxy amino acids. However, the next 20 amino acids are hydrophobic and rich in alanine and valine (Smeekens *et al.,* 1985b). Similar bipartite N-terminal extensions, which determine intraorganellar sorting, occur on mitochondrial proteins that are encoded by nuclear DNA (van Loon *et al.,* 1986; Hurt and van Loon, 1986). It is possible that this unusual feature of the plastocyanin precursor is involved in guiding this protein to the thylakoid lumen. Similarly, the coding sequence for another membrane protein, which is located on the luminal surface and involved in evolution of oxygen, was recently determined by Mayfield *et al.* (1987). The precursor form of this protein also has a bipartite N-terminal extension. The arginine-rich N-terminal portion is followed by a sequence of 13 nonpolar amino acids.

Maturation of plastocyanin is regulated by its ligand copper. In copper-depleted cells of the alga *Scenedesmus*, an M_r 14,000 precursor of plastocyanin accumulated (Bohner *et al.,* 1981). In *Chlamydomonas*, however, neither the mature nor precursor forms of plastocyanin were detected in copper-depleted cells, although translatable mRNA for the protein was present (Merchant and Bogorad, 1986a). In such cells, the mRNA is translated and the plastocyanin precursor is processed in the normal fashion. However, mature apoplastocyanin is rapidly degraded in copper-deficient cells (Merchant and Bogorad, 1986b). In the absence of copper, synthesis of cytochrome c_{552} was induced to replace plastocyanin in electron transport.

Genomic clones recently were obtained from maize that contained genes for phosphoenol pyruvate carboxylase and pyruvate, orthophosphate dikinase (Hudspeth *et al.,* 1986), two enzymes involved in the C_4 pathway of photosynthesis. The carboxylase is a cytoplasmic enzyme, and the coding sequence of the gene provides for the full length of the protein, without an N-terminal extension. In contrast, the dikinase is a chloroplast enzyme, and *in vitro* translation of mRNA selected by the dikinase cDNA yielded a precursor protein of M_r 110,000, about 16,000 larger than the mature enzyme (Hague *et al.,* 1983; Aoyagi and Bassham, 1984, 1985). It will be interesting to learn whether other precursors contain such a large extension.

At least 43 of the 64 chloroplast ribosomal proteins are synthesized on cytoplasmic ribosomes in *Chlamydomonas,* and 19 or 20 are made in the chloroplast as determined by sensitivity of synthesis to ribosomal inhibitors (Schmidt *et al.,* 1983). Schmidt *et al.* (1984) identified precursor forms of several of the large subunit proteins, which are made in the cytoplasm, by *in vitro* translation of poly(A)-rich RNA. The apparent sizes of the N-terminal extensions in these precursors vary considerably, from less than 1000 to about 6000 daltons. Evidence has recently been obtained that ribosomal protein L-18 (Rpl18) is processed in two steps (Schmidt *et al.,* 1985). The *in vitro* translated precursor is an M_r 18,500 protein, whereas the mature form recovered from the ribosome, or after *in vitro* processing (Schmidt *et al.,*

1984), is an M_r 15,500 protein. A precursor of intermediate size (M_r 17,000) was detected in cells pulse-labeled for 5 min, and was converted to the mature form during a 5-min chase period. The M_r 17,000 polypeptide and a precursor of plastocyanin (Bohner et al., 1981) have been the only precursor forms of chloroplast proteins detected in vivo. Moreover, although some evidence suggests that processing of the RbcS precursor occurs in two steps (see Section IX), the maturation of Rpl18 is the first demonstration of two-step processing in vivo.

Gantt and Key (1986) detected 51 proteins in chloroplast ribosomes from pea, 21 that were specific to the small subunit and 30 that were part of the large subunit. A cDNA library was prepared from pea poly(A)-rich RNA and inserted into lambda phage DNA downstream from the promoter for β-galactosidase. Screening colonies for polypeptides that reacted with antibodies against purified ribosomal proteins allowed detection of sequences for Rpsl6, Rpl6, Rpl12, Rpl13, Rpl18, and Rpl25 proteins. After hybrid-selection of the mRNA and translation in vitro, the products were taken up by isolated chloroplasts and processed to the mature proteins. The estimated size of the N-terminal extensions on this group of proteins ranged from 3500 to 5500 daltons.

D. Regulation of Nuclear Gene Expression

Transcription of nuclear genes for chloroplast proteins is initiated at typical eukaryotic promoters. Within each of the rbcS and cab multigene families, one or a few members maintain higher levels of their mRNA transcripts than others (Coruzzi et al., 1984; Dean et al., 1985a; Dunsmuir, 1985; Tumer et al., 1986). Promoters of the more highly expressed members of both gene families, which have been extensively studied, contain near the "−30" position upstream from the start of transcription an AT-rich sequence such as TATATAAA, TATATATA, TATAAATA, or TAAATAAA (Herrera-Estrella et al., 1984; Morelli et al., 1985; Dean et al., 1985a; Dunsmuir, 1985). Similar sequences occur in the promoter used by RNA polymerase II in eukaryotic systems (Breathnach and Chambon, 1981).

A CCAAT sequence occurs in most eukaryotic genes near the "−90" position from the start of transcription and is thought to be involved in regulation of the level of transcription (Breathnach and Chambon, 1981; McKnight and Kingsbury, 1982). Deletion of this sequence did not diminish expression of an rbcS gene (Morelli et al., 1985), and thus no function could be ascribed to this sequence in the plant gene. Its presence in many genes, however, suggests that the sequence may play an as yet unidentified role. A series of deletion mutations reduced quantitatively the expression of some rbcS and cab genes when the 5'-flanking region was shortened to about 350 bp, but tissue-specificity and light-inducibility of expression were still retained (Nagy *et al.*, 1985; Simpson *et al.*, 1985, 1986; Fluhr *et al.*, 1986). Thus, regulatory elements in the region just upstream from the promoter have been sought that provide tissue-specificity and light-inducibility to these genes.

The "−30" AT-rich sequences allow efficient opening of the DNA helix. Initiation of transcription, as a result, occurs at a specific site. The promoters for the *rbcS* and *cab* genes should act as strong expression elements, and indeed, when expressed, the mRNA species transcribed from these genes are abundant. However, although necessary (Morelli *et al.*, 1985; Simpson *et al.*, 1986), the "−30" sequences are not sufficient for expression. In the dark, the levels of RbcS and Cab mRNAs are very low (Nelson *et al.*, 1984). When plant cells are exposed to light, these mRNA species increase in amount about 100-fold (Nelson *et al.*, 1984; Coruzzi *et al.*, 1984; Tobin and Silverthorne, 1985; Eckes *et al.*, 1985; Lamppa *et al.*, 1985b; Simpson *et al.*, 1985, 1986; Fluhr *et al.*, 1986). This effect of light has become the hallmark of the regulation of expression of these genes for plastid proteins. Inducibility by light of a gene in one plant species is transferred along with the gene to another plant (Nagy *et al.*, 1985, 1986). But the response to light is conditional. The *rbcS* and *cab* genes are expressed only in plant cells that contain, or are capable of developing, functional chloroplasts, such as those in leaves and to a lesser extent in stems. No RNA transcripts of these genes can be detected in roots (Dean *et al.*, 1985a; Lamppa *et al.*, 1985b; Nagy *et al.*, 1985; Simpson *et al.*, 1986). Transcript levels in petunia decrease in the order: leaves > sepals > stem > flower petals > roots, with the amount of transcripts from leaves at least 100- to 500-fold greater than in roots. Also, no expression, particularly of the *cab* genes, can be detected in white, mutant plant tissue (Müller *et al.*, 1980; Herrera-Estrella *et al.*, 1984), or in plants in which carotenoid biosynthesis, and consequently chloroplast development, was arrested by mutation or with herbicides (Mayfield and Taylor, 1984; Batschauer *et al.*, 1986; Oelmüller and Mohr, 1986).

The phenomenon of light-inducibility is known to be a function of the 5′-flanking regions of sensitive genes (Morelli *et al.*, 1985; Herrera-Estrella *et al.*, 1984; Simpson *et al.*, 1985; Nagy *et al.*, 1986). Although a specific light-responsive sequence has not yet been identified, Fluhr *et al.* (1986) have shown that light-inducibility is conferred by several short sequences within a 240-bp region just upstream from the promoter of *rbcS* genes. Fusion of this region to other normally nonresponsive genes causes these genes to become light-inducible. Interestingly, this fragment can be introduced in either orientation and still promote light-sensitive expression. Within this fragment is the sequence $GTGTGG_{TTTT}$, which, along with the lack of dependence on orientation, is reminiscent of eukaryotic enhancer elements (Timko *et al.*, 1985; Mazur and Chui, 1985; Tumer *et al.*, 1986; Fluhr *et al.*, 1986). However, as Fluhr *et al.* (1986) note, other genes that are not sensitive to light also contain similar sequences. Thus, the search is still on for the mechanism of light induction and the sequences that are required for expression of these nuclear genes for plastid proteins.

The steady-state level of nuclear-derived mRNAs for many chloroplast proteins is markedly elevated by light. Transcription run-off experiments with nuclei isolated from leaf tissue have shown that light causes an increase in the rate of transcription (Gallagher and Ellis, 1982; Silverthorne and Tobin, 1984; Mösinger *et al.*, 1985). Nuclei isolated from pea and duckweed plants that were exposed to light-synthesized RNA complementary to *rbcS* and *cab* sequences at rates 10- to 20-fold greater than nuclei isolated from dark-grown plants. In contrast, rRNA synthesis was only twofold greater in light-grown plants (Gallagher and Ellis, 1982), and expression of the genes for NADPH : protochlorophyllide reductase was strongly inhibited (Mösinger *et al.*, 1985). Silverthorne and Tobin (1984) further showed that 2 min of red light every 8 hr was sufficient to maintain expression of these genes. If far-red light was given immediately after each pulse of red light, the effect was partially reversed.

The relatively small amount of the RNAs transcribed in nuclei from dark-grown plants was stable in the *in vitro* systems, and thus these results suggest that the increase in the amount of mRNA for these proteins brought about by light is caused by stimulation of transcription rather than inhibition of degradation (Gallagher and Ellis, 1982). The characteristic red-light promotion, far-red-light inhibition of RNA accumulation is typical evidence for the involvement of phytochrome as the photoreceptor of induction of these nuclear genes. Abundant evidence now exists that conversion of phytochrome to the P_{fr} form by irradiation with red light is one of the factors in induction of these genes by light (Tobin and Silverthorne, 1985; Fluhr *et al.*, 1986; Fluhr and Chua, 1986; Nagy *et al.*, 1986; Simpson *et al.*, 1986). Interestingly, phytochrome seems also to control its own synthesis. Red light caused a dramatic *reduction* in phytochrome mRNA in oat seedlings, but the drop in mRNA levels was inhibited by a succeeding pulse of far-red light (Colbert *et al.*, 1983; Hershey *et al.*, 1984).

The *cab* genes are particularly sensitive to light (Kaufman *et al.*, 1984, 1986). Fluences of red light as low as 10^{-4} μmol/m^2 are sufficient to stimulate transcription of *cab* genes. In contrast, a fluence of 1 μmol/m^2 of red light is required to induce *rbcS* genes. Stimulation of expression by red light was fully reversible by far-red light with the *rbcS* genes, if given immediately after the red-light pulse, but not so with the *cab* genes. Because of their increased sensitivity to light, even far-red light slightly stimulated transcription of the *cab* genes. Thus, a process in addition to photoconversion of phytochrome may function in induction of the *cab* genes. In *Chlamydomonas,* in which the *cab* genes likewise are inducible by light, a more rapid increase in the amount of *cab* transcripts can be achieved simply by an elevation in temperature from 25 to 38°C (Hoober et al., 1982). In synchronously grown cultures of this alga, the level of cab mRNA drops sharply before the end of the light period. Johanningmeier and Howell (1984) suggested that an intermediate in chlorophyll synthesis may repress transcription of cab genes under these conditions. The mechanism of induction under different conditions remains to be determined.

Other photoreceptors, however, seem also to be involved in regulation of genes by light. Although phytochrome promoted transcription of pea rbcS genes in etiolated tissues, in mature leaves a blue-light receptor played a more dominant role (Fluhr and Chua, 1986). Simpson et al. (1986), furthermore, observed that some members of the rbcS gene family are more responsive to blue light than to phytochrome-mediated processes. The interplay between the two photoreceptors, particularly as they relate to the various stages of development, should be a fruitful area of future work. Oelmüller and Mohr (1985) found that a substance that absorbs near-ultraviolet (near-UV)/blue light acts in concert with phytochrome to promote anthocyanin formation in Sorghum. A possible candidate for the blue-light receptor in oat seedlings is a derivative of riboflavin that was characterized by Ghisla et al. (1984). The importance of the reduction in the amounts of phytochrome and its mRNA, upon exposure to continuous light, to about 1% of the amount in etiolated oat seedlings remains to be determined (Colbert et al., 1983; Hershey et al., 1984; Shanklin et al., 1987). Perhaps the blue-light responses simply become apparent when phytochrome effects become lessened.

Why light is required for the transcription of specific genes is a fascinating question. Although the gene promoters would be expected to act as strong expression elements, very little transcription occurs without cooperative sequences upstream from the promoter. Most likely, proteins interact with the 5'-flanking sequences to enhance expression. If the activity of these DNA-associated proteins is, in turn, modulated by a light-sensitive compound, the effect of light could be explained. However, such a mechanism may apply more to blue-light responses than to the red/far-red light effects of phytochrome.

VI. PHYTOCHROME

An extensive literature has accumulated on the structure and function of phytochrome (see, for example, reviews by Pratt, 1982; Quail, 1984; Lagarias, 1985). However, although the phenomonology of phytochrome action has been extensively studied, its mechanism of action has remained elusive. Phytochrome is a chromoprotein that exists in two photointerconvertible forms. One form, P_r, absorbs light maximally in the red region of the visible spectrum at 666 nm. The other, P_{fr}, has an absorbance maximum in the far-red region at 730 nm (Vierstra and Quail, 1983). P_{fr} is considered to be the physiologically active form. The native molecule is a dimer, with each subunit composed of a linear tetrapyrrole covalently attached to a cysteine residue within a polypeptide of M_r 125,000 (Pratt, 1982; Lagarias, 1985). Nuclear genes for phytochrome have been cloned from oat plants and sequenced (Hershey et al., 1985). The genes do not contain introns. Hybridization of cloned cDNA sequences with genomic DNA suggested that possibly four separate phytochrome genes exist in oat. Two of the sequences in the oat genome encode a polypeptide of 1128 amino acids; the calculated molecular weight of one is 124,870 and of the other is 124,949 (Hershey et al., 1985). Although there is no evidence for the existence of a cleavable N-terminal extension, the amino acid sequence of the first 20 positions in the protein strongly resembles a transit sequence in containing an enrichment in serine and arginine residues. No extensive hydrophobic segments exist within the primary structure of the protein, which does not support suggestions for membranes as the site of action of phytochrome.

The increase in transcriptional rates of a number of genes in response to phytochrome activation has attracted great interest. But there is a multitude of changes in physiological states, in addition to regulation of gene expression, that are brought about by the action of phytochrome (Quail, 1984; Lagarias, 1985). The variety of these responses suggests that phytochrome mediates a more basic event, to which these other processes respond. Because gene expression does not respond to light in cells that lack functional plastids, yet contain phytochrome, the chromoprotein apparently does not, in most cases at least, affect genes directly. Rather, phytochrome may bring about a fundamental change in the intracellular environment. In some respects the effects of phytochrome are reminiscent of the responses of groups of genes in prokaryotic cells to a change in the environment. As Smith and Neidhart (1983a,b) showed, one group of genes is induced by an aerobic environment whereas another group is induced by an anaerobic environment. In these cells, genes that require relaxed DNA for expression are active in an aerobic environment, whereas those that are more active in supercoiled DNA are expressed in an anaerobic environment (Yamamoto and Droffner, 1985).

The environmental control of expression in prokaryotic cells involves switching some genes off while others are turned on. In a similar manner, genes do not respond in the same manner to light, because some, such as those that encode NADPH: protochlorophyllide reductase and phytochrome itself, are strongly repressed by light (Colbert et al., 1983; Mösinger et al., 1985; Hershey et al., 1985). Such a complementary on–off nature of groups of genes again may suggest that these genes are responding to a more fundamental change in the plant cell. The requirement for functional plastids suggests that a product of the organelle also may play a role in this mechanism. This aspect has been developed by Oelmüller et al. (1986a, 1986b), who found that treating mustard cotyledons with chloramphenicol, an inhibitor of plastid protein synthesis, abolished phytochrome-mediated expression of the *rbcS* gene. They suggested that a factor is produced by plastids that is required for nuclear gene expression and that phytochrome modulates a process initiated by this factor. The elusive nature of phytochrome action implies that either the mechanism is very complicated or that the question has not yet been asked in the right way.

Speth *et al.* (1986) and McCurdy and Pratt (1986) established by immunogold labeling that, in the P_r form, phytochrome is distributed evenly throughout the cytoplasm as expected for a soluble protein. Within minutes after an exposure to red light, however, phytochrome is sequestered into amorphous aggregates, which explain the pelletability of P_{fr}. These aggregates did not seem to be associated specifically with any cellular structure, although some were adjacent to the vacuole (Speth *et al.*, 1986). A pulse of far-red light immediately following the red light prevented formation of these aggregates. Such precipitation in the cytoplasm argues against a specific action of P_{fr}.

A considerable amount of evidence suggests a relation between phytochrome and Ca^{2+} (Hepler and Wayne, 1985). Oat coleoptile protoplasts released Ca^{2+} when irradiated with red light but took up Ca^{2+} when exposed to far-red light (Hale and Roux, 1980). This observation seems to contradict the concept that phytochrome as P_{fr} increases cytosolic Ca^{2+} levels through a stimulation of Ca^{2+} transport systems (Roux *et al.*, 1981; Wayne and Hepler, 1985), unless a very large, rapid increase in free cytosolic Ca^{2+} occurred. The source for this free Ca^{2+} is not known. It must be considered whether phytochrome itself is the source. Sufficient phytochrome is in the cytosol of etiolated seedlings to provide a concentration in the micromolar range (Colbert *et al.*, 1983), which is roughly the same as the concentration of Ca^{2+} (Wayne and Hepler, 1985). It seems possible, because of its structure, that one form of the phytochrome chromophore acts as a chelating agent. The change in conformation of the chromophore that accompanies the $P_r \rightarrow P_{fr}$ conversion (Rüdiger *et al.*, 1983; Sugimoto *et al.*, 1984) would, of course, dramatically alter this property. If phytochrome indeed has this activity, then P_r would be the chelating form.

Blue-light responses may also be related to the mechanism of action of phytochrome (see discussion in Section V,D). Blue light promotes expression of genes in mature plants that no longer respond to photoconversion of phytochrome (Fluhr and Chua, 1986). The mechanism of this effect is no more obvious than that of the action of phytochrome. However, Zeiger *et al.* (1985) have established that blue light promotes opening of stomata, which would facilitate gas exchange into leaf tissue and consequently alter the intraleaf environment. Blue light could also cause flavin-mediated, photodynamic destruction of regulatory molecules. Elucidation of the mechanism of these effects remains an exciting area of research.

VII. ADVANCES IN CHLOROPHYLL SYNTHESIS

None of the effects of light is more consequential for development of the structure and function of the chloroplast than its requirement in chlorophyll synthesis. This statement, however, applies only to higher plants and a few algal strains, which by natural or induced mutation have lost the ability to synthesize chlorophyll in the dark. Indeed, most wild-type algae are green in the dark. The interest of most investigators has long been captivated by the light-mediated process and, unfortunately, little is known about how these latter cells synthesize chlorophyll in the dark.

All photosynthetic organisms, except the photosynthetic bacteria, which contain bacteriochlorophyll or retinal as chromophores, contain chlorophyll *a*. Higher plants and the green algae (chlorophytes) also contain the oxidized analog chlorophyll *b,* whereas several other classes of algae, particularly the chromophytes, contain chlorophyll *c* (Jackson, 1976). The chlorophyll *a/b* ratio in green algae and higher plants varies between species and under different physiological conditions but usually is between 2 and 4.

Chloroplasts contain the complete pathway for synthesis of chlorophyll from glutamate (Kannangara and Gough, 1977; Fuesler *et al.,* 1984a; Bhaya and Castelfranco, 1985). Considerable information is known about the pathway, which was reviewed recently (Rebeiz and Lascelles, 1982; Rebeiz *et al.,* 1983; Porra and Meisch, 1984; Castelfranco and Beale, 1983). In the context of this chapter, only some of the major recent advances, particularly those that relate to regulation of chloroplast development, will be considered. These advances have been made in knowledge of the beginning of the pathway, that is, in the origin of 5-aminolevulinate, and in the final stages of synthesis of chlorophyll.

A.   Synthesis of 5-Aminolevulinate

Beale and Castelfranco (1973, 1974) made the important discovery that glutamate is efficiently converted to 5-aminolevulinate (ALA) in plant cells. This conversion has become known as the C-5 pathway of ALA synthesis, and it provides ALA for a variety of porphyrin products in photosynthetic cells. Glutamate is the source of ALA for synthesis of factor F_{430}, a nickel-containing tetrapyrrole, in the anaerobic archaebacterium *Methanobacterium thermoautotrophicum* (Gilles *et al.,* 1983); of bacteriochlorophyll *a* in the anaerobic purple photosynthetic bacterium *Chromatium* [(Oh-hama *et al.,* 1986), but not in the photosynthetic bacterium *Rhodopseudomonas spheroides,* which produces ALA from glycine and succinylcoenzyme A by ALA synthase (Oh-hama *et al.,* 1985)]; of chlorophyll, hemes, and phycobilins in cyanobacteria and red algae (Avissar, 1983; Weinstein and Beale, 1984); and of the chlorophylls and hemes in green algae and higher plants (Castelfranco and Jones, 1975; Wang *et al.,* 1984; Weinstein and Beale,

1983, 1985a; Schneegurt and Beale, 1986). Oh-hama *et al.* (1982) and Porra *et al.* (1983) proved by ^{13}C nuclear magnetic resonance spectroscopy that chlorophyll *a* in the green alga *Scenedesmus* and in higher plants, respectively, was derived exclusively from ALA made by the C-5 pathway. Indeed, recent work has indicated that all ALA in plant cells, for synthesis of chlorophyll, phycobilins, and mitochondrial hemes, is produced from glutamate (Schneegurt and Beale, 1986). These observations show that ALA for pyrrole synthesis in these organisms does not arise by condensation of glycine and succinylcoenzyme A in a reaction catalyzed by an enzyme analogous to the ALA synthase present in animal cells. However, this latter enzyme was detected in *Euglena* (Foley *et al.*, 1982; Weinstein and Beale, 1983), which taxonomically is not a typical plant. In *Euglena,* mitochondrial hemes seem to be made by the animal-type ALA synthase.

Soluble extracts of higher plants and algal cells were obtained that converted glutamate to ALA in high yield (Gough and Kannagara, 1977; Wang *et al.,* 1981, 1984; Harel and Ne'eman, 1983; Weinstein and Beale, 1985a). Fractionation of these extracts led to the discovery that an essential component of the system is RNA (Kannangara *et al.,* 1984; Huang *et al.,* 1984; Weinstein and Beale, 1985b). Determination of the nucleotide sequence of the RNA moiety from barley chloroplasts revealed that it has a structure typical of a transfer RNA (Schön *et al.,* 1986). The RNA molecule is acylated with glutamate, but whether the resulting glutamyl-RNA participates in protein synthesis or is dedicated only to ALA synthesis is not established. Recent work has demonstrated that in *Chlamydomonas* the RNA is indeed a glutamate-specific tRNA (Huang and Wang, 1986). Because of its involvement in ALA synthesis, this RNA has been referred to as RNA[DALA]. The other two forms of glutamyl-tRNA that occur in chloroplasts are not substrates for ALA synthesis (Kannangara *et al.,* 1984; Schön *et al.,* 1986).

Conversion of glutamate to ALA, therefore, is initiated by the ATP-dependent synthesis of glutamyl-RNA[DALA]. Synthesis of this intermediate is required for the NADPH-dependent reduction of the glutamate moiety. The product of this reduction was tentatively identified as glutamate 1-semialdehyde (Kannangara *et al.,* 1978; Houen *et al.,* 1983; Wang *et al.,* 1984). The mechanism of this reduction is not known. The final reaction in this pathway is the transfer of the amino group from the 2-carbon of glutamate 1-semialdehyde to the 5-carbon of ALA (formerly the 1-carbon of glutamate). Chemically synthesized glutamate 1-semialdehyde is a substrate for this reaction (Houen *et al.,* 1983). Because addition of no other amino acceptor or donor is required (Weinstein and Beale, 1985a), the aminotransferase apparently catalyzes an intramolecular reaction or an exchange between two molecules of glutamate semialdehyde. This final reaction is strongly inhibited by low concentrations of gabaculine and also by aminooxyacetate (Kannangara and Gough, 1978; Weinstein and Beale, 1985a; Schön *et al.,* 1986), which suggests that pyridoxal-phosphate is a cofactor in this reaction. However, positive identification of the amino acceptor on the enzyme has not been achieved.

Synthesis of ALA is highly regulated in plants cells, such that ALA or other precursors of chlorophyll do not normally accumulate in the dark. Light strongly stimulates synthesis of ALA (Beale, 1971; Harel and Klein, 1972), and cells exposed to light increase severalfold their capacity for ALA synthesis (Kannangara and Gough, 1978; Harel and Ne'eman, 1983; Weinstein and Beale, 1985a). But because the level of ALA-synthesizing activity is appreciable even in dark-grown cells, in which neither ALA nor porphyrins accumulate, the regulation of this activity has attracted significant interest. Work with photosynthetic bacteria and algae has shown that hemin is an efficient inhibitor of chlorophyll synthesis (Burnham and Lascelles, 1963; Hoober and Stegeman, 1973; Hoober, 1981). Hemin also efficiently inhibited the *in vitro* synthesis of ALA, with 50% inhibition at about 1 μM hemin (Wang *et al.*, 1984; Weinstein and Beale, 1985a; Huang and Wang, 1986). The inhibition by hemin seems to occur at the step in which glutamyl-RNA-^{DALA} is reduced. Interestingly, other porphyrins and chlorins, including protoporphyrin IX, protochlorophyllide, chlorophyllide *a,* and chlorophyllide *b,* did not significantly inhibit the reaction. The rapidity with which chlorophyll synthesis is initiated when degreened *Chlamydomonas* cells are exposed to light suggests that a direct precursor of chlorophyll should be involved in the regulation. But Castelfranco and Jones (1975) observed that turnover of heme in greening plants was sufficiently rapid to account for regulation of chlorophyll synthesis. On the other hand, Stobart and Ameen-Bukhari (1986) proposed that protochlorophyllide, in association with protochlorophyllide reductase, may be a direct inhibitor of ALA formation. If this latter suggestion is correct, a consequence of photoreduction of protochlorophyllide would be activation of its biosynthetic pathway through promotion of ALA synthesis.

B. Conversion of Protoporphyrin IX to Protochlorophyllide

Protoporphyrin IX exists at a crucial midpoint in porphyrin synthesis and is shunted toward protoheme by chelation with iron and toward chlorophyll by chelation with magnesium. These reactions, being as they are branch points in the pathway, are highly regulated (Castelfranco and Beale, 1983). The pathway for the subsequent conversion of protoporphyrin IX to protochlorophyllide has been established by work in Castelfranco's laboratory. This conversion is initiated by insertion of Mg^{2+} into the porphyrin by magnesium chelatase, a membrane-bound enzyme that requires ATP for activity (Pardo *et al.*, 1980; Fuesler *et al.*, 1984b). This enzyme rapidly looses activity when plastids are broken, which has prevented its characterization. It can be inhibited in intact chloroplasts by reagents that attack sulfhydryl groups but that are thought not to be able to penetrate membranes, which suggests the enzyme is located on the chloroplast envelope (Fuesler *et al.*, 1984b).

Formation of the fifth isocyclic ring to produce protochlorophyllide occurs by a system referred to as Mg-protoporphyrin IX monomethyl ester (oxidative) cyclase (Wong and Castelfranco, 1984, 1985). Cyclization is initiated by formation of the monomethyl ester of magnesium protoporphyrin IX in a reaction with *S*-adenosyl methionine. The methyl propionate side-chain on pyrrole ring C of the porphyrin then is oxidized by a series of reactions that are analogous to β-oxidation of fatty acids. The β-hydroxy- and the β-keto-propionate methyl ester derivatives were detected as intermediates (Wong *et al.*, 1985). The β-keto derivative subsequently cyclizes with the ring methinyl carbon to produce Mg-2,4-divinyl (or monovinyl) pheophorbide a_5, that is, divinyl (or monovinyl) protochlorophyllide. Reduction of the vinyl side-chain on pyrrole ring B to the ethyl group is apparently independent of isocyclic ring formation and occurs to different extents in different plants and under different environmental conditions (Carey *et al.*, 1985).

The conversion of protoporphyrin IX to protochlorophyllide requires both a membrane fraction and soluble components of the chloroplast stroma (Fuesler *et al.*, 1984b; Wong and Castelfranco, 1985). The system is dependent on molecular oxygen and reduced pyridine nucleotides (Wong *et al.*, 1985). A vinyl group on pyrrole ring A, but not on ring B, seems to be required for substrate activity.

C. Synthesis of Chlorophyllide *a*

Protochlorophyllide is reduced by NADPH to chlorophyllide *a* in higher plants in a light-dependent reaction catalyzed by the enzyme NADPH : protochlorophyllide oxidoreductase (also called protochlorophyllide reductase) (Griffiths, 1978; Oliver and Griffiths, 1982; Griffiths *et al.*, 1984). The enzyme, which has a subunit molecular weight of about 36,000 (Apel *et al.*, 1980; Beer and Griffiths, 1981), forms a ternary complex with the two substrates. No reaction occurs within the complex unless light is absorbed by protochlorophyllide. In illuminated samples, a stereospecific addition of a hydride ion, followed by addition of a proton from the medium, occurs to the 7,8-positions in pyrrole ring D.

Interestingly, protochlorophyllide reductase is present at high levels only in etioplasts that have developed an extensive prolamellar body. The protein is degraded upon exposure to light (Santel and Apel, 1981), but the presence of substrates markedly protects it from proteolysis (Kay and Griffiths, 1983). The rate of degradation of the enzyme within the chloroplast was recently found by *in situ* measurements to be relatively slow, with a half-life of several hours (Dehesh *et al.*, 1986a, 1986b), rather than the rapid breakdown observed earlier in extracts of leaves (Santel and Apel, 1981). However, the level of the mRNA for the enzyme is markedly reduced by light in a manner that suggests that phytochrome is involved in suppressing the gene (Mösinger *et al.,* 1985). The enzyme is encoded in nuclear DNA and is synthesized in the cytoplasm (Batschauer *et al.,* 1982; Griffiths and Beer, 1982).

Addition of the 20-carbon alcohol to the propionate side-chain of pyrrole ring D occurs by condensation of chlorophyllide with geranylgeranyl pyrophosphate (Rüdiger *et al.,* 1980). Geranylgeranyl pyrophosphate is synthesized *in vitro* in reconstituted systems that contain the chloroplast stroma and a membrane fraction (envelope or thylakoid), which suggests that part of the pathway is soluble and part is membrane-bound (Block *et al.,* 1980; Soll *et al.,* 1983). The enzyme that catalyzes condensation of geranylgeranyl pyrophosphate with chlorophyllide, referred to as chlorophyll synthase, is located primarily in thylakoid membranes (Soll *et al.,* 1983; Rüdiger, 1986). Subsequent reduction of geranylgeranyl to the phytol moiety occurs by NADPH-dependent reactions. Chlorophyllide *b,* but not protochlorophyllide, also is a substrate for chlorophyll synthase.

Chlorophyll *a* occurs in two major forms, one with vinyl sidechains on pyrrole rings A and B (divinyl) and the other with a vinyl group on ring A and an ethyl group on ring B (monovinyl). Tripathy and Rebeiz (1986) demonstrated that these forms originate at the stage of protoporphyrin. Mono- or divinyl protochlorophyllide is converted to the corresponding form of chlorophyll. Different plants can be characterized by their production of predominantly one form. In some species, the major form is different in light and dark. Algal cells, such as *Chlamydomonas* and *Scenedesmus,* seem to produce monovinyl protochlorophyllide in the dark (Bednarik and Hoober, 1985b; Senger and Brinkmann, 1986).

D. Chlorophyll RC I

Recently Dörnemann and Senger (1982, 1986) discovered a derivative of chlorophyll *a* that is associated only with photosystem 1. This derivative occurs in a one-to-one molar ratio with P700, the reaction-center complex in green plants. The structure of chlorophyll RC I was determined by Dörnemann and Senger (1986) and confirmed by Scheer *et al.* (1986) as $13^2(S)$-hydroxy-20-chloro-chlorophyll *a.* The presence of the chlorine atom in this molecule is particularly interesting. Elucidation of the function of this derivative should provide important information about the reaction center of photosystem 1.

E. Synthesis of Chlorophyllide *b*

The existence of divinyl and monovinyl forms of chlorophyll *b* in plant tissue, and confirmation of their structure by proton nuclear magnetic resonance spectroscopy, was recently established by Wu and Rebeiz (1985). The biosynthesis of chlorophyll(ide) *b* has been a long-standing enigma. Although extensive evidence exists that chlorophyll(ide) *a* is oxidized to chlorophyll(ide) *b* in greening seedlings (Shlyk, 1971; Argyroudi-Akoyunoglou *et al.*, 1976), this reaction occurs slowly and cannot be demonstrated *in vitro*. Thus, this *in vivo* evidence does not satisfactorily explain the synthesis of chlorophyll *b* during rapid chlorophyll accumulation (Benarik and Hoober, 1986).

The isolation and analysis of chlorophyll *b*-less mutants of *Chlamydomonas* (Michel *et al.*, 1983; Maroc *et al.*, 1983; Chunayev *et al.*, 1984; Picaud and Dubertret, 1986), *Euglena* (Cunningham and Schiff, 1986), and higher plants (Markwell *et al.*, 1985a,b) demonstrates that the product of a specific nuclear gene is required for synthesis of this chlorin. In those organisms in which more than one such mutant strain was obtained, genetic analysis indicated that all are allelic. The important problem now is identification of this gene product. Recently an interesting system was developed by Bednarik and Hoober (1985a,b), in which synthesis solely of chlorophyllide *b* was stimulated in *Chlamydomonas* cells treated in the dark with phenanthrolines. The activity was also demonstrated *in vitro*, in which protochlorophyllide was converted nearly quantitatively to chlorophyllide *b* by a membrane fraction obtained from degreened cells (Bednarik and Hoober, 1985b). This was the first demonstration of highly efficient synthesis of this compound *in vitro*. Yet this conversion occurred only in the presence of phenanthrolines. Moreover, phenanthroline acted directly in this reaction, but the mechanism has not been determined. It is intriguing to think that phenanthroline simply mimics a required, endogenous effector of the reaction. Because chlorophyll *a* synthesis normally preceeds, and seems to be required for, chlorophyll *b* synthesis, chlorophyllide *a* perhaps is the endogenous promoter of this reaction.

VIII. THE CHLOROPLAST ENVELOPE

There is a growing appreciation for the role of the chloroplast envelope, the double-membrane structure that surrounds the organelle, in chloroplast development (Douce *et al.*, 1984). It is more than simply a structural barrier between the cytosol and stroma. Indeed, evidence is beginning to accumulate that suggests the envelope plays a pivotal role in development of the structure and function of the organelle. Many soluble proteins are actively transported through the envelope into the stroma by a mechanism not yet understood. It is not clear whether proteins destined for thylakoid membranes pass through the envelope or become integrated first into the inner membrane. The inner membrane, in particular, may be a staging area for assembly of thylakoid membranes. The envelope is a major site of synthesis of lipids, including those of thylakoid membranes. It seems possible that thylakoid membranes develop by the accretion of small vesicles that transport membrane material from the envelope to thylakoids, a proposal that reiterates a long-standing concept in chloroplast development (Menke, 1962; Hoober, 1976; Douce *et al.*, 1984). It is important to consider, within a discussion of chloroplast biogenesis, the properties of the chloroplast envelope.

Several laboratories are intensively studying isolated envelope membranes in order to determine the function of this structure. Procedures were developed a number of years ago to purify the envelope free of thylakoid membranes (Mackender and Leech, 1970; Douce *et al.*, 1973; Poincelot, 1980). Subsequently, procedures were refined to permit separation of inner and outer membranes (Cline *et al.*, 1981; Block *et al.*, 1983b; Douce *et al.*, 1984; Keegstra *et al.*, 1984; Keegstra and Yousif, 1986). Separation of the membranes was afforded by their different buoyant densities: 1.08 g/cm^3 for the outer membrane and 1.13 g/cm^3 for the inner membrane (Cline *et al.*, 1981; Block *et al.*, 1983b). These differences in density reflect marked differences in composition. The outer membrane is richer in lipid, with an acyl lipid/protein weight ratio of about 3, whereas in the inner membrane this ratio is near 1 (Block *et al.*, 1983c). The acyl lipid fraction of the outer membrane consists of galactolipids and phospholipids in a ratio of approximately 1. Greater than 80% of the major phospholipid, phosphatidyl choline, in the envelope is in the outer membrane, and, as judged by its digestion with phospholipase C, is confined to the outer, cytosolic leaflet of the outer membrane bilayer (Dorne *et al.*, 1985). The acyl lipid fraction of the inner membrane consists mainly (80%) of galactolipids, with the remainder as phospholipids. In this respect, the inner membrane closely resembles thylakoid membranes (Cline *et al.*, 1981; Block *et al.*, 1983c). Both envelope membranes contain carotenoids, among which the xanthophyll, violaxanthin, predominates. Inner membranes, which have 7.2 μg carotenoid/mg protein, contain nearly three times more of these pigments than the outer membrane (Block *et al.*, 1983c).

The polypeptide and enzymatic constituents of the two membranes of the envelope are quite different (Cline *et al.*, 1981; Block *et al.*, 1983b; Werner-Washburne *et al.*, 1983), indicating that the two have very different functions. The outer membrane apparently is quite porous to small molecules but does not allow large, globular protein molecules to penetrate (Joyard *et al.*, 1983; Block *et al.*, 1983b; Keegstra *et al.*, 1984). The outer surface probably contains receptors for precursors of chloroplast-destined proteins (Dobberstein *et al.*, 1977; Schmidt *et al.*, 1981; Pfisterer *et al.*, 1982; Cline *et al.*, 1985; Bitsch and Kloppstech, 1986; see also Section IX). Precursors of chloroplast proteins that are synthesized on cytoplasmic ribosomes bind to the outer surface of the chloroplast, but, in the absence of energy, remain susceptible to digestion by added proteases (Cline *et al.*, 1985). After ATP-driven uptake of the precursors into the chloroplast, the proteins are protected from digestion by proteases (Grossman *et al.*, 1980; Cline *et al.*,

1985). Several polypeptides are exposed on the outer surface of the outer membrane, particularly those of M_r 10,000 and 24,000, which are reactive in intact chloroplast to proteases and antibodies (Joyard *et al.*, 1983; Block *et al.*, 1983b). Several other high-molecular-weight polypeptides also are digested by proteases in intact chloroplasts, which marks them as outer-membrane polypeptides (Cline *et al.*, 1981). Whether any of these serve as receptors is not known, although such proteolytic treatment of chloroplasts destroys their ability to take up the precursor forms of chloroplast proteins.

The inner membrane of the envelope is a permeability barrier to large and small molecules. Consequently, transport systems occur in this membrane to maintain metabolic communication with the cytosol (Heber and Heldt, 1981). A major polypeptide of M_r 29,000–30,000 in the inner membrane (Block *et al.*, 1983b) has been identified as the phosphate translocator, which is important for exchange of triose phosphates from the chloroplast with inorganic phosphate from the cytoplasm (Flügge and Heldt, 1986; Heldt and Flügge, 1986). The inner membrane also contains several other important transport activities, including those for nucleotides, sugars, amino acids, and dicarboxylic acids such as malate, oxaloacetate, and α-ketoglutarate (Heber and Heldt, 1981; Keegstra *et al.*, 1984; Heldt and Flügge, 1986). Apparently none of the polypeptides in the envelope, or thylakoid, membranes occur as glycoproteins (Keegstra and Cline, 1982).

Several of the more than 20 enzymes that are known to occur in the envelope have been localized to one or other of the membranes (Douce and Joyard, 1981; Douce *et al.*, 1984; Keegstra and Yousif, 1986). Particularly intriguing is the presence in the inner membrane of an active nucleoside triphosphatase (Douce *et al.*, 1973; Joyard and Douce, 1975; McCarty *et al.*, 1984), whose function unfortunately is unknown. The activity seems to be associated with an M_r 37,000 polypeptide that was partially purified, after detergent solubilization, by McCarty and Selman (1986). A vanadate-insensitive ATPase also was present in the extracts but lost activity upon fractionation. Thus, more than one ATPase may occur in the envelope. Muto and Shimogawara (1985) studied a Ca^{2+}- and phospholipid-dependent protein kinase in envelopes of spinach chloroplasts and suggested that such an enzyme may be involved in transport of proteins through the membrane. Soll and Buchanan (1983) observed that RbcS polypeptides that were associated with the envelope, and possibly in the process of transport into the organelle, were phosphorylated.

The envelope is a major site of lipid synthesis. Dorne *et al.* (1985) suggested that phosphatidyl choline, which is localized to the outer leaflet of the outer membrane, was possibly derived from lipid synthesis in the cytoplasmic endoplasmic reticulum (Douce and Joyard, 1981). Indeed, among the various chloroplast membranes, this phospholipid is a major constituent only of the outer envelope membrane (Block *et al.*, 1983b). The envelope, however, is involved in the synthesis of the remaining major polar lipids of the other membranes (inner envelope and thylakoid membranes), such as the mono- and digalactosyl diacylglycerol, sulfoquinovosyl diacylglycerol, and phosphatidyl glycerol. The galactolipids comprise more than two-thirds of the total lipid in thylakoid membranes.

Most of the final steps in the biosynthetic pathways for the glycolipids occur within the organelle (Andrews *et al.*, 1985; Kleppinger-Sparace *et al.*, 1985). A key substrate, lysophosphatidic acid (1-acyl-*sn*-glycerol-3-phosphate), is synthesized by a soluble enzyme in the stroma (Bertrams and Heinz, 1981) from glycerol-3-phosphate and fatty acyl acyl-carrier protein. The stroma, in fact, is apparently the sole site of fatty acid synthesis in the leaf cell of higher plants (Shimakata and Stumpf, 1983a,b; Stumpf, 1984). The chloroplast fatty acid synthase system has properties very similar to the system in prokaryotic cells (Stumpf, 1984). The second acyl group is then added to lysophosphatidic acid by an acyl transferase within the inner envelope membrane (Andrews *et al.*, 1985). Evidence also exists for the presence of this enzyme on thylakoid membranes (Michaels *et al.*, 1983). The resulting phosphatidic acid is the substrate for the major thylakoid phospholipid, phosphatidyl glycerol, which is synthesized entirely within the chloroplast (Bishop *et al.*, 1985). Andrews and Mudd (1985) established that the reactions phosphatidic acid → CDP-diacylglycerol → phosphatidyl glycerol all occur in the inner envelope membrane.

The source of the diacylglycerol moiety of the glycoplipids varies among different plant species. A phosphatidic acid phosphatse is located in the inner membrane of spinach chloroplast envelopes (Joyard and Douce, 1979; Block *et al.*, 1983a) and presumably catalyzes generation of diacylglycerol, the substrate for glycolipid synthesis. This phosphatase also is localized to the inner membrane in pea chloroplasts. Its activity, however, is much less in pea than in spinach chloroplasts (Andrews *et al.*, 1985). Gardiner and Roughan (1983) demonstrated that the phosphatase activity is correlated directly with the content of 16:3 fatty acids in the *sn*-2 position of diacylgalactosyl diglycerides of the membrane; plants that are low in this activity (such as pea) contain 18:3 fatty acids in this position. On the grounds that the presence of 16:3 fatty acids at the *sn*-2 position represents a prokaryotic-type biosynthetic pathway, which occurs in the chloroplast, and that insertion of an 18:3 fatty acid at this position is characteristic of a eukaryotic-type pathway in the cytoplasm (Heinz and Roughan, 1983), Bishop *et al.* (1985) recently showed that in 18:3 plants such as wheat and cucumber (and presumably pea), the diacylglycerol portion of the galactolipids was derived entirely from the cytoplasm. Apparently, a system exists in the cytoplasm to elongate the 16-carbon fatty acids that are synthesized by the chloroplast to yield the 18-carbon fatty acids. Even in 16:3 plants such as spinach and tobacco, the diacylglycerol moiety of monogalactosyl diacylglycerol was derived about equally from the chloroplast and cytoplasm, but less than 20% was supplied by the chloroplast for synthesis of digalactosyl diacy-lglycerol. Interestingly, the diacylglycerol portion of phosphatidyl glycerol, whose synthesis does not require phosphatase activity, is provided entirely by the chloroplast (Bishop *et al.*, 1985).

The enzymes UDP-Gal : diacylglycerol galactosyl transferase and galacto-lipid : galactolipid galactosyl transferase, which synthesize respectively mono- and digalactosyl diacylglycerides, are located on the outer membrane of the envelope (Dorne *et al.*, 1982a; Cline and Keegstra, 1983). This orienta-tion apparently is required because the substrate for the first enzyme, UDP-galactose, is synthesized in the cytoplasm of the plant cell (Bertrams *et al.*, 1981). However, in etioplasts isolated from wheat, more than half of the galactolipid biosynthetic activity was found in the prothylakoid–prolamellar body fraction (Sandelius and Selstam, 1984). This distribution may be ex-pected if the prolamellar body serves as a depository of excess lipid made by the envelope during growth in the dark (Lütz, 1981). Perhaps during devel-opment of the etioplast, excess lipid-rich envelope membrane, which con-tains some of the envelope enzymes, is transferred to internal membranes.

The location of the enzymes involved in lipid synthesis implies that a transfer of the glycolipid products occurs between the two envelope mem-branes. Since the galactolipids are specific markers for envelope and thyla-koid membranes, a mechanism must also exist for transfer from their site of synthesis on the outer membrane to thylakoids during chloroplast develop-ment. Sites of fusion between the outer and inner membranes of the enve-lope have been detected (Douce *et al.*, 1984; Keegstra *et al.*, 1984), which may allow diffusion of these lipids into the inner membrane. Whether the rate of such a process, if it occurs, is adequate to support rapid thylakoid membrane biogenesis is not known. The mechanism of subsequent transfer to thylakoid membranes also is unknown.

Kosmac and Feierabend (1985) found that the rate of glycerolipid synthe-sis was drastically reduced when chlorophyll synthesis was inhibited in rye plants rendered deficient in chloroplast ribosomes by growth at 32°C or with the use of inhibitors of these ribosomes. The presence of the enzymes for glycerolipid synthesis in 32°C-grown plants, however, indicated that their synthesis does not occur in the chloroplast. These results confirm the earlier suggestion of Dorne *et al.* (1982b) that the enzymes involved in galactolipid synthesis are encoded by nuclear genes and synthesized on cytoplasmic ribosomes. Janero and Barnett (1981) observed that in *Chlamydomonas* cells the rate of glycerolipid synthesis closely paralleled that of chlorophyll syn-thesis during the cell cycle. Thus, these two processes may be regulated in a coordinate fashion, by mechanisms still to be resolved. Because these lipids constitute about 50% of thylakoid membranes, this area of research remains a very important part of the investigation into the processes and mechanisms of the biogenesis of this membrane.

IX. PROCESSING OF CYTOPLASMICALLY MADE PRECURSORS OF PLASTID PROTEINS

All chloroplast proteins studied thus far that are synthesized on cytoplasmic 80 S ribosomes are made as larger precursors (Schmidt and Mishkind, 1986). Dobberstein *et al.* (1977) and Highfield and Ellis (1978) discovered that the RbcS polypeptide is initially made as a precursor longer than the mature polypeptide. The N-terminal extension is required for entry of the precursor into the chloroplast. In recognition of this function, the extension was designated the "transit sequence" (Chua and Schmidt, 1979). Since then, precursor forms have also been identified for Cab polypeptides, ferredoxin, plastocyanin, the Rieske iron–sulfur protein, ribosomal proteins, and several stromal enzymes (Schmidt and Mishkind, 1986). Very shortly after the N-terminus of the precursor polypeptides enters the stroma, the transit sequence is cleaved in a processing step to yield the mature protein.

Precursor polypeptides made by *in vitro* translation of poly(A)-rich RNA have been used to study their transport into isolated chloroplasts and the subsequent processing steps. Pretreatment of chloroplasts with proteases greatly reduced their ability to take up polypeptides, which suggested that the envelope may contain receptor-like proteins (Chua and Schmidt, 1978; Pfisterer *et al.*, 1982). Pfisterer *et al.* (1982) showed that precursor forms, but not the mature forms, of chloroplast proteins were specifically bound to envelope membranes. Several envelope polypeptides were identified as candidates for receptors by the transfer of a photoactive cross-linking agent [^{35}S]cysteamine-*N*-succinimidyl-3-[(2-nitro-4-azidophenyl)-2-aminoethyldithio]propionate from precursor proteins to envelope preparations (Kloppstech and Bitsch, 1986). In these experiments, the reagent was first attached to precursors through a disulfide bond. The complex was incubated with envelope membranes, irradiated, and then treated with reducing agents to release the precursor polypeptides, leaving the ^{35}S-labeled reagent attached to envelope proteins. Bitsch and Kloppstech (1986) also found that a group of proteins could be removed from envelope membranes by mild treatment with detergents, which destroyed receptor activity. Most of the activity, however, could be restored upon reconstitution with the detergent-solubilized proteins. Cline *et al.* (1985) showed that binding of RbcS and Cab precursors to the outer membrane of the envelope is specific but does not require a source of energy. Thus, the way is now open for the isolation and characterization of these binding proteins, and for an evaluation of the function of receptors in the transport process.

The transit sequences on precursors of plastid proteins are variable in length and show no close homology. A common pattern, however, is a marked abundance in serine, threonine, and the basic amino acids lysine and arginine; acidic amino acids are rare. Karlin-Neumann and Tobin (1986) analyzed the amino acid sequences of a number of the known N-terminal extensions and concluded that three short blocks of homology indeed may exist, which lie at the beginning, middle, and end of transit sequences. The homology applies in particular to positions containing serine, proline, and basic amino acids. Karlin-Neumann and Tobin (1986) proposed that differences in length of transit peptides, which vary from 800 to 16,000 daltons, are caused by differing lengths of nonhomologous sequences between the blocks. If the value of 800 daltons, the shortest known N-terminal extension as estimated by gel electrophoresis, is correct, a sequence as short as about eight amino acids may suffice for uptake by the chloroplast. This short sequence exists in the precursor of the large ribosomal subunit protein Rp12 in *Chlamydomonas* (Schmidt *et al.,* 1984). Presequences shorter than the first 12 amino acids in the precursor of cytochrome oxidase subunit IV of yeast are insufficient for polypeptide import into mitochondria (Hurt *et al.,* 1985).

As with the presequences that direct polypeptides into mitochondria (Hay *et al.,* 1984; Hurt and van Loon, 1986; Horwich *et al.,* 1986), the basic amino acids apparently are essential for transport into chloroplasts, because substitution of canavanine for arginine and *S*-(2-aminoethyl)-cysteine for lysine markedly inhibited import of RbcS and Cab precursors into the organelle (Robinson and Ellis, 1985). Substitution of azetidine-2-carboxylate for proline also prevented uptake by chloroplasts. Transit sequences generally contain, but are not enriched in, proline, which suggests that the secondary structure as well as the charge of the N-terminal extension is necessary for function. Similarity between the chloroplast and mitochondrial transport systems was demonstrated by the ability of the first 31 amino acids of the 45-residue-long N-terminal extension in the *Chlamydomonas* RbcS precursor, when attached to a cytosolic protein, to transport the protein into mitochondria (Hurt *et al.,* 1986).

In mitochondria, a transmembrane electrochemical potential is required for import of proteins (Hay *et al.,* 1984; Hurt and van Loon, 1986; Randall, 1986). Presumably, since the matrix of the mitochondria is negative with respect to the cytoplasm, the transmembrane potential may be responsible for pulling the positively charged presequence into the matrix. It is not clear whether a significant potential exists across the chloroplast envelope. Experiments have indicated that energy directly in the form of ATP, either added exogenously or produced endogenously by photophosphorylation, is required for import of precursors into chloroplasts; ionophores that would be expected to dissipate ion gradients, and thus transmembrane potentials, did not block import in the presence of ATP (Grossman *et al.,* 1980; Cline *et al.,* 1985; Flügge and Hinz, 1986). Flügge and Hinz (1986) also demonstrated that uptake of proteins was dependent upon ATP on the outside surface of the envelope. In illuminated chloroplasts, which generated ATP by photophosphorylation, inhibition of ATP export severely inhibited uptake of RbcS precursors. They suggested that protein transport involves phosphorylation and dephosphorylation. A phosphorylated form of RbcS associated with the envelope was detected by Soll and Buchanan (1983). The ability to import and process precursor proteins exists even in etioplasts. In this case, transport of proteins is completely dependent on added ATP (Schindler and Soll, 1986). Interestingly, the transport was found to be inhibited by the phosphatase inhibitor NaF, which implies that phosphorylation of the protein may occur during transport.

Along with the requirement for transport, the highly polar transit sequence may also be required to confer water-solubility to membrane proteins such as Cab. The transit sequence is cleaved from the precursor either during transport of the polypeptide into the organelle or immediately thereafter. The cleavage step may occur as soon as the N-terminus of the precursor extends sufficiently into the stroma to be recognized by the processing protease. Sorting of the mature forms to their proper compartment within the chloroplast, and integration into functional complexes, also occurs rapidly. Precursor forms generally cannot be found in any chloroplast subfraction after import (Schmidt *et al.*, 1981; Kohorn *et al.*, 1986), nor can precursor forms for most proteins be detected *in vivo* in pulse-labeled cells (Slovin and Tobin, 1982; Hoober *et al.*, 1982; Mishkind and Schmidt, 1983; Marks *et al.*, 1985, 1986). However, a polypeptide intermediate in size between the precursor and the mature form of a ribosomal protein has been detected in *Chlamydomonas* after a short-term label (Schmidt *et al.*, 1985). In a heterologous system, the Cab protein precursor, translated *in vitro* from *Lemna* poly(A)-rich RNA, was taken up by barley and maize chloroplasts, but processing occurred sufficiently slowly that both the precursor and processed forms were recovered in chlorophyll–protein complexes (Chitnis *et al.*, 1986). In another heterologous system, Mishkind *et al.* (1985) observed that precursors of RbcS from *Chlamydomonas* were cleaved *within* the transit sequence after uptake by higher plant chloroplasts to yield a polypeptide larger than the mature subunit. These results suggest that the recognition site, and specificity of the processing enzyme that produces the mature subunit, may have diverged between the alga and higher plants.

Processing of precursors of proteins destined for the thylakoid lumen is more complex than that of the examples cited above. The precursor of plastocyanin contains a bipartite N-terminal extension, a basic, polar sequence typical of other transit sequences followed by a hydrophobic segment of 20 amino acids (Smeekens *et al.*, 1985b). Evidence was recently reported that plastocyanin is processed in two steps (Smeekens *et al.*, 1986). In the first, the terminal polar segment is cleaved by the processing protease in the stroma. A second cleavage then occurs by a protease apparently located on thylakoid membranes. Smeekens *et al.* (1986) proposed that precursors of luminal proteins, such as plastocyanin, are transported into the stroma, where the first cleavage occurs. They further suggest that the hydrophobic portion of the N-terminal extension then serves as the signal for transport of the protein across thylakoid membranes, where the second cleavage generates the mature protein within the lumen. A similar two-step processing was found for luminal proteins involved in the evolution of oxygen by photosystem 2 (Chia and Arntzen, 1986).

A soluble protease that processes that RbcS precursor was partially purified from pea chloroplasts (Robinson and Ellis, 1984a). The estimated molecular weight of the processing enzyme from pea was about 180,000. The features of the transit sequence that are recognized by the enzyme are not known, but the introduction of analogs that distort the conformation and charge of the sequence, in particular azetidine carboxylate, an analog of proline, strongly inhibits cleavage (Robinson and Ellis, 1985). The enzyme has a broad pH optimum but is maximally active near pH 9. It is highly specific for transit sequences and does not show general proteolytic activity. The processing protease apparently is a metalloenzyme, because it is inhibited by chelators such as ethylenediamine tetraacetate and 1,10-phenanthroline, but it is not inhibited by inhibitors of serine or thiol proteases.

A polypeptide of M_r 18,000, intermediate in size between the RbcS precursor from pea (M_r 20,000) and the mature RbcS (M_r 14,500; Bedbrook *et al.*, 1980), appeared during incubation of this precursor with the enzyme purified from pea, which suggested that processing occurred in two steps (Robinson and Ellis, 1984b). Processing could be arrested at the intermediate stage by treating the RbcS precursor with iodoacetate, which carboxymethylated the cysteine residue adjacent to the cleavage site that produces the mature subunit. Uptake of RbcS precursors, translated from *Chlamydomonas* poly(A)-rich RNA, also resulted in cleavage only to a polypeptide intermediate in size between the precursor (M_r about 22,000) and the mature RbcS (M_r 16,500) (Mishkind *et al.*, 1985).

In extracts of *Chlamydomonas* cells, processing activity is present that cleaves the precursor of RbcS (Dobberstein *et al.*, 1977; Chua and Schmidt, 1978; Marks *et al.*, 1986), ribosomal proteins (Schmidt *et al.*, 1984), and Cab polypeptides (Marks *et al.*, 1985). No significant general proteolytic degradation occurs with these extracts (Marks *et al.*, 1985, 1986). In contrast, *in vitro* processing of the RbcS precursor occurs very rapidly and is highly specific. No evidence of intermediate forms can be detected in a homologous *Chlamydomonas* system. However, when cleavage at the primary site, which yields the mature RbcS, is hindered by antibodies bound to the precursor polypeptide, cleavage at a secondary site is detected (Marks *et al.*, 1986). The product of this cleavage is an intermediate (M_r 18,500) that is similar in size to the polypeptide produced when the *Chlamydomonas* RbcS precursor is taken up by pea chloroplasts (Mishkind *et al.*, 1985). These results indicate that a secondary cleavage site may occur in the transit sequence. The significance of the second site, which is most clearly observed when the "primary" site is hindered, is not known. Whether processing does in fact proceed *in vivo* in two steps, or whether the second site is cleaved only when the "primary," kinetically more favorable, site is blocked, also is not known.

In vitro experiments, in which cell-free translates are added to intact chloroplasts of higher plants, demonstrated that RbcS enters the stroma, where it is recovered as part of the RuBPC holoenzyme (Chua and Schmidt, 1978; Smith and Ellis, 1979). In contrast, Cab polypeptides were recovered in thylakoid membranes; no detectable amounts of Cab polypeptides were found in envelope or stromal fractions (Schmidt *et al.,* 1981; Bellemare *et al.,* 1982; Kohorn *et al.,* 1986). Thus the posttranslational transport of these polypeptides culminates in correct localization within the functional compartments of the chloroplast.

No information exists on the fate of transit sequences after they are removed from proteins by the processing step. Presumably these peptides are rapidly degraded by proteases within the organelle. A variety of proteolytic activities, in addition to the processing enzyme, have been detected in chloroplast preparations. Recently, Liu and Jagendorf (1986a,b) have characterized several endo- and aminopeptidases that are optimally active at neutral pH values. Whereas the peptidases are soluble enzymes in the stroma, these investigators have also detected ATP-dependent proteolytic activity associated with thylakoid membranes.

X. ASSEMBLY OF COMPLEXES IN THYLAKOID MEMBRANES

A. The Complexes

An understanding of the biogenesis of the chloroplast rests on the underlying knowledge of gene structures and the interaction of the gene products with other chloroplast components. Both the structure and the function of the chloroplast are dependent on the proper assembly of elaborate yet stringently defined complexes. Most of these complexes are composed of products of both nuclear and chloroplast genes. The search for knowledge of how these gene products are processed, sorted among plastid compartments, interact with each other and with chlorophyll and other lipids, and associate to form the thylakoid membrane has been a major impetus for current research. But the assembly process is not confined to the membrane, for ribulose bisphosphate carboxylase and ribosomes are complexes that reside in the stroma whose structure and function also are dependent on the proper coordination between nuclear and plastid gene products.

In this last section of this chapter, aspects of the assembly of several of the complexes in thylakoid membranes are considered. Eight such complexes have been identified, which comprise 45–50 different proteins. Three of the complexes are associated with photosystem 2: the photosystem reaction-center complex, the water-oxidation complex, and the major light-harvesting complex (LHC-2). Three other complexes are associated with photosystem 1: the reaction center complex, a light-harvesting complex (LHC-1), and the complex involved in the transfer of electrons from photosystem 1 to $NADP^+$. The seventh is the cytochrome b_6/f complex, which functionally connects the two photosystems. The eighth is the ATP synthase, which comprises both the integral membrane complex CF_0 (the proton channel) and the attached peripheral complex CF_1 that contains the active site for ATP synthesis. The discussion that follows will emphasize knowledge of the assembly of the cytochrome b_6/f complex.

B. The Cytochrome b_6/f Complex

The cytochrome b_6/f complex of higher plant chloroplasts contains a minimum of one molecule of cytochrome f (285 amino acids, M_r 31,800 including the heme), one of cytochrome b_{563} (b_6, with two heme groups, 208 amino acids, M_r 23,000), one Rieske iron–sulfur (Fe_2S_2) protein (M_r 19-20,000), and one polypeptide of M_r 17,500 (spinach) or 15,200 (pea) (Hurt and Hauska, 1982; Alt et al., 1983; Hauska et al., 1983; Widger et al., 1984a,b; Willey et al., 1984a,b; Ortiz and Malkin, 1985; Mansfield and Anderson, 1985). The M_r 17,500 (or 15,200) polypeptide has been designated "subunit IV." In some preparations, an additional polypeptide of M_r 5000 (Hurt and Hauska, 1982) and one of M_r 37,000 (Clark and Hind, 1983; Alt et al., 1983; Coughlan et al., 1985; Ortiz and Malkin, 1985) are present. The M_r 37,000 polypeptide has been identified as ferredoxin-$NADP^+$ oxidoreductase (Clark et al., 1984) and is not an authentic member of the complex.

As an aside, it must be pointed out that estimates of relative molecular mass (M_r) by polyacrylamide gel electrophoresis are subject to considerable error. For most of the components of the cytochrome complex, the primary sequences of the polypeptides, and thus the molecular weights, are known as a result of nucleotide sequence analyses of the genes. Previous estimates by electrophoresis of the molecular weight of cytochrome f provided a value of 33,000–37,000 (Alt et al., 1983; Ortiz and Malkin, 1985; Coughlan et al., 1985; Rothstein et al., 1985), whereas the gene contains information for a polypeptide of 31,096 (Willey et al., 1984a). Another example is the photosystem 2 reaction-center apoprotein (P_{680}), which by electrophoresis was estimated to have a M_r of 51,000; the gene encodes a protein of molecular weight 56,246 (Morris and Herrmann, 1984). The reaction-center apoproteins of photosystem 1 migrate during electrophoresis as M_r 66,000–68,000 components, although their plastid genes encode proteins of 83,200 and 82,500 in molecular weight (Fish et al., 1985; Lehmbeck et al., 1986). These differences do not seem to reflect cleavage of precursor forms. Thus, until the gene for a protein is sequenced and the amino acid composition is known, values for molecular weight estimated by electrophoresis must be used with reservation.

The M_r 15,200 polypeptide in the cytochrome b_6/f complex from pea (subunit IV) contains three hydrophobic regions within its sequence of 139 amino acids (Phillips and Gray, 1984). Widger *et al.* (1984b) found that the gene for the M_r 17,500 subunit IV in spinach was somewhat homologous to the C-terminal half of the mitochondrial cytochrome b, a protein of M_r 42,000. On this basis they suggested that because the coding sequence for the M_r 17,500 polypeptide is contiguous with that for cytochrome b_{563} in the chloroplast genome, the chloroplast gene may have became split during evolution. The M_r 17,500 remnant does not bear a heme group and is not directly involved in electron transport through the complex. Yet it is readily cross-linked to the Rieske iron–sulfur protein (Lam, 1986) and thus is an authentic component of the complex. Although a function has not been assigned to subunit IV, it is an integral membrane protein as a consequence of its hydrophobicity, and possibly plays a structural role in maintaining close aposition of the other electron carriers.

The cytochrome b_6/f complex also contains one molecule of a quinone, probably plastoquinone C, and acts as a plastoquinol–plastocyanin oxidoreductase (Hurt and Hauska, 1982; Hauska *et al.*, 1983). Thus, its function in photosynthesis is to transfer electrons from plastoquinone, which is reduced on the stromal side of the membrane by the action of photosystem 2, to plastocyanin, the donor of electrons to photosystem 1 on the luminal side of the membrane (Hauska *et al.*, 1983; Lam and Malkin, 1985a). Lam and Malkin (1985a) demonstrated that the purified cytochrome complex transferred electrons *in vitro* from an oxygen-evolving photosystem 2 complex to plastocyanin. Electron transport was maximal when three to four cytochrome complexes were present per photosystem 2 reaction center. In thylakoid membranes, these units would be connected by a common plastoquinone pool. The order of electron transfer within the cytochrome complex seems to be plastoquinone → cytochrome b_{563} → Rieske iron-sulfur protein → cytochrome f → plastocyanin.

Cross-linking studies have shown that the cytochrome complex in thylakoid membranes is closely associated with both photosystem 2 and photosystem 1 (Lam and Malkin, 1985b). Because results from membrane fractionation indicated that photosystem 2 is primarily located with stacked, granal membranes, whereas photosystem 1 is primarily located with nonappressed, stromal membranes (Anderson and Andersson, 1982; Andersson, 1986), the cytochrome complex apparently is distributed between both membrane subtypes. Allred and Staehelin (1985) confirmed this distribution in spinach chloroplasts. This arrangement would suggest that separate units of the complex are associated with either photosystem 1 or photosystem 2. In contrast, cytochrome b_{559}, a large (M_r 111,000) protein that is associated with the oxidizing side of photosystem 2, is localized exclusively in granal thylakoids in spinach (Rao *et al.*, 1986).

The location of the genes and the site of synthesis of each of the proteins of the cytochrome b_6/f complex are known. The cytochromes and subunit IV are coded by chloroplast DNA (Alt *et al.*, 1983; Willey *et al.*, 1984a,b; Phillips and Gray, 1984) and are synthesized within the chloroplast from poly(A)-minus mRNA (Doherty and Gray, 1979; Alt *et al.*, 1983) (see Section II). However, the mRNA for the Rieske iron–sulfur protein is found in the poly(A)-rich fraction (Alt *et al.*, 1983) and is transcribed from nuclear DNA and translated in the cytoplasm (Tittgen *et al.*, 1986). The nucleotide sequences and the deduced amino acid sequences for cytochrome *f* from pea and wheat (Willey *et al.*, 1984a,b) and for cytochrome b_{563} (Widger *et al.*, 1984b) and subunit IV from pea (Phillips and Gray, 1984) have been determined.

Based on results of hydropathy analyses and studies in which topological arrangements of these proteins in thylakoid membranes were probed with proteases, the evidence is clear that (1) the bulk of cytochrome *f*, including the N-terminal heme-binding domain, is exposed on the luminal surface of thylakoid membranes (Willey *et al.*, 1984a,b; Mansfield and Anderson, 1985). The C-terminus is exposed on the stromal surface, where it is accessible to attack by carboxypeptidase, trypsin, proteinase K, and pronase E (Willey *et al.*, 1985a; Mansfield and Anderson, 1985; Ortiz and Malkin, 1985). Trinitrobenzene sulfonate, a membrane-impermeant reagent, forms adducts with cytochrome *f* (Ortiz and Malkin, 1985), possibly by reacting with the cluster of lysine residues near the C-terminus. The protein appears to be anchored to the membrane by a single hydrophobic sequence of 20 amino acids near the C-terminus (Willey *et al.*, 1984a). (2) The C-terminus of cytochrome b_{563} also is accessible to carboxypeptidase and pronase E digestion on the stromal surface of the membrane (Mansfield and Anderson, 1985; Ortiz and Malkin, 1985). The model developed by Widger *et al.* (1984b) from the amino acid sequence suggests that the N-terminus is on the opposite side of the membrane. Five hydrophobic, potentially membrane-spanning segments are present in the polypeptide. Although the sequence suggests that the bulk of cytochrome b_{563} is embedded within the membrane, trypsin digests the protein when the luminal surface is exposed on inside-out vesicles (Mansfield and Anderson, 1985). Thus, this cytochrome is exposed on both surfaces of the membrane, as predicted by the model of Widger *et al.* (1984b). (3) Subunit IV is highly hydrophobic (Phillips and Gray, 1984) and is embedded within the membrane. Its C-terminus is accessible to digestion by carboxypeptidase in right-side-out but not in inside-out vesicles (Mansfield and Anderson, 1985).

Cytochromes b_{563} and *f* and subunit IV, therefore, are encoded by chloroplast DNA, synthesized within the organelle, and span the thylakoid membrane with their C-termini on the stromal surface. The N-termini, although definitely known only for cytochrome f, are on the luminal surface. Messenger RNAs for cytochrome b_{563} and subunit IV were translated *in vitro* into products the same size as the mature proteins, but cytochrome f was initially made as a precursor about 4000 daltons larger than the mature form (Alt *et al.*, 1983). Cytochrome f also appears to be made by ribosomes attached to the membrane (Willey *et al.*, 1984a). The nucleotide sequence of the gene for cytochrome f indicates that the coding region extends further upstream from the sequence that codes for the N-terminus of the mature protein (Willey *et al.*, 1984a). The N-terminal extension, possibly 35 amino acids long, has the hydrophobic characteristics of a signal sequence (von Heijne, 1985). This notion was established by Rothstein *et al.* (1985), who fused the 5'-end of the cytochrome f gene to the *lacZ* gene. The resulting fused protein became associated with the cell membrane of the host bacterial cells, which demonstrated that the chloroplast "signal sequence," the N-terminus of the cytochrome f precursor, indeed was recognized by the bacterial secretory system and initiated export of β-galactosidase, which is a cytoplasmic enzyme in bacterial cells. The mode of synthesis and insertion into thylakoid membranes of these polypeptides, therefore, may be similar to the mode that occurs on ribosomes bound to the endoplasmic reticulum of animal cells and the cell membrane of bacteria (for reviews, see Sabatini *et al.*, 1982; Wickner and Lodish, 1985).

The orientation in the membrane of the Rieske iron–sulfur protein is less certain than for the other members of this complex. Mansfield and Anderson (1985) found that digestion of the stromal surface of thylakoid membranes with carboxypeptidase markedly decreased the apparent antigenicity of the protein but did not cause a detectable change in size. However, pronase E, when added to the stromal surface, digested the protein to a fragment of lower molecular weight (Ortiz and Malkin, 1985). Thus, a portion of the protein is exposed on the stromal surface. The bulk of the protein, however, including the Fe_2S_2 center, resides on the luminal surface of the membrane, where it functions in the transfer of electrons to cytochrome f (Hauska *et al.*, 1983). *In vitro* translation of mRNA for the Rieske iron–sulfur protein produced a polypeptide about 7000 daltons larger than the mature protein (Alt *et al.*, 1983). If this extension is at the N-terminus, as is the case for other cytoplasmically made chloroplast proteins (see Section IX), this presequence presumably functions to transfer this end of the molecule into the stroma, where the processing activity is located. Consequently, it is possible that this protein spans the membrane with its N-terminus exposed in the stroma and its C-terminus within the thylakoid lumen.

There is no information at present to predict how the components of this or any other complex organize themselves into specific multiprotein assemblies. Light markedly stimulates synthesis of most of the components by causing an increase in the level of mRNAs, but the mRNAs are not induced to the same extent or at the same rate (Takabe *et al.*, 1986). Furthermore, an interruption in the synthesis of one component of the complex causes a coordinate reduction in the amounts of the other components such that the correct stoichiometric relationships are maintained (Barkan *et al.*, 1986). This exquisite supervision over the accumulation of membrane proteins occurs at a posttranscriptional level and indicates that the process of assembly is closely monitored, possibly by proteolytic enzymes.

Because the various aggregates, such as the cytochrome b_6/f complex, the photosystem complexes, and the light-harvesting complexes, can be isolated, the interaction between the constituents apparently is relatively strong. Elucidation of the characteristics of these interactions perhaps will be the next phase in the analysis of membrane assembly.

C. Plastocyanin and Other Luminal Proteins

Plastocyanin resides on the luminal surface of thylakoid membranes (Hauska *et al.*, 1983) and is easily washed from inside-out membranes. It is not an integral membrane component. The structure of the protein, which contains a single copper atom and consequently forms intensely blue crystals, has been extensively studied (Church *et al.*, 1986). The protein is quite soluble in water and has its nonpolar amino acid side-chains oriented towards the core of the protein; polar side-chains are on the surface (Draheim *et al.*, 1986), an arrangement typical of globular proteins. The gene for plastocyanin is located in nuclear DNA, and the mRNA is recovered with the poly(A)-rich fraction (Grossman *et al.*, 1982). Smeekens *et al.* (1985b) cloned a full-length cDNA and determined its nucleotide sequence. The mature protein consists of 99 amino acids (M_r 10,400), but the gene encodes a precursor polypeptide that is 66 amino acids longer. This is an unusually long and complex presequence for such a small protein and it possibly is involved in directing the protein to its site of function (see below).

Assembly of the cytochrome b_6/f complex and association with its cognate, plastocyanin, requires a joint effort between the nuclear-cytoplasmic and chloroplast compartments. The cytochromes and subunit IV are synthesized within the plastid and are inserted into thylakoid membranes from the stromal surface, with an orientation predicted if the ribosomes are attached to the membrane. The path traveled by the Rieske iron–sulfur protein and plastocyanin to their sites of function on the membrane, after synthesis in the cytoplasmic compartment, is, however, less clear. Of considerable interest is how plastocyanin, a water-soluble protein, achieves its location within the thylakoid lumen. Evidence was presented recently in support of the suggestion that such proteins pass through the two membranes of the envelope, enter the stroma, and subsequently cross the thylakoid membrane into the lumen (see Section IX). The bipartite N-terminal extension in the plastocyanin precursor is processed in two steps, one in the stroma and one in thylakoid membranes (Smeekens *et al.*, 1986).

Whether the path described above adequately explains transport of proteins into the thylakoid lumen remains to be established. Such a pathway seems highly unfavorable from a thermodynamic standpoint. Usually, hydrophobic segments serve as "stop-transfer" signals and halt translocation of a polypeptide through a membrane at the point where these segments maximally interact with the hydrophobic core of the membrane (Blobel, 1983; Davis and Model, 1985; Wickner and Lodish, 1985). Thus, it seems unlikely that the hydrophobic sequence within the N-terminal extension of the plastocyanin precursor can pass completely through the inner membrane of the envelope. For the same reason, it should be thermodynamically favorable for thylakoid membrane proteins, which also contain hydrophobic sequences (Fig. 3), to become integrated into the inner membrane of the envelope during import into the organelle, by a process dictated by these sequences. In contrast, proteins that lack extensive hydrophobic segments apparently are not restrained in this manner and pass through the envelope into the stroma. Once in the stroma and processed, these soluble proteins are trapped in this compartment.

There are other possible means by which a protein such as plastocyanin reaches the thylakoid lumen, which would not require crossing three membranes. The presequence for plastocyanin resembles that for mitochondrial proteins such as cytochrome *c* peroxidase (Hurt and van Loon, 1986). Processing of mitochondrial proteins that contain such long and bipartite presequences also occurs in two steps, one in the matrix that is similar to the process for other imported proteins, and the second on the outside of the inner membrane. The result is that the protein becomes localized within the mitochondrial intermembrane space. By analogy to this process in mitochondria, it is possible that the precursor of plastocyanin associates with the cytosolic surface of the inner envelope membrane. The N-terminal "transit sequence" would permit entrance of this end of the polypeptide into the stroma, where processing by the stromal protease would occur. However, transfer across the membrane would be halted by the contiguous hydrophobic segment. Cleavage of the remainder of the N-terminal extension then may occur on the cytoplasmic surface of the inner membrane of the envelope, by a reaction analogous to the second processing step for mitochondrial proteins. The consequence of this processing would be an association of mature plastocyanin initially with the cytosolic surface of the inner envelope membrane.

Several extrinsic thylakoid proteins of M_r 33,000, 23,000, and 16,000, which are involved with the oxidation of water by photosystem 2, are also localized to the luminal surface of thylakoid membranes (Liveanu *et al.*, 1986) but synthesized in the cytoplasm (Westhoff *et al.*, 1985b; Tittgen *et al.*, 1986). Chia and Arntzen (1986) described an interesting defect in processing of these proteins, which normally occurs in two steps just as that of plasto-cyanin. In a chloroplast genome mutant of tobacco, *lutescens-1,* the second processing activity is lost as the chloroplasts mature. The result is the integration of the intermediate forms of these proteins into the luminal side of thylakoid membranes, held onto the membrane apparently by the hydrophobic portion of the N-terminal extension. It seems likely that these proteins achieved this orientation because this segment served as a stop-transfer signal at the level of the chloroplast envelope.

Mayfield *et al.* (1987) cloned from *Chlamydomonas* a cDNA and a genomic fragment that encode a M_r 20,000 protein that is analogous to the M_r 23,000 protein in the oxygen-evolving complex of higher plants. The gene for this protein is located in nuclear DNA and encodes a precursor protein with a 57-amino acid, bipartite, N-terminal extension similar to the presequence for plastocyanin. Such complex N-terminal extensions seem to be characteristic of luminal proteins that are synthesized on cytoplasmic ribosomes.

Therefore, as illustrated in Fig. 4 (pathway 1), thylakoid membranes perhaps expand as the result of invaginations of the inner membrane, fission to form vesicles, and then fusion of these vesicles with thylakoid membranes. This process would result in the transfer of membrane material, as vesicles that also enclose cytoplasmically derived, luminal proteins, to growing thylakoids (Blobel, 1983). The cytosolic surface of the inner envelope membrane, with its associated proteins, then becomes the luminal surface of the

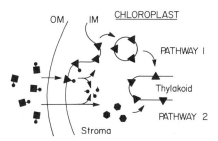

Fig. 4. Possible paths for transport of proteins from their site of synthesis in the cytoplasm to thylakoid membranes. Precursors of stromal proteins are transported across the chloroplast envelope and are processed to the mature forms by a protease located in the stroma. One possible path for membrane proteins, and those destined for the thylakoid lumen, involves an initial processing step in the stroma but an additional step at the level of the thylakoid membrane (pathway 2). This second step may involve a conformational change and/or an additional proteolytic step to achieve the mature form (▲). Another possible path (pathway 1) would involve integration of precursor forms into the envelope inner membrane. In this pathway, the N-termini of the proteins would be transported into the stroma, where the initial processing step would occur. Pathway 1 implies that extended hydrophobic sequences within the proteins would act to stop-transfer signals, thus preventing further transfer of the proteins into the stroma. This hypothesis requires that membranous elements would transfer such proteins, along with lipids synthesized in the envelope, to thylakoid membranes.

thylakoid. In this scheme, only the outer membrane of the envelope needs to be crossed by membrane-bound and luminal proteins, a feat that every other chloroplast-destined protein also accomplishes. Possibly some associated envelope enzymes could also be carried to the thylakoid membrane, where they may tend to accumulate. It remains to be established whether thylakoid and luminal proteins indeed are transferred in from the envelope by a flow of vesicles. Resolution of this question most likely will also require an understanding of the transfer of lipids from their site of synthesis in the envelope to thylakoids.

D. The Photosystems

Several recent studies have advanced our knowledge of the composition and assembly of photosynthetic units. Two closely related proteins occur in reaction centers of photosystem 1. The proteins from maize have molecular weights of 83,200 and 82,500, as deduced from the gene sequences (Fish *et al.*, 1985). In pea, the molecular weights of these proteins are 84,100 and 82,400 (Lehmbeck *et al.*, 1986). These proteins bind chlorophyll *a*, or chlorophyll RC I (see Section VII,D), to form a complex that constitutes P700, the functional reaction center for photosystem 1. Both proteins are encoded in chloroplast DNA, and the genes for these proteins (*psaA* and *psaB*) lie adjacent to each other (Fig. 1). Expression of these genes is dependent on light (Vierling and Alberte, 1983; Fish *et al.*, 1985; Takabe *et al.*, 1986), although synthesis of the polypeptides continues after illuminated leaves are transferred to the dark (Vierling and Alberte, 1983).

Reaction-center complexes of photosystem 1 are associated with a light-harvesting chlorophyll-*a*/*b* complex (LHC-1), which contains several (generally four) polypeptides of M_r 19,000–24,000 in pea (Haworth *et al.*, 1983) or of M_r 14,000–22,000 in algae (Ish-Shalom and Ohad, 1983). Five polypeptides ranging from M_r 19,000 to 24,500 were found in LHC-1 from the siphonaeous alga *Codium* (Chu and Anderson, 1985). The chlorphyll *a*/*b* ratio in LHC-1 is 2–4 (Haworth *et al.*, 1983; Melis and Anderson, 1983; Chu and Anderson, 1985). The light-harvesting complexes surround, and transfer absorbed energy to, the reaction centers where the photochemical reactions occur at high efficiency.

Two homologous proteins also are present in the reaction center of photosystem 2. Several years ago, the reaction center activity was thought to reside on proteins that by electrophoresis were estimated to have molecular weights of 47,000–51,000 and 43,000–44,000 (Delepelaire and Chua, 1981; Camm and Green, 1983; Satoh *et al.*, 1983; Delepelaire, 1984; Minami *et al.*, 1986). The genes for both proteins were located on chloroplast DNA (Westhoff *et al.*, 1983; Westhoff, 1985). The gene encoding the larger protein was sequenced, which provided a deduced amino acid sequence for a protein of M_r 56,246 (Morris and Herrmann, 1984). More recently, evidence has mounted for D1, the *psbA* gene product, and D2, a protein homologous to D1 and encoded by *psbD,* as the reaction center components. Nanba and Satoh (1987) isolated a complex that exhibited a light-induced absorbance change at 682 nm, a characteristic of the photosystem 2 reaction center pigment designated P_{680}. This complex contained only D1, D2, cytochrome b_{559}, five molecules of chlorophyll *a,* two of pheophytin *a,* and one of β-carotene.

Assembly of functional photosystem 2 reaction centers from proteins synthesized in isolated spinach chloroplasts has been demonstrated (Minami *et al.,* 1986). In *Chlamydomonas,* nuclear gene products are required to achieve proper assembly of the reaction center complex (Jensen *et al.,* 1986). Mutant strains were isolated in which arrested synthesis of a M_r 51,000, chlorophyll *a*-binding protein seemed to result in inhibition within the chloroplast of translation of the abundant mRNA for D1. Consequently, D1 did not accumulate and newly synthesized D2 was rapidly degraded. Assembly of these complexes seemed to be tightly coordinated, particularly at the translational and post-translational levels. The requirement of the M_r 51,000 protein for assembly of the complex may have caused the difficulty in identifying the reaction center components.

The photosystem 2 reaction centers are surrounded by light-harvesting chlorophyll *a/b* complexes, among which the major complex is referred to as LHC-2. Each LHC-2 includes several proteins of M_r 26,000–29,000, but one member of the group usually is predominant (Delepelaire and Chua, 1981; Schmidt *et al.,* 1981; Kaplan and Arntzen, 1982; Mullet, 1983). The heterogeneity of the apoproteins may partly result from the fact that they are encoded by a gene family but also from posttranslational modifications (Kohorn *et al.,* 1986; Leutwiler *et al.,* 1986; Darr *et al.,* 1986). The chlorophyll *a/ b* ratio in this complex is near 1, with probably four molecules of chlorophyll *a* and three molecules of chlorophyll *b* associated with each polypeptide (Melis and Anderson, 1983; Zuber, 1985). Each LHC-2 also contains one to two xanthophylls, which may be required for assembly of the complex (Siefermann-Harms, 1985; Zuber, 1985). Plumley and Schmidt (1987) developed a method for reconstituting LHC-2 from purified lipids and delipidated, dodecyl sulfate-denatured, thylakoid polypeptides. Successful reconstitution required chlorophyll *a,* chlorophyll *b,* and xanthophylls; maximal assembly occurred only when all three xanthophylls, lutein, violaxanthin, and neoxanthin, were present. The chlorophyll *a/b*-binding (Cab) proteins are synthesized on cytoplasmic ribosomes, and the nuclear genes for several of these proteins have been cloned and sequenced (see Section V,B). These proteins are the predominant protein components of thylakoid membranes and have attracted great interest and attention (for reviews, see Bennett, 1983; Zuber, 1985; Gounaris *et al.,* 1986; Thornber, 1986).

The core proteins of the reaction centers are encoded within the chloroplast genome and are synthesized within the organelle. In contrast, the apoproteins of the light-harvesting complexes, and the Cab proteins of LHC-2 in particular, are encoded by nuclear DNA and synthesized in the cytoplasm. Interestingly, the LHC-1 apoproteins may be related to those of LHC-2, because they have recently been found to exhibit immunological cross-reactivity (Evans and Anderson, 1986; Darr et al., 1986). Models of the photosystems place the chloroplast-derived, core proteins at the center of the photosystem complexes, with the light-harvesting and electron donor and acceptor complexes surrounding the reaction centers (Camm and Green, 1983; Gounaris et al., 1986). However, the LHCs can integrate into thylakoid membranes without the core proteins, but in this case the absorbed energy is not trapped and is emitted as fluorescent light (Delepelaire, 1984; Delepelaire and Wollman, 1985).

A major redistribution of chlorophyll occurs when leaves that contain chloroplasts still in the process of development are returned to the dark. Under these conditions, photosystem 1 and 2 reaction centers continue to be formed but at the expense of light-harvesting complexes (Argyroudi-Akoyunoglou et al., 1982; Akoyunoglou and Akoyunoglou, 1985). As a result, chlorophyll a is conserved, whereas chlorophyll b and the LHC-1 and LHC-2 apoproteins are degraded. Synthesis of the LHC apoproteins also continues to some extent in the dark, but they do not accumulate. The explanation given for these phenomena is that the affinity of reaction center apoproteins for chlorophyll a is greater than that of the LHC apoproteins. Consequently, continued assembly of the reaction center complexes apparently causes dissociation of the LHCs, which results in degradaton of the LHC apoproteins and chlorophyll b (Akoyunoglou and Argyroudi-Akoyunoglou, 1986). These data again demonstrate that pigments and polypeptides that are not associated with other membrane components in proper complexes are degraded. Perhaps the LHCs, which contain much more chlorophyll than the reaction centers, serve not only as energy antennae but also as a reserve of chlorophyll a for the reaction centers.

E.   Entry of Cab Proteins into Chloroplasts

A major question still calling for an answer is the path of transport of imported chloroplast proteins to their site of function. Possible schemes were proposed above (see Section X,C) in relation to the transport of plastocyanin to the lumen of thylakoids. An argument was developed in favor of initiation of assembly of complexes at the level of the inner membrane of the envelope, such that growth of thylakoids occurs by a flow of membrane from the envelope. This scheme is simply an elaboration of suggestions proposed years ago (Mühlethaler and Frey-Wyssling, 1959; Menke, 1962; Hoober, 1976; Douce et al., 1984). One feature of proteins that may provide a clue to the transport mechanism, albeit an equivocal one, is the orientation of proteins in thylakoid membranes. For example, the major LHC-2 (Cab) proteins span the thylakoid membrane such that the N-termini of some are exposed on the stromal surface. Karlin-Neumann *et al.* (1985) proposed a model for the orientation of the Cab polypeptides in membranes based on hydropathy plots of the amino acid sequence. The model suggests that a large portion of the protein, including the N-terminal region and a highly polar internal segment, is exposed on the stromal surface (see Fig. 3). However, as discussed in Section V,B, sufficient evidence to verify this model is still lacking.

In *Chlamydomonas,* three major Cab polypeptides exist (Chua and Blomberg, 1979; Delepelaire and Chua, 1981). Trysin and thermolysin cleave about 15 amino acids from two of the polypeptides but not the third (Bar-Nun *et al.,* 1977; Chua and Bomberg, 1979; Delepelaire and Wollman, 1985). The two that are sensitive to protease also contain blocked N-termini, but the third does not (Hoober *et al.,* 1980). Furthermore, the former two polypeptides also are readily phosphorylated *in vivo,* whereas the latter is not (Owens and Ohad, 1983; Delepelaire and Wollman, 1985). Although immunochemically and structurally very similar (Chua and Blomberg, 1979; Hoober *et al.,* 1982; Marks *et al.,* 1985), these polypeptides may be positioned differently within the membrane.

It seems clear that transfer of the N-terminus of Cab precursor proteins through the inner envelope membrane is required for processing to the mature form. The processing enzyme would cleave the transit sequence, leaving the N-terminus exposed to the stroma. Subsequently, internal hydrophobic sequences may serve as stop-transfer sequences. This orientation of the N-terminus is established by its sensitivity of proteolytic digestion (Mullet, 1983). The position of the remainder of the protein remains in doubt. Kühlbrandt (1984) found that the Cab proteins form an asymmetric transmembrane arrangement, with large surface areas exposed on one side of the membrane. But the membrane surfaces were not identified in this study.

Whether the Cab proteins associate with the inner envelope membrane, as suggested above, or pass through into the stroma is not known. Cline (1986) developed a system that integrated Cab precursors directly into thylakoid, but not envelope, membranes. Integration into thylakoid membranes required ATP and the chloroplast stromal fraction. Surprisingly, in this system very little processing to the mature Cab polypeptides occurred, but the integrated precursors were proteolytically cleaved to a fragment characteristic of the membrane-bound form. During studies of the import of precursors into intact chloroplasts, soluble forms of the processed Cab proteins could not be detected (Schmidt *et al.,* 1981; Bellemare *et al.,* 1982; Kohorn *et al.,* 1986). Moreover, the polypeptides were recovered in the thylakoid fraction in these studies. The suggestion has been made that the presence of chlorophyll, particularly chlorophyll *b,* is required for stable integration of the proteins into developing membranes (Apel and Kloppstech, 1980; Bennett, 1981; Schmidt *et al.,* 1981; Bellemare *et al.,* 1982; Kohorn *et al.,* 1986). In its absence, the proteins are rapidly broken down. On the other hand, in *Chlam-*

ydomonas, which differs from higher plants in that chlorophyll *b*-less mutants still retain nearly full complements of the Cab proteins (Michel *et al.,* 1983; Picaud and Dubertret, 1986), this breakdown system seems less active. When degreened cells of this alga were induced to synthesize the Cab proteins in the light under conditions that prevented chlorophyll synthesis, the Cab proteins were integrated into membranes and achieved the same orientation as in control cells in which chlorophyll synthesis was allowed (J. K. Hoober, unpublished observations). These experiments have not established the membrane type into which the proteins were integrated, but they showed that integration of the proteins into membranes is independent of chlorophyll. However, even without chlorophyll synthesis, a soluble form of the proteins could not be detected.

The site of association of the Cab proteins with chlorophyll also is not known. Purified envelope membranes are essentially devoid of chlorophyll (Block *et al.,* 1983c; Pineau *et al.,* 1986). But confounding this problem is a lack of definitive evidence of where the terminal steps in chlorophyll synthesis occur. Fuesler *et al.* (1984b) suggested from studies of the inhibition of magnesium chelatase with *p*-chloromercuribenzene sulfonate, which is thought to be unable to penetrate the envelope of intact chloroplasts, that this step occurs in the envelope membranes. This suggestion is supported by the detection of protochlorophyllide in purified envelope membranes but not in thylakoid membranes of mature spinach chloroplasts (Pineau *et al.,* 1986). Furthermore, conversion of protochlorophyllide to chlorophyllide *a* in the light, and to chlorophyllide *b* in phenanthroline-treated cells in the dark, occurs at maximal rates even in extensively degreened *Chlamydomonas* cells that are nearly depleted of thylakoid membranes (Bednarik and Hoober, 1985a,b).

These results indicate that a high-capacity system for chlorophyll synthesis exists in nonthylakoid membranes and presumably in the envelope. If this location can be established, then chlorophyll(ide)s may associate with Cab proteins in the envelope inner membrane. The complexes then may travel to thylakoids along with the lipids that are synthesized in the envelope (see Section VIII). The question then arises as to what triggers expulsion of thylakoid components from the envelope. Development of a well-defined system is necessary to study these questions in the assembly of the light-harvesting complexes.

The scenario above predicts that thylakoid membranes expand by random incorporation of products of the cytoplasmic and plastid protein-synthesizing systems. The cytoplasmic proteins achieve their site of function by the expression of information within the proteins themselves. Interestingly, most of the plastid products seem to be made by ribosomes attached to thylakoid membranes (reviewed by Margulies, 1986). The products of membrane-bound ribosomes include the photosystem 1 reaction center apoprotein (Margulies, 1986), polypeptide D1, the herbicide-binding protein (Herrin and Michaels, 1985a), the α and β subunits of ATP synthase, CF_1 (Margulies, 1983; Bhaya and Jagendorf, 1985; Herrin and Michaels, 1985b), and cytochrome f (Willey *et al.,* 1984a,b). This site of synthesis of cytochrome f, in particular, seems to be determined by a hydrophobic N-terminal extension and results in transfer of the bulk of the protein across the membrane. Of interest also is the observation that the major portion of RbcL mRNA is associated with membrane-bound polysomes (Margulies, 1986; Hattori and Margulies, 1986). These studies have set the stage for an analysis of the distribution of these proteins after synthesis.

XI. CONCLUSION

The amount of knowledge on the molecular basis of chloroplast development is increasing at a rapid pace. Many of the membrane-bound and stroma proteins have been identified, their sites of synthesis established, and their genes isolated and sequenced. The information for import of plastid proteins after synthesis in the cytoplasm and for achievement of their proper functional location is contained entirely within the amino acid sequence of the proteins themselves. The specificity of integration of plastid-derived proteins into complexes also is dictated by information contained within each protein. The recognition of this fact has provided a great impetus to the current emphasis on the molecular biology of plastid genes and proteins. How the protein sequence information is read, and definition of the physical–chemical interactions between proteins, will provide a rich area of research for some time to come.

Just as the phenomenological descriptions of chloroplast development at the morphological level provided the base for biochemical descriptions of this process, and then as the biochemical information provided the base for the progress in understanding mechanisms at the level of the gene, so this current era of molecular biology will allow future investigations into the actual mechanisms of assembly of chloroplast components into the functional structures of the organelle. This latter area perhaps may require different techniques than are currently available. Nevertheless, those who are interested in, and provide support for, the processes in development of chloroplast structure and function, should be well satisfied with the progress in this area.

ACKNOWLEDGMENTS

I wish to express appreciation for the valuable assistance given by Maria E. Hoober in the preparation of this chapter. Research in my laboratory is supported by a grant from the National Science Foundation.

REFERENCES

Akoyunoglou, A., and Akoyunoglou, G. (1985). *Plant Physiol.* **79**, 425–431.
Akoyunoglou, G., and Argyroudi-Akoyunoglou, J. (1986). *In* "Regulation of Chloroplast Differentiation" (G. Akoyunoglou and H. Senger, eds.), pp. 571–582. Alan R. Liss, Inc., New York.
Akoyunoglou, G., and Senger, H., eds. (1986). "Regulation of Chloroplast Differentiation." Alan R. Liss, Inc., New York.
Allred, D. R., and Staehelin, L. A. (1985). *Plant Physiol.* **78**, 199–202.
Alt, J., Westhoff, P., Sears, B. B., Nelson, N., Hurt, E., Hauska, G., and Herrmann, R. G. (1983). *EMBO J.* **2**, 979–986.
Amann, E., Brosius, J., and Ptashne, M. (1983). *Gene* **25**, 167–178.
Anderson, J. M., and Andersson, B. (1982). *Trends Biochem. Sci.* **7**, 288–292.
Andersson, B. (1986). *In* "Methods in Enzymology" (A. Weissbach and H. Weissbach, eds.), Vol. 118, pp. 325–338. Academic Press, New York.
Andrews, J., and Mudd, J. B. (1985). *Plant Physiol.* **79**, 259–265.
Andrews, J., Ohlrogge, J. B., and Keegstra, K. (1985). *Plant Physiol.* **78**, 459–465.
Aoyagi, K., and Bassham, J. A. (1984). *Plant Physiol.* **76**, 278–280.
Aoyagi, K., and Bassham, J. A. (1985). *Plant Physiol.* **78**, 807–811.
Apel, K., and Kloppstech, K. (1980). *Planta* **150**, 426–430.
Apel, K., Santel, H.-J., Redlinger, T. E., and Falk, H. (1980). *Eur. J. Biochem.* **111**, 251–258.
Argyroudi-Akoyunoglou, J. H., Kondylaki, S., and Akoyunoglou, G. (1976). *Plant Cell Physiol.* **17**, 939–954.
Argyroudi-Akoyunoglou, J. H., Akoyunoglou, A., Kalosakas, K., and Akoyunoglou, G. (1982). *Plant Physiol.* **70**, 1242–1248.
Avissar, Y. J. (1983). *Plant Physiol.* **72**, 200–203.
Barkan, A., Miles, D., and Taylor, W. C. (1986). *EMBO J.* **5**, 1421–1427.
Bar-Nun, S., Schantz, R., and Ohad, I. (1977). *Biochim. Biophys. Acta* **459**, 451–467.
Bassi, R., Peruffo, A. d. B., Barbato, R., and Ghisi, R. (1985). *Eur. J. Biochem.* **146**, 589–595.
Batschauer, A., Santel, H.-J., and Apel, K. (1982). *Planta* **154**, 459–464.
Batschauer, A., Mösinger, E., Kreuz, K., Dörr, I., and Apel, K. (1986). *Eur. J. Biochem.* **154**, 625–634.
Beale, S. I. (1971). *Plant Physiol.* **48**, 316–319.
Beale, S. I., and Castelfranco, P. A. (1973). *Biochem. Biophys. Res. Commun.* **52**, 143–149.
Beale, S. I., and Castelfranco, P. A. (1974). *Plant Physiol.* **53**, 297–303.
Bedbrook, J. R., and Bogorad, L. (1976). *Proc. Natl. Acad. Sci. U.S.A.* **73**, 4309–4313.
Bedbrook, J. R., Link, G., Coen, D. M., Bogorad, L., and Rich, A. (1978). *Proc. Natl. Acad. Sci. U.S.A.* **75**, 3060–3064.
Bedbrook, J. R., Smith, S. M., and Ellis, R. J. (1980). *Nature (London)* **287**, 692–697.
Bednarik, D. P., and Hoober, J. K. (1985a). *Arch. Biochem. Biophys.* **240**, 369–379.
Bednarik, D. P., and Hoober, J. K. (1985b). *Science* **230**, 450–453.
Bednarik, D. P., and Hoober, J. K. (1986). *In* "Regulation of Chloroplast Differentiation" (G. Akoyunoglou and H. Senger, eds.), pp. 105–114. Alan R. Liss, Inc., New York.
Beer, N. S., and Griffiths, W. T. (1981). *Biochem. J.* **195**, 83–92.
Belford, H. S., Offner, G. D., and Troxler, R. F. (1983). *J. Biol. Chem.* **258**, 4503–4510.
Bellemare, G., Bartlett, S. G., and Chua, N. H. (1982). *J. Biol. Chem.* **257**, 7762–7767.
Bennett, J. (1981). *Eur. J. Biochem.* **118**, 61–70.
Bennett, J. (1983). *Biochem. J.* **212**, 1–13.
Bennoun, P., Spierer-Herz, M., Erickson, J., Girard-Bascou, J., Pierres, Y., Delosme, M., and Rochaix, J.-D. (1986). *Plant Mol. Biol.* **6**, 151–160.
Berends, T., Kubicek, Q., and Mullet, J. E. (1986). *Plant Mol. Biol.* **6**, 125–134.
Bergmann, P., Seyer, P., Burkard, G., and Weil, J.-H. (1984). *Plant Mol. Biol.* **3**, 29–36.
Berk, A. J., and Sharp, P. A. (1977). *Cell (Cambridge, Mass.)* **12**, 721–732.

Berry, J. O., Nikolau, B. J., Carr, J. P., and Klessig, D. F. (1986). *Mol. Cell. Biol.* **6**, 2347–2353.

Bertrams, M., and Heinz, E. (1981). *Plant Physiol.* **68**, 653–657.

Bertrams, M., Wrage, K., and Heinz, E. (1981). *Z. Naturforsch., C: Biosci.* **36**, 62–70.

Bhaya, D., and Castelfranco, P. A. (1985). *Proc. Natl. Acad. Sci. U.S.A.* **82**, 5370–5374.

Bhaya, D., and Jagendorf, A. T. (1985). *Arch. Biochem. Biophys.* **237**, 217–223.

Biekmann, S., and Feierabend, J. (1985). *Eur. J. Biochem.* **152**, 529–535.

Bird, C. R., Koller, B., Auffret, A. D., Huttly, A. K., Howe, C. J., Dyer, T. A., and Gray, J. C. (1985). *EMBO J.* **4**, 1381–1388.

Bishop, D. G., Sparace, S. A., and Mudd, J. B. (1985). *Arch. Biochem. Biophys.* **240**, 851–858.

Bitsch, A., and Kloppstech, K. (1986). *Eur. J. Cell Biol.* **40**, 160–166.

Blobel, G. (1983). *In* "Methods in Enzymology" (S. Fleischer and B. Fleischer, eds.), Vol. 96, pp. 663-682. Academic Press, New York.

Block, M. A., Joyard, J., and Douce, R. (1980). *Biochim. Biophys. Acta* **631**, 210–219.

Block, M. A., Dorne, A.-J., Joyard, J., and Douce, R. (1983a). *FEBS Lett.* **164**, 111–115.

Block, M. A., Dorne, A.-J., Joyard, J., and Douce, R. (1983b). *J. Biol. Chem.* **258**, 13273–13280.

Block, M. A., Dorne, A.-J., Joyard, J., and Douce, R. (1983c). *J. Biol. Chem.* **258**, 13281–13286.

Bogorad, L., Gubbins, E. J., Jolly, S. O., Krebbers, E. T., Larrinua, I. M., Muskavitch, K. M. T., Rodermel, S. R., Subramanian, A., and Steinmetz, A. (1983). *In* "Gene Structure and Regulation in Development" (S. Subtelny and F. C. Kafatos, eds.), pp. 13–32. Alan R. Liss, Inc., New York.

Bohner, H., Böhme, H., and Böger, P. (1981). *FEBS Lett.* **131**, 386–388.

Bottomley, W., Smith, H. J., and Bogorad, L. (1971). *Proc. Natl. Acad. Sci. U.S.A.* **68**, 2412-2416.

Breathnach, R., and Chambon, P. (1981). *Annu. Rev. Biochem.* **50**, 349–383.

Briat, J.-F., and Mache, R. (1980). *Eur. J. Biochem.* **111**, 503–509.

Broglie, R., Bellemare, G., Bartlett, S. G., Chua, N.-H., and Cashmore, A. R. (1981). *Proc. Natl. Acad. Sci. U.S.A.* **78**, 7304–7308.

Broglie, R., Coruzzi, G., Lamppa, G., Keith, B., and Chua, N.-H. (1983). *Bio Technology* **1**, 55–61.

Broglie, R., Coruzzi, G., Keith, B., and Chua, N.-H. (1984). *Plant Mol. Biol.* **3**, 431–444.

Burgess, R. R. (1976). *In* "RNA Polymerase" (R. Losick and M. Chamberlin, eds.), pp. 69–100. Cold Spring Harbor Lab., Cold Spring Harbor, New York.

Burnham, B. F., and Lascelles, J. (1963). *Biochem. J.* **87**, 462–472.

Camm, E. L., and Green, B. R. (1983). *J. Cell. Biochem.* **23**, 171–179.

Carey, E. E., Tripathy, B. C., and Rebeiz, C. A. (1985). *Plant Physiol.* **79**, 1059–1063.

Cashmore, A. R. (1984). *Proc. Natl. Acad. Sci. U.S.A.* **81**, 2960–2964.

Castelfranco, P. A., and Beale, S. I. (1983). *Annu. Rev. Plant Physiol.* **34**, 241–278.

Castelfranco, P. A., and Jones, O. T. G. (1975). *Plant Physiol.* **55**, 485–490.

Chia, C. P., and Arntzen, C. J. (1986). *J. Cell Biol.* **103**, 725–731.

Chitnis, P. R., Harel, E., Kohorn, B. D., Tobin, E. M., and Thornber, J. P. (1986). *J. Cell Biol.* **102**, 982–988.

Chu, N. M., and Tewari, K. K. (1982). *Mol. Gen. Genet.* **186**, 23–32.

Chu, N. M., Shapiro, D. R., Oishi, K. K., and Tewari, K. K. (1985). *Plant Mol. Biol.* **4**, 65–80.

Chu, Z.-X., and Anderson, J. M. (1985). *Biochim. Biophys. Acta* **806**, 154–160.

Chua, N.-H., and Blomberg, F. (1979). *J. Biol. Chem.* **254**, 215–223.

Chua, N.-H., and Schmidt, G. W. (1978). *Proc. Natl. Acad. Sci. U.S.A.* **75**, 6110-6114.

Chua, N.-H., and Schmidt, G. W. (1979). *J. Cell Biol.* **81**, 461–483.

Chunayev, A. S., Ladygin, V. G., Mirnaya, O. N., Semyonov, E. P., Gayevsky, N. A., and Boldina, O. N. (1984). *Genetika (Moscow)* **20**, 775–781.

Church, W. B., Guss, J. M., Potter, J. J., and Freeman, H. C. (1986). *J. Biol. Chem.* **261**, 234–237.

Cifferi, O., and Dure, L., III, eds. (1983). "Structure and Function of Plant Genomes." Plenum, New York.

Clark, R. D., and Hind, G. (1983). *J. Biol. Chem.* **258,** 10348–10354.

Clark, R. D., Hawkesford, M. J., Coughlan, S. J., Bennett, J., and Hind, G. (1984). *FEBS Lett.* **174,** 137–142.

Cline, K. (1986). *J. Biol. Chem.* **261,** 14804–14810.

Cline, K., and Keegstra, K. (1983). *Plant Physiol.* **71,** 366–372.

Cline, K., Andrews, J., Mersey, B., Newcomb, E. H., and Keegstra, K. (1981). *Proc. Natl. Acad. Sci. U.S.A.* **78,** 3595–3599.

Cline, K., Werner-Washburne, M., Lubben, T. H., and Keegstra, K. (1985). *J. Biol. Chem.* **260,** 3691–3696.

Colbert, J. T., Hershey, H. P., and Quail, P. H. (1983). *Proc. Natl. Acad. Sci. U.S.A.* **80,** 2248–2252.

Coruzzi, G., Broglie, R., Cashmore, A., and Chua, N.-H. (1983). *J. Biol. Chem.* **258,** 1399–1402.

Coruzzi, G., Broglie, R., Edwards, C., and Chua, N.-H. (1984). *EMBO J.* **3,** 1671–1679.

Coughlan, S., Matthijs, H. C. P., and Hind, G. (1985). *J. Biol. Chem.* **260,** 14891–14893.

Cozens, A. L., Walker, J. E., Phillips, A. L., Huttly, A. K., and Gray, J. C. (1986). *EMBO J.* **5,** 217–222.

Crouse, E. J., Mubumbila, M., Stummann, B. M., Bookjans, G., Michalowski, C., Bohnert, H. J., Weil, J.-H., and Henningsen, K. W. (1986). *Plant Mol. Biol.* **7,** 143–149.

Cunningham, F. X., Jr., and Schiff, J. A. (1986). *Plant Physiol.* **80,** 223–230.

Dairi, T., Inokuchi, K., Mizuno, T., and Mizushima, S. (1985). *J. Mol. Biol.* **184,** 1–6.

Darr, S. C., Somerville, S. C., and Arntzen, C. J. (1986). *J. Cell Biol.* **103,** 733–740.

Davis, N. G., and Model, P. (1985). *Cell (Cambridge, Mass.)* **41,** 607–614.

Dean, C., van den Elzen, P., Tamaki, S., Dunsmuir, P., and Bedbrook, J. (1985a). *EMBO J.* **4,** 3055–3061.

Dean, C., van den Elzen, P., Tamaki, S., Dunsmuir, P., and Bedbrook, J. (1985b). *Proc. Natl. Acad. Sci. U.S.A.* **82,** 4964–4968.

Dehesh, K., Klaas, M., Hauser, I., and Apel, K. (1986a). *Planta* **169,** 162–171.

Dehesh, K., van Cleve, B., Ryberg, M., and Apel, K. (1986b). *Planta* **169,** 172–183.

Delepelaire, P. (1984). *EMBO J.* **3,** 701–706.

Delepelaire, P., and Chua, N.-H. (1981). *J. Biol. Chem.* **256,** 9300–9307.

Delepelaire, P., and Wollman, F.-A. (1985). *Biochim. Biophys. Acta* **809,** 277–283.

Dobberstein, B., Blobel, G., and Chua, N.-H. (1977). *Proc. Natl. Acad. Sci. U.S.A.* **74,** 1082–1085.

Doherty, A., and Gray, J. C. (1979). *Eur. J. Biochem.* **98,** 87–92.

Dorne, A.-J., Block, M. A., Joyard, J., and Douce, R. (1982a). *FEBS Lett.* **145,** 30–34.

Dorne, A.-J., Carde, J.-P., Joyard, J., Borner, T., and Douce, R. (1982b). *Plant Physiol.* **69,** 1467–1470.

Dorne, A.-J., Joyard, J., Block, M. A., and Douce, R. (1985). *J. Cell Biol.* **100,** 1690–1697.

Dörnemann, D., and Senger, H. (1982). *Photochem. Photobiol.* **35,** 821–826.

Dörnemann, D., and Senger, H. (1986). *Photochem. Photobiol.* **43,** 573–581.

Douce, R., and Joyard, J. (1981). *Trends Biochem. Sci.* **6,** 237–239.

Douce, R., Holtz, R. B., and Benson, A. A. (1973). *J. Biol. Chem.* **248,** 7215–7222.

Douce, R., Block, M. A., Dorne, A.-J., and Joyard, J. (1984). *Subcell. Biochem.* **10,** 1–84.

Draheim, J. E., Anderson, G. P., Duane, J. W., and Gross, E. L. (1986). *Biophys. J.* **49,** 891–900.

Dron, M., Rahire, M., and Rochaix, J.-D. (1982). *J. Mol. Biol.* **162,** 775–793.

Dunsmuir, P. (1985). *Nucleic Acids Res.* **13,** 2503–2518.

Dunsmuir, P., Smith, S., and Bedbrook, J. (1983). *Nucleic Acids Res.* **11,** 4177–4183.

Eckes, P., Schell, J., and Willmitzer, L. (1985). *Mol. Gen. Genet.* **199,** 216–224.

Ellis, R. J. (1977). *Biochim. Biophys. Acta* **463,** 185–215.

Ellis, R. J. (1981). *Annu. Rev. Plant Physiol.* **32**, 111–137.

Eneas-Filho, J., Hartley, M. R., and Mache, R. (1981). *Mol. Gen. Genet.* **184**, 484–488.

Erickson, J. M., Rahire, M., and Rochaix, J.-D. (1984). *EMBO J.* **3**, 2753–2762.

Erickson, J. M., Rahire, M., Rochaix, J.-D., and Mets, L. (1985). *Science* **228**, 204–207.

Erickson, J. M., Rahire, M., Malnoe, P., Girard-Bascou, J., Pierre, Y., Bennoun, P., and Rochaix, J.-D. (1986). *EMBO J.* **5**, 1745–1754.

Evans, P. K., and Anderson, J. M. (1986). *FEBS Lett.* **199**, 227–233.

Fawley, M. W., and Grossman, A. R. (1986). *Plant Physiol.* **81**, 149–155.

Fish, L. E., and Jagendorf, A. T. (1982). *Plant Physiol.* **70**, 1107–1114.

Fish, L. E., Kück, U., and Bogorad, L. (1985). *J. Biol. Chem.* **260**, 1413–1421.

Flügge, U. I., and Heldt, H. W. (1986). *In* "Methods in Enzymology" (S. Fleischer and B. Fleischer, eds.), Vol. 125, pp. 716–730.

Flügge, U. I., and Hinz, G. (1986). *Eur. J. Biochem.* **160**, 563–570.

Fluhr, R., and Chua, N.-H. (1986). *Proc. Natl. Acad. Sci. U.S.A.* **83**, 2358–2362.

Fluhr, R., Kuhlmeier, C., Nagy, F., and Chua, N.-H. (1986). *Science* **232**, 1106–1112.

Foley, T., Dzelzkalns, V., and Beale, S. I. (1982). *Plant Physiol.* **70**, 219–226.

Friedman, A. L., and Alberte, R. S. (1986). *Plant Physiol.* **80**, 43–51.

Fromm, H., Devic, M., Fluhr, R., and Edelman, M. (1985). *EMBO J.* **4**, 291–295.

Fromm, H., Edelman, M., Koller, B., Goloubinoff, P., and Galun, E. (1986). *Nucleic Acids Res.* **14**, 883–898.

Fuesler, T. P., Castelfranco, P. A., and Wong, Y.-S. (1984a). *Plant Physiol.* **74**, 928–933.

Fuesler, T. P., Wong, Y.-S., and Castelfranco, P. A. (1984b). *Plant Physiol.* **75**, 662–664.

Fukuzawa, H., Kohchi, T., Shirai, H., Ohyama, K., Umesono, K., Inokuchi, H., and Ozeki, H. (1986). *FEBS Lett.* **198**, 11–15.

Gallagher, T. F., and Ellis, R. J. (1982). *EMBO J.* **1**, 1493–1498.

Gantt, J. S., and Key, J. L. (1986). *Mol. Gen. Genet.* **202**, 186–193.

Gardiner, S. E., and Roughan, P. G. (1983). *Biochem. J.* **210**, 949–952.

Ghisla, S., Mack, R., Blankenhorn, G., Hemmerich, P., Krienitz, E., and Kuster, T. (1984). *Eur. J. Biochem.* **138**, 339–344.

Gibbs, S. P. (1981). *Int. Rev. Cytol.* **72**, 49–99.

Gilles, H., Jaenchen, R., and Thauer, R. K. (1983). *Arch. Microbiol.* **135**, 237–240.

Goldschmidt-Clermont, M. (1986). *Plant Mol. Biol.* **6**, 13–21.

Goldschmidt-Clermont, M., and Rahire, M. (1986). *J. Mol. Biol.* **191**, 421–432.

Gordon, K. H. J., Crouse, E. J., Bohnert, and Herrmann, R. G. (1982). *Theor. Appl. Genet.* **61**, 373–384.

Gough, S. P., and Kannangara, C. G. (1977). *Carlsberg Res. Commun.* **42**, 459–464.

Gounaris, K., Barber, J., and Harwood, J. L. (1986). *Biochem. J.* **237**, 313–326.

Gray, P. W., and Hallick, R. B. (1977). *Biochemistry* **16**, 1665–1671.

Gray, P. W., and Hallick, R. B. (1978). *Biochemistry* **17**, 284–290.

Griffiths, W. T. (1978). *Biochem. J.* **174**, 681–692.

Griffiths, W. T., and Beer, N. S. (1982). *Plant Physiol.* **70**, 1014–1018.

Griffiths, W. T., Oliver, R. P., and Kay, S. A. (1984). *In* "Protochlorophyllide Reduction and Greening" (C. Sironval and M. Brouers, eds.), pp. 19–29. Martinus Nijhoff/Dr. W. Junk Publ., The Hague, The Netherlands.

Grossman, A. R., Bartlett, S. G., and Chua, N.-H. (1980). *Nature (London)* **285**, 625–628.

Grossman, A. R., Bartlett, S. G., Schmidt, G. W., Mullet, J. E., and Chua, N.-H. (1982). *J. Biol. Chem.* **257**, 1558-1563.

Gruissem, W., and Zurawski, G. (1985). *EMBO J.* **4**, 1637–1644.

Gruissem, W., Narita, J. O., Greenberg, B. M., Prescott, D. M., and Hallick, R. B. (1983). *J. Cell. Biochem.* **22**, 31–46.

Hagemann, R. (1986). *In* "Regulation of Chloroplast Differentiation" (G. Akoyunoglou and H. Senger, eds.), pp. 455-466. Alan R. Liss, Inc., New York.

Hauge, D. R., Uhler, M., and Collins, P. D. (1983). *Nucleic Acids Res.* **11**, 4853–4865.

Hale, C. C., and Roux, S. J. (1980). *Plant Physiol.* **65**, 658–662.
Hallick, R. B., Hollingsworth, M. J., and Nickoloff, J. A. (1984). *Plant Mol. Biol.* **3**, 169–176.
Hanley-Bowdoin, L., and Chua, N.-H. (1987). *Trends Biochem. Sci.* **12**, 67–70.
Harel, E., and Klein, S. (1972). *Biochem. Biophys. Res. Commun.* **49**, 364–370.
Harel, E., and Ne'eman, E. (1983). *Plant Physiol.* **72**, 1062–1067.
Hattori, T., and Margulies, M. M. (1986). *Arch. Biochem. Biophys.* **244**, 630–640.
Hauska, G., Hurt, E., Gabellini, N., and Lockau, W. (1983). *Biochim. Biophys. Acta* **726**, 97–133.
Hawley, D. K., and McClure, W. R. (1983). *Nucleic Acids Res.* **11**, 2237–2255.
Haworth, P., Watson, J. L., and Arntzen, C. J. (1983). *Biochim. Biophys. Acta* **724**, 151–158.
Hay, R., Bohni, P., and Gasser, S. (1984). *Biochim. Biophys. Acta* **779**, 65–87.
Heber, U., and Heldt, H. W. (1981). *Annu. Rev. Plant Physiol.* **32**, 139–168.
Heinz, E., and Roughan, P. G. (1983). *Plant Physiol.* **72**, 273–279.
Heldt, H. W., and Flügge, U. I. (1986). *In* "Methods in Enzymology" (S. Fleischer and B. Fleischer, eds.), Vol. 125, pp. 705–716. Academic Press, New York.
Hennig, J., and Herrmann, R. G. (1986). *Mol. Gen. Genet.* **203**, 117–128.
Hepler, P. K., and Wayne, R. O. (1985). *Annu. Rev. Plant Physiol.* **36**, 397–439.
Herrera-Estrella, L., Van den Broeck, G., Maenhaut, R., Van Montagu, M., Schell, J., Timko, M., and Cashmore, A. (1984). *Nature (London)* **310**, 115–120.
Herrin, D., and Michaels, A. (1985a). *Arch. Biochem. Biophys.* **237**, 224–236.
Herrin, D., and Michaels, A. (1985b). *FEBS Lett.* **184**, 90–95.
Herrin, D. L., Michaels, A. S., and Paul, A.-L. (1986). *J. Cell Biol.* **103**, 1837–1845.
Herrmann, R. G., Bohnert, H.-J., Driesel, A., and Hobom, G. (1976). *In* "Genetics and Biogenesis of Chloroplasts and Mitochondria" (T. Bücher, W. Neupert, W. Sebald, and S. Werner, eds.), pp. 351–359. Elsevier/North-Holland Biomedical Press, Amsterdam.
Hershey, H. P., Colbert, J. T., Lissemore, J. L., Barker, R. F., and Quail, P. H. (1984). *Proc. Natl. Acad. Sci. U.S.A.* **81**, 2332–2336.
Hershey, H. P., Barker, R. F., Idler, K. B., Lissemore, J. L., and Quail, P. H. (1985). *Nucleic Acids Res.* **13**, 8543–8559.
Highfield, P. E., and Ellis, R. J. (1978). *Nature (London)* **271**, 420–424.
Hird, S. M., Willey, D. L., Dyer, T. A., and Gray, J. C. (1986). *Mol. Gen. Genet.* **203**, 95–100.
Hollingsworth, M. J., Johanningmeier, U., Karabin, G. D., Stiegler, G. L., and Hallick, R. B. (1984). *Nucleic Acids Res.* **12**, 2001–2017.
Hoober, J. K. (1976). *In* "Genetics and Biogenesis of Chloroplasts and Mitochondria" (T. Bücher, W. Neupert, W. Sebald, and S. Werner, eds.), pp. 87–94. Elsevier/North-Holland Biomedical Press, Amsterdam.
Hoober, J. K. (1981). *In* "Photosynthesis V. Chloroplast Development" (G. Akoyunoglou, ed.), pp. 859–866. Balaban Int. Sci. Serv., Philadelphia, Pennsylvania.
Hoober, J. K., and Stegeman, W. J. (1973). *J. Cell Biol.* **56**, 1–12.
Hoober, J. K., Millington, R. H., and D'Angelo, L. P. (1980). *Arch. Biochem. Biophys.* **202**, 221–234.
Hoober, J. K., Marks, D. B., Keller, B. J., and Margulies, M. M. (1982). *J. Cell Biol.* **95**, 552–558.
Horwich, A. L., Kalousek, F., Fenton, W. A., Pollock, R. A., and Rosenberg, L. E. (1986). *Cell (Cambridge, Mass.)* **44**, 451–459.
Houen, G., Gough, S. P., and Kannangara, C. G. (1983). *Carlsberg Res. Commun.* **48**, 567–572.
Huang, D.-D., and Wang, W.-Y. (1986). *J. Biol. Chem.* **261**, 13451–13455.
Huang, D.-D., Wang, W.-Y., Gough, S. P., and Kannangara, C. G. (1984). *Science* **225**, 1482–1484.
Hudspeth, R. L., Glackin, C. A., Bonner, J., and Grula, J. W. (1986). *Proc. Natl. Acad. Sci. U.S.A.* **83**, 2884–2888.
Hurt, E., and Hauska, G. (1982). *J. Bioenerg. Biomembr.* **14**, 405–424.
Hurt, E. C., and van Loon, A. P. G. M. (1986). *Trends Biochem. Sci.* **11**, 204–207.

Hurt, E. C., Pesold-Hurt, B., Suda, K., Opplinger, W., and Schatz, G. (1985). *EMBO J.* **4**, 2061–2068.
Hurt, E. C., Soltanifar, N., Goldschmidt-Clermont, M., Rochaix, J.-D., and Schatz, G. (1986). *EMBO J.* **5**, 1343–1350.
Inamine, G., Nash, B., Weissbach, H., and Brot, N. (1985). *Proc. Natl. Acad. Sci. U.S.A.* **82**, 5690–5694.
Ish-Shalom, D., and Ohad, I. (1983). *Biochim. Biophys. Acta* **722**, 498–507.
Jackson, A. H. (1976). *In* "Chemistry and Biochemistry of Plant Pigments" (T. W. Goodwin, ed.), Vol. 1, pp. 1–63. Academic Press, New York.
Janero, D. R., and Barnett, R. (1981). *J. Cell Biol.* **91**, 126–134.
Jenni, B., and Stutz, E. (1979). *FEBS Lett.* **102**, 95–99.
Jensen, K. H., Herrin, D. L., Plumley, F. G., and Schmidt, G. W. (1986). *J. Cell Biol.* **103**, 1315–1325.
Johanningmeier, U., and Howell, S. H. (1984). *J. Biol. Chem.* **259**, 13541–13549.
Jolly, S. O., and Bogorad, L. (1980). *Proc. Natl. Acad. Sci. U.S.A.* **77**, 822–826.
Jolly, S. O., McIntosh, L., Link, G., and Bogorad, L. (1981). *Proc. Natl. Acad. Sci. U.S.A.* **78**, 6821–6825.
Joyard, J., and Douce, R. (1975). *FEBS Lett.* **51**, 335–340.
Joyard, J., and Douce, R. (1979). *FEBS Lett.* **102**, 147–150.
Joyard, J., Billecocq, A., Bartlett, S. G., Block, M. A., Chua, N.-H., and Douce, R. (1983). *J. Biol. Chem.* **258**, 10000–10006.
Kannangara, C. G., and Gough, S. P. (1977). *Carlsberg Res. Commun.* **42**, 444–457.
Kannangara, C. G., and Gough, S. P. (1978). *Carlsberg Res. Commun.* **43**, 185–194.
Kannangara, C. G., Gough, S. P., Oliver, R. P., and Rasmussen, S. K. (1984). *Carlsberg Res. Commun.* **49**, 417–437.
Kaplan, S., and Arntzen, C. J. (1982). *In* "Photosynthesis" (Govindjee, ed.), Vol. 1, pp. 65–151. Academic Press, New York.
Karlin-Neumann, G. A., and Tobin, E. M. (1986). *EMBO J.* **5**, 9–13.
Karlin-Neumann, G. A., Kohorn, B. D., Thornber, J. P., and Tobin, E. M. (1985). *J. Mol. Appl. Genet.* **3**, 45–61.
Kaufman, L. S., Thompson, W. F., and Briggs, W. R. (1984). *Science* **226**, 1447–1449.
Kaufman, L. S., Roberts, L. L., Briggs, W. R., and Thompson, W. F. (1986). *Plant Physiol.* **81**, 1033-1038.
Kay, S. A., and Griffiths, W. T. (1983). *Plant Physiol.* **72**, 229–236.
Keegstra, K., and Cline, K. (1982). *Plant Physiol.* **70**, 232–237.
Keegstra, K., and Yousif, A. (1986). *In* "Methods in Enzymology" (A. Weissbach and H. Weissbach, eds.), Vol. 118, pp. 316–325. Academic Press, New York.
Keegstra, K., Werner-Washburne, M., Cline, K., and Andrews, J. (1984). *J. Cell. Biochem.* **24**, 55–68.
Kidd, G. H., and Bogorad, L. (1979). *Proc. Natl. Acad. Sci. U.S.A.* **76**, 4890–4892.
Kidd, G. H., and Bogorad, L. (1980). *Biochim. Biophys. Acta* **609**, 14–30.
Kirk, M. M., and Kirk, D. L. (1985). *Cell (Cambridge, Mass.)* **41**, 419–428.
Kleppinger-Sparace, K. F., Mudd, J. B., and Bishop, D. G. (1985). *Arch. Biochem. Biophys.* **240**, 859–865.
Kloppstech, K., and Bitsch, A. (1986). *In* "Regulation of Chloroplast Differentiation" (G. Akoyunoglou and H. Senger, eds.), pp. 235–240. Alan R. Liss, Inc., New York.
Kohorn, B. D., Harel, E., Chitnis, P. R., Thornber, J. P., and Tobin, E. M. (1986). *J. Cell Biol.* **102**, 972–981.
Koller, B., and Delius, H. (1980). *Mol. Gen. Genet.* **178**, 261–269.
Koller, B., and Delius, H. (1982). *EMBO J.* **1**, 995–998.
Koller, B., Delius, H., and Helling, R. B. (1984a). *Plant Mol. Biol.* **3**, 127–136.
Koller, B., Gingrich, J. C., Stiegler, G. L., Farley, M. A., Delius, H., and Hallick, R. B. (1984b). *Cell (Cambridge, Mass.)* **36**, 545–553.
Koller, B., Clarke, J., and Delius, H. (1985). *EMBO J.* **4**, 2445-2450.

Kornberg, A. (1983). *Eur. J. Biochem.* **137**, 337–382.
Kosmac, U., and Feierabend, J. (1985). *Plant Physiol.* **79**, 646–652.
Kühlbrandt, W. (1984). *Nature (London)* **307**, 478–480.
Kühsel, M., and Kowallik, K. V. (1987). *Mol. Gen. Genet.* **207**, 361–368.
Kung, S. D., and Lin, C. M. (1985). *Nucleic Acids Res.* **13**, 7543–7549.
Kyle, D. J., Ohad, I., and Arntzen, C. J. (1984). *Proc. Natl. Acad. Sci. U.S.A.* **81**, 4070–4074.
Lagarias, J. C. (1985). *Photochem. Photobiol.* **42**, 811–820.
Lam, E. (1986). *Biochim. Biophys. Acta* **848**, 324–332.
Lam, E., and Malkin, R. (1985a). *Arch. Biochem. Biophys.* **242**, 58–63.
Lam, E., and Malkin, R. (1985b). *Arch. Biochem. Biophys.* **242**, 64–71.
Lamppa, G. K., Morelli, G., and Chua, N.-H. (1985a). *Mol. Cell. Biol.* **5**, 1370–1378.
Lamppa, G. K., Nagy, F., and Chua, N.-H. (1985b). *Nature (London)* **316**, 750–752.
Lehmbeck, J., Rasmussen, O. F., Bookjans, G. B., Jepsen, B. R., Stummann, B. M., and Henningsen, K. W. (1986). *Plant Mol. Biol.* **7**, 3–10.
Lemaux, P. G., and Grossman, A. R. (1985). *EMBO J.* **4**, 1911–1919.
Lerbs, S., Briat, J.-F., and Mache, R. (1983). *Plant Mol. Biol.* **2**, 67–74.
Lerbs, S., Bräutigam, E., and Parthier, B. (1985). *EMBO J.* **4**, 1661–1666.
Leto, K. J., Bell, E., and McIntosh, L. (1985). *EMBO J.* **4**, 1645–1653.
Leutwiler, L. S., Meyerowitz, E. M., and Tobin, E. M. (1986). *Nucleic Acids Res.* **14**, 4051–4064.
Lindenhahn, M. M., Metzlaff, M., and Hagemann, R. (1985). *Mol. Gen. Genet.* **200**, 503–505.
Link, G. (1984). *EMBO J.* **3**, 1697–1704.
Link, G., Coen, D. M., and Bogorad, L. (1978). *Cell (Cambridge, Mass.)* **15**, 725–731.
Liu, X.-Q., and Jagendorf, A. T. (1986a). *In* "Regulation of Chloroplast Differentiation" (G. Akoyunoglou and H. Senger, eds.), pp. 597–606. Alan R. Liss, Inc., New York.
Liu, X.-Q., and Jagendorf, A. T. (1986b). *Plant Physiol.* **81**, 603–608.
Liveanu, B., Yocum, C. F., and Nelson, N. (1986). *J. Biol. Chem.* **261**, 5296–5300.
Ludwig, M., and Gibbs, S. P. (1985). *Protoplasma* **127**, 9–20.
Lütz, C. (1981). *Protoplasma* **109**, 99–115.
McCarty, D. R., and Selman, B. R. (1986). *Arch. Biochem. Biophys.* **248**, 523–531.
McCarty, D. R., Keegstra, K., and Selman, B. R. (1984). *Plant Physiol.* **76**, 584–588.
McClure, W. R. (1985). *Annu. Rev. Biochem.* **54**, 171–204.
McCurdy, D. W., and Pratt, L. H. (1986). *J. Cell Biol.* **103**, 2541–2550.
Mackender, R. O., and Leech, R. M. (1970). *Nature (London)* **228**, 1347–1349.
McKnight, S. L., and Kingsbury, R. (1982). *Science* **217**, 316–324.
McKown, R. L., and Tewari, K. K. (1984). *Proc. Natl. Acad. Sci. U.S.A.* **81**, 2354–2358.
Mansfield, R. W., and Anderson, J. M. (1985). *Biochim. Biophys. Acta* **809**, 435–444.
Marder, J.B., Goloubinoff, P., and Edelman, M. (1984). *J. Biol. Chem.* **259**, 3900–3908.
Margulies, M. M. (1983). *Eur. J. Biochem.* **137**, 241–248.
Margulies, M. M. (1986). *In* "Regulation of Chloroplast Differentiation" (G. Akoyunoglou and H. Senger, eds.), pp. 171–180. Alan R. Liss, Inc., New York.
Marks, D. B., Keller, B. J., and Hoober, J. K. (1985). *Plant Physiol.* **79**, 108–113.
Marks, D. B., Keller, B. J., and Hoober, J. K. (1986). *Plant Physiol.* **81**, 702–704.
Markwell, J. P., Webber, A. N., Danko, S. J., and Baker, N. R. (1985a). *Biochim. Biophys. Acta* **808**, 156–163.
Markwell, J. P., Webber, A. N., and Lake, B. (1985b). *Plant Physiol.* **77**, 948–951.
Maroc, J., Guyon, D., and Garnier, J. (1983). *Plant Cell Physiol.* **24**, 1217–1230.
Mattoo, A. K., Hoffman-Falk, H., Marder, J. B., and Edelman, M. (1984). *Proc. Natl. Acad. Sci. U.S.A.* **81**, 1380–1384.
Mayfield, S. P., and Taylor, W. C. (1984). *Eur. J. Biochem.* **144**, 79–84.
Mayfield, S. P., Rahire, M., Frank, G., Zuber, H., and Rochaix, J.-D. (1987). *Proc. Natl. Acad. Sci. U.S.A.* **84**, 749–753.
Mazur, B. J., and Chui, C.-F. (1985). *Nucleic Acids Res.* **13**, 2373–2386.

Melis, A., and Anderson, J. M. (1983). *Biochim. Biophys. Acta* **724**, 473–484.
Menke, W. (1962). *Annu. Rev. Plant Physiol.* **13**, 27–44.
Merchant, S., and Bogorad, L. (1986a). *Mol. Cell. Biol.* **6**, 462–469.
Merchant, S., and Bogorad, L. (1986b). *J. Biol. Chem.* **261**, 15850–15853.
Michaels, A. S., Jelsema, C. L., and Barrnett, R. J. (1983). *J. Ultrastruct. Res.* **82**, 35–51.
Michel, H., Tellenbach, M., and Boschetti, A. (1983). *Biochim. Biophys. Acta* **725**, 417–424.
Minami, E.-I., Shinohara, K., Kuwabara, T., and Watanabe, A. (1986). *Arch. Biochem. Biophys.* **244**, 517–527.
Mishkind, M. L., and Schmidt, G. W. (1983). *Plant Physiol.* **72**, 847–854.
Mishkind, M. L., Wessler, S. R., and Schmidt, G. W. (1985). *J. Cell Biol.* **100**, 226–234.
Montandon, P.-E., and Stutz, E. (1984). *Nucleic Acids Res.* **12**, 2851–2859.
Moore, P. D. (1982). *Nature (London)* **295**, 647–648.
Morelli, G., Nagy, F., Fraley, R. T., Rogers, S. G., and Chua, N.-H. (1985). *Nature (London)* **315**, 200–204.
Morgenthaler, J. J., and Mendiola-Morgenthaler, L. (1976). *Arch. Biochem. Biophys.* **172**, 51–58.
Morris, J., and Hermann, R. G. (1984). *Nucleic Acids Res.* **12**, 2837–2850.
Mösinger, E., Batschauer, A., Schäfer, E., and Apel, K. (1985). *Eur. J. Biochem.* **147**, 137–142.
Mubumbila, M., Bowman, C. M., Droog, F., Dyer, T., Kuntz, M., and Weil, J.-H. (1985). *Plant Mol. Biol.* **4**, 315–322.
Mühlethaler, K., and Frey-Wyssling, A. (1959). *J. Biochem. Biophys. Cytol.* **6**, 507–512.
Müller, M., Viro, M., Balke, C., and Kloppstech, K. (1980). *Planta* **148**, 444–447.
Mullet, J. E. (1983). *J. Biol. Chem.* **258**, 9941–9948.
Mullet, J. E., Klein, R. R., and Grossman, A. R. (1986). *Eur. J. Biochem.* **155**, 331–338.
Mulligan, B., Schultes, N., Chen, L., and Bogorad, L. (1984). *Proc. Natl. Acad. Sci. U.S.A.* **81**, 2693–2697.
Muto, S., and Shimogawara, K. (1985). *FEBS Lett.* **193**, 88–92.
Nagy, F., Morelli, G., Fraley, R. T., Rogers, S. G., and Chua, N.-H. (1985). *EMBO J.* **4**, 3063–3068.
Nagy, F., Kay, S. A., Boutry, M., Hsu, M.-Y., and Chua, N.-H. (1986). *EMBO J.* **5**, 1119–1124.
Nanba, O., and Satoh, K. (1987). *Proc. Natl. Acad. Sci. U.S.A.* **84**, 109–112.
Nelson, T., Harpster, M. H., Mayfield, S. P., and Taylor, W. C. (1984). *J. Cell Biol.* **98**, 558–564.
Nierzwicki-Bauer, S. A., Curtis, S. E., and Haselkorn, R. (1984). *Proc. Natl. Acad. Sci. U.S.A.* **81**, 5961–5965.
Nivison, H. T., and Jagendorf, A. T. (1984). *Plant Physiol.* **75**, 1001-1008.
Oelmüller, R., and Mohr, H. (1985). *Proc. Natl. Acad. Sci. U.S.A.* **82**, 6124–6128.
Oelmüller, R., and Mohr, H. (1986). *Planta* **167**, 106–113.
Oelmüller, R., Dietrich, G., Link, G., and Mohr, H. (1986a). *Planta* **169**, 260–266.
Oelmüller, R., Levitan, I., Bergfeld, R., Rajasekhar, V. K., and Mohr, H. (1986b). *Planta* **168**, 482–492.
Ohad, I., Kyle, D. J., and Arntzen, C. J. (1984). *J. Cell Biol.* **99**, 481–485.
Ohad, I., Kyle, D. J., and Hirschberg, J. (1985). *EMBO J.* **4**, 1655–1659.
Oh-hama, T., Seto, H., Otake, N., and Miyachi, S. (1982). *Biochem. Biophys. Res. Commun.* **105**, 647–652.
Oh-hama, T., Seto, H., and Miyachi, S. (1985). *Arch. Biochem. Biophys.* **237**, 72–79.
Oh-hama, T., Seto, H., and Miyachi, S. (1986). *Arch. Biochem. Biophys.* **246**, 192–198.
Ohme, M., Tanaka, M., Chunwongse, J., Shinozaki, K., and Sugiura, M. (1986). *FEBS Lett.* **200**, 87–90.
Ohyama, K., Fukuzawa, H., Kohchi, T., Shirai, H., Sano, T., Sano, S., Umesono, K., Shiki, Y., Takeuchi, M., Chang, Z., Aota, S., Inokuchi, H., and Ozeki, H. (1986). *Nature (London)* **322**, 572–574.
Oliver, R. P., and Griffiths, W. T. (1982). *Plant Physiol.* **70**, 1019–1025.

Ortiz, W., and Malkin, R. (1985). *Biochim. Biophys. Acta* **808**, 164–170.
Owens, G. C., and Ohad, I. (1983). *Biochim. Biophys. Acta* **722**, 234–241.
Palmer, J. D. (1983). *Nature (London)* **301**, 92–93.
Palmer, J. D. (1985). *Annu. Rev. Genet.* **19**, 325–354.
Palmer, J. D., and Thompson, W. F. (1981). *Proc. Natl. Acad. Sci. U.S.A.* **78**, 5533–5537.
Palmer, J. D., and Thompson, W. F. (1982). *Cell (Cambridge, Mass.)* **29**, 537–550.
Palmer, J. D., Nugent, J. M., and Herbon, L. A. (1987). *Proc. Natl. Acad. Sci. U.S.A.* **84**, 769–773.
Pardo, A. D., Chereskin, B. M., Castelfranco, P. A., Franceschi, V. R., and Wezelman, B. E. (1980). *Plant Physiol.* **65**, 956–960.
Passavant, C. W., Stiegler, G. L., and Hallick, R. B. (1983). *J. Biol. Chem.* **258**, 693–695.
Pfisterer, J., Lachmann, P., and Kloppstech, K. (1982). *Eur. J. Biochem.* **126**, 143–148.
Phillips, A. L., and Gray, J. C. (1984). *Mol. Gen. Genet.* **194**, 477–484.
Picaud, A., and Dubertret, G. (1986). *Photosynth. Res.* **7**, 221–236.
Pichersky, E., Bernatzky, R., Tanksley, S. D., Breidenbach, R. B., Kausch, A. P., and Cashmore, A. R. (1985). *Gene* **40**, 247–258.
Piechulla, B., Imlay, K. R. C., and Gruissem, W. (1985). *Plant Mol. Biol.* **5**, 373–384.
Pineau, B., Dubertret, G., Joyard, J., and Douce, R. (1986). *J. Biol. Chem.* **261**, 9210–9215.
Plumley, F. G., and Schmidt, G. W. (1987). *Proc. Natl. Acad. Sci. U.S.A.* **84**, 146–150.
Plumley, F. G., Kirchman, D. L., Hodson, R. E., and Schmidt, G. W. (1986). *Plant Physiol.* **80**, 685–691.
Poincelot, R. P. (1980). *In* "Methods in Enzymology" (A. San Pietro, ed.), Vol. 69, pp. 121–128. Academic Press, New York.
Pollans, N. O., Weeden, N. F., and Thompson, W. F. (1985). *Proc. Natl. Acad. Sci. U.S.A.* **82**, 5083–5087.
Porra, R. J., and Meisch, H.-U. (1984). *Trends Biochem. Sci.* **9**, 99–104.
Porra, R. J., Klein, O., and Wright, P. E. (1983). *Eur. J. Biochem.* **130**, 509–516.
Posno, M., van Vliet, A., and Groot, G. S. P. (1986). *Nucleic Acids Res.* **14**, 3181–3195.
Pratt, L. H. (1982). *Annu. Rev. Plant Physiol.* **33**, 557–582.
Quail, P. H. (1984). *Trends Biochem. Sci.* **9**, 450–453.
Randall, L. L. (1986). *In* "Methods in Enzymology" (S. Fleischer and B. Fleischer, eds.), Vol. 125, pp. 129–138. Academic Press, New York.
Rao, L. V. M., Usharani, P., Butler, W. L., and Tokuyasu, K. T. (1986). *Plant Physiol.* **80**, 138–141.
Ravel-Chapuis, P., Heizmann, P., and Nigon, V. (1982). *Nature (London)* **300**, 78–81.
Rawson, J. R. Y., Kushner, S. D., Vapnek, D., Alton, V. N. K., and Boerma, C. L. (1978). *Gene* **3**, 191–209.
Rebeiz, C. A., and Lascelles, J. (1982). *In* "Photosynthesis" (Govindjee, ed.), Vol. 1, pp. 699–780. Academic Press, New York.
Rebeiz, C. A., Wu, S. M., Kuhadja, M., Daniell, H., and Perkins, E. J. (1983). *Mol. Cell. Biochem.* **57**, 97–125.
Reith, M. E., and Cattolico, R. A. (1985a). *Biochemistry* **24**, 2556–2561.
Reith, M. E., and Cattolico, R. A. (1985b). *Plant Physiol.* **79**, 231–236.
Reith, M., and Cattolico, R. A. (1986). *Proc. Natl. Acad. Sci. U.S.A.* **83**, 8599–8603.
Robinson, C., and Ellis, R. J. (1984a). *Eur. J. Biochem.* **142**, 337–342.
Robinson, C., and Ellis, R. J. (1984b). *Eur. J. Biochem.* **142**, 343–346.
Robinson, C., and Ellis, R. J. (1985). *Eur. J. Biochem.* **152**, 67–73.
Rochaix, J.-D. (1978). *J. Mol. Biol.* **126**, 597–617.
Rochaix, J.-D., Dron, M., Rahire, M., and Malnoe, P. (1984a). *Plant Mol. Biol.* **3**, 363–370.
Rochaix, J. D., Erickson, J., Schneider, M., Vallet, J. M., Dron, M., Masson, A., and Bennoun, P. (1984b). *In* "Biosynthesis of the Photosynthetic Apparatus: Molecular Biology, Development and Regulation" (J. P. Thornber, L. A. Staehelin, and R. B. Hallick, eds.), pp. 285–294. Alan R. Liss, Inc., New York.

Rochaix, J. D., Rahire, M., and Michel, F. (1985). *Nucleic Acids Res.* **13,** 975–984.
Rodermel, S. R., and Bogorad, L. (1985). *J. Cell Biol.* **100,** 463–476.
Rosenberg, M., and Court, D. (1979). *Annu. Rev. Genet.* **13,** 319–353.
Rothstein, S. J., Gatenby, A. A., Wiley, D. L., and Gray, J. C. (1985). *Proc. Natl. Acad. Sci. U.S.A.* **82,** 7955-7959.
Roux, S. J., McEntire, K., Slocum, R. D., Cedel, T. E., and Hale, C. C. (1981). *Proc. Natl. Acad. Sci. U.S.A.* **78,** 283–287.
Rüdiger, W. (1986). *In* "Regulation of Chloroplast Differentiation" (G. Akoyunoglou and H. Senger, eds.), pp. 3–12. Alan R. Liss, Inc., New York.
Rüdiger, W., Benz, J., and Guthoff, C. (1980). *Eur. J. Biochem.* **109,** 193–200.
Rüdiger, W., Thümmler, F., Cmiel, E., and Schneider, S. (1983). *Proc. Natl. Acad. Sci. U.S.A.* **80,** 6244-6248.
Ryrie, I. J., and Fuad, N. (1982). *Arch. Biochem. Biophys.* **214,** 475–488.
Sabatini, D. D., Kreibich, G., Morimoto, T., and Adesnik, M. (1982). *J. Cell Biol.* **92,** 1–22.
Sandelius, A. S., and Selstam, E. (1984). *Plant Physiol.* **76,** 1041–1046.
Santel, H.-J., and Apel, K. (1981). *Eur. J. Biochem.* **120,** 95–103.
Satoh, K., Nakatani, H. Y., Steinback, K. E., Watson, J., and Arntzen, C. J. (1983). *Biochim. Biophys. Acta* **724,** 142–150.
Scheer, H., Gross, E., Nitsche, B., Cmiel, E., Schneider, S., Schäfer, W., Schiebel, H.-M., and Schulten, H.-R. (1986). *Photochem. Photobiol.* **43,** 559–571.
Schindler, C., and Soll, J. (1986). *Arch. Biochem. Biophys.* **247,** 211–220.
Schmidt, G. W., and Mishkind, M. L. (1983). *Proc. Natl. Acad. Sci. U.S.A.* **80,** 2632–2636.
Schmidt, G. W., and Mishkind, M. L. (1986). *Annu. Rev. Biochem.* **55,** 879–912.
Schmidt, G. W., Devillers-Thiery, A., Desruisseaux, H., Blobel, G., and Chua, N.-H. (1979). *J. Cell Biol.* **83,** 615–622.
Schmidt, G. W., Bartlett, S. G., Grossman, A. R., Cashmore, A. R., and Chua, N.-H. (1981). *J. Cell Biol.* **91,** 468–478.
Schmidt, R. J., Richardson, C. B., Gillham, N. W., and Boynton, J. E. (1983). *J. Cell Biol.* **96,** 1451–1463.
Schmidt, R. J., Myers, A. M., Gillham, N. W., and Boyton, J. E. (1984). *J. Cell Biol.* **98,** 2011-2018.
Schmidt, R. J., Gillham, N. W., and Boynton, J. E. (1985). *Mol. Cell. Biol.* **5,** 1093–1099.
Schneegurt, M. A., and Beale, S. I. (1986). *Plant Physiol.* **81,** 965–971.
Schneider, M., and Rochaix, J.-D. (1986). *Plant Mol. Biol.* **6,** 265–270.
Schneider, M., Darlix, J.-L., Erickson, J., and Rochaix, J.-D. (1985). *Nucleic Acids Res.* **13,** 8531–8541.
Schön, A., Krupp, G., Gough, S., Berry-Lowe, S., Kannangara, C. G., and Söll, D. (1986). *Nature (London)* **322,** 281–284.
Schuster, G., Ohad, I., Martineau, B., and Taylor, W. C. (1985). *J. Biol. Chem.* **260,** 11866–11873.
Schuster, G., Pecker, I., Hirschberg, J., Kloppstech, K., and Ohad, I. (1986). *FEBS Lett.* **198,** 56–60.
Schweiger, H.-G., de Groot, E. J., Leible, M. B., and Tymms, M. J. (1986). *In* "Regulation of Chloroplast Differentiation" (G. Akoyunoglou and H. Senger, eds.), pp. 467–476. Alan R. Liss, Inc., New York.
Senapathy, P. (1986). *Proc. Natl. Acad. Sci. U.S.A.* **83,** 2133–2137.
Senger, H., and Brinkmann, G. (1986). *Physiol. Plant.* **68,** 119–124.
Shanklin, J., Jabben, M., and Vierstra, R. D. (1987). *Proc. Natl. Acad. Sci. U.S.A.* **84,** 359–363.
Shepherd, H.S., Ledoigt, G., and Howell, S. H. (1983). *Cell (Cambridge, Mass.)* **32,** 99–107.
Shimakata, T., and Stumpf, P. K. (1983a). *Arch. Biochem. Biophys.* **220,** 39–45.
Shimakata, T., and Stumpf, P. K. (1983b). *J. Biol. Chem.* **258,** 3592–3598.
Shinozaki, K., and Sugiura, M. (1985). *Mol. Gen. Genet.* **200,** 27–32.

Shinozaki, K., Sun, C.-R., and Sugiura, M. (1984). *Mol. Gen. Genet.* **197**, 363–367.
Shinozaki, K., Deno, H., Sugita, M., Kuramitsu, S., and Sugiura, M. (1986a). *Mol. Gen. Genet.* **202**, 1–5.
Shinozaki, K., Ohme, M., Tanaka, M., Wakasugi, T., Hayashida, N., Matsubayashi, T., Zaita, N., Chunwongse, J., Obokata, J., Yamaguchi-Shinozaki, K., Ohto, C., Torazawa, K., Meng, B. Y., Sugita, M., Deno, H., Kamogashira, T., Yamada, K., Kusuda, J., Takaiwa, F., Kato, A., Tohdoh, N., Shimada, H., and Sugiura, M. (1986b). *EMBO J.* **5**, 2043–2049.
Shlyk, A. A. (1971). *Annu. Rev. Plant Physiol.* **22**, 169–184.
Siefermann-Harms, D. (1985). *Biochim. Biophys. Acta* **811**, 325–355.
Sijben-Müller, G., Hallick, R. B., Alt, J., Westhoff, P., and Herrmann, R. G. (1986). *Nucleic Acids Res.* **14**, 1029–1044.
Silverthorne, J., and Tobin, E. M. (1984). *Proc. Natl. Acad. Sci. U.S.A.* **81**, 1112–1116.
Simpson, J., Timko, M. P., Cashmore, A. R., Schell, J., van Montagu, M., and Herrera-Estrella, L. (1985). *EMBO J.* **4**, 2723–2729.
Simpson, J., van Montagu, M., and Herrera-Estrella, L. (1986). *Science* **233**, 34–38.
Slovin, J. P., and Tobin, E. M. (1982). *Planta* **154**, 456–472.
Smeekens, S., van Binsbergen, J., and Weisbeek, P. (1985a). *Nucleic Acids Res.* **13**, 3179–3194.
Smeekens, S., de Groot, M., van Binsbergen, J., and Weisbeek, P. (1985b). *Nature (London)* **317**, 456–458.
Smeekens, S., Bauerle, C., Hageman, J., Keegstra, K., and Weisbeek, P. (1986). *Cell (Cambridge, Mass.)* **46**, 365–375.
Smith, J. H., and Bogorad, L. (1974). *Proc. Natl. Acad. Sci. U.S.A.* **71**, 4839–4842.
Smith, M. W., and Neidhardt, F. C. (1983a). *J. Bacteriol.* **154**, 336–343.
Smith, M. W., and Neidhardt, F. C. (1983b). *J. Bacteriol.* **154**, 344–350.
Smith, S. M., and Ellis, R. J. (1979). *Nature (London)* **278**, 662–664.
Soll, J., and Buchanan, B. B. (1983). *J. Biol. Chem.* **258**, 6686–6689.
Soll, J., Schultz, G., Rüdiger, W., and Benz, J. (1983). *Plant Physiol.* **71**, 849–854.
Speth, V., Otto, V., and Schäfer, E. (1986). *Planta* **168**, 299–304.
Starnes, S. M., Lambert, D. H., Maxwell, E. S., Stevens, S. E., Jr., Porter, R. D., and Shively, J. M. (1985). *FEMS Microbiol. Lett.* **28**, 165–169.
Steinback, K. E., Burke, J. J., and Arntzen, C. J. (1979). *Arch. Biochem. Biophys.* **195**, 546–557.
Steinmetz, A. A., and Weil, J.-H. (1986). In "Methods in Enzymology" (A. Weissbach and H. Weissbach, eds.), Vol. 118, pp. 212–231. Academic Press, New York.
Steinmetz, A. A., Castroveijo, M., Sayre, R. T., and Bogorad, L. (1986). *J. Biol. Chem.* **261**, 2485–2488.
Steinmüller, K., Kaling, M., and Zetsche, K. (1983). *Planta* **159**, 308–313.
Stobart, K., and Ameen-Buckari, I. (1986). In "Regulation of Chloroplast Differentiation" (G. Akoyunoglou and H. Senger, eds.), pp. 71–80. Alan R. Liss, Inc., New York.
Stumpf, P. K. (1984). In "Biosynthesis of the Photosynthetic Apparatus: Molecular Biology, Development and Regulation" (J. P. Thornber, L. A. Staehelin, and R. B. Hallick, eds.), pp. 225–235. Alan R. Liss, Inc., New York.
Sugimoto, T., Inoue, Y., Suzuki, H., and Furuya, M. (1984). *Photochem. Photobiol.* **39**, 697–702.
Sugita, M., Kato, A., Shimada, H., and Sugiura, M. (1984). *Mol. Gen. Genet.* **194**, 200–205.
Surzycki, S. J., and Shellenbarger, D. L. (1976). *Proc. Natl. Acad. Sci. U.S.A.* **73**, 3961–3965.
Takabe, T., Takabe, T., and Akazawa, T. (1986). *Plant Physiol.* **81**, 60–66.
Tewari, K. K. (1986). In "Methods in Enzymology" (A. Weissbach and H. Weissbach, eds.), Vol. 118, pp. 186–201. Academic Press, New York.
Tewari, K. K., and Goel, A. (1983). *Biochemistry* **22**, 2142–2148.
Tewari, K. K., Kolodner, R. D., and Dobkin, W. (1976). In "Genetics and Biogenesis of Chloroplasts and Mitochondria" (T. Bücher, W. Neupert, W. Sebald, and S. Werner, eds.), pp. 379–386. Elsevier/North-Holland Biomedical Press, Amsterdam.

Thomasson, J. R., Nelson, M. E., and Zakrzewski, R. J. (1986). *Science* **233**, 876–878.
Thornber, J. P. (1986). *In* "Encyclopedia of Plant Physiology, New Series" (L. A. Staehelin and C. J. Arntzen, eds.), Vol. 19, pp. 98–142. Springer-Verlag, Berlin and New York.
Timko, M. P., Kausch, A. P., Castresana, C., Fassler, J., Herrera-Estrella, L., Van den Broeck, G., van Montagu, M., Schell, J., and Cashmore, A. R. (1985). *Nature (London)* **318**, 579–582.
Tittgen, J., Hermans, J., Steppuhn, J., Jansen, T., Jansson, C., Andersson, B., Nechushtai, R., Nelson, N., and Hermann, R. (1986). *Mol. Gen. Genet.* **204**, 258–265.
Tobin, E. M. (1981). *Plant Physiol.* **67**, 1078–1083.
Tobin, E. M., and Silverthorne, J. (1985). *Annu. Rev. Plant Physiol.* **46**, 569–593.
Torazawa, K., Hayashida, N., Obokata, J., Shinozaki, K., and Sugiura, M. (1986). *Nucleic Acids Res.* **14**, 3143.
Trebst, A. (1986). *Z. Naturforsch., C:Biosci.* **41C**, 240–245.
Tripathy, B. C. and Rebeiz, C. A. (1986). *J. Biol. Chem.* **261**, 13556-13564.
Tumer, N. E., Clark, W. G., Tabor, G. J., Hironaka, C. M., Fraley, R. T., and Shah, D. M. (1986). *Nucleic Acids Res.* **14**, 3325–3342.
Vallet, J.-M., Rahire, M., and Rochaix, J.-D. (1984). *EMBO J.* **3**, 415–421.
van Loon, A. P. G. M., Brändl, A. W., and Schatz, G. (1986). *Cell (Cambridge, Mass.)* **44**, 801–812.
van Vloten-Doting, L., Groot, G. S. P., and Hall, T. C., eds. (1985). "Molecular Form and Function of the Plant Genome." Plenum, New York.
Vierling, E., and Alberte, R. S. (1983). *J. Cell Biol.* **97**, 1806–1814.
Vierstra, R. D., and Quail, P. H. (1983). *Plant Physiol.* **72**, 264–267.
Viro, M., and Kloppstech, K. (1980). *Planta* **150**, 41–45.
von Heijne, G. (1985). *J. Mol. Biol.* **184**, 99–105.
von Heijne, G. (1986a). *FEBS Lett.* **198**, 1–4.
von Heijne, G. (1986b). *EMBO J.* **5**, 1335–1342.
Waddell, J., Wang., X.-M., and Wu, M. (1984). *Nucleic Acids Res.* **12**, 3843–3856.
Wang, W.-Y., Gough, S. P., and Kannangara, C. G. (1981). *Carlsberg Res. Commun.* **46**, 243–257.
Wang, W.-Y., Huang, D.-D., Stachon, D., Gough, S. P., and Kannangara, C. G. (1984). *Plant Physiol.* **74**, 569–575.
Watson, J. C., and Surzycki, S. J. (1983). *Curr. Genet.* **7**, 201–210.
Wayne, R., and Hepler, P. K. (1985). *Plant Physiol.* **77**, 8–11.
Weaver, R. F., and Weissmann, C. (1979). *Nucleic Acids Res.* **7**, 1175–1193.
Weinstein, J. D., and Beale, S. I. (1983). *J. Biol. Chem.* **258**, 6799–6807.
Weinstein, J. D., and Beale, S. I. (1984). *Plant Physiol.* **74**, 146–151.
Weinstein, J. D., and Beale, S. I. (1985a). *Arch. Biochem. Biophys.* **237**, 454–464.
Weinstein, J. D., and Beale, S. I. (1985b). *Arch. Biochem. Biophys.* **239**, 87–93.
Werner-Washburne, M., Cline, K., and Keegstra, K. (1983). *Plant Physiol.* **73**, 569–575.
Westhoff, P. (1985). *Mol. Gen. Genet.* **201**, 115–123.
Westhoff, P., Alt, J., and Herrmann, R. G. (1983). *EMBO J.* **2**, 2229-2237.
Westhoff, P., Alt, J., Nelson, N., and Herrmann, R. G. (1985a). *Mol. Gen. Genet.* **199**, 290–299.
Westhoff, P., Jansson, C., Klein-Hitpass, L., Berzborn, R., Larsson, C., and Bartlett, S. G. (1985b). *Plant Mol. Biol.* **4**, 137–146.
Whitfeld, P. R., and Bottomley, W. (1983). *Annu. Rev. Plant Physiol.* **34**, 279–310.
Whitfeld, P. R., Herrmann, R. G., and Bottomley, W. (1978). *Nucleic Acids Res.* **5**, 1741–1751.
Wickner, W. T., and Lodish, H. F. (1985). *Science* **230**, 400–407.
Widger, W. R., Cramer, W. A., Hermodson, M., Meyer, D., and Gullifor, M. (1984a). *J. Biol. Chem.* **259**, 3870–3876.
Widger, W. R., Cramer, W. A., Herrmann, R. G., and Trebst, A. (1984b). *Proc. Natl. Acad. Sci. U.S.A.* **81**, 674–678.

Willey, D. L., Auffret, A. D., and Gray, J. C. (1984a). *Cell (Cambridge, Mass.)* **36,** 555–562.
Willey, D. L., Howe, C. J., Auffret, A. D., Bowman, C. M., Dyer, T. A., and Gray, J. C. (1984b). *Mol. Gen. Genet.* **194,** 416–422.
Wong, Y.-S., and Castelfranco, P. A. (1984). *Plant Physiol.* **75,** 658–661.
Wong, Y.-S., and Castelfronco, P. A. (1985). *Plant Physiol.* **79,** 730–733.
Wong, Y.-S., Castelfranco, P. A., Goff, D. A., and Smith, K. M. (1985). *Plant Physiol.* **79,** 725–729.
Wu, S.-M., and Rebeiz, C. A. (1985). *J. Biol. Chem.* **260,** 3632–3634.
Yamada, T., and Shimaji, M. (1986). *Nucleic Acids Res.* **14,** 3827–3839.
Yamada, T., Shimaji, M., and Fukuda, Y. (1986). *Plant Mol. Biol.* **6,** 245–252.
Yamamoto, N., and Droffner, M. L. (1985). *Proc. Natl. Acad. Sci. U.S.A.* **82,** 2077–2081.
Zeiger, E., Iino, M., and Ogawa, T. (1985). *Photochem. Photobiol.* **42,** 759–763.
Zhu, Y. S., Kung, S. D., and Bogorad, L. (1985). *Plant Physiol.* **79,** 371–376.
Zuber, H. (1985). *Photochem. Photobiol.* **42,** 821–844.

Composition, Organization, and Dynamics of the Thylakoid Membrane in Relation to Its Function

2

J. BARBER

I. Introduction
II. Thylakoid Lipids
    A. Types and Levels
    B. *In Vitro*
    C. *In Vivo*
III. Thylakoid Proteins
    A. Photosystem Two Complex
    B. Photosystem One Complex
    C. Light-Harvesting Chlorophyll *a*/*b* Complex
    D. Cytochrome b_6/f Complex
    E. ATP-Synthase (CF_0–CF_1) Complex
IV. Thylakoid-Membrane Organization
    A. Electron Microscopy
    B. Lateral Distribution of Proteins
    C. Physical Factors Responsible for Thylakoid Membrane Organization
    D. Stoichiometry of Complexes
V. Thylakoid Membrane Dynamics
    A. Fluidity
    B. Platoquinol/Plastocyanin Diffusion
    C. LHC-2 Diffusion
    D. Rotational Diffusion
VI. Final Comment
    References

The Biochemistry of Plants, Vol. 10

I. INTRODUCTION

The transfer of reducing equivalents to NAD^+ or $NADP^+$, together with the net synthesis of ATP, represents the first major step in the process of photosynthesis, whether it be in anoxygenic (photosynthetic bacteria) or oxygenic (green plants, algae and cyanobacteria) organisms. In the case of anoxygenic photosynthetic bacteria, the reducing equivalents can be obtained from oxidising a range of compounds including H_2S and organic acids. In contrast, the prokaryotic cyanobacteria are able to utilize H_2O as an electron/proton source, and in this respect they are similar to the eukaryotic systems of green plants and algae. In all cases, however, the reactions giving rise to the net synthesis of the above mentioned energy rich compounds are located in membrane systems, while the carbon fixation processes occur in aqueous environments. A feature that has been emerging during the past few years is that, despite the morphological, physiological, and ecological differences between various types of photosynthetic organisms, at the molecular level there are striking similarities. Such similarities are particularly evident in the case of electron transport, a fact that is now being widely exploited as more information becomes available from gene analyses and from the application of techniques of molecular biology.

In this chapter I overview our present understanding of electron flow in photosynthesis, especially as it relates to membrane structure and function. With the constraints of space, it will be impossible to discuss all details or to cover the wide range of different types of organisms. Rather, I have chosen to restrict my discussions to those organisms that contain chlorophyll *b*, that is, higher plants and green algae, and to concentrate on new developments that have bearings on structural-functional relationships.

For some time it has been obvious that a complete understanding of the mechanisms that underlie light interception, electron transfer, proton pumping, and photophosphorylation in higher plants and green algae would require an appreciation of the structure and properties of the chloroplast thylakoid membrane (Menke, 1962; Boardman, 1970; Park and Sane, 1971; Boardman *et al.*, 1978; Arntzen, 1978). Despite this, it was not until the 1980s that a detailed organizational picture started to emerge as emphasized in recent reviews (Hiller and Goodchild, 1981; Barber, 1983a, 1985; Staehelin and Arntzen, 1983; Staehelin, 1986; Murphy, 1986). This is not to say that efforts to relate the structure of the thylakoid membrane with its functions had not begun several years before, starting with the pioneering work of Boardman and colleagues (Boardman and Anderson, 1964; Boardman, 1968, 1970, 1977). Indeed, in 1975, Anderson published detailed and thoughtful discussions on the molecular organization of the thylakoid membrane and its relationship with functional activities. To my mind, Anderson's review was an outstanding contribution in that it heralded a new era in which for the first time the fluid-mosaic model of Singer and Nicolson (1972) was adopted to describe the thylakoid system. Since then there have been tremendous de-

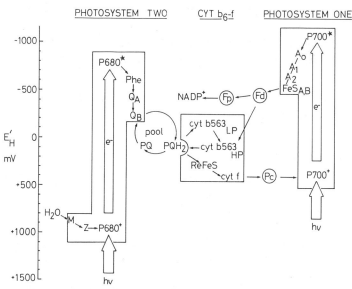

Fig. 1. A scheme for electron transfer in oxygen evolving photosynthetic organisms based on midpoint redox potentials at pH 7.0. The symbols are: M, components that can accumulate oxidizing potential for water splitting; Z, primary electron donor to P680, the reaction-center chlorophyll of PS2; Phe, phaeophytin; Q_A, bound plastoquinone (one-electron acceptor); Q_B, plastoquinone (two-electron acceptor) able to exchange with plastoquinone (PQ) pool; ReFeS, Rieske iron–sulfur center; cyt, cytochrome; Pc, plastocyanin; P700, reaction-center chlorophyll of PS1; A_0, A_1, A_2, primary electron acceptors of PS1; $FeS_{A,B}$, bound iron–sulfur centers A and B; Fd, soluble ferredoxin; F_P, flavoprotein having $NADP^+$-ferredoxin oxidoreductase activity (FNR); $NADP^+$, oxidized nicotinamide adenine dinucleotide phosphate. The boxed-in sections represent three distinct intrinsic membrane protein complexes, while other components are water-soluble except the PQ pool, which is soluble in membrane lipids.

velopments in our understanding of the structural implications of photosynthetic membranes, with the most outstanding being the recent elucidation, by x-ray crystallographic analysis, of the structure of the reaction center of *Rhodopseudomonas viridis* (Deisenhofer *et al.,* 1985a).

Before embarking on the details of the structural and functional aspects of electron transport in chloroplasts, it is appropriate to emphasise that the transfer of electrons and protons from water to $NADP^+$ is driven by a series of redox reactions powered with light energy absorbed by two photosystems, one (PS1) and two (PS2), which interact cooperatively (Fig. 1). This concept was first suggested by Hill and Bendall (1960) and quickly confirmed by a number of elegant experiments conducted by Duysens and his colleagues in the Netherlands (see Duysens, 1964). A modern version of the Hill and Bendall Z-scheme is given in Fig. 1, which indicates how PS2 and PS1 interact via a number of redox components and also shows that under some circumstances electrons may also cycle around PS1. This scheme is based on potential energies of the redox components and gives no information about the structures involved or how such structures are related to thylakoid membrane composition and organization. It is therefore the purpose of this chapter to discuss the structural implications of the Z-scheme in terms of the lipoprotein and dynamic nature of the membrane in which it is sited.

II. THYLAKOID LIPIDS

A. Types and Levels

The lipids of the thylakoid membrane can be divided into those that are saponifiable and those that are not (see Harwood, 1980; Quinn and Williams, 1983). The saponifiable lipids are the diacylglycerolipids, which make up the matrix of the membrane, while the nonsaponifiable lipids are the various forms of pigments (chlorophylls and carotenoids) and quinones. The pigments are bound within protein complexes, and the quinones (mainly plastoquinone-9) are located in the hydrophobic lipid matrix. The major classes of saponifiable lipids are the electroneutral galactolipids, monogalactosyldiglyceride (MGDG) and digalactosyldiglyceride (DGDG), which together can

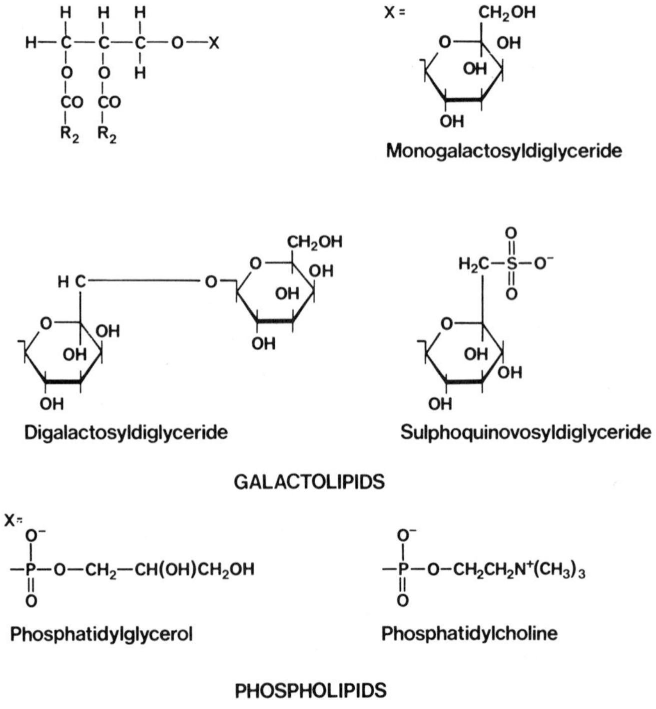

Monogalactosyldiglyceride

Digalactosyldiglyceride   Sulphoquinovosyldiglyceride

GALACTOLIPIDS

Phosphatidylglycerol   Phosphatidylcholine

PHOSPHOLIPIDS

Fig. 2. Chemical structure of the main classes of lipids found in chloroplast thylakoid membranes.

TABLE I
Polar Lipid Composition and Fatty Acid Composition of Pea Chloroplast Thylakoids[a]

Lipid class	Relative proportion (mol %)	Fatty acid (mol %)							Double-bond index
		16:0	16:1	16:3	18:0	18:1	18:2	18:3	
MGDG	42	2	0	—	1	1	4	92	5.7
DGDG	28	8	0	—	2	1	2	86	5.2
SQDG	11	25	0	—	5	2	6	62	3.8
PG	12	29	17	—	3	7	10	35	3.0
PC	8	24	0	—	6	4	33	33	3.3

[a] For this preparation the lipid to chlorophyll ratio was 2.99 mol/mol. Data from Chapman *et al.* (1985).

represent as much as 75% of the total polar lipid composition of the thylakoid membrane. The remaining polar lipids are sulfoquinovosyldiglyceride (SQDG), phosphatidylglycerol (PG), and phosphatidylcholine (PC). The chemical structure of these various lipid classes is presented in Fig. 2, while Table I gives their relative levels for the thylakoids of pea. Table I also emphasizes that the polar lipids of the thylakoids are extremely unsaturated, with a double-bond index (average number of double bonds per molecule) in the region of five or six. The dominating unsaturated fatty acid is linolenic acid, which has a hydrocarbon tail of 18 carbons and three double bonds (18:3 fatty acid; octadecatrienoic acid) and can comprise 90% of the total acyl chains. In the so-called "16:3 plants" (e.g., spinach), hexadecatrienoic acid also occurs in MGDG. A very important feature of chloroplast PG is that it contains the unusual fatty acid 16:1 *trans*-Δ3-hexadecenoic acid. Also worthy of note is that the thylakoid membranes possess little or no sterols and that 20% or less of the lipids (SQDG + PG) carry net negative electrical charge.

B. *In Vitro*

The polar lipids of the thylakoids can be isolated and analyzed by a wide range of procedures as recently reviewed (Chapman and Barber, 1987). When isolated, the total lipid extract or the majority of the lipid classes (DGDG, SQDG, PG, and PC) forms bilayers when dispersed in water or salt solutions at physiological temperatures. In contrast, the naturally occurring MGDG does not form bilayers in aqueous dispersions but arranges itself into nonbilayer structures, called the hexagonal type II phase (Hex-II) (Rivas and Luzzati, 1969; Shipley *et al.*, 1973; Sen *et al.*, 1981; Gounaris *et al.*, 1983b; see also Quinn and Williams, 1983; Gounaris and Barber, 1983). The formation of the Hex-II lattice only occurs with polyunsaturated MGDG when it is in the liquid crystalline state. Below the phase transition temperature, when it is in the gel state, the preferred organization is in the form of lamellar sheets (Sen *et al.*, 1983). To obtain the gel state with naturally occurring polyunsaturated species of MGDG, the temperature must be lowered to −30°C (Shipley *et al.*, 1973). Decreasing the level of unsaturation by catalytic hydrogenation, however, dramatically raises the phase transition to well above room temperature (Gounaris *et al.*, 1983b). Aqueous dispersion of binary mixtures of MGDG and conventional bilayer-forming lipids can give rise to a variety of structures intermediate between bilayers and Hex-II depending on the molar ratio of MGDG to the bilayer lipid (Sen *et al.*, 1981; Sprague and Staehelin, 1984). Typical structures formed include those shown diagrammatically in Fig. 3 and in the electron micrographs of Fig. 4. With the total lipid extract, the conformational states adopted depend on conditions such as temperature, pH, electrolyte levels, and the presence of cryoprotectants. For example, the bilayer configuration of the total lipid extract is lost when the pH is lowered or the cation levels raised (Gounaris *et al.*, 1983a). Under these conditions, neutralization or electrostatic screening of the charges of SGDG and PG facilitates their phase separation from MGDG and thus the formation of nonbilayer structures occur where the relative level of MGDG is high.

C. *In Vivo*

Despite the fact that polyunsaturated MGDG is the dominant polar lipid of the thylakoids, there is no evidence that this membrane, under normal conditions, contains nonbilayer structures (Gounaris *et al.*, 1983c). Therefore, as for other biological membranes, it can be assumed that the thylakoid polar lipids exist predominantly as a bilayer *in vivo*. Only under extreme conditions, such as heat stress (Gounaris *et al.*, 1984a) or protein denaturation (Machold *et al.*, 1977), have nonbilayer structures been detected in the natural membrane. The precise reason for the existence of high levels of nonbilayer-forming lipid in the thylakoid membrane is unclear, although it has been suggested that the molecular shape of MGDG makes it ideal for packaging large multipeptide complexes into the membrane (Murphy, 1982; W. P. Williams *et al.*, 1984). Therefore locally perturbed bilayers may exist, possibly as "half" nonbilayer structures.

Several studies have been initiated to investigate whether the lipid composition is symmetrical between the two leaflets of the bilayer. The main techniques employed have been antibody labeling (e.g., Radunz, 1979, 1980) and selective attack by lipases (e.g., Rawyler and Siegenthaler, 1981a; Unitt and Harwood, 1982, 1985). Unfortunately, these approaches have not yet produced a clear picture. A trend that emerged from these studies is that PG is more abundant in the outer half of the bilayer while SQDG is preferentially located in the inner leaflet. It has also been reported by Unitt and Harwood (1985) that while palmitate in PG is evenly distributed, linolenate and particularly *trans*-Δ3-hexadecenoate are preferentially found in the outer leaflet. These results are in accordance with the earlier work of Duval *et al.* (1980).

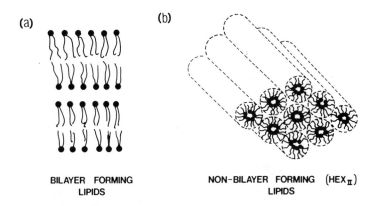

(a)

(b)

BILAYER FORMING
LIPIDS

NON-BILAYER FORMING (HEX$_{II}$)
LIPIDS

(c)

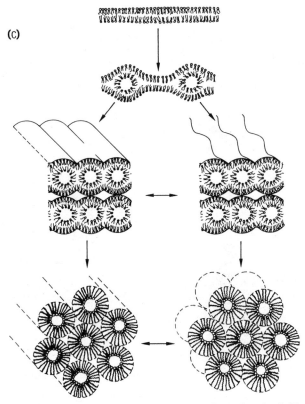

Fig. 3. Structures formed in aqueous dispersions by (a) bilayer-forming lipids, (b) pure non-bilayer-forming lipids, and (c) mixtures of bilayer- and non-bilayer-forming lipids. Structures seen in mixed systems vary from a pure bilayer, through intermediate structures consisting of spherical and cylindrical inverted micelles sandwiched within bilayers, to quasi-crystalline arrays of inverted micelles. When isolated, polar lipids of the thylakoid membrane will form these types of structure, depending on conditions and degree of purification and unsaturation. [Drawing taken from Quinn and Williams (1983).]

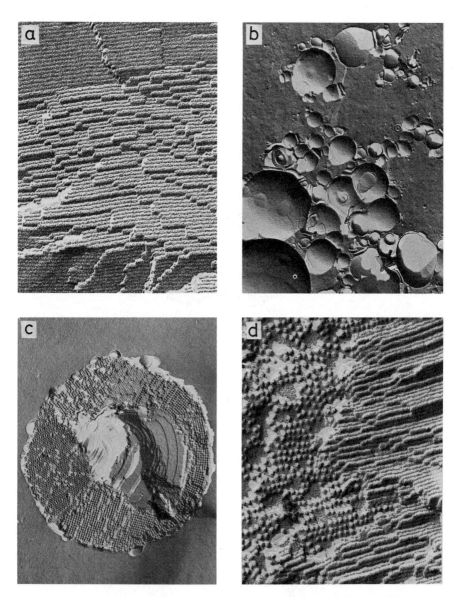

Fig. 4. Electron micrographs of freeze-fracture replicas prepared from (a) MGDG in its unsaturated form showing Hex-II type configuration (×200,000), (b) phosphatidylglycerol (PG) isolated from thylakoids showing normal bilayer organization (×54,000), (c) thylakoid total lipid extract in the presence of cations showing aggregated inverted micelles and cylindrical Hex-II configuration (×47,000), and (d) a binary mixture (2 : 1) of MGDG and DGDG, demonstrating the coexistence of Hex II and inverted micelles sandwiched within a bilayer (×190,000). (Pictures kindly supplied by Dr. K. Gounaris.)

Inconsistencies in data exist regarding the distribution of PC and the two electroneutral galactolipids. Earlier Radunz (1980) had concluded that galactolipids were enriched at the inner surface, while more recently Sundby and Larsson (1985) labeled MGDG and DGDG with tritiated sodium borohydride and concluded that they are more abundant at the outer surface. The work of Gounaris et al. (1984b), using monoclonal antibodies, also suggested that the level of MGDG in the outer leaflet was higher than the inner, as did the chemical labeling and hydrolase studies of Rawyler et al. (1986) and Rawyler and Siegenthaler (1985). These latter workers, however, did not confirm the conclusions of Sundby and Larsson concerning the distribution of DGDG.

Since the thylakoid membranes of higher plants normally exist as interconnecting granal and stromal lamellae, there is a possibility that a lateral, as well as a transverse, heterogeneity of lipid distribution may occur. From investigations that have been made, a general picture has emerged that there is no extreme lateral separation of lipid classes between the two membrane regions (see Chapman et al., 1985). The only consistent difference found was that the granal, compared with stromal, lamellae had a higher MGDG to DGDG ratio. This latter conclusion is supported by analyses of various membrane preparations derived from the appressed regions of the grana (Gounaris et al., 1983d; Murphy and Woodrow, 1983; Henry et al., 1983). Fatty acid analyses did not reveal any major differences between the two membrane regions, although the granal membranes were found to be slightly more unsaturated (Chapman et al., 1985; Ford et al., 1982). The finding that PG with its 16:1 trans-Δ3-hexadecenoic acid occurred both in granal and stromal lamellae (Chapman et al., 1985) contrasts with the views of Tuquet et al. (1977), who postulated that PG with this fatty acid was involved in membrane stacking. In fact, the elegant studies of McCourt et al. (1985) using a trans-Δ3-hexadecenoic acid deficient mutant of Arabidopsis thaliana (L.) has clearly indicated that this fatty acid is not required for normal structural or functional integrity of the thylakoids.

Although this is a rather unexplored area, there is every reason to believe that the various intrinsic protein complexes of the thylakoid membrane interact with specific lipids in order to attain optimal organization and functional activities (see Siegenthaler and Rawyler, 1986). It has been suggested that PG may play a structural role in the organization of the light-harvesting chlorophyll a/b (LHC-2) complex (Trémolières et al., 1981; Remy et al., 1982, 1985). Using lipases and acyl hydrolases with different substrate specificities, Rawyler and Siegenthaler (1980, 1981a,b) concluded that PC is closely associated with photosystem two (PS2) and PG with photosystem one (PS1). The former conclusion was also supported by the recent work of Jordan et al. (1983). Unfortunately, most of these studies have been indirect, relying on the action of various enzymes, although studies on developing systems are starting to yield additional information (Siegenthaler and Giroud, 1986). Recently two thylakoid-membrane protein complexes have been isolated and subjected to lipid analyses following treatments to remove loosely bound lipid. Interestingly, both studies found SQDG to be tightly bound to the isolated complexes, the implications of which have been discussed by Barber and Gounaris (1986). One such study was made by Pick *et al.* (1985) on isolated CF_0–CF_1 ATP-synthase, while the other involved the analysis of a PS2 core complex (Gounaris and Barber, 1985). In addition to the above approaches, experiments with isolated proteins reconstituted into different lipid environments can also help elucidate the functional significance of protein-lipid interactions. From such studies, Siefermann-Harms *et al.* (1982) concluded that unsaturated MGDG is required to restore energy transfer between isolated LHC-2 and PS2 complexes, while Gounaris *et al.* (1983e) and Matsuda and Butler (1983) demonstrated that DGDG enhances or activates PS2 reactions. Akabori *et al.* (1984) and Imoaka *et al.* (1986) also reported that lipids were required for PS2 activity, but the study was limited to thylakoid total lipid extract. A detailed study has been made involving the reconstitution of the CF_0–CF_1 ATP-synthase complex with thylakoid lipids (Pick *et al.*, 1984) that also emphasized the importance of galactolipids, especially MGDG, for optimal activities.

III. THYLAKOID PROTEINS

There are five functionally distinct protein complexes within the thylakoid membrane: photosystem two (PS2), photosystem one (PS1), light-harvesting chlorophyll *a/b* (LHC-2), cytochrome b_6/f (cyt b_6/f), and ATP-synthase (CF_0–CF_1). These supramolecular complexes interact together to promote the conversion of light energy into NADPH and ATP. They are embedded in the lipid matrix of the membrane with a protein to lipid ratio of about 1.7. In addition to these main complexes, there are other protein components such as chlorophyllases, kinases, and phosphatases. Except for LHC-2, the other main complexes are composed of several different polypeptides, which are either nucleus or chloroplast gene products. The transcription of the nuclear genes occurs on 80 S cytoplasmic ribosomes so that the resulting polypeptides must be transferred to the chloroplast before insertion into the thylakoid membrane. Such a transfer is commonly but not always aided by a "transit peptide" attached to the polypeptide. For chloroplast-encoded polypeptides, the transcription occurs on 70 S ribosomes within the organelle. No matter whether a polypeptide is a product of the nucleus or chloroplast genome, its native molecular weight and its functional conformation will not be obtained until its final processing after insertion into the membrane. Nevertheless, the elucidation of primary structures from gene sequencing has already become a valuable tool for predicting the organizational and functional properties of many thylakoid polypeptides. Coupled with this important growth in the application of molecular genetics have been considerable improvements in protein separation and assaying techniques, including the application of immunology. However, perhaps the most important step in the last few years has been the development of procedures to crystallize membrane proteins (Michel, 1982, 1983; Garavito *et al.*, 1983). In the case of the reaction centers of *Rhodopseudomonas viridis*, the crystals obtained were sufficiently ordered to allow x-ray diffraction analyses to be carried out to a resolution of atomic distances (Deisenhofer *et al.*, 1984, 1985a,b). This outstanding study has tremendous implications for photosynthesis research, particularly because it has revealed striking analogies between bacterial and higher plant systems.

As a consequence of these developments, our appreciation of the structural–functional relationships of the five main complexes is rapidly advancing, and below I summarize how our knowledge stands at present.

A. Photosystem Two Complex

1. Function

This complex functions as a water–plastoquinone oxidoreductase. A discussion of the kinetics of the electron transfer processes involved is complicated by the requirement for the reaction center to turn over four times in order to produce one oxygen molecule from two molecules of water (Joliot and Kok, 1975). It is generally agreed that the charge storage involves various valency states of manganese, and several models have been proposed (see Renger and Govindjee, 1985). The primary photochemical reactions that generate the necessary redox potential to oxidize water and reduce plastoquinone occur in the PS2 reaction center (see Van Gorkom, 1985) and are remarkably similar to those that take place in the reaction centers of nonsulfur purple bacteria (Rutherford, 1985). When photoexcited, the reaction-center chlorophyll P680 reduces a bound quinone (Q_A) within a microsecond. This electron donation is facilitated by a very rapid reduction of a phaeophytin molecule, probably in the time domain of 10 psec. The electron on Q_A is passed, within a millisecond, to a plastoquinone molecule designated Q_B. This quinone, unlike Q_A, is a two-electron acceptor and when fully reduced is converted to plastoquinol by the addition of two H^+ from the aqueous medium. Unlike the fully oxidized or semiquinone form of Q_B, this fully reduced species no longer binds to the reaction centre and is replaced by an oxidized species. If any of these electron-transfer reactions are blocked, then back reactions will occur with half-times as indicated in Fig. 5(a). Normally, however, P680$^+$ can be rereduced in about 10 nsec by a primary donor Z, which has recently been suggested to be a quinone (O'Malley *et al.*, 1984). However, the kinetics of P680$^+$ reduction are dependent on the charge accumulation state of the water splitting process and the redox condition of Z (Brettel *et al.*, 1984; Schlodder *et al.*, 1985). Even so, at

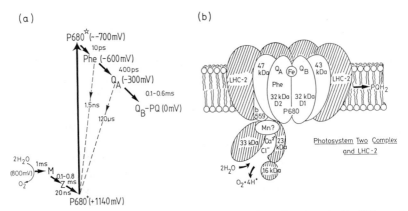

Fig. 5. (a) Kinetics of electron transport processes that take place in the PS2 reaction-center complex, giving approximate half-time for the reaction rates and suggested midpoint redox potential for the components. For symbols, see Fig. 1. (b) A schematic model of the composition and organization of the PS2 surrounded by LHC-2 complexes. Shaded components are probably nucleus-encoded, while unshaded are chloroplast-encoded.

normal light intensities the electron flow to $P680^+$ from water is very efficient and cannot be improved by addition of artificial electron donors, such as diphenylcarbazide (DPC). If the water oxidizing system is partially or totally inhibited, DPC can effectively reduce $P680^+$, as can cytochrome b_{559}, which is an intrinsic component of the PS2 complex.

2. Polypeptides

The PS2 complex consists of several polypeptides of which the minimum number are shown in Fig. 5(b). As can be seen, some of the polypeptides are intrinsic components of the membrane while others are extrinsic, being bound to the lumenal surface. In higher plant chloroplasts there are at least three extrinsic proteins, having apparent molecular weights of about 33 kDa, 23 kDa, and 16 kDa, that play a role in the water-splitting process (for reviews, see Ghanotakis and Yocum, 1985; Murata and Miyao, 1985; Andersson and Akerlund, 1986). All three proteins are encoded in the nucleus (Westhoff et al., 1985b) and have been well characterized in terms of their amino acid contents and physical properties (Jansson, 1984; Kuwabara and Murata, 1984). Recently the complete amino acid sequence of the 33-kDa protein was published (Oh-oka et al., 1986).

The 23-kDa and 16-kDa proteins are readily removed by washing membranes, which have their lumenal surface exposed, with high concentrations of NaCl (Åkerlund et al., 1982; Miyao and Murata, 1983; Ghanotakis et al., 1984b). Removal of the 33-kDa protein requires other washes and treatments, such as 0.8 M Tris (Yamamoto et al., 1981a), high pH (Sandusky et al., 1983), 2.5 M urea (Miyao and Murata, 1984), and heating after deoxycholate extraction (Franzen and Andreasson, 1984). Except for urea and heating, these treatments also remove the 23-kDa and 16-kDa polypeptides and the Mn involved in charge accumulation states. However, under appropriate conditions washing with 1 M CaCl$_2$ or MgCl$_2$ removes the 33–kDa, 23-kDa, and 16-kDa polypeptides without removing Mn (Ono and Inoue, 1984a; Imaoka et al., 1986). In addition to removing the 33-kDa polypeptide without extracting Mn, as mentioned above, several reductants such as H$_2$O$_2$, hydroquinone, and benzidine will extract most of the Mn without removing the 33-kDa unit (Ghanotakis et al., 1984c). A variety of reconstitution studies, whereby the proteins have been rebound to the membrane in order to activate oxygen evolution, have led to general conclusions about their functions (see Murata and Miyao, 1985). It is generally agreed that the 23-kDa and 16-kDa polypeptides function to maintain elevated levels of calcium and chloride at the site of water oxidation while the role of the 33-kDa unit is more closely linked to the stability of the manganese involved in the accumulation of oxidation states. Therefore the inhibition of water oxidation brought about by the removal of the 23-kDa and 16-kDa polypeptide can be overcome by raising the calcium and chloride levels in the reaction medium (Andersson et al., 1984; Ghanotakis et al., 1984a). On the other hand, as stated above, the removal of the 33-kDa protein is often accompanied by the loss of the manganese necessary for the catalytic breakdown of water. If precautions are taken to prevent the loss of manganese (e.g., by using 1 M calcium chloride), then the removal of the 33-kDa protein does not totally inhibit water oxidation as long as high concentrations of calcium and chloride are present in the assay medium (Miyao and Murata, 1984; Kuwabara et al., 1985; Ono and Inoue, 1984b; Tang and Satoh, 1986).

Rebinding of the 23-kDa protein occurs under low salt conditions and is enhanced if the 33-kDa protein is present. Similarly, the rebinding of the 16-kDa protein only occurs when the 23-kDa protein is present. It is possible, however, to remove the 33 kDa with 2.5 M urea without totally removing the 23- or 16-kDa protein. Depending on chloride and calcium levels, the binding of the 23- and 16-kDa proteins can stimulate oxygen evolution (Åkerlund et al., 1982; Toyoshima and Fakutaka, 1982; Ghanotakis et al., 1984d; Andersson et al., 1984; Imoaka et al., 1986), with the degree of stimulation being enhanced with some types of preparation if extracted thylakoid membrane lipids are also added (Akabori et al., 1984). Although it has been established that the 16-kDa and 23-kDa units do not contain metal ions or show redox properties, there are two reports that the 33-kDa unit can be isolated under some conditions with manganese bound to it (Abramowicz and Dismukes, 1984; Yamamoto et al., 1984). Despite these latter claims, it must be concluded for the time being that the 33-kDa unit protects, but does not harbor, the four-manganese cluster thought to be necessary for water oxidation and that this catalytic site has not yet been located precisely.

When the PS2 core is isolated free of LHC-2 and of extrinsic proteins, it appears to consist of five polypeptides readily detectable by sodium dodecyl sulfate–polyacrylamide gel electrophoresis (SDS–PAGE) (Satoh, 1979; Satoh *et al.,* 1983; Gounaris and Barber, 1985). These components have apparent molecular weights of about 47 kDa, 43 kDa, 10 kDa, and two in the region of 34–31 kDa. The latter two are visualized as diffuse bands designated D1 and D2, which do not stain well with coomassie blue. In addition to these components there is evidence, based on ^{14}C labeling (Delepelaire, 1984), that there are one or more lower-molecular-weight polypeptides, including a 4-kDa polypeptide, that can be resolved on SDS–PAGE under special conditions (Ljungberg *et al.,* 1986). Until recently it had been argued that the 47-kDa unit is the PS2 reaction-center polypeptide containing P680, phaeophytin, and Q_A (Nakatani *et al.,* 1984; Yamagishi and Katoh, 1984). However, it now seems likely that this chlorophyll-binding polypeptide has no photochemical activity and functions only as a light-harvesting system as does the chlorophyll-binding 43-kDa polypeptide (Camm and Green, 1983). As explained in the next section (Section III,A,3), based on amino acid sequence analogies with the L and M subunits of the purple bacterial reaction center, it is highly likely that the D1 and D2 proteins are the reaction-center polypeptides of PS2 (Deisenhofer *et al.,* 1985a,b; Trebst, 1986). D1 has convincingly been shown to be the site for Q_B and herbicide binding. Compared with other thylakoid membrane proteins, D1 turns over remarkably rapidly (Edelman *et al.,* 1984), possibly because of its susceptibility to photodamage (Kyle *et al.,* 1984a). At present there is no biochemical evidence that it binds chlorophyll or is directly involved in primary photochemistry. Its strong homology with the L and M subunits of the bacterial reaction center had been noted, however (Youvan *et al.,* 1984; J. C. Williams *et al.,* 1984), and led to speculations regarding the involvement of a conserved histidine–methionine pair in quinone binding (Hearst and Sauer, 1984). The D2, unlike the D1 protein, contains lysine and does not rapidly turn over (Millner *et al.,* 1986; Pick *et al.,* 1986). Nevertheless, its amino acid sequence has close homologies with D1 and with the L and M subunits, which suggests that it is also a quinone binding protein. As explained below, the structural elucidation of the reaction-center crystals of *Rhodopseudomonas viridis* (Deisenhofer *et al.,* 1985a) strongly indicate that D1 and D2 are indeed structurally and functionally comparable with L and M subunits. Therefore a working model for PS2 organization should place D1 and D2 as the reaction-center polypeptides and assume that the 47-kDa and 43-kDa units act only as core antennae systems [see Fig. 5(b)]. This new concept has to be proven and as yet no photochemically active chlorophyll-binding complex, consisting of D1 and D2 only, has been isolated. However, a close examination of the existing literature does not exclude this new model. For example, Yamagishi and Katoh (1984) used lauryldimethylamine *N*-oxide (LDAO) to dissociate a PS2 core preparation isolated from *Synechococcus* into two chlorophyll binding complexes. One of these consisted almost entirely of a 40-kDa polypeptide (equivalent to the 43-kDa polypeptide in higher plants) and showed no photochemical activity. The other complex was photochemically active and therefore was assumed to contain the PS2 reaction center. SDS–PAGE of this active complex using denaturing conditions revealed it to consist of the 47-kDa polypeptide but also to contain components at about 32 kDa and 10 kDa. Indeed, when the 47-kDa unit has been purified as a chlorophyll-binding protein, free of other components, it no longer shows photochemical activity (Tang and Satoh, 1984). The low-molecular-weight polypeptides of the PS2 core, notably the 10-kDa and 4-kDa units, are thought to be the apoproteins of cytochrome b_{559} (Widger *et al.*, 1984). These polypeptides, like D1, contain no lysine and probably form a heterodimer (see below), of which there may be two per PS2 complex (Cramer *et al.*, 1985, 1986).

3. *Molecular Biology*

As indicated in Fig. 5(b), all the known polypeptides that make up the core of PS2 are encoded on the chloroplast genome. In the past few years the genes for these polypeptides have been located and sequenced for a wide variety of photosynthetic tissue (see reviews by Dyer, 1985; Barber and Marder, 1986). Sequences for the nuclear genes of the 17-kDa, 23-kDa, and 33-kDa polypeptides have not yet been published, although they are being actively investigated (Westhoff *et al.*, 1985b). The genes for the 47-kDa and 43-kDa polypeptides are known as *psbB* and *psbC,* respectively, adopting the notation of Hallick and Bottomley (1984). The *psbB* gene from spinach codes for a hydrophobic protein of 508 amino acids having a molecular mass of 56,246 (Morris and Herrmann, 1984). The spinach *psbC* gene product is anticipated to have 473 residues and a molecular weight of 51,800 (Holschuh *et al.*, 1984; Alt *et al.*, 1984). Despite the fact that, in the case of the *psbB* gene, there is some debate about the starting point of the open reading frame and that both the *psbB* and *psbC* gene sequences are for the unprocessed products, there is a remarkable similarity between the two encoded polypeptides (compare Figs. 6 and 7). Hydropathy plots for the two sequences based on Kyte and Doolittle (1982) indices suggest seven possible transmembrane helical spans, as shown in Figs. 6 and 7. To be noted in these models is that both polypeptides have a very large hydrophobic domain between helices VI and VII and that both have conserved histidine residues strategically located in transmembrane segments III, IV, and VII. Since both polypeptides non-covalently bind chlorophyll *a,* it seems likely that these histidines are vital for stabilizing the porphyrin heads within the protein. As indicated in Figs. 6 and 7, the large hydrophobic domains may be exposed on the lumenal side of the thylakoid and play a role in the binding of the extrinsic proteins and catalysing the water oxidation process. The comparability of the two polypeptides is consistent with their functional role as core antennae systems for the PS2 reaction center as depicted in Fig. 5(b).

Although the precise function of cytochrome b_{559} is not known, the genes

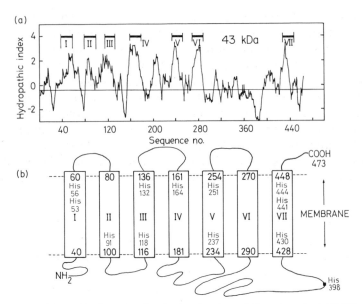

Fig. 6. (a) A plot of hydropathy distribution for the *psbC* gene product (43-kDa protein) presented by Alt *et al.* (1984) using Kyte and Doolittle indices (1982) and an amino acid window of 11 units. (b) A folding model based on the hydropathy profile and the strategic placing of histidines in transmembrane segments. This model has been drawn in order to have symmetry with the model for the *psbB* gene product shown in Fig. 7b.

for the 10-kDa and 4-kDa polypeptides, *psbE* and *psbF,* respectively, have been located side by side on the spinach chloroplast genome (Herrmann *et al.,* 1984; Westhoff *et al.,* 1985a). The *psbE* product has 83 amino acids while the *psbF* gene codes for 39 residues. Hydropathy considerations of the two sequences indicate that each polypeptide has one transmembrane segment containing a single histidine residue. As indicated above, it therefore seems likely that each histidine residue is involved in heme coordination giving rise to an alpha–beta type heterodimer (see Cramer *et al.,* 1985, 1986). The two histidines are localized toward one end of the hydrophobic membrane segment, so that if there were two heterodimers per PS2 complex they could be arranged such that the two hemes facilitate electron transfer across the membrane (see Cramer *et al.,* 1985).

Perhaps the greatest impact that molecular biology has had on our understanding of the structure and function of PS2 has come from locating and sequencing the *psbA* and *psbD* genes. The former gene codes for the D1 herbicide-binding protein and was first sequenced for spinach and tobacco (Zurawski *et al.,* 1982), but now sequences have been obtained for *psbA* genes from a wide variety of organisms (Marder, 1985). A remarkable feature is that these sequences are highly conserved, predicting 352–360 amino acid residues (353 in higher plants) and a nonprocessed molecular mass of

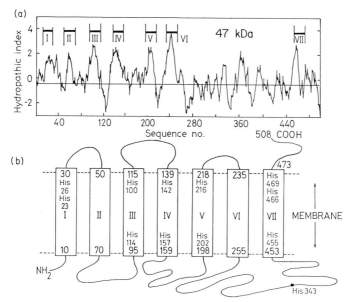

Fig. 7. (a) A plot of hydropathy distribution for the *psbB* gene product (47-kDa protein) as presented by Alt *et al.* (1984) using Kyte and Doolittle (1982) indices and an amino acid window of 11 units. (b) A folding model based on the hydropathy profile and the strategic placing of histidine residues in transmembrane segments. This model has been drawn in order to have symmetry with the model for the *psbC* product shown in Fig. 6b.

about 39 kDa. There are arguments, however, that translation actually starts at 37 codons upstream, giving a smaller translation product of molecular weight 34.6 kDa (Hirschberg and McIntosh, 1983; Cohen *et al.*, 1984) although recent studies from the laboratory of Edelman indicate that this is very unlikely (M. Edelman, personal communication). Several point mutations have been found giving rise to amino acid residue changes that interfere with the ability of the D1 protein to bind triazine or urea herbicides (Hirschberg and McIntosh, 1983; Erickson *et al.*, 1985). The *psbD* has been studied far less than the *psbA* gene, but sequences exist for *Chlamydomonas* (Rochaix *et al.*, 1984), pea (Rasmussen *et al.*, 1984), and spinach (Holschuh *et al.*, 1984; Alt *et al.*, 1984). The sequences are highly conserved, giving a nonprocessed molecular weight of 39.5 kDa. Hydropathy plots of the D1 and D2 sequences indicate similar folding structures, while the occurrence of highly conserved regions emphasizes their close relationship with the L and M subunits of the bacterial reaction center. If D1 and D2 are assumed to have five transmembrane segments (Trebst, 1986; Barber and Marder, 1986), then it is segment IV from the N-terminus that has the most striking analogy with the corresponding transmembrane segments of the L and M polypeptides (see Fig. 8). In this transmembrane region, histidines 198 and 215 are conserved in D1 and D2 and correspond to histidines 200 and 217 in M and

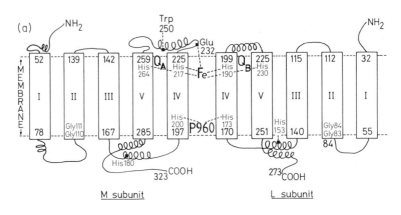

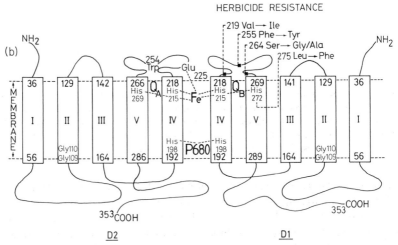

Fig. 8. (a) A diagrammatic representation of the organization of the L and M polypeptides of the reaction center of *Rhodopseudomonas viridis* based on details obtained by x-ray crystallography (Deisenhofer *et al.*, 1985a). The figure indicates how the helical transmembrane segments IV and V of both L and M polypeptides are involved in the binding of Fe and the special pair of bacteriochlorophyll-*b* molecules that form P960. (b) A diagrammatic representation of the organization of D2 and D1 polypeptides of PS2 based on analogies with the bacterial reaction-center structure shown in (a). The model has been contrived especially to emphasize homologies of segment IV and V with the bacterial system so as to suggest possible binding arrangements for Fe and P680. Other key conserved amino acids between the PS2 and bacterial polypeptides are also shown, together with residue positions 219, 255, and 264 in D1, which are mutation sites for conferring resistance to the triazine herbicides (e.g., atrazine) and the substituted urea herbicides [e.g., DCMU, 3-(3′,4′-dichlorophenyl)-1,1-dimethylurea].

173 and 190 in L. From the crystal structure, Deisenhofer *et al.* (1985a) have shown that histidines 200 M and 173 L bind the bacteriochlorophyll pair that corresponds to the primary donor P960 in *Rhodopseudomonas viridis* and that histidines 217 M and 190 L coordinate with the iron atom in the reaction center. Therefore it can be reasonably argued that histidines 198 on D1 and

D2 bind the P680 chlorophyll and that the 218 residues interact with iron. There are several other conserved residues between the higher plant and bacterial systems that further support the concept that D1 and D2 are the reaction-center polypeptides of PS2 (see Trebst, 1986). For example, in the crystal structure it can be seen that the menaquinone molecule, which acts as Q_A, is firmly bound by interacting with a tryptophan residue on the M subunit at position 250. A tryptophan can also be found on the D2 sequence at position 254, suggesting that this is the site for Q_A binding in the PS2 complex. Similar arguments can be developed for the binding of Q_B involving phenylalanine 255 on the D1 protein (Trebst, 1986). Figure 8(b) gives a possible arrangement of the D1 and D2 as reaction-center polypeptides based on analogy with organization of the L and M subunits of *R. viridis* as recently published by Deisenhofer *et al.* (1985a).

B. Photosystem One Complex

1. *Function*

Photosystem one (PS1) acts as a light-stimulated plastocyanin–ferredoxin oxidoreductase. The oxidation of the PS1 reaction-center chlorophyll (P700) has been characterised extensively by absorption bleaching at 700 nm and by an electron paramagnetic resonance (EPR) signal at $g = 2.005$ (see reviews by Malkin, 1982; Rutherford and Heathcote, 1985). The midpoint redox potential of $P700^+$ is about $+450$ mV and its formation occurs in less than 20 psec (Il'ina *et al.*, 1984). Although the line width of the EPR signal had suggested that P700 was a dimeric form of chlorophyll (Norris *et al.*, 1971), there are new arguments based on resonance spectroscopy to suggest that it is not (Wasielewski *et al.*, 1981). Recently, Dornemann and Senger (1984) isolated a chlorinated derivative of chlorophyll *a* from a wide range of organisms and proposed that this species is P700. The formation of $P700^+$ coincides with the reduction of a primary electron acceptor designated A_0. As Fig. 9(a) shows, subsequent electron flow is to the bound iron–sulfur centers Fe–S_A and Fe–S_B via two intermediate electron acceptors, A_1 and A_2. The precise chemical identity of A_0 and A_1 is unknown, although it is widely agreed that they are not phaeophytin (Damm *et al.*, 1984). A_0 seems to be a monomeric form of chlorophyll *a* (Gest and Favinger, 1983; Shuvalov *et al.*, 1986), while A_1 does show spectroscopic properties indicative of a quinone (Thurnauer and Gast, 1985), possibly menaquinone (R. Malkin, personal communication) or vitamin K (P. Mathis, personal communication). A_2 has EPR characteristics of a bound iron–sulfur center (McIntosh and Bolton, 1976; Evans *et al.*, 1975), although other identities for this acceptor have been suggested (see Bolton, 1978). The electron donation to A_1 or A_2 seems to occur with a half-time of about 200 psec, while the reduction time of FeS_A and FeS_B is in the region of 170 nsec. Whether FeS_A and FeS_B are involved

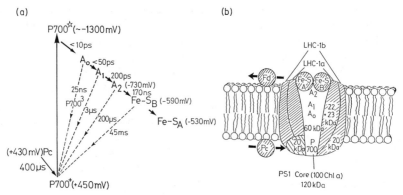

Fig. 9. (a) Kinetics of the electron-transport process that takes place in the PS1 reaction-center complex, giving approximate half-times for the reaction rates and suggested midpoint redox potentials for the components. For symbols see Fig. 1. (b) A schematic model for the composition and organization of the PS1 complex. Shaded components are probably nuclear-encoded, while unshaded are chloroplast-encoded.

in linear electron flow or are on a branched pathway is still a matter for debate (Rutherford and Heathcote, 1985). Nevertheless there is evidence that the presence of $Fe-S_B$ is obligatory for $Fe-S_A$ reduction (Malkin *et al.*, 1984), suggesting that the electron-transfer relationship shown in Fig. 9(a) is a useful working model. If the forward reactions are blocked by selective chemical reduction, it is possible to estimate the back reaction rates as indicated in Fig. 9(a) (see Rutherford and Heathcote, 1985). Normally, however, the reduction of $P700^+$ occurs in the submillisecond time domain, with the electron being donated by reduced plastocyanin. The kinetics of the reduction process can be complex, being governed by the diffusion of plasto-cyanin and its binding to the PS1 reaction-center complex (Haehnel, 1984), and is achieved in less than 1 msec.

2. Polypeptides

For higher plants grown under normal light conditions, approximately 30% of the total chlorophyll is bound in the PS1 complex, with each complex containing about 200 chlorophyll molecules (Anderson, 1980). The complex has a diameter of about 10.6 nm (Mullet *et al.*, 1980) and seems to consist of an inner core surrounded by outer layers of light-harvesting polypeptides. The inner reaction center core probably binds half the total chlorophyll and consists of two polypeptides in the region of 67-kDa and several lower-molecular-weight components. Presumably these lower-molecular-weight polypeptides bind the secondary electron acceptors, such as $Fe-S_A$ and $Fe-S_B$, although recent evidence suggests that A_2, as well as A_0 and A_1, is located within the P700 binding protein (Golbeck and Cornelius, 1986).

Treatment with SDS will separate the P700-binding protein from the lower-molecular-weight components to yield a photochemically active preparation containing about 100 chlorophyll *a* molecules and some beta-carotene per reaction center. This basic unit of PS1 is known as CP1 and seems to be ubiquitous among a wide range of photosynthetic organisms (Thornber, 1975). A stoichiometric analysis indicates that there may be two 67-kDa polypeptides per CP1.

3. Molecular Biology

Of the several polypeptides that make up the PS1 complex, only the genes for the chlorophyll-binding 67-kDa core polypeptides have been located and sequenced. These genes, of which there are two (*psaA* and *psaB*), are located side by side in the maize chloroplast genome (Fish *et al.*, 1985) and code for polypeptides consisting of 752 and 736 amino acids, respectively. They are light-activated and have sequence homologies of 45%. Hydropathy plots indicate that they may have as many as 11 transmembrane helical segments. The *psaA* gene product contains 42 histidine residues, while 38 histidines are coded for by the *psaB* gene. The majority of these histidines are probably located in membrane-buried segments and, as already suggested for the PS2 polypeptides, are likely to be involved in chlorophyll binding. Since SDS–PAGE indicates that CP1 contains only the 67-kDa polypeptides, it is assumed that either or both the *psaA* or *psaB* products bind the reaction-center chlorophyll that constitutes P700 and the primary electron acceptors (Malkin, 1986). Nothing is known about the location of the genes that code for the smaller polypeptides of the PS1 core; however, the LHCP-1a and LHCP-1b genes are almost certainly encoded in the nucleus.

C. Light-Harvesting Chlorophyll *a/b* Complex

1. Function

This complex, referred to as LHC-2, is a major chlorophyll-binding protein of the thylakoid membrane of higher plants and green algae. With plants grown in normal light intensities it binds 40–60% of the total chlorophyll (Thornber, 1975, 1986). It acts as an antenna system to PS2, although there does seem to be a pool of LHC-2 that is mobile and can redirect its absorbed energy in favor of PS1 (see Section IV,C). The presence of the complex in the membrane is necessary for the normal stacking of thylakoids to form grana (Arntzen, 1978).

2. Proteins

The LHC-2 complex is composed of two or more polypeptides of approximate molecular weights between 24 and 27 kDa, has a chlorophyll *a/b* ratio in the region of about 1.0–1.4, contains little or no beta-carotene, but binds xanthophylls (Thornber, 1975; Braumann *et al.*, 1982; Menger *et al.*, 1984). Since LHC-2 is not homogeneous, there may be some variation in the polypeptide and pigment ratios for each particular type (Larsson and Andersson, 1985; Remy *et al.*, 1985). On average, the LHC-2 complexes create an antenna of about 200 chlorophyll molecules per P680 (Anderson, 1980) and do so by forming aggregates. From freeze-fracture electron microscopy it seems that the LHC-2–PS2 unit has a diameter of 15–18 nm (Armond and Arntzen, 1977; Mullet and Arntzen, 1980), while free and mobile forms of LHC-2 appear as 8-nm-diameter particles (Kyle *et al.*, 1983a). Just how the individual polypeptides of molecular weights between 24 and 27 kDa aggregate is uncertain. However, recent structural studies on semicrystalline isolated LHC-2 using electron microscopy and image processing have revealed a two-dimensional array of trimers (Kuehlbrandt, 1984). The image analysis was possible to 1.6 nm resolution and showed that LHC-2 is an asymmetric membrane-spanning protein of about 6 nm in length [see Fig. 10(a)]. The precise number of chlorophylls bound to each polypeptide is a matter for debate but is in the region of 6–13, indicating that each trimeric aggregate contains 20–40 chlorophylls. If this trimeric structure exists *in vivo*, then there would be 5–10 of them associated with each PS2 core. However, each trimer has a width of about 6.4 nm, so that the 8-nm particles observed in freeze-fracture micrographs attributed to LHC-2 complexes free of PS2 (Mullet and Arntzen, 1980) may only represent one trimeric structure. Since the diameter of the PS2 core is thought to be about 8.5 nm (Sprague *et al.*, 1985), it seems unlikely that more than four or five LHC-2 trimeric complexes are associated with it, indicating that the upper value for chlorophyll binding to each LHC-2 polypeptide is more likely.

3. *Molecular Biology*

The polypeptides of LHC-2 are encoded by the nuclear *Cab* gene family, of which several have now been sequenced: for pea (see Cashmore, 1984), for petunia (Dunsmuir, 1985), and for *Lemna gibba* (Tobin *et al.*, 1984; Kohorn *et al.*, 1986). They contain 228, 232, and 229 amino acids, respectively, and are highly conserved except for those near to the N-terminus. Two important developments are emerging from this work. First, from hydropathy considerations it has been possible to formulate folding models of the type shown in Fig. 10(b) (Karlin-Neumann *et al.*, 1985); second, it has led to a series of very interesting studies aimed at understanding the uptake and processing of the precursor polypeptide by the chloroplast (Chitnis *et al.*, 1986). The model shown in Fig. 10(b) of the LHC-2 polypeptide suggests that there are three transmembrane segments. A striking feature of this model is that a considerable portion of the polypeptide is exposed at the outer surface, which is in line with the conclusion drawn from the image analysis work of Kuehlbrandt [see Fig. 10(a)]. Another interesting facet is the lack of histidine residues often assumed to be necessary for binding

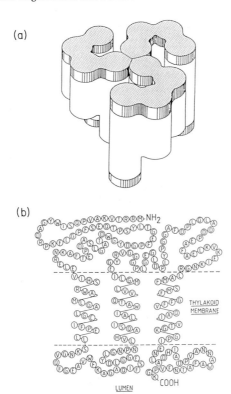

Fig. 10. (a) A representation of the trimeric structure of isolated LHC-2 as elucidated from the image analysis of electron micrographs of two-dimensional crystals (Kuehlbrandt, 1984). The resolution is 1.6 nm and the shaded areas show those parts of the trimers that probably protrude from the membrane. (b) A folding model of a single polypeptide of LHC-2 based on the amino acid sequence obtained from the gene of *Lemna* chromosome (Tobin *et al.*, 1984).

chlorophyll (see comments in Sections III,A and III,B). Good support for the folding model presented in Fig. 10(b) comes from polarized infrared spectroscopy and ultraviolet circular dichroism (Nebedryk *et al.*, 1984) and from the elegant gene deletion studies of Kohorn *et al.* (1986).

D. Cytochrome b_6/f Complex

1. Function

This complex functions in higher plants and green algae as a plastoquinol–plastocyanin oxidoreductase and thereby catalyzes the electron/proton exchange between PS2 and PS1. It is also required for cyclic electron flow supported by PS1 and is probably capable of operating a Q-cycle type mechanism (Hind *et al.*, 1984). In many different ways it resembles the ubiquinol–cytochrome c oxidoreductase complexes of mitochondria and photosynthetic bacteria. Several reviews detailing its structure and function have recently been published: Bendall (1982), Hauska *et al.* (1983), Rich (1984), Hauska (1986), and Joliot and Joliot (1986). All the cytochrome (cyt) *b/c* complexes contain a two-heme cyt *b*, one-heme cyt c_1 or *f*, and one high-potential $Fe_2–S_2$ Rieske center. A bound quinone may also be associated with the isolated complex. In the chloroplast cyt b_6/f complex, the midpoint potentials for the two *b*-type hemes are about -50 and -150 mV at pH 7.0, although a wide discrepancy in these values exists in the literature. The redox potential of cyt *f* is pH-independent, having a value of 340 mV. The Rieske Fe–S center has a midpoint potential of about 285 mV, which becomes pH-dependent above pH 8.0. The precise way in which these redox centers functionally interact is a matter of considerable debate (see Rich, 1984; Joliot and Joliot, 1986), but the most popular models are based on the Q-cycle scheme of Mitchell (1975) (see Bendall, 1982). The general idea is that the cyt b_6/f complex contains two plastoquinone-binding sites known as site C and site Z, which are located toward the two surfaces of the membrane. It seems that the first electron is passed to the Rieske center, which then reduces cyt *f*. (Cyt *f* acts as a reductant for oxidized plastocyanin.) The second electron can be passed either to the reoxidized Rieske center or donated to the low potential form of cyt b_6, since the semiquinone species is a strong reductant. If the latter occurs, the electron on cyt b_{6LP} ($E_m \sim -150$ mV) is probably passed to the high-potential cyt b_{6HP} ($E_m \sim -50$ mV) and then used to reduce a bound plastoquinone to a semiquinone at site C. The further reduction of this semiquinone bound at the C site could occur in several different ways. Either another plastoquinol molecule is reduced at the Z site, making a second electron available via the reduction of cyt b_{6HP}, or electron donation may come directly from PS2 or indirectly from PS1 via reduced ferredoxin [as indicated in Fig. 11(a)]. If site C is near the outer surface, then the effect of the cyt b_6-catalyzed Q cycle is to pump protons across the membrane and shift the H^+/e stoichiometry from 1 to 2 for plastoquinol oxidation.

2. Proteins

The cyt b_6/f complex, when isolated from spinach chloroplasts (Hauska *et al.*, 1983), consists of four major polypeptides as revealed by SDS–PAGE: cyt *f* corresponds to a doublet having apparent molecular weights of 34 and 31 kDa, cyt b_6 is a 22-kDa polypeptide, the Rieske FeS protein runs at a molecular mass of 20 kDa, and the fourth subunit of unknown function is a 17-kDa subunit [see Fig. 11(b)]. Some small hydrophobic polypeptides below 6 kDa are also present and can be observed by silver staining (Hurt and Hauska, 1982). In addition, Clark *et al.* (1984) found that under some circumstances a 37-kDa flavoprotein, corresponding to ferredoxin–NADP$^+$ oxidoreductase (FNR), was associated with the isolated complex. However, this extrinsic component is not necessary for plastoquinol oxidation and

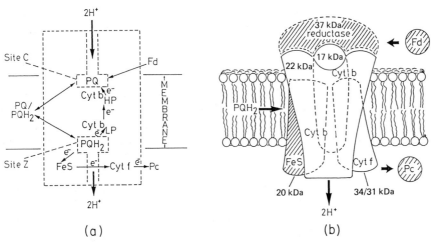

(a) (b)

Fig. 11. (a) A scheme of electron transfer through the cyt b_6/f complex, including a Q-cycle mechanism involving electron donation from ferredoxin (Fd) reduced by PS1. Cyt b is cytochrome b-563 in its low-potential (LP, $E_m \sim -150$ mV) and high-potential (HP, $E_m \sim -50$ mV) forms. It is a matter of debate whether PQ/PQH$_2$ exchanges with the quinone pool in the bulk lipid matrix of the membrane or remains within the protein complex during the Q-cycling process. (b) A schematic model of the cytochrome b_6/f complex where cyt b indicates cytochrome b_{563} in its high- or low-potential forms. Under some conditions the complex may bind the 37-kDa ferredoxin-NADP oxidoreductase (Clark and Hind, 1983) possibly at a 17-kDa polypeptide (Vallejos *et al.*, 1984). The shaded components are probably nucleus-encoded, and unshaded are chloroplast-encoded.

presumably plays a role in ferredoxin-mediated electron flow powered by PS1 photochemistry. The function of the 17-kDa polypeptide, often called subunit IV, is unknown. The stoichiometry of these subunits has been shown by Hurt and Hauska (1982) to be approximately $1:1:1:1$, but their organization within the cyt b_6/f complex is not yet known. Nevertheless, Fig. 11(b) attempts to provide a schematic structural model for the complex based on present knowledge, but it does not include a low-potential form of cytochrome b_{559} that may be associated with it *in vivo* (Nelson and Neumann, 1972; Peters *et al.*, 1983; Hodges *et al.*, 1984), nor does it include any low-molecular-weight components.

3. Molecular Biology

Three of the four major polypeptides that make up the cyt b_6/f complex have their genes located on the chloroplast genome. The remaining polypeptide, which binds the Rieske Fe–S center, is almost certainly encoded in the nucleus, although as yet its gene has not been cloned and sequenced.

The gene that codes for cyt f is known as *petA* and has been sequenced for pea (Willey *et al.*, 1984a), wheat (Willey *et al.*, 1984b), and spinach (Alt and Herrmann, 1984). It is translated into a precursor polypeptide 4 kDa larger than the mature polypeptide consisting of 285 amino acids. Hydropathy plots indicate only one transmembrane segment near the C-terminus, so that a large hydrophilic domain of 250 amino acids is expected to be exposed at the membrane surface. Functional considerations, data from proteolytic digestion, and immunological studies suggest that, on balance, it is likely that this hydrophobic region is located on the lumenal surface and is the site for heme binding (Willey *et al.*, 1984a; Mansfield and Bendall, 1984; Mansfield and Anderson, 1985). A histidine residue at position 25 may well act as one of the ligands for the heme, while lysine residue 145 or 222 could act as the other. It has also been speculated that 10 highly conserved basic amino acids give rise to a positively charged region between residues 58 and 154 that may be important for binding plastocyanin (Cramer *et al.*, 1985). In higher plant chloroplasts, the gene for cyt b_6 is called *petB* and is adjacent to the *petD* gene that codes for the 17-kDa subunit IV of the cyt b_6/f complex (Heinemeyer *et al.*, 1984; Phillips and Gray, 1984). They are transcribed into a bicistronic mRNA (Alt *et al.*, 1983). In the case of *petB*, the translated protein contains 211 amino acids corresponding to a molecular mass of 23.4 kDa. Hydropathy plots indicate five transmembrane segments with four strategically located histidines (residues 82 and 96 in segment II and residues 183 and 196 in segment V) that could bind the two "*b*-type" hemes to opposite sides of the membrane (Widger *et al.*, 1984). In the case of the mitochondrial cyt *b* polypeptide, the same four histidines are conserved despite the fact that the protein is larger than that in the chloroplast. Cramer *et al.* (1985) have therefore proposed a common arrangement for cyt *b* polypeptides of all types of *b/c* complexes, in which the two hemes are orientated perpendicular to the plane of the membrane with their iron centers spaced at about 2 nm. The mitochondrial polypeptide contains 380–385 amino acid residues and has a molecular mass of 40 kDa. Hydropathy considerations suggest that it has nine transmembrane segments, with the first five from the N-terminus having considerable homology with the 23-kDa cyt b_6 polypeptide and containing the four highly conserved histidines mentioned above. Interestingly, the remaining C-terminal portion, with its proposed four transmembrane segments, is similar to the 17-kDa polypeptide of the cyt b_6/f complex. This latter polypeptide is the product of the chloroplast *petD* gene, which has been located and sequenced by Phillips and Gray (1984) and Heinemeyer *et al.* (1984). It could be, therefore, that the single cyt *b* gene of mitochondria is replaced by a split gene in the chloroplast genome.

E. ATP Synthase (CF$_0$–CF$_1$) Complex

1. Function

This complex consists of two parts: the CF_0, which is an intrinsic component of the membrane, and CF_1, which is an extrinsic portion attached to the CF_0 on its outer surface (see Fig. 12). The CF_0–CF_1 complex functions as a

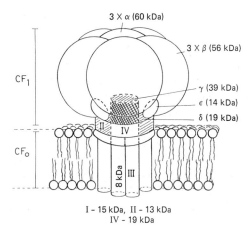

3 × α (60 kDa)

3 × β (56 kDa)

CF$_1$

γ (39 kDa)
ε (14 kDa)
δ (19 kDa)

CF$_o$

8 kDa

I - 15 kDa, II - 13 kDa
IV - 19 kDa

Fig. 12. A schematic model of the composition and organization of the ATP-synthase (CF$_0$–CF$_1$) complex. The shaded components are probably nucleus-encoded, unshaded are chloroplast-encoded.

proton translocating enzyme able to utilize the electrochemical potential gradient acting on protons across the thylakoid membrane to drive the net synthesis of ATP. Recent reviews covering the structure and function of the chloroplast ATP synthase complex have been written by Schlodder *et al.* (1982), Nelson (1982), McCarty and Moroney (1985), Merchant and Selman (1985), McCarty and Nalin (1986), and Strotmann (1986). The CF$_0$ portion probably has a molecular mass of about 100 kDa and acts as a passive proton-conducting channel. On the other hand, CF$_1$ is a 400-kDa multipeptide complex containing the catalytic site responsible for ATP synthesis or hydrolysis. The substrates for CF$_1$ are metal–nucleotide complexes and inorganic phosphate (Pi). It has been proposed that the metal ion may be necessary for positioning the beta-phosphoryl group of ADP for nucleotide attack on Pi (Frasch and Selman, 1982). Normally CF$_1$ shows a very low ATP hydrolysis activity, and its ability to synthesize ATP only occurs when the membrane is energized (see Schlodder *et al.,* 1982). However, it is possible to enhance ATPase activity of the native or purified enzyme by a wide range of treatments (heating, alcohols, proteases, reducing agents, and certain detergents—see McCarty and Moroney, 1985). Although Mg^{2+} is the preferred cation for ATP hydrolysis and synthesis activity of the membrane-bound CF$_1$, when isolated, other divalent cations can replace the requirement for Mg^{2+} depending on the mode of ATPase activation. Optimizing the ATPase and ATP synthetase activity of isolated CF$_0$–CF$_1$ also appears to require the presence of certain chloroplast thylakoid lipids as emphasized by the recent studies of Pick and colleagues (1984, 1985, 1987). For ATP synthesis it seems that a minimum of three H$^+$ are required to produce one ATP molecule, although the precise stoichiometry is probably governed by the size and nature of the electrochemical potential gradient acting on the protons (see Schlodder *et al.*, 1982).

The fact that CF_0–CF_1 will show ATPase activity under reducing conditions suggests that this property may be used for regulating the activity of the enzyme in the intact chloroplast. Indeed, several studies (see, for example, Mills and Mitchell, 1982) have indicated that reduced thioredoxin could function to facilitate the activation of the enzyme in the light so as to allow ATP synthesis to occur at suboptimal $\Delta\mu_{H^+}$ conditions. In the dark, thioredoxin is no longer reduced by electron flow and the CF_0–CF_1 complex reverts back to a low ATPase activity, thus avoiding the futile dissipation of stromal and cytosol ATP levels. However, the presence of reduced thioredoxin is not an absolute requirement, since the ATP synthase can be activated in its thiooxidized state as long as the $\Delta\mu_{H^+}$ is sufficiently large.

2. *Proteins*

CF_0 has not been isolated as a complete entity but, as indicated in Fig. 12, it is now thought to consist of four different polypeptides. Three of these have been well documented (e.g., see Nelson, 1982) and are denoted subunits I, II, and III, having apparent molecular masses of about 15, 13, and 8 kDa, respectively. The fourth subunit was detected some years ago by Pick and Racker (1979) as a polypeptide of 19-kDa, but its existence has only recently been confirmed by gene mapping of the pea chloroplast chromosome using a probe corresponding to the gene of subunit *a* of the bacterial and mitochondrial F_0 complex (Cozens *et al.*, 1986). The relative stoiochiometry of the various subunits of CF_0 is still a matter for debate, but it is often assumed that six copies of subunit III form a hexagonal-shaped proton-conducting channel across the membrane with subunit II helping to maintain this structural organisation. Subunit III reacts covalently with dicyclohexylcarbodiimide (DCCD), the effect of which is to block proton translocation through the channel. The roles of subunits I and IV are not clear but they could function to allow a tight association of CF_1 with the proton conducting channel formed by the subunit III hexamer.

CF_1 is a spherical-shaped complex, which, by electron microscopy, can be observed on the outer surface of the membrane as a knob-like structure of 13–14 nm diameter. It consists of five subunits, designated alpha, beta, gamma, delta, and epsilon, with molecular masses of about 60, 56, 39, 19, and 14 kDa, respectively (see Fig. 12). As with the mitochondrial and bacterial F_1 complex, the subunit stoichiometry of CF_1 is probably three alpha, three beta, one gamma, one delta, and one epsilon.

It seems that ADP and ATP binding occurs at three functionally equivalent sites on the beta subunits in the close vicinity of the alpha subunits (Bruist and Hammes, 1981). It is not known precisely how these three sites interact during the process of ATP synthesis or hydrolysis, although a cyclic cooperative model has been proposed by Boyer and Kohlbrenner (1981) (see also Strotmann, 1986). Activation of CF_1 seems to involve both the gamma and epsilon subunits. Alcohols, heat, and detergents displace the latter subunit, while the gamma subunit contains the disulfide bond, which is broken during activation with reducing agents. Despite the fact that a considerable amount of work has been carried out on the structure and function of the CF_0–CF_1 complex, it is still a mystery how the energy of the proton flux induces the conformational changes that give rise to the activation of the enzyme and the concomitant synthesis or hydrolysis of ATP.

3. Molecular Biology

The structural and functional similarity between the CF_0–CF_1 complex and the ATP synthase of mitochondria (F_0–F_1) and bacteria (BF_0–BF_1) is also maintained at the level of molecular genetics, as emphasized in the review of Walker et al. (1984) and the recent experimental paper of Cozens et al. (1986). Of the nine known polypeptides that constitute the CF_0–CF_1 complex, six are encoded on the chloroplast DNA. The alpha subunit is the product of the atpA gene and has been located on the chloroplast genome of several different plants, including tobacco (Deno et al., 1983; Fluhr et al., 1983) and wheat (Howe et al., 1985). Using a variety of different plants, the chloroplast genes for the beta subunit (atpB) and epsilon subunit (atpE) have been located and sequenced: for example, barley (Zurawski and Clegg, 1984), maize (Krebbers et al., 1982), wheat (Howe et al., 1985), tobacco (Shinozaki et al., 1983), and pea (Howe et al., 1985). The stop codon of atpB gene overlaps the initiation codon of the atpE and the genes are co-transcribed to give a single polycistronic mRNA. The genes for the gamma (atpC) and delta (atpD) subunits are probably located in the nucleus (Bouthyette and Jagendorf, 1978; Nechushtai et al., 1981), as is the gene for subunit II (atpG) of the CF_0 complex. Subunits I and III are, however, chloroplast-encoded and again their genes, atpF and atpH, respectively (particularly the latter), have been located and sequenced for several different plants ranging from spinach (Alt et al., 1983), wheat (Howe et al., 1982; Bird et al., 1985), and tobacco (Deno et al., 1984; Shinozaki et al., 1986) to Euglena (Passavant and Hallick, 1985). The atpA, atpF, and atpH genes are arranged in a single cluster. The cluster has a fourth open reading frame, which has been designated atpI. This gene has recently been sequenced in pea and is homologous to the gene for subunit a in the proton-translocating portion of the mitochondrial and E. coli ATPase (Cozens et al., 1986).

A comparison of the amino acid sequences of the alpha and beta subunits reveals a high degree of homology, especially at sites that may be involved in nucleotide binding (Howe et al., 1985; Deno and Sugiura, 1984). There seem to be four critical regions where this homology is particularly strong, not only between the chloroplast polypeptides but also when compared with the equivalent proteins of mitochondria and bacteria. In tobacco the alpha and beta subunits consist of 507 and 498 amino acids, respectively, and some key residues on the beta-polypeptide, glutamic acids 203, 209, and tyrosine 385, may be intimately involved in ATP synthase/ATPase activity (Deno and Sugiura, 1984).

The gene for the epsilon subunit (*atpE*) in maize consists of 138 amino acids residues (Krebbers *et al.*, 1982). In this case the sequence shows a lower homology with the epsilon subunit of *E. coli* ATPase complex (23%).

The sequencing of the gene for subunit III (*atpH*) indicates that the unprocessed polypeptide contains 81 amino acids corresponding to a molecular mass of about 8 kDa (Howe *et al.*, 1982; Deno *et al.*, 1984). The polypeptide is very hydrophobic and has about 30% homology with the corresponding subunit of the *E. coli* ATP synthase (subunit *c*) (Walker *et al.*, 1984). Hydropathy plots of both sequences, however, are very similar and indicate that there are two transmembrane-spanning segments. A conserved glutamate residue, at position 62 in subunit III and at position 66 in the subunit of *E. coli* and the related subunit in mitochondrial ATP synthase, is thought to be crucial for H^+ translocation and the site for the binding of DCCD (Walker *et al.*, 1984).

Nucleotide sequencing and analyses of RNA–DNA hybrids have indicated that the gene of the CF_0 subunit I (*atpF*) is very unusual for higher-plant plastid DNA, since it is interrupted by an 823-bp intron (Bird *et al.*, 1985; Shinozaki *et al.*, 1986). The deduced amino acid sequence of CF_0 subunit I indicates a polypeptide of 183 residues. This polypeptide is probably processed to 166 amino acids from its N-terminus (Bird *et al.*, 1985) so that its native molecular weight is about 19 kDa. A hydropathy profile of the amino acid sequence of the processed polypeptide indicates a short hydrophobic region at the N-terminus, which spans the membrane from residue 9 to 28. The remaining 138 amino acids are hydrophilic (33% charged residues) and are likely to be exposed at the outer membrane surface and function to bind CF_1 to the CF_0 complex. Although the amino acid sequence for CF_0 subunit I is not highly homologous with the related subunit *b* of the *E. coli* ATP synthase, the hydropathy plots, as for subunits III and *c*, are very similar (Walker *et al.*, 1984; Bird *et al.*, 1985). Because of this it is possible that subunit I forms a homodimer as postulated for the mitochondrial *b* subunit (Cox *et al.*, 1984).

The fourth subunit of CF_0 (called subunit IV in this chapter), corresponding to the product of the bacterial gene coding for subunit *a* of BF_0, consists of 247 amino acids. The homology between the bacterial and chloroplast amino acid sequences is restricted to the C-terminal half of the polypeptides, although hydropathy plots of the two proteins are remarkably similar. Such plots indicate six or seven possible transmembrane segments (see Cozens *et al.*, 1986).

IV. THYLAKOID MEMBRANE ORGANIZATION

A. Electron Microscopy

The thylakoids of higher-plant chloroplasts are observed by the electron microscope as complex folded membranes having stacked (granal) and non-stacked (stromal) regions. There are many reviews that have dealt with the organization of thylakoids, including those of Kirk and Tilney-Bassett (1967), Thomson (1974), Gunning and Steer (1975), Coombs and Greenwood (1976), and more recently the excellent article of Staehelin (1986). Unfortunately, many of the models presented are too precise and static and do not reflect the dynamic nature of this membrane system. A picture that is often given, based on electron-microscopy studies using tangential and serial sections, shows the stromal and granal lamellae connected by a tilted network of spiral frets that follow a right-to-left helix around the surface of cylindrical grana (Paolillo, 1970; Brangeon and Mustardy, 1979). Such a rigid model must, however, be reconciled with the fact that it is relatively easy to un-stack and restack isolated thylakoids by simply manipulating the ionic environment in which they are suspended (Izawa and Good, 1966; Barber, 1976; Barber and Chow, 1979). Concomitant with the salt-induced membrane-stacking changes are characteristic changes in the lateral distribution of intrathylakoid proteinaceous particles, as observed by freeze-fracture electron microscopy (Wang and Packer, 1973; Ojakian and Satir, 1974; Staehelin, 1976; Staehelin and Arntzen, 1983). In contrast with isolated membranes, the thylakoids within mature intact chloroplasts are usually differentiated into stacked and unstacked lamellae, although the relative extents vary with growth conditions (Boardman, 1977; Barber, 1985) and may fluctuate on a short time scale in response to reversible phosphorylation of LHC-2 (see Section V,C). It should also be noted that extensive grana formation is not a feature of bundle-sheath chloroplasts of C_4 plants, which are deficient in LHC-2 and PS2 (Coombs and Greenwood, 1976).

B. Lateral Distribution of Proteins

The first hints of a lateral separation of different protein complexes along the plane of the thylakoid membrane of higher-plant chloroplasts stem from the pioneering work of Boardman and Anderson (1964), in which they used the detergent digitonin to obtain membrane fragments enriched in either PS1 or PS2. This initial study was followed by the work of many others involving the use of detergents (Vernon *et al.*, 1966; Sane *et al.*, 1970), mechanical fragmentation (Michel and Michel-Wolwertz, 1969), and freeze-fracture electron microscopy (Staehelin, 1976, 1986; Armond *et al.*, 1977; Arntzen, 1978), which supported the idea that PS2 is mainly associated with stacked membranes (grana) while PS1 is located both in stacked and unstacked membrane regions. A more precise picture of thylakoid membrane organization, however, has emerged from the use of the phase partition technique developed and exploited by Albertsson and co-workers (1982) for separating mixtures of mechanically derived membrane fragments. With this technique, Andersson (1978) and Andersson and Anderson (1980) were able to show that the tightly interacting (appressed) membranes of the granal stacks contained mainly PS2 and LHC-2 while CF_0–CF_1 and PS1 were found to be located in the nonappressed regions. The restricted localization of CF_0–CF_1 to the nonappressed membranes, including end granal lamellae, had been advocated earlier by Miller and Staehelin (1976) based on freeze-fracture electron microscopy, but the extreme lateral separation of PS2 and PS1 had not been generally accepted previously, even though there was good evidence that ferredoxin–NADP oxidoreductase (FNR) was restricted to the external surface of the nonappressed membranes (Berzborn, 1969; Jennings *et al.*, 1979). According to the kinetic absorption studies of Haehnel (1982), the cyt b_6/f complexes should be located close to PS1. Despite this conclusion, several independent studies have consistently produced evidence that this cytochrome containing complex is evenly distributed between appressed and nonappressed membranes (Cox and Andersson, 1981; Anderson, 1982; Anderson *et al.*, 1985; Allred and Staehelin, 1985). As mentioned in Section III,D,1, and cyt b_6/f complex has been isolated with FNR bound to it. Since FNR is only found associated with the outer surface of the nonappressed membranes, then it could be that the partitioning of cyt b_6/f complexes between appressed and nonappressed membranes is dictated by whether or not the oxidoreductase is bound to it (see Fig. 13).

Although appressed membranes normally contain most of the PS2 and LHC-2 complexes, a variable fraction of these two complexes can be found in nonappressed regions. In the case of those PS2 units located in stromal lamellae their light-harvesting antenna size is small, showing little or no energy transfer to other PS2 complexes (Melis and Thielen, 1980). These independent units correspond to beta centers (Melis and Homann, 1976) that may not be fully functional in linear electron transport. They contrast with the PS2 alpha centers of appressed membranes, which readily show interunit energy transfer. Some LHC-2 complexes are also found in the nonappressed regions, particularly after their surfaces have been phosphorylated (see Section V,C).

C. Physical Factors Responsible for Thylakoid Membrane Organization

A simple model showing the lateral distribution of protein complexes is given in Fig. 13. An important consequence of this model is that it is necessary to discard the use of the old terminology of granal and stromal lamellae

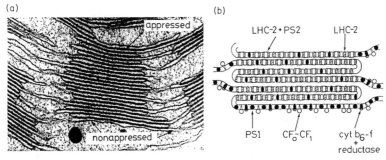

(a)   (b)

Fig. 13. (a) Electron micrograph of a typical stacked granal region of the thylakoid membranes of higher plant chloroplasts showing that this membrane system can be divided into appressed and nonappressed regions. (b) A diagrammatic "static" representation of the possible lateral distribution of supramolecular complexes between the appressed and nonappressed regions of the thylakoid membranes.

to distinguish between stacked and unstacked thylakoids. Rather, it is more satisfactory to use the terms appressed and nonappressed membranes, where the former are the grana partition regions having the outer surfaces of adjacent membranes in close contact and not directly exposed to the aqueous stromal phase of the chloroplast, while the latter constitute the end granal and stromal lamellae with their outer surfaces exposed to the stroma. Clearly such a distinction emphasizes differences that must exist in the physical and chemical properties of the outer surfaces of the two membrane regions. An understanding of this difference has been possible by analyzing the salt-induced changes in membrane stacking mentioned in Section IV,A above and from monitoring concomitant chlorophyll fluorescence yield changes (Barber *et al.,* 1977; Barber and Chow, 1979; Barber, 1980a). Such studies revealed the important conclusion that the salt-induced changes were controlled by the concentration of cations rather than anions and that for those cations that did not strongly bind to the surface, the relationship was dependent on their valency rather than their chemical nature (Mills and Barber, 1978; Barber *et al.,* 1980). These findings indicated that the membrane conformational and chlorophyll fluorescence changes were under the control of electrostatic forces resulting from the net surface charge density on the membrane (Barber, 1980b, 1982a). Further experimental and theoretical investigations of the two related phenomena revealed that the degree of membrane interaction was regulated by the balance between double-layer repulsive and van der Waals attractive forces (Sculley *et al.,* 1980; Rubin *et al.,* 1981b; Thorne and Duniec, 1983), comparable with the concepts of the Derjaguin–Landau–Verwey–Overbeek (DLVO) theory for describing the coagulation of electrically charged colloidal particles (Overbeek, 1978). Indeed, a particularly impressive relationship between experiment and theory showed that it was possible to explain the observation of Gross and Prasher

(1974) that isolated thylakoids remained stacked in very-low-salt-containing medium and unstacked on the introduction of low levels of monovalent cations (Barber *et al.*, 1977; Duniec *et al.*, 1979; Rubin and Barber, 1980). However, as emphasized by Sculley *et al.* (1980), there was one serious problem when applying classical DLVO theory to thylakoid stacking. The problem was to account for the close distance of approach of adjacent membranes in the appressed regions (about 4 nm), since calculations of van der Waals attractive forces indicated them to be too weak to overcome the electrical double-layer force of repulsion. In order to explain this anomaly, Duniec *et al.* (1981) proposed that ion binding occurred in such a way as to reduce the surface net charge density and allow very close membrane–membrane interactions to occur. An alternative possibility was developed by Barber (1980a), which visualized that the stabilization of appressed membrane regions occurred by lateral migration of electrical charge components away from the region where membrane appression occurred (see Fig. 14). In this way an electroneutral domain would be formed with little or no electrical double-layer repulsion. Because the formation of appressed and nonappressed membrane regions was paralleled by an increase in chlorophyll fluorescence yield, indicative of a decrease in energy transfer between PS2 and PS1, it was concluded that the two photosystems partitioned into different domains: PS2 and LHC-2 into the appressed regions, and PS1 into the nonappressed regions (Barber, 1979, 1980a,b, 1982a,b). As mentioned in Section IV,A, such movements have been detected by freeze-fracture electron microscopy and are possible because of the fluid nature of the membrane. An important outcome of these analyses is that in normal appressed/nonappressed configuration the lateral asymmetry in the distribution of components given in Fig. 13 is based on their surface electrical charge properties. It is concluded therefore that the PS2/LHC-2 domain of the appressed region carries a low net electrical charge on its outer surface, while PS1 and other components in the nonappressed region are sufficiently charged on their outer surfaces to maintain significant electrical double-layer repulsion even in the presence of a relatively high level of screening cations.

In support of the lateral charge displacement model is the fact that isolated thylakoids can also be induced to form stacked regions by adding cationic species that bind to the surface (e.g., protons and polyvalent cations). In this case, electrostatic neutralization rather than screening occurs, and gives rise to appressed membranes that are different from those found *in vivo* (Barber, 1980a,b; Scoufflaire *et al.*, 1983). This type of "nonphysiological" stacking does not require long-distance lateral movements of protein complexes in order to form electroneutral domains and therefore is less dependent on membrane fluidity (Barber *et al.*, 1980). Incomplete neutralization of surface charges—for example, by decreasing pH to 5.4 rather than to the isoelectric point at pH 4.3, or addition of nonsaturating levels of polyvalent cations—seems to bring about conformational changes more like those observed by

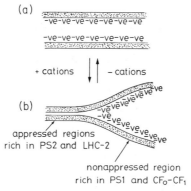

Fig. 14. A diagrammatic representation of the lateral charge displacement model to explain the randomization of supramolecular complexes in unstacked thylakoids and their subsequent ability to partition into discrete domains under ionic conditions that induce membrane stacking. (a) Poor electrostatic screening leads to unstacking and randomization of electrically charged components due to increased coulombic repulsion; (b) increased electrostatic screening lowers coulombic repulsion and leads to lateral charge movement and the establishment of appressed regions with low net surface charge density and nonappressed regions with high surface charge.

cation screening (Mills and Barber, 1978; Jennings *et al.,* 1981; Scoufflaire *et al.,* 1983). Thus the possibility of some cation binding occurring during normal grana formation cannot be totally dismissed, although to date very little selectivity has been observed between a wide range of ionic species within a valency group (e.g., alkaline metal cations or alkali metal cations—see Mills and Barber, 1978).

D. Stoichiometry of Complexes

The fact that electron transport and photophosphorylation involves the cooperative interaction of several discrete macromolecular complexes linked by diffusible components indicates that there is probably no fixed stoichiometric relationship between them. Indeed, it is well documented that growth at low light intensities or in shade conditions brings about a relative increase in LHC-2 that is accompanied by an increase in (1) the ratio of appressed to nonappressed membranes and (2) the total amount of chlorophyll and membrane area per chloroplast (Boardman, 1977; Lichtenthaler *et al.,* 1981; Barber, 1985; Anderson, 1986). On the other hand, based on total chlorophyll concentrations, low light or shade conditions decrease the levels of ATP synthase and cyt b_6/f, leading to a lower maximum rate of full-chain electron flow measured in saturating light. At limiting light intensities, however, the quantum yield for photosynthesis measured with plants grown either in low (shade) or high (sun) light are about the same (Boardman, 1977). This important observation indicates that the stoichiometry of the PS2 and PS1 complexes and the sizes of their light-harvesting antennae have been optimized. A part of this optimization probably involves the state 1–state 2 transitions, which act as a fine-tuning mechanism, termed the "carburetor effect" by Myers (1971). At maximum, this short-term regulatory mechanism will probably not adjust the quantum yield by more than 15–20% (Canaani and Malkin, 1984). It is therefore necessary for plants to adjust the composition of their thylakoid membranes. In the canopy, shade leaves are subjected not only to low light intensities, but also to radiation, which is poor in blue and red wavelengths and relatively rich in green and far-red (beyond 700 nm) (see Bjorkman and Ludlow, 1972). If no adjustment of antenna size occurred and the concentration of PS2 and PS1 were the same as in sun plants, the consequences of shading would be an overexcitation of PS1 relative to PS2. This undesirable situation could be overcome either by raising the PS2/PS1 ratio or by increasing the antenna size of PS2. To investigate these possibilities, experiments have been carried out on a wide variety of plant species, the results of which have recently been reviewed (Barber, 1985; Anderson, 1986). Unfortunately there is still considerable controversy, with some arguing that the stoichiometry of PS2 and PS1 is fixed (Whitmarsh and Ort, 1984) and others claiming variations in the PS2/PS1 ratio with little change in antenna sizes (Melis, 1984). In actual fact it seems likely that in response to growth in shade conditions, there is an increase in both the PS2/PS1 ratio and in the antenna size of PS2 (see Leong and Anderson, 1984) with the relevant extents depending on whether the regulation is in response to light intensity or quality. In any event these studies emphasize that plants have a remarkable ability to adjust the content of their thylakoid membranes in a manner that is beneficial to their growth and development depending on their illumination conditions and that the old concept of a fixed stoichiometry of PS2 and PS1 is not an absolute requirement.

V. THYLAKOID MEMBRANE DYNAMICS

A. Fluidity

The separation of various complexes between appressed and nonappressed lamellae implies that lateral movement of components must occur to facilitate electron flow and photophosphorylation. As discussed in the subsections below, some of these diffusional processes occur within the lipid matrix. Recognition of this fact places emphasis on the fluidity properties of the thylakoid membrane. As already mentioned in Section II, the acyl chains of the thylakoid polar lipids are very unsaturated, indicating that they are likely to give rise to a bilayer having a hydrophobic interior with a high degree of motion (Stubbs and Smith, 1984). Indeed, experimental evidence for this contention has been obtained from time-resolved anisotropy measurements using the fluorescence probe 1,6-diphenyl-1,2,5-hexatriene (DPH)

(Millner *et al.*, 1984a). Using the same technique, Millner and colleagues (1984b) went on to show that the reconstitution of protein complexes (namely, CF_0–CF_1) into this bilayer decreased acyl chain motion. Since the thylakoid membrane has a relatively high protein to lipid ratio (1.7 w/w), its fluidity must be decreased significantly from that of the pure lipids. Such a decrease can be detected by DPH fluorescence (Ford and Barber, 1981) and EPR spin label studies (Hiller and Raison, 1980; Ford *et al.*, 1982; Aronson *et al.*, 1983). Time-dependent fluorescence anisotropy data obtained with DPH indicated that despite the presence of protein the thylakoid lipid matrix was relatively fluid with an estimated microviscosity of 0.34 P at 25°C (Ford and Barber, 1983). This value contrasts with 0.42 P at 35°C for the cell membrane of human erythrocytes (which contain cholesterol) and 0.82 P at 35°C for the purple membrane of *Halobacterium halobium* (which has an unusually high protein to lipid ratio) (see Kinosita *et al.*, 1981). The value for the thylakoid membrane compares well, however, with 0.29 P at 35°C for rat liver mitochondrial membranes (Kinosita *et al.*, 1981). It should be noted that all these values involve measurement of the rotational diffusion of DPH and may not necessarily be indicative of viscosity factors controlling lateral movements. Moreover, the probe may selectively partition into more fluid microdomains in the membrane. Nevertheless, Shinitzky and Barenholz (1978), Shinitzky and Inbar (1976) and Shinitzky and Yuli (1982) have repeatedly argued that DPH fluorescence anisotropy is an adequate measure of membrane fluidity at the submacroscopic level. They have divided membrane lipid dynamics into three distinct levels of resolution: microscopic, macroscopic, and submacroscopic. The microscopic level provides information about individual atoms or molecular segments at various depths within the lipid bilayer and is best investigated using nuclear magnetic and electron spin resonance techniques. Shinitzky and colleagues argue that the macroscopic level deals with lipid domains in terms of bulk thermodynamics which can be investigated using calorimetry. However, the submacroscopic level provides low molecular resolution of lipid fluidity and can be investigated by nonspecific fluorescence probes such as DPH. They make the point that as far as general physiological relevance is concerned, the submacroscopic level of resolution of membrane fluidity correlates well with a wide range of functional processes (Shinitzky and Yuli, 1982).

According to the DPH technique, there is no obvious phase change in bulk lipids of pea chloroplast thylakoids over the temperature range 55 to −20°C (Barber *et al.*, 1984). This does not exclude the possibility of phase changes and lipid reorganization occurring in localized membrane regions that are not detected by this fluorescence probe; therefore there is a need to extend this study to other plant species before a general conclusion can be drawn. An interesting observation made with thylakoids isolated from peas grown either at 7 or 17°C was that the lower growth temperature resulted in thylakoids having a more fluid membrane at a particular measuring temperature than did those from warm grown plants. As a consequence of this adjustment, the fluidity of the thylakoids was about the same at their respective growth temperatures, indicative of the operation of a homeostatic mechanism (Barber *et al.*, 1984). Therefore it seems important to the higher plant that the bulk lipid matrix of the thylakoid membrane is maintained in a liquid cystalline state at a certain level of fluidity.

Bearing in mind the organization of chloroplasts, the question arises as to whether there is a fluidity difference between appressed and nonappressed membranes. In fact, several lines of evidence indicate that appressed lamellae are less fluid than nonappressed (Ford *et al.*, 1982; Aronson *et al.*, 1983). This physical difference is not due to differences in the level of acyl chain unsaturation but seems to be a consequence of a higher protein to lipid ratio in appressed compared with nonappressed membranes (Ford *et al.*, 1982; Dunahay *et al.*, 1984). The EPR spin-label studies of Aronson *et al.* (1983) concluded that appressed membranes isolated with Triton X-100 were almost rigid, an observation that is consistent with the considerable delipidation of these preparations leading to a very high protein to lipid ratio (Gounaris *et al.*, 1983d). The high protein to lipid ratio of the appressed membranes is consistent with theoretical considerations of the role of van der Waals forces in stabilizing close intermembrane distances (Sculley *et al.*, 1980; Rubin *et al.*, 1981a). Recent lipid analyses of thylakoid fragments enriched in appressed or nonappressed lamellae confirm that there is no striking difference in the fatty acid unsaturation levels of the two membrane types even when individual lipid classes are considered (Chapman *et al.*, 1986). The importance of the protein to lipid ratio in controlling thylakoid membrane fluidity also seems to underlie the adaptation to growth temperature mentioned above for peas. With this chill-resistant plant, growth temperatures of 7 or 17°C had little effect on the fatty acid unsaturation level but did affect the protein to lipid ratio in a manner consistent with the fluidity changes (Chapman *et al.*, 1983).

B. Plastoquinol/Plastocyanin Diffusion

Figure 15 indicates how the various intrinsic and extrinsic proteins of the thylakoid membrane of higher-plant chloroplasts interact to facilitate electron transfer from H_2O to $NADP^+$, catalyze electron flow around PS1, and pump protons from the outer to the inner membrane surface so as to generate the electrochemical potential gradient necessary to drive the synthesis of ATP at the CF_0–CF_1 complex. According to this model, intercommunication between complexes is achieved by long-range diffusional processes. The likely candidates for these diffusion steps are PQ/PQH_2 between PS2 and cyt b_6/f complexes, plastocyanin between cyt b_6/f and PS1, and ferredoxin between PS1 and cyt b_6/f. The latter candidate is a water-soluble low-molecular-weight protein of about 11 kDa that contains a 2Fe–2S active center of

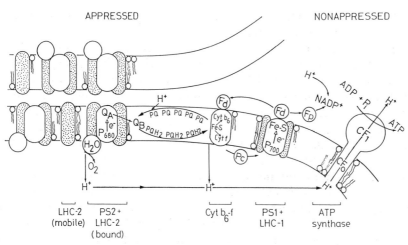

Fig. 15. A diagram indicating the spatial and functional relationships between five types of intrinsic membrane complexes that cooperate to capture light energy, promote electron and proton transport, and bring about the synthesis of ATP. The lateral separation of these complexes requires diffusion of mobile components (PQ/PQH_2, Pc, and Fd) and gives rise to the vectorial pumping of protons across the membrane so that the electrochemical potential gradient created can be used to synthesize ATP. Symbols as for Fig. 1, together with LHC-2, light-harvesting complex containing a high level of chlorophyll b usually located close to PS2; LHC-1, light-harvesting complex with a low level of chlorophyll b that forms a part of the PS1 complex; ADP or ATP, adnosine di- or triphosphate; CF_0 and CF_1, intrinsic and extrinsic portions of ATP-synthase complex; Fe–S, iron–sulfur centers; cyt b_6, two redox forms of cytochrome b_{563}.

95–100 amino acid residues (see Hall and Rao, 1977). Its diffusion between PS1 and cyt b_6/f is required for cyclic electron flow, but since there is no precise information about the kinetics of the cyclic process we cannot be certain of diffusion rates. It has been reported that electron donation to FNR occurs with a half-time of 1 μsec (Bouges-Bocquet, 1980), indicating that soluble ferredoxin does not diffuse over large distances. Such a conclusion is consistent with the location of FNR close to PS1 in the nonappressed membranes and the involvement of only those cyt b_6/f complexes in this region with cyclic electron flow. Regarding noncyclic electron flow, it is known that the oxidation of PQH_2 is the rate-limiting step having a $t_{1/2}$ of 10–20 msec (see Witt, 1971; Haehnel, 1984) while the oxidation of the Rieske Fe–S center and the reduction of $P700^+$ occurs in less than 1 msec (Haehnel, 1984).

The main species of PQ in the thylakoid membrane is plastoquinone-9. It is a hydrophobic molecule with two methyl groups attached to its quinone ring and a side-chain of nine isoprenic groups. It is located in the lipid matrix at a sufficiently high concentration for there to be a pool of six to seven molecules for each electron-transport chain (Witt, 1971; Chapman and Barber, 1986). In all likelihood its preferred location is at the midplane of the bilayer, since the bulky isoprenoid side-chain methyl groups would not be expected to pack effectively between the acyl chains of the thylakoid lipid matrix (Millner and Barber, 1984). Such a conclusion is supported by fluorescence quenching and NMR studies of other quinone systems (Katsikas and Quinn, 1982; Kingsley and Feigenson, 1981). When compared with the level of polar lipid, the concentration of PQ is equal in the appressed and nonappressed regions, indicating that it is freely diffusible throughout the complete thylakoid membrane system (Chapman and Barber, 1986). However, it is worth noting that because the appressed membranes have a higher protein to lipid ratio than nonappressed membranes (Ford $et\ al.$, 1982), the level of PQ on a chlorophyll basis is higher in the former compared to the latter regions (see Jennings $et\ al.$, 1983). The lateral diffusion of PQ and PQH$_2$ is presumably dependent on the fluidity of the lipid matrix, particularly at the midplane of the bilayer. As emphasized by Millner and Barber (1984), the midplane viscosity may be very low, allowing PQ to have a diffusion coefficient as high as 10^{-6} cm^2 s^{-1}. Such high values are indicated by the work of Marcus and Hawley (1970) and more recently by the fluorescence quenching studies of Fato $et\ al.$ (1985). In contrast, values of $2-5 \times 10^{-9}$ cm^2 s^{-1} were estimated for the room-temperature diffusion coefficient of a quinone-fluorescence analog in mouse liver megamitochondria based on measurements using the technique of fluorescence recovery after photobleaching (FRAP) (Gupta $et\ al.$, 1984). The interpretation of this work, however, is open to question because the fluorescence analog must be constructed such that its fluorophore is exposed at the membrane surface, meaning that the quinone cannot reside in its preferred midplane location. Blackwell $et\ al.$ (1986) have attempted to obtain diffusion coefficients of PQ and PQH$_2$ in liposomes composed of phosphatidylcholine by monitoring pyrene fluorescence quenching. Room-temperature values ranging from 10^{-7} to 3×10^{-8} cm^2 s^{-1} were obtained with PQH$_2$ consistently showing a slower diffusion rate than PQ. A knowledge of PQ and PQH$_2$ diffusion coefficients is vital for estimating diffusion pathlengths, which can be calculated using the two-dimensional Einstein equation $\langle x \rangle^2 = 4Dt$ [where $\langle x \rangle$ is the mean distance of diffusion in time t when the diffusion coefficient is D (Einstein, 1956)]. Thus a D value of 3×10^{-8} cm^2 s^{-1} would give a diffusion path of about 110 nm in 1 msec and 350 nm in 10 msec while a D value of 10^{-7} cm^2 s^{-1} gives 200 nm and 630 nm for the same two time intervals. The radius of a typical appressed region is 250 nm, and if cyt b_6/f complexes occur in this region, as well as in the nonappressed lamellae, the diffusion path between PS2 and nearby cyt b_6/f complexes could be as short as 20 nm (Whitmarsh, 1986). If cyt b_6/f complexes are only localized in the nonappressed region, then the minimum pathlength increases to about 70–100 nm (Millner and Barber, 1984; Haehnel, 1984). In either case it seems unlikely that long-distance diffusion of PQH$_2$, under normal circumstances and at room temperature, restricts the rate of noncyclic electron flow. Since the oxidation of PQH_2 is the slowest process in the overall electron-transfer chain, it must be concluded, for the time being, that the reaction is limited within the cyt b_6/f complex (see Mauro *et al.*, 1986).

Plastocyanin is a one-copper-containing protein that is encoded in the nucleus (Grossman *et al.*, 1982) and is readily lost from isolated chloroplasts by sonication. It appears blue in its oxidized state and has a molecular weight of 10.5 kDa. There seem to be two monomeric molecules per PS1 (Haehnel, 1986) and, according to x-ray crystallography, these molecules are ellipsoidal with the shape of a flattened barrel with dimensions of 0.4 × 3.2 × 2.8 nm (Colman *et al.*, 1978). The core of the molecule is hydrophobic, consisting mainly of aromatic residues, while the polar exterior surface has net negative charge clustered at one end. The walls of the barrel are made up of eight segments of extended beta-pleated sheets and the Cu atom is located near one end of the molecule with approximately tetrahedral-coordinate geometry. Consideration of plastocyanin diffusion within the lumen of the intact thylakoids is problematical. First, the distance between the two interior surfaces of the thylakoid lumen decreases from about 7 nm in the dark to about 3 nm in the light (Murakami and Packer, 1970). Considering the size of the monomeric molecule, the movement of plastocyanin within the lumenal compartment is likely to be very restricted. Moreover, restrictions of movement may also be expected because of proteins that are exposed at the inner surface, including those associated with the water-splitting processes of PS2. That the lumenal space does restrict diffusional motion was shown by EPR spin label studies, which indicated that the viscosity of this compartment is severalfold higher than that of free water (Berg *et al.*, 1979). Second, Haehnel (1982) carried out flash absorption studies to investigate the effect of partially blocking noncyclic electron transport, on the light-induced redox kinetics of P700 and cyt f. From his results he was forced to conclude that there is no freely diffusing pool of plastocyanin. The third problem is that according to the FRAP studies of Fragata *et al.* (1984) using lipid vesicles, the diffusion coefficient of plastocyanin is about 5×10^{-8} cm^2 s^{-1}. In fact, for the reasons given above, the diffusion coefficient for plastocyanin within the lumen of stacked thylakoids is likely to be much lower than 5×10^{-8} cm^2 s^{-1}. This conclusion is difficult to reconcile with the reaction kinetics for the oxidation of cyt f and the Rieske Fe–S center and a diffusion pathlength from appressed to nonappressed regions of 10 nm or more. If all the cyt b_6/f complexes are located in nonappressed membranes and thus in the close proximity of PS1, the problem is simplified. However, at present there seem to be good data to indicate that such an organization does not exist (see Section IV,B). With unstacked membranes or with PS1-enriched subchloroplast particles, the situation is clearer and the kinetics of P700$^+$ re-reduction are second-order and consistent with platocyanin acting as a freely diffusing component (Haehnel *et al.*, 1980; Olsen and Cox, 1982).

Despite the above difficulties, it seems likely that at room temperature photosynthetic electron flow is not rate-limited by diffusion processes (see Mauro *et al.,* 1986), although conditions that alter the physical characteristics of membranes (e.g., reduction of the fluidity) may change this situation (see Yamamoto *et al.,* 1981b).

C. LHC-2 Diffusion

As mentioned previously (Section IV,C), isolated thylakoid membranes can be made to unstack and restack simply by changing the ionic composition of the suspending medium. Concomitant with these gross conformational changes is the lateral movement of protein complexes in the membrane. Such lateral movements alter the extent of energy transfer between complexes and result in changes in the yield of chlorophyll fluorescence. Many different types of experiments have been performed to establish these interrelationships as reviewed by Barber (1980a, 1982a) and Staehelin (1986). Thus from these studies it can be concluded that the thylakoid membrane is sufficiently fluid to allow large membrane complexes to diffuse freely along the plane of the membrane. The fact that the changes in chlorophyll fluorescence yield reflect changes in the spatial distances between PS2 and PS1 allows the possibility of estimating lateral diffusion coefficients for these complexes. With this concept in mind, Rubin *et al.* (1981a) and Briantais *et al.* (1983) were able to calculate diffusion coefficients for the lateral diffusion of PS1 and PS2 complexes to be in the region of 10^{-10} to 10^{-11} cm^2 s^{-1}, which are similar to those reported for a wide range of protein complexes in different membrane systems (see Cherry, 1979; Barber, 1982b; Gupta *et al.,* 1984).

The physiological relevance of the potential for lateral diffusion of protein complexes in the thylakoid membrane did not become obvious until the discovery that some of the LHC-2 complexes can undergo a reversible phosphorylation on their outer surface (Bennett, 1977). The importance of this phosphorylation/dephosphorylation process in relation to the regulation of energy distribution between PS2 and PS1, and to the state 1–state 2 phenomenon observed in intact tissue, has been extensively reviewed (Allen, 1983; Barber, 1982a, 1983b; Bennett, 1983; Horton, 1983; Staehelin and Arntzen, 1983), most recently by Barber (1985, 1986a,b). The phosphorylation occurs at one or two threonyl residues at the N-terminal portion of the LHC-2 polypeptide and is catalyzed by a membrane-bound kinase activated under conditions when the plastoquinone pool is overreduced (either by excess PS2 light or artificially by adding appropriate reductants). When the kinase is not activated, for example, in the dark under oxidizing conditions, or in excess PS1 light, a membrane-bound phosphatase brings about the dephosphorylation of LHC-2. Bennett *et al.* (1980) and Horton and Black (1980) were the first to show that the phosphorylation/dephosphorylation processes were associated with changes in energy distribution between PS2 and PS1

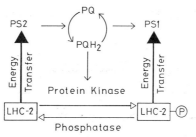

Fig. 16. The role of LHC-2 phosphorylation in controlling energy redistribution processes. Imbalance of the excitation rate in favor of PS2 reduces the plastoquinone pool (PQ → PQH$_2$), which activates a membrane-bound protein kinase. This leads to a decrease in the excitation rate of PS2 in favor of PS1 and, under limiting light conditions, an increase in the efficiency of photosynthetic electron flow. Imbalance of excitation rate in favor of PS1 reoxidizes plastoquinone (PQH$_2$ → PQ), thereby "switching off" the protein kinase and leading to a dephosphorylation of LHC-2 by a membrane-bound phosphatase. Again, under these conditions the overall effect is to increase the quantum efficiency of electron transport.

(also see Allen *et al.*, 1981). From the work of Bonaventura and Myers (1969) it was known that changes in energy distribution between PS2 and PS1 occurred within intact tissue so as to maximize photosynthetic efficiency under conditions of limiting light intensities (see Myers, 1971). Such changes are known as state 1–state 2 transitions and are linked to the LHC-2 phosphorylation process by comparing chlorophyll fluorescence yield changes induced by a combination of PS2 and PS1 light with changes in the incorporation of radioactive phosphate in LHC-2 (Chow *et al.*, 1981; Telfer and Barber, 1981; Telfer *et al.*, 1983a). From these and other experiments, it has become well accepted that the redox control of phosphorylation and dephosphorylation of LHC-2 underlies the molecular mechanism of the state transitions along the lines shown in Fig. 16.

How can phosphorylation of LHC-2 regulate energy distribution between PS2 and PS1? Bearing in mind the concepts presented in Section IV,C, it was logical to postulate that the phosphorylation/dephosphorylation of LHC-2 would alter its surface electrical charge properties and thus dictate its preference to partition into the appressed or nonappressed regions of the membrane (Barber, 1980a,b, 1982a,b, 1983b). In the nonphosphorylated condition, the mobile LHC-2 would be closely associated with PS2 complexes in the appressed regions while the introduction of negative charge onto its surface by phosphorylation would induce coulombic repulsive forces leading to lateral diffusion into the PS1 enriched nonappressed membranes. In this way the phosphorylatable and mobile form of LHC-2 can act either as an antenna for PS2 or PS1 (see Fig. 17). Support for this mobile antenna model has been obtained from a variety of experiments ranging from freeze-fracture electron microscopy (Kyle *et al.*, 1983b; Staehelin, 1986) and biochemical analyses (Chow *et al.*, 1981; Andersson *et al.*, 1982; Kyle *et al.*,

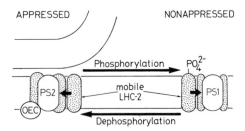

Fig. 17.   The concept of a mobile pool of LHC-2 that can act as a light-harvesting antenna for PS1 when phosphorylated and for PS2 when nonphosphorylated. The lateral movements between the appressed and nonappressed regions are in response to the surface electrical charge changes induced by the phosphorylation/dephosphorylation processes.

1984a,b), to energy-transfer and quantum-yield studies (Farchaus *et al.*, 1982; Krause and Behrend, 1983; Kyle *et al.*, 1983a; Telfer *et al.*, 1984a, 1986). However, as yet the quantitative details of this regulatory process are not firmly established. We do not know precisely how much of the LHC-2 in the membrane is involved or to what degree the LHC-2 becomes functionally linked to PS1. Moreover, it is unclear whether LHC-2 moves alone or whether some PS2-LHC-2 complexes also laterally diffuse into the nonappressed regions. If the latter occurs then there will be changes in energy transfer between PS2 and PS1 (spillover) as well as changes in the absorption cross section of the two photosystems. Indeed, with isolated thylakoids the degree of LHC-2 and PS2-LHC-2 diffusion in response to phosphorylation is, as would be expected, sensitive to the level of cations in the suspension medium (Hodges and Barber, 1984). At suboptimal levels of cations (e.g., below 5 mM) the effect of LHC-2 phosphorylation is to cause a greater lateral mixing of complexes as monitored by changes in the extent of chlorophyll fluorescence quenching (Horton and Black, 1983; Telfer *et al.*, 1983b) and the degree of membrane unstacking (Telfer *et al.*, 1984b). Such a dependence on the background level of cations is in agreement with the concept that the regulatory process relies on changes in electrostatic forces within the intact chloroplast. Indeed, changes in cation levels within the stroma, as well as changes in LHC-2 phosphorylation, may occur in response to changes in light quality and intensity. However, those experiments that have set out to investigate this additional influence on the state transitions have concluded that in mature tissue the dominating process is a change in absorption cross section indicative of LHC-2 migration alone (Hodges and Barber, 1983; Canaani and Malkin, 1984; Malkin *et al.*, 1986).

The lateral shuffling of LHC-2 complexes in the thylakoid membrane in response to surface phosphorylation requires the lipid matrix to be fluid, as demonstrated by the experiment of Haworth (1983). Thus the maintenance of a fluid lipid matrix is important not only for the diffusional processes of electron transport but also for the regulation of energy distribution between the two photosystems.

D. Rotational Diffusion

It seems that many intrinsic membrane protein complexes can rotate within the lipid matrix (see Cherry, 1979; Thomas, 1986). The functional significance of this rotational diffusion is, however, not always clear. Very few studies of this type have been conducted on thylakoids. Wagner and Junge (1980) have studied the rotation of the CF_0-CF_1 complex, while Rousselet and Wollman (1986) monitored the rotational mobility of the PS1 and LHC-2 complex. The former workers detected the rotational motion by labeling CF_1 with eosin isothiocyanate and following linear dichroism of flash-induced absorption change (i.e., by photoselection). These workers also used this approach to monitor the rotation of ferredoxin–NADP oxidoreductase (FNR) (Wagner *et al.*, 1981). It was found in the case of CF_1 that its rotational correlation time was affected by the existence of an electrochemical potential gradient of protons across the membrane or by the presence of ATP, indicating that significant conformational changes occur in this protein complex under these conditions. Similar changes in rotational correlation times were also observed for FNR bound to the membrane surface when ferredoxin interacted with the enzyme (a change from 1 to 40 μsec measured to 10°C). Wagner *et al.* (1982) argued that this slowing down of rotation was due to the formation of a complex between ferredoxin, the PS1 complex, and FNR, a result that may suggest that soluble ferredoxin may not normally diffuse over large distances. In the case of the work of Rousselet and Wollman (1986), the rotational movements were monitored by labeling polypeptides with a paramagnetic analog of *N*-ethylmaleimide. Using thylakoids isolated from *Chlamydomonas reinhardtii*, their EPR saturation transfer spectra indicated that the PS1 complex undergoes slow rotation compared to other labeled proteins. In contrast the LHC-2 complex was found to rotate rapidly in unstacked membranes but to become far less mobile after restacking by the addition of Mg^{2+}.

VI. FINAL COMMENT

Despite the fact that I have restricted the discussions of this chapter almost entirely to the structural–functional properties of the thylakoid membranes of higher plants and green algae, there is considerable growth in the understanding of the comparable system of the non-chlorophyll *b*-containing eukaryotic and prokaryotic organisms. Clearly the advances in the organizational details of the purple bacterial membrane, and especially the reaction-center complex itself, are an outstanding example of this. But substantial progress is also being made with other major classes of organisms, including the cyanobacteria and red algae. These photosynthetic organisms contain phycobilisomes, the structure of which is being rapidly elucidated (see Glazer, 1981, 1984; Deisenhofer *et al.*, 1985b; Zuber, 1987). New organisms are being discovered that have great potential for experimental work, such as *Heliobacterium chlorum* (Gest and Favinger, 1983), which represents a completely new class of organism linking the simple anoxygenic photosynthetic prokaryotes with the more complex oxygenic organisms, and the recently isolated new member of the chlorophyll *b*-containing Prochlorophyta (Burger-Wiersma *et al.*, 1986). Although the various classes of organisms at first sight do seem to have distinctly different photosynthetic systems, it is remarkable that at the molecular level they are often very similar. Who for example, 5 years ago, could have guessed that functional and structural studies on the reaction centers of purple bacteria would be so important for understanding the organizational and mechanistic properties of PS2? Without any doubt, a continuing effort to study and compare a wide range of photosynthetic organisms will continue to be a valuable approach. Such comparative studies will be particularly important because we can look forward to using the techniques of *in vitro* genetic manipulation to obtain detailed molecular information about the photosynthetic processes. Indeed, these techniques are fully operational in prokaryotic organisms, and the information that will arise from, for example, site-directed mutagenesis studies should be extrapolated to the eukaryotic systems. At present, genetic engineering of eukaryotic photosynthetic organisms is problematic due to the division of DNA between nuclear and plastid chromosomes, but presumably new procedures will arise so that these systems will be manipulatable in a manner comparable with their prokaryotic cousins.

I also feel that we will come to appreciate in more depth the dynamics of photosynthetic membranes and how they regulate their organization in response to changes in environmental conditions. For example, in the higher-plant thylakoid system we still need to understand much more about why certain polypeptides become reversibly phosphorylated and how this affects membrane conformation. The assembly and disassembly of membrane complexes is also an intriguing problem, especially in the case of the PS2 complex where the D1 protein is known to turn over relatively rapidly. These areas, and others, will form the focus of experimental work in the future with the overall aim to elucidate fully the structure of the photosynthetic apparatus. With this information we will be able to appreciate how nature has constructed molecular machinery that is sufficiently malleable to maintain a high quantum efficiency over a wide range of environmental conditions.

ACKNOWLEDGMENTS

I acknowledge the financial support of the Science and Engineering Research Council and the Agricultural and Food Research Council. The preparation of the manuscript involved the help and patience of Lyn Barber and Julie Westcott, to whom I am most thankful. I must also thank my research colleagues for their support and for valuable discussions, particularly Dr.

Alison Telfer, Dr. David Chapman, Dr. Niki Gounaris, and Dr. Jonathan Marder, and Dr. Christa Critchley for critically reading the manuscript.

REFERENCES

Abramowicz, D. A., and Dismukes, C. G. (1984). *Biochim. Biophys. Acta* **765,** 309–318.

Akabori, K., Imaoka, A., and Toyoshima, Y. (1984). *FEBS Lett.* **173,** 36–40.

Åkerlund, H.-E., Jansson, C., and Andersson, B. (1982). *Biochim. Biophys. Acta* **681,** 1–10.

Albertsson, P.-A., Andersson, B., Larsson, C., and Akerlund, H. E. (1982). *Methods Biochem. Anal.* **28,** 115–10.

Allred, D. R., and Staehelin, L. A. (1985). *Plant Physiol.* **78,** 199–202.

Allen, J. F. (1983). *Trends Biochem. Sci.* **8,** 369–373.

Allen, J. F., Bennett, J., Steinback, K. E., and Arntzen, C. J. (1981). *Nature (London)* **291,** 21–25.

Alt, J., and Herrmann, R. G. (1984). *Curr. Genet.* **8,** 551–557.

Alt, J., Winter, P., Sebalt, W., Moses, J. G., Schedel, R., Westhoff, P., and Herrmann, R. G. (1983). *Curr. Genet.* **7,** 129–138.

Alt, J., Morris, J., Westhoff, P., and Herrmann, R. G. (1984). *Curr. Genet.* **8,** 597–606.

Anderson, J. M. (1975). *Biochim. Biophys. Acta* **416,** 191–235.

Anderson, J. M. (1980). *FEBS Lett.* **117,** 327–331.

Anderson, J. M. (1982). *FEBS Lett.* **138,** 62–66.

Anderson, J. M. (1986). *Annu. Rev. Plant Physiol.* **37,** 93–136.

Anderson, J. M., Goodchild, D. J., and Andersson, B. (1985). *Cell Biol. Int. Rep.* **9,** 715–721.

Andersson, B. (1978). Ph.D. Thesis, University of Lund, Sweden.

Andersson, B., and Akerlund, H.-E. (1986). *Top. Photosynth.* **8,** 379–420.

Andersson, B., and Anderson, J. M. (1980). *Biochim. Biophys. Acta* **593,** 427–440.

Andersson, B., Akerlund, H.-E., Jergil, B., and Larsson, C. (1982). *FEBS Lett.* **149,** 181–185.

Andersson, B., Critchley, C., Ryrie, I. J., Jansson, C., Larsson, C., and Anderson, J. M. (1984). *FEBS Lett.* **168,** 113–117.

Armond, P. A., and Arntzen, C. J. (1977). *Plant Physiol.* **59,** 398–404.

Armond, P. A., Staehelin, L. A., and Arntzen, C. J. (1977). *J. Cell Biol.* **73,** 400–418.

Arntzen, C. J. (1978). *Curr. Top. Bioenerg.* **8,** Part B, 111–160.

Aronson, H., Waggoner, C., More, J., and Berg, S. P. (1983). *Biochim. Biophys. Acta* **725,** 519–528.

Barber, J. (1976). *Top. Photosynth.* **1,** 89–134.

Barber, J. (1979). *Ciba Found Symp.* (N.S.) **61,** 283–304.

Barber, J. (1980a). *FEBS Lett.* **118,** 1–10.

Barber, J. (1980b). *Biochim. Biophys. Acta* **594,** 253–308.

Barber, J. (1982a). *Annu. Rev. Plant Physiol.* **33,** 261–295.

Barber, J. (1982b). *BioSci. Rep.* **2,** 1–13.

Barber, J. (1983a). *Plant, Cell Environ.* **6,** 311–312.

Barber, J. (1983b). *Photobiochem. Photobiophys.* **5,** 181–190.

Barber, J. (1985). *Top. Photosynth.* **6,** 91–134.

Barber, J. (1986a). *In* "Encyclopedia of Plant Physiology, New Series" (L. A. Staehelin and C. J. Arntzen, eds.), Vol. 19, pp. 653–664. Springer-Verlag, Berlin and New York.

Barber, J. (1986b). *Photosynth. Res.* (Butler Memorial Issue) **10,** 243–253.

Barber, J., and Chow, W. S. (1979). *FEBS Lett.* **105,** 5–10.

Barber, J., and Gounaris, K. (1986). *Photosynth. Res.* **9,** 239–249.

Barber, J., and Marder, J. B. (1986). *Biotechnol. Genet. Eng. Rev.* **4,** 355–404.

Barber, J., Mills, J. D., and Love, A. (1977). *FEBS Lett.* **74**, 174–181.

Barber, J., Chow, W. S., Scoufflaire, C., and Lannoye, R. (1980). *Biochim. Biophys. Acta* **591**, 92–103.

Barber, J., Ford, R. C., Mitchell, R. A. C., and Millner, P. A. (1984). *Planta* **161**, 375–380.

Bendall, D. (1982). *Biochim. Biophys. Acta* **683**, 119–152.

Bennett, J. (1977). *Nature (London)* **269**, 344–346.

Bennett, J. (1983). *Biochem. J.* **212**, 1–13.

Bennett, J., Steinback, K. E., and Arntzen, C. J. (1980). *Proc. Natl. Acad. Sci. U.S.A.* **77**, 5253–5257.

Berg, S. P., Lusczakoski, D. M., and Morse, P. D. (1979). *Arch. Biochem. Biophys.* **194**, 138–148.

Berzborn, R. J. (1969). *Z. Naturforsch., B: Anorg. Chem., Org. Chem., Biochem., Biophys., Biol.* **24B**, 436–446.

Bird, C. R., Koller, B., Auffret, A. D., Huttly, A. K., Howe, C. J., Dyer, T. A., and Gray, J. C. (1985). *EMBO J.* **4**, 1381–1388.

Bjorkman, O., and Ludlow, M. (1972). *Year Book—Carnegie Inst. Washington* **71**, 85–94.

Blackwell, M. F., Gounaris, K., and Barber, J. (1986). *Biophys. J.* **858**, 221–234.

Boardman, N. K. (1968). *Adv. Enzymol.* **30**, 1–79.

Boardman, N. K. (1970). *Annu. Rev. Plant Physiol.* **26**, 369–401.

Boardman, N. K. (1977). *Annu. Rev. Plant Physiol.* **28**, 355–377.

Boardman, N. K., and Anderson, J. M. (1964). *Nature (London)* **203**, 166–167.

Boardman, N. K., Anderson, J. M., and Goodchild, D. J. (1978). *Curr. Top. Bioenerg.* **8**, Part B, 35–109.

Bolton, J. R. (1978). *In* "The Photosynthetic Bacteria" (R. K. Clayton and W. R. Sistrom, eds.), pp. 419–429. Plenum, New York.

Bonaventura, C., and Myers, J. (1969). *Biochim. Biophys. Acta* **189**, 366–383.

Bouges-Bocquet, B. (1980). *Biochim. Biophys. Acta* **590**, 223–233.

Bouthyett, P. Y., and Jagendorf, A. T. (1978). *Plant Cell Physiol.* **19**, 1169–1174.

Boyer, P. D., and Kohlbrenner, W. E. (1981). *In* "Energy Coupling in Photosynthesis" (B. R. Selman and S. Selman-Reimer, eds.), pp. 231–240. Elsevier, Amsterdam.

Brangeon, J., and Mustardy, L. A. (1979). *Biol. Cell.* **36**, 71–80.

Braumann, T., Weber, G., and Grimme, L. H. (1982). *Photobiochem. Photobiophys.* **4**, 1–8.

Brettel, K., Schlodder, E., and Witt, H. T. (1984). *Biochim. Biophys. Acta* **766**, 403–415.

Briantais, J.-M., Vernotte, C., Lavorel, J., Olive, J., and Wollman, F. A. (1983). *Proc. Int. Congr. Photosynth. 6th, 1983,* Abstr. 1, p. 284.

Bruist, M. F., and Hammes, G. G. (1981). *Biochemistry* **20**, 6298–6305.

Burger-Wiersma, T., Veenhuis, M., Korthals, H. J., van de Wiel, C. C., and Mur, L. R. (1986). *Nature (London)* **320**, 262–264.

Camm, E. L., and Green, B. R. (1983). *Biochim. Biophys. Acta* **724**, 291–293.

Canaani, O., and Malkin, S. (1984). *Biochim. Biophys. Acta* **766**, 513–524.

Cashmore, A. R. (1984). *Proc. Natl. Acad. Sci. U.S.A.* **81**, 2960–2964.

Chapman, D. J., and Barber, J. (1986). *Biochim. Biophys. Acta* **850**, 170–172.

Chapman, D. J., and Barber, J. (1987). *In* "Methods in Enzymology" (R. Douce and L. Packer, eds.). Vol. 148, pp. 294–319. Academic Press, New York.

Chapman, D. J., Millner, P. A., and Barber, J. (1983). *Biochem. Soc. Trans.* **11**, 387–388.

Chapman, D. J., DeFelice, J., and Barber, J. (1985). *Dev. Plant Biol.* **9**, 457–464.

Chapman, D. J., DeFelice, J., and Barber, J. (1986). *Photosynth. Res.* **8**, 257–265.

Cherry, R. J. (1979). *Biochim. Biophys. Acta* **559**, 289–327.

Chitnis, P. R., Harel, E., Kohorn, B. D., Tobin, E. M., and Thornber, J. P. (1986). *J. Cell Biol.* **102**, 982–988.

Chow, W. S., Telfer, A., Chapman, D. J., and Barber, J. (1981). *Biochim. Biophys. Acta* **638**, 60–68.

Clark, R. D., and Hind, G. (1983). *J. Biol. Chem.* **258**, 10348–10354.

Clark, R. D., Hawkesford, M. J., Coughlan, S. J., Bennett, J., and Hind, G. (1984). *FEBS Lett.* **174**, 137–142.

Cohen, B. N., Coleman, T. A., Schmitt, J. J., and Weissbach, H. (1984). *Nucleic Acids Res.* **12**, 6221–6230.

Colman, P. M., Freeman, H. C., Guss, J. M., Murata, M., Norris, V. A., Romshaw, J. A. M., and Venkatapp, M. P. (1978). *Nature (London)* **272**, 319–324.

Coombs, J., and Greenwood, A. D. (1976). *Top. Photosynth.* **1**, 1–51.

Cox, G. B., Jans, D. A., Fimmel, A. L., Gibson, F., and Hatch, L. (1984). *Biochim. Biophys. Acta* **768**, 201–208.

Cox, R. P., and Andersson, B. (1981). *Biochem. Biophys. Res. Commun.* **103**, 1336–1342.

Cozens, A. L., Walker, J. E., Phillips, A. L., Huttly, A. K., and Gray, J. C. (1986). *EMBO J.* **5**, 217–222.

Cramer, W. A., Widger, W. R., Herrmann, R. G., and Trebst, A. (1985). *Trends Biochem. Sci.* **10**, 125–129.

Cramer, W. A., Theg, S. M., and Widger, W. R. (1986). *Photosynth. Res.* **10**, 393–404.

Damm, I., Gropper, T., Branmann, T., and Grimme, L. H. (1984). *In* "Advances in Photosynthesis Research" (C. Sybesma, ed.), Vol. 1, 109–112. Martinus Nijhoff/Dr. W. Junk Publ., The Hague, The Netherlands.

Deisenhofer, J., Epp, O., Miki, K., Huber, R., and Michel, H. (1984). *J. Mol. Biol.* **180**, 385–398.

Deisenhofer, J., Epp, O., Miki, K., Huber, R., and Michel, H. (1985a). *Nature (London)* **318**, 618–624.

Deisenhofer, J., Michel, H., and Huber, R. (1985b). *Trends Biochem. Sci.* **10**, 243–248.

Delepelaire, P. (1984). *EMBO J.* **3**, 701–706.

Deno, H., and Sugiura, M. (1984). *FEBS Lett.* **172**, 209–211.

Deno, H., Shinozaki, K., and Sugiura, M. (1983). *Nucleic Acids Res.* **11**, 2185–2191.

Deno, H., Shinozaki, K., and Sugiura, M. (1984). *Gene* **32**, 195–201.

Dornemann, D., and Senger, H. (1984). *In* "Advances in Photosynthesis Research" (S. Sybesma, ed.), Vol. 2, pp. 77–80. Martinus Nijhoff/Dr. W. Junk Publ., The Hague, The Netherlands.

Dunahay, T. G., Staehelin, L. A., Seibert, M., Ogilvi, P. D., and Berg, S. P. (1984). *Biochim. Biophys. Acta* **764**, 179–193.

Duniec, J. T., Sculley, M. J., and Thorne, S. W. (1979). *J. Theor. Biol.* **79**, 473–484.

Duniec, J. T., Israelachvili, J. N., Ninham, B. W., Pashley, R. M., and Thorne, S. W. (1981). *FEBS Lett.* **129**, 193–196.

Dunsmuir, P. (1985). *Nucleic Acids Res.* **13**, 2503–2518.

Duval, J. C., Dubacq, J. P., and Trémolières, A. (1980). *In* "Biogenesis and Function of Plant Lipids" (P. Mazliak, P. Beneveniste, C. Costes, and R. Douce, eds.), pp. 91–94. Elsevier, Amsterdam.

Duysens, L. N. M. (1964). *Prog. Biophys. Mol. Biol.* **14**, 1–104.

Dyer, T. A. (1985). *Oxford Surv. Plant Mol. Cell Biol.* **2**, 147–177.

Edelman, M., Mattoo, A. K., and Marder, J. B. (1984). *In* "Chloroplast Biogenesis" (R. J. Ellis, ed.), pp. 283–302. Cambridge Univ. Press, London and New York.

Einstein, A. (1956). *In* "Investigations on the Theory of Brownian Movement" (R. Furth, ed.). Dover, New York.

Erickson, J. M., Rahire, M., Rochaix, J.-D., and Mets, L. (1985). *Science* **228**, 204–207.

Evans, M. C. W., Sihra, C. K., and Cammack, R. (1975). *Biochem. J.* **158**, 71–77.

Farchaus, J. W., Widger, W. R., Cramer, W. A., and Dilley, R. A. (1982). *Arch. Biochem. Biophys.* **217**, 362–367.

Fato, R., Battino, M., Parenti Castelli, G., and Lenaz, G. (1985). *FEBS Lett.* **179**, 238–242.

Fish, L. E., Kuck, U., and Bogorad, L. (1985). *J. Biol. Chem.* **260**, 1413–1421.

Fluhr, R., Fromm, H., and Edelman, M. (1983). *Gene* **25,** 271–280.

Ford, R. C., and Barber, J. (1981). *Photobiochem. Photobiophys.* **1,** 263–270.

Ford, R. C., and Barber, J. (1983). *Biochim. Biophys. Acta* **722,** 341–348.

Ford, R. C., Chapman, D. J., Barber, J., Pedersen, J. Z., and Cox, R. P. (1982). *Biochim. Biophys. Acta* **681,** 145–151.

Fragata, M., Ohnishi, S., Adada, K., Ito, T., and Takahashi, M. (1984). *Biochemistry* **23,** 4044–4051.

Franzen, L.-G., and Andreasson, L.-E. (1984). *Biochim. Biophys. Acta* **765,** 166–170.

Frasch, W. D., and Selman, B. R. (1982). *Biochemistry* **21,** 3636–3643.

Garavito, R. M., Jenkins, J., Jansonius, J. N., Karlsson, R., and Rosenbusch, J. P. (1983). *J. Mol. Biol.* **164,** 313–327.

Gest, H., and Favinger, J. (1983). *Arch. Microbiol.* **136,** 11–16.

Ghanotakis, D. F., and Yocum, C. F. (1985). *Photosynth. Res.* **7,** 97–114.

Ghanotakis, D. F., Babcock, G. T., and Yocum, C. F. (1984a). *FEBS Lett.* **167,** 127–130.

Ghanotakis, D. F., Babcock, G. T., and Yocum, C. F. (1984b). *Biochim. Biophys. Acta* **765,** 388–398.

Ghanotakis, D. F., Topper, J. N., and Yocum, C. F. (1984c). *Biochim. Biophys. Acta* **767,** 524–531.

Ghanotakis, D. F., Topper, J. N., Babcock, G. T., and Yocum, C. F. (1984d). *FEBS Lett.* **170,** 169–173.

Glazer, A. N. (1981). *In* "The Biochemistry of Plants" (M. D. Hatch and N. K. Boardman, eds.), Vol. 8, pp. 51–96. Academic Press, New York.

Glazer, A. N. (1984). *Biochim. Biophys. Acta* **768,** 29–51.

Golbeck, J. H., and Cornelius, J. M. (1986). *Biochim. Biophys. Acta* **849,** 16–24.

Gounaris, K., and Barber, J. (1983). *Trends Biochem. Sci.* **8,** 378–381.

Gounaris, K., and Barber, J. (1985). *FEBS Lett.* **188,** 68–72.

Gounaris, K., Sen, A., Brain, A. P. R., Quinn, P. J., and Williams, W. P. (1983a). *Biochim. Biophys. Acta* **728,** 129–139.

Gounaris, K., Mannock, D. A., Sen, A., Brain, A. P. R., Williams, W. P., and Quinn, P. J. (1983b). *Biochim. Biophys. Acta* **732,** 229–242.

Gounaris, K., Brain, A. P. R., Quinn, P. J., and Williams, W. P. (1983c). *FEBS Lett.* **153,** 47–52.

Gounaris, K., Sundby, C., Andersson, B., and Barber, J. (1983d). *FEBS Lett.* **157,** 170–174.

Gounaris, K., Whitford, D., and Barber, J. (1983e). *FEBS Lett.* **163,** 230–234.

Gounaris, K., Brain, A. P. R., Quinn, P. J., and Williams, W. P. (1984a). *Biochim. Biophys. Acta* **766,** 198–208.

Gounaris, K., Lambillotte, M., Barber, J., Muehlethaler, K., and Jay, F. (1984b). *Dev. Plant Biol.* **9,** 485–488.

Gross, E. L., and Prasher, S. H. (1974). *Arch. Biochem. Biophys.* **164,** 460–468.

Grossman, A. R., Bartlett, S. G., Schmidt, G., Mullet, J. E., and Chua, N.-H. (1982). *J. Biol. Chem.* **257,** 1558–1563.

Gunning, B. E. S., and Steer, M. W. (1975). "Ultrastructure and the Biology of Plant Cells." Arnold, London.

Gupta, S., Wu, E.-S., Hoechli, M., Jacobson, K., Sowers, A. E., and Hackenbrock, C. R. (1984). *Proc. Natl. Acad. Sci. U.S.A.* **81,** 2606–2610.

Haehnel, W. (1982). *Biochim. Biophys. Acta* **682,** 245–257.

Haehnel, W. (1984). *Annu. Rev. Plant Physiol.* **35,** 659–693.

Haehnel, W. (1986). *In* "Encyclopedia of Plant Physiology, New Series" (L. A. Staehelin and C. J. Arntzen, eds.), Vol. 19, pp. 547–559. Springer-Verlag, Berlin and New York.

Haehnel, W., Hesse, V., and Propper, A. (1980). *FEBS Lett.* **111,** 79–82.

Hall, D. O., and Rao, K. K. (1977). *In* "Encyclopedia of Plant Physiology, New Series" (A. Trebst and M. Avron, eds.), Vol. 5, pp. 206–221. Springer-Verlag, Berlin and New York.

Hallick, R. B., and Bottomley, W. (1984). *Plant Mol. Biol. Rep.* **1,** 38–43.

Harwood, J. L. (1980). *In* "Biochemistry of Plants" (P. K. Stumpf and E. E. Conn, eds.), Vol. 4, pp. 1–55. Academic Press, New York.

Hauska, G. (1986). *In* "Encyclopedia of Plant Physiology, New Series" (L. A. Staehelin and C. J. Arntzen, eds.), Vol. 19, pp. 496–507. Springer-Verlag, Berlin and New York.

Hauska, G., Hurt, E., Gabellini, N., and Lokau, W. (1983). *Biochim. Biophys. Acta* **726,** 97–133.

Haworth, P. (1983). *Arch. Biochem. Biophys.* **226,** 145–154.

Hearst, J. E., and Sauer, K. (1984). *Z. Naturforsch.* **39C,** 421–424.

Heinemeyer, W., Alt, J., and Herrmann, R. G. (1984). *Curr. Genet.* **8,** 543–549.

Henry, L. E. A., Mikkelsen, J. D., and Moller, B. L. (1983). *Carlsberg Res. Commun.* **48,** 131–148.

Herrmann, R. G., Alt, J., Schiller, B., Widger, W. R., and Cramer, W. A. (1984). *FEBS Lett.* **176,** 239–244.

Hill, R., and Bendall, F. (1960). *Nature (London)* **186,** 136–137.

Hiller, R. G., and Goodchild, D. J. (1981). *In* "The Biochemistry of Plants" (M. D. Hatch and N. K. Boardman, eds.), Vol. 8, pp. 1–49. Academic Press, New York.

Hiller, R. G., and Raison, J. K. (1980). *Biochim. Biophys. Acta* **599,** 63–72.

Hind, G., Clark, R. D., and Houchins, J. P. (1984). *In* "Advances in Photosynthesis Research" (C. Sybesma, ed.), Vol. 1, pp. 529–536. Martinus Nijhoff/Dr. W. Junk, The Hague, The Netherlands.

Hirschberg, J., and McIntosh, L. (1983). *Science* **222,** 1346–1349.

Hodges, M., and Barber, J. (1983). *Plant Physiol.* **72,** 1119–1122.

Hodges, M., and Barber, J. (1984). *Biochim. Biophys. Acta* **767,** 102–107.

Hodges, M., Packham, N. K., and Barber, J. (1984). *Photobiochem. Photobiophys.* **7,** 311–317.

Holschuh, K., Bottomley, W., and Whitfeld, P. R. (1984). *Nucleic Acids Res.* **12,** 8819–8834.

Horton, P. (1983). *FEBS Lett.* **152,** 47–52.

Horton, P., and Black, M. T. (1980). *FEBS Lett.* **119,** 141–144.

Horton, P., and Black, M. T. (1983). *Biochim. Biophys. Acta* **772,** 214–218.

Howe, C. J., Auffret, A. D., Doherty, A., Bowman, C. M., Dyer, T. A., and Gray, J. C. (1982). *Proc. Natl. Acad. Sci. U.S.A.* **79,** 6903–6907.

Howe, C. J., Fearnley, I., Walker, J. E., Dyer, T. A., and Gray, J. C. (1985). *Plant Mol. Biol.* **4,** 333–345.

Hurt, E., and Hauska, G. (1982). *Biochim. Biophys. Acta* **682,** 466–473.

Il'ina, M. D., Krasanskas, V. V., Rotomskis, R. J., and Borisov, A. Yu. (1984). *Biochim. Biophys. Acta* **767,** 501–506.

Imoaka, A., Akabori, K., Yanagi, M., Izumi, K., Toyoshima, Y., Kawamori, A., Nakayama, H., and Sato, J. (1986). *Biochim. Biophys. Acta* **848,** 201–211.

Izawa, S., and Good, N. E. (1966). *Plant Physiol.* **41,** 533–543.

Jansson, C. (1984). *In* "Advances in Photosynthesis Research" (C. Sybesma, ed.), Vol. 1, pp. 375–378. Martinus Nijhoff/Dr. W. Junk, The Hague, The Netherlands.

Jennings, R. C., Garlaschi, F. M., Gerola, P. D., and Forti, G. (1979). *Biochim. Biophys. Acta* **546,** 207–219.

Jennings, R. C., Garlaschi, F. M., Gerola, P. D., Etzion-Katz, R., and Forti, G. (1981). *Biochim. Biophys. Acta* **638,** 100–107.

Jennings, R. C., Garlaschi, F. M., and Gerola, P. D. (1983). *Biochim. Biophys. Acta* **722,** 144–149.

Joliot, P., and Joliot, A. (1986). *In* "Encyclopedia of Plant Physiology, New Series" (L. A. Staehelin and C. J. Arntzen, eds.), Vol. 19, pp. 528–538. Springer-Verlag, Berlin and New York.

Joliot, P., and Kok, B. (1975). *In* "Bioenergetics of Photosynthesis" (Govindjee, ed.), pp. 387–412. Academic Press, New York.

Jordan, B. R., Chow, W. S., and Baker, A. J. (1983). *Biochim. Biophys. Acta* **725,** 77–86.

Karlin-Neumann, G. A., Kohorn, B. D., Thornber, J. P., and Tobin, E. M. (1985). *J. Mol. Appl. Genet.* **3**, 45–61.

Katsikas, H., and Quinn, P. J. (1982). *Eur. J. Biochem.* **124**, 165–169.

Kingsley, P. B., and Feigenson, G. W. (1981). *Biochim. Biophys. Acta* **635**, 602–618.

Kinosita, K., Kataoka, R., Kimura, Y., Gotch, O., and Ikegami, A. (1981). *Biochemistry* **20**, 4270–4277.

Kirk, J. T. O., and Tilney-Bassett, R. A. E. (1967). "The Plastids: Their Chemistry, Structure, Growth and Inheritance." Freeman, San Francisco, California.

Kohorn, B. D., Harel, E., Chitnis, P. R., Thornber, J. P., and Tobin, E. M. (1986). *J. Cell Biol.* **102**, 972–981.

Krause, G. H., and Behrend, U. (1983). *Biochim. Biophys. Acta* **723**, 176–181.

Krebbers, E. T., Larrinua, I. M., McIntosh, L., and Bogorad, L. (1982). *Nucleic Acids Res.* **10**, 4985–5002.

Kuehlbrandt, W. (1984). *Nature (London)* **307**, 478–480.

Kuwabara, T., and Murata, N. (1984). In "Advances in Photosynthesis Research" (C. Sybesma, ed.), Vol. 1, pp. 371–374. Martinus Nijhoff/Dr. W. Junk, The Hague, The Netherlands.

Kuwabara, T., Miyao, M., Murata, T., and Murata, N. (1985). *Biochim. Biophys. Acta* **806**, 283–289.

Kyle, D. J., Staehelin, L. A., and Arntzen, C. J. (1983a). *Arch. Biochem. Biophys.* **222**, 527–541.

Kyle, D. J., Baker, N. R., and Arntzen, C. J. (1983b). *Photobiochem. Photobiophys.* **5**, 79–85.

Kyle, D. J., Ohad, I., and Arntzen, C. J. (1984a). *Proc. Natl. Acad. Sci. U.S.A.* **81**, 4070–4074.

Kyle, D. J., Kuang, T.-Y., Watson, J. L., and Arntzen, C. J. (1984b). *Biochim. Biophys. Acta* **765**, 89–96.

Kyte, J., and Doolittle, R. F. (1982). *J. Mol. Biol.* **157**, 105–132.

Larsson, C., and Andersson, B. (1985). *Biochim. Biophys. Acta* **809**, 396–402.

Leong, T.-Y., and Anderson, J. M. (1984). *Photosynth. Res.* **5**, 105–115.

Lichtenthaler, H. K., Buschmann, C., Doll, M., Fietz, H.-J., Bach, T., Kozel, H., Meier, D., and Rahmsdorf, U. (1981). *Photosynth. Res.* **2**, 115–141.

Ljungber, U., Henrysson, T., Rochester, C. P., Akerlund, H.-E., and Andersson, B. (1986). *Biochim. Biophys. Acta* **849**, 112–120.

McCarty, R. E., and Moroney, J. V. (1985). *Enzymes Biol. Membr.* **4**, 383–414.

McCarty, R. E., and Nalin, C. M. (1986). In "Encyclopedia of Plant Physiology, New Series" (L. A. Staehelin and C. J. Arntzen, eds.), Vol. 19, pp. 576–583. Springer-Verlag, Berlin and New York.

McCourt, P., Browse, J., Watson, J., Arntzen, C. J., and Somerville, C. R. (1985). *Plant Physiol.* **78**, 853–858.

Machold, O., Simpson, D. J., and Hayer-Hanson, G. H. (1977). *Carlsberg Res. Commun.* **42**, 499–516.

McIntosh, A. R., and Bolton, J. R. (1976). *Biochim. Biophys. Acta* **430**, 555–559.

Malkin, R. (1982). *Annu. Rev. Plant Physiol.* **33**, 455–479.

Malkin, R. (1986). *Photosynth. Res.* (in press).

Malkin, R., Lam, E., and Ortiz, W. (1984). In "Advances in Photosynthesis Research" (C. Sybesma, ed.), Vol. 1, pp. 179–185. Martinus Nijhoff/Dr. W. Junk, The Hauge, The Netherlands.

Malkin, S., Telfer, A., and Barber, J. (1986). *Biochim. Biophys. Acta* **848**, 42–48.

Mansfield, R. W., and Anderson, J. M. (1985). *Biochim. Biophys. Acta* **809**, 435–444.

Mansfield, R. W., and Bendall, D. S. (1984). *Biochim. Biophys. Acta* **766**, 62–69.

Marcus, M. F., and Hawley, M. D. (1970). *Biochim. Biophys. Acta* **201**, 1–8.

Marder, J. B. (1985). Ph.D. Thesis, Weizmann Institute, Israel.

Matsuda, H., and Butler, W. L. (1983). *Biochim. Biophys. Acta* **725**, 330–324.

Mauro, S., Lannoye, R., Vandeloise, R., and Vander Donckt, E. (1986). *Photobiochem. Photobiophys.* **11,** 83–94.

Melis, A. (1984). *J. Cell. Biochem.* **24,** 271–285.

Melis, A., and Homann, P. (1976). *Photochem. Photobiol.* **23,** 343–350.

Melis, A., and Thielen, A. P. G. M. (1980). *Biochim. Biophys. Acta* **589,** 275–286.

Menger, W., Knotzel, J., Braumann, T., and Grimme, L. H. (1984). *In* "Advances in Photosynthesis Research" (C. Sybesma, ed.), Vol. 2, pp. 145–148. Martinus Nijhoff/Dr. W. Junk, The Hague, The Netherlands.

Menke, W. (1962). *Annu. Rev. Plant Physiol.* **13,** 27–44.

Merchant, S., and Selman, B. R. (1985). *Photosynth. Res.* **6,** 3–31.

Michel, H. (1982). *J. Mol. Biol.* **158,** 567–572.

Michel, H. (1983). *Trends Biochem. Sci.* **8,** 56–59.

Michel, J. M., and Michel-Wolwertz, M. R. (1969). *Prog. Photosynth. Res.* **1,** 115–127.

Miller, K. R., and Staehelin, L. A. (1976). *J. Cell Biol.* **68,** 30–37.

Millner, P. A., and Barber, J. (1984). *FEBS Lett.* **169,** 1–6.

Millner, P. A., Mitchell, R. A. C., Chapman, D. J., and Barber, J. (1984a). *Photosynth. Res.* **5,** 63–76.

Millner, P. A., Chapman, D. J., and Barber, J. (1984b). *Biochim. Biophys. Acta* **765,** 282–287.

Millner, P. A., Marder, J. B., Gounaris, K., and Barber, J. (1986). *Biochim. Biophys. Acta* **852,** 30–37.

Mills, J. D., and Barber, J. (1978). *Biophys. J.* **21,** 257–272.

Mills, J. D., and Mitchell, P. (1982). *Biochim. Biophys. Acta* **679,** 75–83.

Mitchell, P. (1975). *FEBS Lett.* **59,** 137–139.

Miyao, M., and Murata, N. (1983). *Biochim. Biophys. Acta* **725,** 87–93.

Miyao, M., and Murata, N. (1984). *FEBS Lett.* **170,** 350–354.

Morris, J., and Herrmann, R. G. (1984). *Nucleic Acids Res.* **12,** 2837–2850.

Mullet, J. E., and Arntzen, C. J. (1980). *Biochim. Biophys. Acta* **589,** 100–117.

Mullet, J. E., Burke, J. J., and Arntzen, C. J. (1980). *Plant Physiol.* **65,** 814–822.

Murakami, S., and Packer, L. (1970). *J. Cell Biol.* **47,** 332–351.

Murata, N., and Miyao, M. (1985). *Trends Biochem. Sci.* **10,** 122–124.

Murphy, D. J. (1982). *FEBS Lett.* **150,** 19–26.

Murphy, D. J. (1986). *Biochim. Biophys. Acta* **864,** 33–94.

Murphy, D. J., and Woodrow, I. E. (1983). *Biochim. Biophys. Acta* **725,** 104–112.

Myers, J. (1971). *Annu. Rev. Plant Physiol.* **22,** 289–312.

Nakatani, H. Y., Ke, B., Dolan, E., and Arntzen, C. J. (1984). *Biochim. Biophys. Acta* **756,** 347–352.

Nebedryk, E., Andriaambinitsoa, S., and Breton, J. (1984). *Biochim. Biophys. Acta* **765,** 380–387.

Nechushtai, R., Nelson, N., Mattoo, A. K., and Edelman, M. (1981). *FEBS Lett.* **125,** 115–119.

Nelson, N. (1982). *Top. Photosynth.* **4,** 81–104.

Nelson, N., and Neumann, J. (1972). *J. Biol. Chem.* **247,** 1917–1924.

Norris, J. R., Uphaus, R. A., Crespi, H. G., and Katz, J. J. (1971). *Proc. Natl. Acad. Sci. U.S.A.* **68,** 625–629.

Oh-oka, H., Tanaka, S., Wada, K., Kuwabara, T., and Murata, N. (1986). *FEBS Lett.* **197,** 63–66.

Ojakian, G. K., and Satir, P. (1974). *Proc. Natl. Acad. Sci. U.S.A.* **21,** 2052–2056.

Olsen, L. F., and Cox, R. P. (1982). *Biochim. Biophys. Acta* **679,** 436–443.

O'Malley, P. J., Babcock, G. T., and Prince, R. C. (1984). *Biochim. Biophys. Acta* **766,** 283–288.

Ono, T.-A., and Inoue, Y. (1984a). *FEBS Lett.* **166,** 381–384.

Ono, T.-A., and Inoue, Y. (1984b). *FEBS Lett.* **168,** 281–286.

Overbeek, J. T. G. (1978). *J. Colloid Interface Sci.* **58,** 408–422.

Paolillo, D. J. (1970). *J. Cell Sci.* **6**, 234–255.

Park, R. B., and Sane, P. V. (1971). *Annu. Rev. Plant Physiol.* **22**, 395–430.

Passavant, C. W., and Hallick, R. B. (1985). *Plant Mol. Biol.* **4**, 347–354.

Peters, F. A. L. J., van Wielink, J. E., Sang, H. W. W. F., De Vries, S., and Kraayenhoff, F. R. (1983). *Biochim. Biophys. Acta* **722**, 460–470.

Phillips, A. L., and Gray, J. C. (1984). *Mol. Gen. Genet.* **194**, 477–484.

Pick, U., and Racker, E. (1979). *J. Biol. Chem.* **254**, 2793–2799.

Pick, U., Gounaris, K., Admon, A., and Barber, J. (1984). *Biochim. Biophys. Acta* **765**, 12–20.

Pick, U., Gounaris, K., Weiss, M., and Barber, J. (1985). *Biochim. Biophys. Acta* **808**, 415–420.

Pick, U. Weiss, M., Gounaris, K., and Barber, J. (1987). *Biochim. Biophys. Acta* **891**, 128–139.

Quinn, P. J., and Williams, W. P. (1983). *Biochim. Biophys. Acta* **737**, 223–266.

Radunz, A. (1979). *Z. Naturforsch. C: Biosci.* **34C**, 1199–1204.

Radunz, A. (1980). *Z. Naturforsch., C: Biosci.* **35C**, 1024–1031.

Rasmussen, O. F., Brookjans, G., Stunmann, B. M., and Henningsen, K. W. (1984). *Plant Mol. Biol.* **3**, 191–199.

Rawyler, A., and Siegenthaler, P. A. (1980). *Eur. J. Biochem.* **110**, 179–187.

Rawyler, A., and Siegenthaler, P. A. (1981a). *Biochim. Biophys. Acta* **635**, 348–358.

Rawyler, A., and Siegenthaler, P. A. (1981b). *Biochim. Biophys. Acta* **638**, 30–39.

Rawyler, A., and Siegenthaler, P. A. (1985). *Biochim. Biophys. Acta* **815**, 287–298.

Rawyler, A., Unitt, M. D., Giroud, C., Davies, H., Mayor, J.-P., Harwood, J. L., and Siegenthaler, P.-A. (1987). *Photosynth. Res.* **11**, 3–13.

Remy, R., Trémolières, A., Duval, J. C., Ambard-Bretteville, F., and Dubacq, J. P. (1982). *FEBS Lett.* **137**, 271–275.

Remy, R., Ambard-Bretteville, F., and Dubertret, G. (1985). *FEBS Lett.* **188**, 43–47.

Renger, G., and Govindjee (1985). *Photosynth. Res.* **6**, 33–35.

Rich, P. (1984). *Biochim. Biophys. Acta* **768**, 53–79.

Rivas, E., and Luzzati, V. C. (1969). *J. Mol. Biol.* **41**, 261–275.

Rochaix, J.-D., Dron, M., Rahire, M., and Malnoe, P. (1984). *Plant Mol. Biol.* **3**, 363–370.

Rousselet, A., and Wollman, F.-A. (1986). *Arch. Biochem. Biophys.* **246**, 321–331.

Rubin, B. T., and Barber, J. (1980). *Biochim. Biophys. Acta* **592**, 87–102.

Rubin, B. T., Barber, J., Paillotin, G., Chow, W. S., and Yamamoto, Y. (1981a). *Biochim. Biophys. Acta* **638**, 69–76.

Rubin, B. T., Chow, W. S., and Barber, J. (1981b). *Biochim. Biophys. Acta* **634**, 174–190.

Rutherford, A. W. (1985). *Biochem. Soc. Trans.* **14**, 15–17.

Rutherford, A. W., and Heathcote, P. (1985). *Photosynth. Res.* **6**, 295–316.

Sandusky, P. O., DeRoo, C. L. S., Hicks, D. B., Yocum, C. F., Ghanotakis, D. F., and Babcock, G. T. (1983). *In* "The Oxygen Evolving System of Photosynthesis" (Y. Inoue, A. R. Crofts, Govindjee, N. Murata, G. Renger, and K. Satoh, eds.), pp. 123–133. Academic Press, New York.

Sane, P. V., Goodchild, D. J., and Park, R. B. (1970). *Biochim. Biophys. Acta* **216**, 162–178.

Satoh, K. (1979). *Biochim. Biophys. Acta* **546**, 84–92.

Satoh, K., Nakatani, H. Y., Steinback, K. E., Watson, J., and Arntzen, C. J. (1983). *Biochim. Biophys. Acta* **724**, 142–150.

Schlodder, E., Graber, P., and Witt, H. T. (1982). *Top. Photosynth.* **4**, 105–176.

Schlodder, E., Brettel, K., and Witt, H. T. (1985). *Biochim. Biophys. Acta* **808**, 123–131.

Scoufflaire, C., Lannoye, R., and Barber, J. (1983). *Photobiochem. Photobiophys.* **4**, 249–256.

Sculley, M. J., Duniec, J. T., Thorne, S. W., Chow, W. S., and Boardman, N. K. (1980). *Arch. Biochem. Biophys.* **201**, 339–346.

Sen, A., Williams, W. P., Brain, A. P. R., Dickins, M. J., and Quinn, P. J. (1981). *Nature (London)* **293**, 488–490.

Sen, A., Mannock, D. A., Collings, D. J., Quinn, P. J., and Williams, W. P. (1983). *Proc. R. Soc. London, Ser. B* **218**, 349–364.

Shinitzky, M., and Barenholz, Y. (1978). *Biochim. Biophys. Acta* **515**, 367–394.

Shinitzky, M., and Inbar, M. (1976). *Biochim. Biophys. Acta* **433**, 133–149.

Shinitzky, M., and Yuli, I. (1982). *Chem. Phys. Lipids* **30**, 261–282.

Shinozaki, K., Deno, H., Kato, A., and Sugiura, M. (1983). *Gene* **24**, 137–155.

Shinozaki, K., Deno, H., Wakasugi, T., and Sugiura, M. (1986). *Curr. Genet.* **10**, 421–423.

Shipley, G. G., Green, J. P., and Nichols, B. W. (1973). *Biochim. Biophys. Acta* **311**, 531–544.

Shuvalov, V. A., Nuijs, A. M., van Gorkom, H. J., Smit, H. W. J., and Duysens, L. N. M. (1986). *Biochim. Biophys. Acta* **850**, 319–323.

Siefermann-Harms, D., Ross, J. W., Kaneshiro, K. H., and Yamamoto, H. Y. (1982). *FEBS Lett.* **149**, 191–196.

Siegenthaler, P. A., and Giroud, C. (1986). *FEBS Lett.* **201**, 215–220.

Siegenthaler, P. A., and Rawyler, A. (1986). *In* "Encyclopedia of Plant Physiology, New Series" (L. A. Staehelin and C. J. Arntzen, eds.), Vol. 19, pp. 693–705. Springer-Verlag, Berlin and New York.

Singer, S. L., and Nicolson, G. L. (1972). *Science* **175**, 720–731.

Sprague, S. G., and Staehelin, L. A. (1984). *Biochim. Biophys. Acta* **777**, 306–322.

Sprague, S. G., Camm, E. L., Green, B. R., and Staehelin, L. A. (1985). *J. Cell Biol.* **100**, 552–557.

Staehelin, L. A. (1976). *J. Cell Biol.* **71**, 136–158.

Staehelin, L. A. (1986). *In* "Encyclopedia of Plant Physiology, New Series" (L. A. Staehelin and C. J. Arntzen, eds.), Vol. 19, pp. 1–84. Springer-Verlag, Berlin and New York.

Staehelin, L. A., and Arntzen, C. J. (1983). *J. Cell Biol.* **97**, 1327–1337.

Strotmann, H. (1986). *In* "In Encyclopedia of Plant Physiology, New Series" (L. A. Staehelin and C. J. Arntzen, eds.), Vol. 19, pp. 584–594. Springer-Verlag, Berlin and New York.

Stubbs, C. D., and Smith, A. D. (1984). *Biochim. Biophys. Acta* **779**, 80–137.

Sundby, C., and Larsson, C. (1985). *Biochim. Biophys. Acta* **813**, 61–67.

Tang, X.-S., and Satoh, K. (1984). *Plant Cell Physiol.* **25**, 935–945.

Tang, X.-S., and Satoh, K. (1986). *FEBS Lett.* **201**, 221–224.

Telfer, A., and Barber, J. (1981). *FEBS Lett.* **129**, 161–165.

Telfer, A., Allen, J. F., Barber, J., and Bennett, J. (1983a). *Biochim. Biophys. Acta* **722**, 176–181.

Telfer, A., Hodges, M., and Barber, J. (1983b). *Biochim. Biophys. Acta* **724**, 167–175.

Telfer, A., Bottin, H., Barber, J., and Mathis, P. (1984a). *Biochim. Biophys. Acta* **764**, 324–330.

Telfer, A., Hodges, M., Millner, P. A., and Barber, J. (1984b). *Biochim. Biophys. Acta* **766**, 554–562.

Telfer, A., Whitelegge, J., Bottin, H., and Barber, J. (1986). *J. Chem. Soc., Faraday Trans. 2* (*Spec. Ed.*) **82**, 2207–2215.

Thomas, D. (1986). *In* "Techniques for the Analysis of Membrane Proteins" (C. I. Ragan and R. J. Cherry, eds.), pp. 377-431. Chapman & Hall, London.

Thomson, W. W. (1974). *In* "Dynamic Aspects of Plant Development" (A. W. Robards, ed.), pp. 138–177. McGraw-Hill, New York.

Thornber, J. P. (1975). *Annu. Rev. Plant Physiol.* **26**, 127–158.

Thornber, J. P. (1986). *In* "Encyclopedia of Plant Physiology, New Series" (L. A. Staehelin and C. J. Arntzen, eds.), Vol. 19, pp. 98–142. Springer-Verlag, Berlin and New York.

Thorne, S. W., and Duniec, J. T. (1983). *Q. Rev. Biophys.* **16**, 197–278.

Thurnauer, M. C., and Gast, P. (1985). *Photobiochem. Photobiophys.* **9**, 29–38.

Tobin, E. M., Wimpee, C. F., Silverthorne, J., Stiekema, W. J., Neumann, G. A., and Thornber, J. P. (1984). *In* "Biosynthesis of the Photosynthetic Apparatus: Molecular Biology, Development and Regulation". (J. P. Thornber, L. A. Staehelin, and R. B. Hallick, eds.), pp. 325–334. Alan R. Liss, Inc., New York.

Toyoshima, Y., and Fakutaka, E. (1982). *FEBS Lett.* **150**, 223–227.

Trebst, A. (1986). *Z. Naturforsch., C. Biosci.* **41C**, 240–245.

Trémolières, A., Dubacq, J. P., Ambard-Bretteville, F., and Remy, R. (1981). *FEBS Lett.* **130**, 27–31.

Tuquet, C., Guillot-Salomon, T., de Lubac, M., and Signol, M. (1977). *Plant Sci. Lett.* **8**, 59–64.

Unitt, M. D., and Harwood, J. L. (1982). In "Biochemistry and Metabolism of Plant Lipids" (J. F. G. Wintermans and P. J. C. Kuiper, eds.), pp. 359–362. Elsevier, Amsterdam.

Unitt, M. D., and Harwood, J. L. (1985). *Biochem. J.* **228**, 707–711.

Vallejos, R. H., Ceccarelli, E. A., and Chan, R. L. (1984). *J. Biol. Chem.* **259**, 8048–8051.

Van Gorkom, H. J. (1985). *Photosynth. Res.* **6**, 97–112.

Vernon, L. P., Shawf, E. R., and Ke, B. (1966). *J. Biol. Chem.* **241**, 4101–4109.

Wagner, R., and Junge, W. (1980). *FEBS Lett.* **114**, 327–333.

Wagner, R., Carrillo, N., Junge, W., and Vallejos, R. H. (1981). *FEBS Lett.* **131**, 335–340.

Wagner, R., Carrillo, N., Junge, W., and Vallejos, R. H. (1982). *Biochim. Biophys. Acta* **680**, 317–330.

Walker, J. E., Saraste, M., and Gay, W. J. (1984). *Biochim. Biophys. Acta* **768**, 164–200.

Wang, A. Y. I., and Packer, L. (1973). *Biochim. Biophys. Acta* **305**, 488–492.

Wasielewski, M. R., Norris, J. R., Shipman, L. L., Lin, C. P., and Svec, W. A. (1981). *Proc. Natl. Acad. Sci. U.S.A.* **78**, 2957–2961.

Westhoff, P., Alt, J., Widger, W. R., Cramer, W. A., and Hermann, R. G. (1985a). *Plant Mol. Biol.* **4**, 103–110.

Westhoff, P., Jansson, C., Klein-Hitpass, L., Berzborn, R. J., Larsson, C., and Bartlett, S. (1985b). *Plant Mol. Biol.* **4**, 137–144.

Whitmarsh, J. (1986). In "Encyclopedia of Plant Physiology, New Series" (L. A. Staehelin and C. J. Arntzen, eds.), Vol. 19, pp. 508–527. Springer-Verlag, Berlin and New York.

Whitmarsh, J., and Ort, D. R. (1984). *Arch. Biochem. Biophys.* **231**, 378–389.

Widger, W. R., Cramer, W. A., Herrmann, R. G., and Trebst, A. (1984). *Proc. Natl. Acad. Sci. U.S.A.* **81**, 674–678.

Willey, D. L., Auffret, A. D., and Gray, J. C. (1984a). *Cell (Cambridge, Mass.)* **36**, 555–562.

Willey, D. L., Howe, C. J., Auffret, A. D., Bowman, C. M., Dyer, T. A., and Gray, J. C. (1984b). *Mol. Gen. Genet.* **194**, 416–422.

Williams, J. C., Steiner, L. A., Feher, G., and Simon, M. I. (1984). *Proc. Natl. Acad. Sci. U.S.A.* **81**, 7303–7307.

Williams, W. P., Gounaris, K., and Quinn, P. J. (1984). In "Advances in Photosynthesis Research" (C. Sybesma, ed.), Vol. 3, pp. 123–130. Martinus Nijhoff/Dr. W. Junk, The Hague, The Netherlands.

Witt, H. T. (1971). *Q. Rev. Biophys.* **4**, 365–477.

Yamagishi, A., and Katoh, S. (1984). *Biochim. Biophys. Acta* **765**, 118–124.

Yamamoto, Y., Doi, M., Tamura, N., and Nishimura, M. (1981a). *FEBS Lett.* **133**, 265–268.

Yamamoto, Y., Ford, R. C., and Barber, J. (1981b). *Plant Physiol.* **67**, 1069–1072.

Yamamoto, Y., Shinkai, H., Isogai, Y., Matsuura, K., and Nishimura, M. (1984). *FEBS Lett.* **175**, 429–432.

Youvan, D. C., Bylina, E. J., Alberti, M., Bugusch, H., and Hearst, J. E. (1984). *Cell (Cambridge, Mass.)* **37**, 949–957.

Zuber, H. (1987). *Top. Photosynth.* **8**, 197–259.

Zurawski, G., and Clegg, M. T. (1984). *Nucleic Acids Res.* **12**, 2549–2559.

Zurawski, G., Bohnert, H. J., Whitfeld, P. R., and Bottomley, W. (1982). *Proc. Natl. Acad. Sci. U.S.A.* **79**, 7699–703.

Rubisco: Structure, Mechanisms, and Prospects for Improvement

3

T. JOHN ANDREWS
GEORGE H. LORIMER

I. Introduction
II. The Catalytic Bifunctionality of Rubisco: A Fixed Constraint in the Architecture of Life?
   A. An Unexpected Oxygenation
   B. A Unique Catalyst
   C. Biospheric Role
   D. A Fixed Constraint?
   E. The Costs of Oxygenation
III. Structure and Synthesis
   A. Quaternary Structure
   B. Primary Structure
   C. Structure Determination by Crystallography
   D. Synthesis and Assembly
IV. Mechanisms of Carbamylation and Catalysis
   A. Carbamylation of the Activator Lysine by $^ACO_2$ and Me^{2+}
   B. Binding of RuBP, 2CABP, and Other Effector Ligands
   C. Order of Addition of Substrates
   D. Reaction Intermediates and Their Properties
   E. Stereochemistry
   F. Partial Reactions
   G. Role of the Divalent Metal Ion in Carboxylation
   H. Mechanism of Oxygenation
   I. Relative Specificity
   J. Identification of Groups within the Catalytic Site
   K. Site-Directed Mutagenesis
V. Mechanisms for Modulating Activity
   A. Does Rubisco Need to Be Regulated?
   B. Possible Regulatory Mechanisms

 VI. Subunit Interactions
     A. Why Two Subunits?
     B. Reversible Dissociation of Small Subunits from the Large-Subunit Octamer
     C. Capabilities of the Large-Subunit Octamer
     D. Hybridization of Rubisco Subunits Derived from Different Species
     E. Possible Roles for the Small Subunit
 VII. Evolution of Rubisco
     A. A Single Phylogeny
     B. Adaptive Responses to the Declining CO_2/O_2 Ratio
VIII. Prospects for a Better Rubisco
     A. Is a Better Rubisco Possible?
     B. How Might a Better Rubisco Be Achieved?
     References

I. INTRODUCTION*

Much of the current diverse interest in D-ribulose 1,5-bisphosphate carboxylase–oxygenase (Rubisco) can be attributed to its pivotal function in
photosynthesis and photorespiration. It is clearly a rate-limiting factor in
both processes. Yet, curiously, it appears to be a grossly inefficient catalyst.
How can it be that an enzyme that must have been subject to the most
intense kind of selection for catalytic specificity and turnover rate is unable

* Abbreviations: $^ACO_2$, activator CO_2; $^SCO_2$, substrate CO_2; Me^{2+}, generic divalent metal
ion; RuBP, D-ribulose 1,5-bisphosphate; XuBP, D-xylulose 1,5-bisphosphate; u-3-P-glycerate,
upper 3-phospho-D-glycerate (the product formed from C-1 and C-2 of RuBP and $^SCO_2$);
l-3-P-glycerate, lower 3-phospho-D-glycerate (the product formed from C-3, C-4, and C-5 of
RuBP); 2CABP, 2'-carboxy-D-arabinitol 1,5-bisphosphate; 2CRBP, 2'-carboxy-D-ribitol 1,5-
bisphosphate; 2CXBP, 2'-carboxy-D-xylitol 1,5-bisphosphate; 4CABP, 4'-carboxy-D-arabinitol
1,5-bisphosphate (equivalent to 2'-carboxy-D-lyxitol 1,5-bisphosphate); 3-keto-2CABP, 2'-
carboxy-3-keto-D-arabinitol 1,5-bisphosphate; 3-keto-2CRBP, 2'-carboxy-3-keto-D-ribitol 1,5-
bisphosphate; 2PPBP, 2-peroxy-D-pentitol 1,5-bisphosphate; 3-keto-2PABP, 2-peroxy-3-keto-
D-arabinitol 1,5-bisphosphate; Rubisco, D-ribulose 1,5-bisphosphate carboxylase–oxygenase;
Pi, inorganic phosphate.
     Amino acid nomenclature: The single-letter code will be used. When referring to a specific
residue in the *Rhodospirillum rubrum* Rubisco, the single letter specifying the amino acid will be
followed by the superscript L2 and the amino acid number. For the hexadecameric enzyme, the
superscripts L8 and S8 will be used to designate the large and small subunits, respectively.
Thus, the activator lysine of the *R. rubrum* Rubisco is designated $K^{L2}166$, while the corresponding residue of the spinach enzyme will be designated $K^{L8}201$. The N-terminal methionine of the
mature small subunit is thus $M^{S8}1$.
     Carbamylation state: The terms deactivated and activated, previously used to refer, respec-
tively, to the catalytically incompetent enzyme lacking the carbamate–Me^{2+} complex and the
catalytically competent enzyme containing the carbamate–Me^{2+} complex, create confusion.
Not all "activated" forms of the enzyme are catalytically competent. The quaternary complex
of enzyme–$^ACO_2$–Me^{2+}-2CABP is an example. We shall therefore refer to decarbamylated
enzyme as that lacking the carbamate–Me^{2+} complex and carbamylated enzyme as that containing the carbamate–Me^{2+} complex.
     Kinetic constants: K_c, Michaelis constant for CO_2 in carboxylation (M); V_c, maximal rate of
carboxylation (sec^{-1}); v_c, actual rate of carboxylation (sec^{-1}); K_o, Michaelis constant for O_2 in
oxygenation (M); V_o, maximal rate of oxygenation (sec^{-1}); v_o, actual rate of oxygenation (s^{-1});
K_{Rc}, Michaelis constant for RuBP in carboxylation (M); K_{Ro}, Michaelis constant for RuBP in
oxygenation (M); K_{iR}, kinetically determined inhibition constant for RuBP (M) (Cleland, 1963);
C, [CO_2] (M); O, [O_2] (M); R, [RuBP] (M); $s_{rel} = [V_c/K_c]/[V_o/K_o]$.

to distinguish between the substrate of photosynthesis, CO_2, and its product, O_2, and is such a slow catalyst that photosynthetic cells must invest one quarter or more of their precious nitrogen budget in this one enzyme? The dubious distinction of being the world's most abundant protein is simply a consequence of its catalytic ineffectiveness. So far, little has emerged to satisfy this curiosity and, in the absence of exact knowledge of the details of the catalytic mechanism responsible for this apparently slow evolutionary progress, the prospect of large increases in photosynthetic productivity enabled by an engineered improvement of Rubisco remains a tantalizing possibility.

Equally obscure are the reasons for the complex subunit structure of Rubisco. While single-subunit Rubiscos are known, these appear to be restricted to the purple, nonsulfur bacteria. All others so far studied appear to be complex hexadecamers composed of eight copies each of a 50- to 55-kDa large subunit and a 12- to 18-kDa small subunit. The catalytic sites are present on the large subunits, but do not express catalytic activity unless the small subunits are also bound. The manner of the small subunit's involvement in the mechanism is particularly unclear. Further interest is added by the complex mechanisms for synthesis and assembly of the subunits in eukaryotes. One is left with the feeling that the role of the small subunit in Rubisco's mechanism must be of fundamental importance and that any attempts at improvement must take account of both subunits and the interactions between them.

Rubisco's central role in photosynthesis and photorespiration makes it a likely candidate for regulation, though whether it is more or less regulated than other photosynthetic enzymes remains to be seen. Rubisco's activity *in vivo* certainly seems to be tightly controlled, very probably by a multiplicity of mechanisms. These have been extensively investigated in recent years.

In this chapter, we summarize recent advances in understanding of Rubisco, its mechanisms of catalysis and regulation, the synthesis and assembly of its subunits, and the role of interactions between them. We pay particular attention to those aspects that seem most in need of "improvement," that is, the relative specificity for CO_2 as opposed to O_2 and the rate of catalytic throughput at limiting CO_2.

II. THE CATALYTIC BIFUNCTIONALITY OF RUBISCO: A FIXED CONSTRAINT IN THE ARCHITECTURE OF LIFE?

A. An Unexpected Oxygenation

Nearly two decades separated the discoveries of Rubisco's carboxylase (Quayle *et al.*, 1954; Weissbach *et al.*, 1954) and oxygenase functions (Bowes *et al.*, 1971; Andrews *et al.*, 1973; Lorimer *et al.*, 1973) (Fig. 1). This may reflect a tendency to discover the expectable. Characterization of the

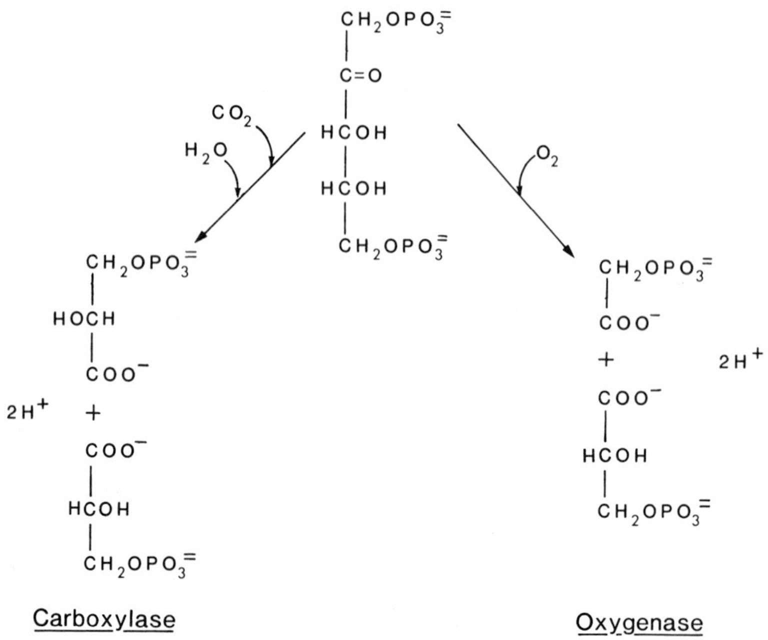

$$
\begin{array}{c}
CH_2OPO_3^= \\
| \\
C=O \\
| \\
HCOH \\
| \\
HCOH \\
| \\
CH_2OPO_3^=
\end{array}
$$

Carboxylase (left, with CO_2, H_2O):

$$
\begin{array}{c}
CH_2OPO_3^= \\
| \\
HOCH \\
| \\
COO^- \\[4pt]
2H^+ \;\; + \\[4pt]
COO^- \\
| \\
HCOH \\
| \\
CH_2OPO_3^=
\end{array}
$$

Oxygenase (right, with O_2):

$$
\begin{array}{c}
CH_2OPO_3^= \\
| \\
COO^- \\[4pt]
+ \qquad 2H^+ \\[4pt]
COO^- \\
| \\
HCOH \\
| \\
CH_2OPO_3^=
\end{array}
$$

<u>Carboxylase</u>   <u>Oxygenase</u>

Fig. 1.   The carboxylation and oxygenation of RuBP.

enzyme that catalyzed the carboxylation of RuBP to produce two molecules of phosphoglycerate fulfilled expectations raised by observations of the $^{14}CO_2$-labeling patterns produced by C_3 photosynthesis (Calvin and Benson, 1948; Benson et al., 1950; Bassham et al., 1950). However, no such rational expectations predicted the oxygenation of RuBP. Indeed, all of the reactions that constitute the photorespiratory glycolate pathway (Fig. 2) were pieced together, and their subcellular locations determined (Tolbert, 1971), before the RuBP oxygenase reaction that produced its substrate, phosphoglycolate (together with another molecule of phosphoglycerate), came to light (Bowes et al., 1971; Andrews et al., 1973; Lorimer et al., 1973).

Even now, more than a decade later, we are still puzzled by the oxygenase reaction. No essential function for its attendant photorespiration has emerged. Indeed, the only function the glycolate pathway seems to serve is to salvage three-quarters of the carbon diverted from photosynthesis by RuBP oxygenase as phosphoglycolate. In so doing, it consumes energy in the form of ATP and reducing equivalents (Fig. 2). Such energy consumption may be advantageous in some circumstances. For example, it may dissipate excess photosynthetic reductant under photoinhibitory conditions associated with CO_2 limitation (Osmond, 1981). But it is clear that RuBP oxygenation cannot have arisen in response to selection pressure for an energy-wasting mechanism, because oxygenation is a property of all Rubiscos, even those from anaerobic organisms.

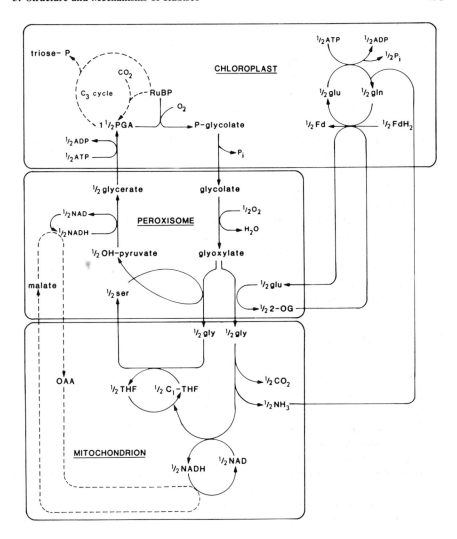

Fig. 2. The stoichiometry and energetics of the photorespiratory pathways. Net Result: RuBP + $1\frac{1}{2}O_2$ + ATP + $\frac{1}{2}FdH_2 \rightarrow 1\frac{1}{2}PGA$ + $\frac{1}{2}CO_2$ + ADP + $1\frac{1}{2}P_i$ + $\frac{1}{2}Fd$ + H_2O [Adapted from Somerville and Ogren (1982).]

B. A Unique Catalyst

Rubisco stands at the interface between the inorganic and organic phases of the biosphere's carbon cycle, catalyzing the only reaction by which atmospheric CO_2 may be acquired by living organisms. There are many other carboxylases, but the reactions they catalyze do not result in the net acquisition of carbon for gluconeogenic purposes. For all carboxylases except Rubisco, the carbon they fix is only temporarily resident in the organism's metabolism, being obligatorily released by subsequent decarboxylation reactions. The only other exceptions known are the reductive carboxylations of the coenzyme A derivatives of acetate and succinate. These may enable net carbon acquisition in certain green sulfur bacteria via a reversed citric acid cycle (Evans *et al.*, 1966; Takabe and Akazawa, 1977; but see McFadden, 1973, for a contrary opinion). However, if this pathway functions, it does so only in anaerobiosis and is quantitatively unimportant to the biosphere.

While the enediol chemistry employed by Rubisco (Section IV,D) is reminiscent of other enzymatic mechanisms, for example, triose phosphate isomerase (Fersht, 1985), Rubisco's use of this chemistry for carboxylation and oxygenation is unique. The oxygenase mechanism is particularly unusual. Other oxygenases require a cofactor, such as a transition metal like copper or iron, or a flavin, to assist in overcoming the problem of spin inversion which is associated with the reaction of triplet O_2 (Section IV,H). Rubisco lacks such cofactors and appears to use analogous mechanisms for both carboxylation and oxygenation involving, in both cases, electrophilic attack of the gaseous substrates on the enediol.

C. Bisopheric Role

The fundamental importance of Rubisco to life on this planet can hardly be overstated. Its role in acquiring the carbon substrate of life from the sparse atmospheric resource of CO_2 is readily apparent. Moreover, consideration of the balance between carboxylation and oxygenation (i.e., between photosynthesis and photorespiration) leads to contemplation of the delicate fulcrum on which life is balanced.

Central to these considerations is the parameter that describes the balance between Rubisco's competing (and opposing) activities. This is the relative specificity for CO_2 as opposed to O_2, which we will call s_{rel}. Because CO_2 and O_2 are linear competitive inhibitors with respect to each other (Badger and Andrews, 1974), it may be shown (Laing *et al.*, 1974) that

$$\frac{v_c}{v_o} = \frac{V_c/K_c}{V_o/K_o} \cdot \frac{[CO_2]}{[O_2]}$$

Therefore, V_cK_o/K_cV_o, the ratio between the V_{max}/K_m ratios for the two activities, is the proportionality constant that specifies the ratio of carboxylation to oxygenation that occurs at a particular $[CO_2]/[O_2]$ ratio. This we define as s_{rel}. Alternatively, s_{rel} may be thought of as the ratio between the carboxylase and oxygenase rates when CO_2 and O_2 are at equal concentrations. For higher-plant Rubiscos, s_{rel} values cluster around 100, while, at the other end of the scale, single-subunit Rubiscos have values around 15. Cyanobacterial and algal Rubiscos have intermediate s_{rel} values (Jordan and Ogren, 1981, 1983; Andrews and Lorimer, 1985) (Fig. 3).

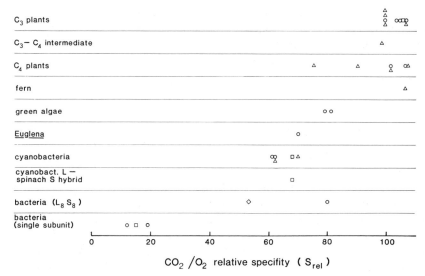

Fig. 3. The relative specificity (s_{rel}) scale; $s_{rel} = [V_c/K_c]/[V_o/K_o]$. Data were taken from: $\bigcirc$, Jordan and Ogren (1981); $\triangle$, Jordan and Ogren (1983); $\square$, Andrews and Lorimer (1985); $\diamond$, Jordan and Chollet (1985). The s_{rel} values from different laboratories vary according to the pK' assumed for the $CO_2 \leftrightarrow HCO_3^-$ equilibrium. This pK' decreases linearly with the logarithm of ionic strength and thus becomes relatively insensitive to ionic strength above ionic strength 0.1. Ionic strengths of approximately 0.1 and above are usual in most carboxylase assay buffers. Therefore, we have recalculated published s_{rel} values in accordance with a pK' of 6.12, which is the value appropriate to an ionic strength of 0.1 at 25°C (Yokota and Kitaoka, 1985).

O_2 is much more abundant in the biosphere than CO_2. The O_2/CO_2 concentration ratio in air-saturated, aqueous solution at 25°C is approximately 25. At the site of Rubisco during active photosynthesis in C_3 leaves, the ratio is considerably greater because of stomatal and other diffusive resistances to CO_2 transport and, to a lesser degree, because of photosynthetic O_2 production in close proximity. If we assume a ratio of 40, we see that the ratio of oxygenation to carboxylation for a typical C_3 leaf will be 0.4. This means that the rate of photorespiratory CO_2 release will be 20% of the gross rate of CO_2 fixation because two molecules of RuBP must be oxygenated to produce one molecule of photorespiratory CO_2. This rate approximates to that observed in natural conditions (i.e., 0.03% CO_2, 21% O_2).

Consider the consequences of a Rubisco with a lower s_{rel} value. Let us take, for example, the single-subunit Rubisco from a rhodopseudomonad that has an s_{rel} value of 12 (Jordan and Ogren, 1981). Under the same conditions considered above, this s_{rel} translates to a rate of oxygenation 3.3-fold faster than that of carboxylation or a rate of photorespiratory CO_2 release 1.7 times faster than gross CO_2 fixation. Such a negative carbon balance is, of course, inconsistent with life under present-day atmospheric conditions, but such a nondiscriminatory Rubisco satisfies this particular organism's anaerobic lifestyle. We can calculate that life in present-day atmospheric conditions, in the absence of ancillary mechanisms to alter the CO_2 or O_2 concentrations, requires a minimum s_{rel} value of approximately 12.5. Of course, diffusional limitations to gaseous transport mean that a much higher s_{rel} value is required in practice.

D. A Fixed Constraint?

Since RuBP oxygenation and photorespiration consume energy for no apparent useful purpose, the continued existence of this wasteful phenomenon is a curious riddle. Why has not RuBP oxygenation simply been selected out of existence? We still do not have a satisfying answer to this question. A decade and a half ago, we suggested that oxygenation of RuBP was unavoidable because the chemistry of carboxylation obligatorily involved an intermediate which was susceptible to attack by O_2 (Lorimer and Andrews, 1973; Andrews and Lorimer, 1978). This would explain the lack of an obvious function for the oxygenase activity and its ubiquitous expression by all Rubiscos. It is now clear that the 2,3-enediol of RuBP is that O_2-sensitive intermediate (Section IV,D). Somerville *et al.* (1984) pointed out that this "inevitability" hypothesis was, if true, a clear example of what Gould and Lewontin (1979) have described as an "evolutionary spandrel." Such phenomena are "inherent architectural constraints in the structure of life" that are not the products of adaptation, nor do they depend on selective pressure for their maintenance.

We feel that the inevitability hypothesis still provides the most convincing explanation for the existence of RuBP oxygenase and photorespiration. However, one important caveat has arisen during the intervening years since its original proposition. The "architectural constraint" is not totally fixed. Selective pressure clearly is able to modify the ratio of oxygenation to carboxylation over the range mentioned in the preceding section (Fig. 3), though not able, at least in the time that has elapsed to date, to eliminate oxygenation completely. Indeed, the variation in s_{rel} values displayed by Rubiscos so far studied (Fig. 3), while enormously significant at the physiological level, is very limited when viewed in the context of the specificity of many other enzymes. Nature has, in other circumstances, been able to find the means for extraordinary selectivity. To take an isolated example, tyrosyl-tRNA synthetase, a precision tool on the assembly line of proteins, has a relative specificity for tyrosine, as compared to phenylalanine, in excess of 10^5 (Fersht *et al.*, 1985). Here the only difference between the two substrates is a hydroxyl group, and even then, it is the absence of the group that must be recognized. Therefore, while any oxygenation of RuBP remains at all, we will continue to regard it as inevitable.

E. The Costs of Oxygenation

The physiological costs of this inherent inefficiency in the acquisition of carbon by photosynthetic organisms are severe. However, they are not directly related to the CO_2 that is lost in photorespiration. Although CO_2 is a sparse resource, it is nevertheless virtually infinite as far as an individual plant is concerned. The major costs are threefold, affecting the efficiencies of use of light energy, water, and nutrient N.

The energy costs of RuBP oxygenation are incurred in the recovery of three-quarters of the carbon lost from the photosynthetic carbon reduction cycle in the form of phosphoglycolate. The reductant and ATP expended in this way (Fig. 2) adversely affects the light (quantum) use efficiency of photosynthesis. Some idea of the size of this effect may be gauged from the beneficial effects of high CO_2 or low O_2 concentrations. Both conditions nearly double the quantum efficiency at 25–30°C, raising it, in red light, to levels that approach the theoretical limit of eight quanta per CO_2, fixed (Ehleringer and Bjorkman, 1976; Bjorkman and Demmig, 1987).

The oxygen inhibition of the carboxylase reaction, together with its inherently feeble catalytic power, limits a plant's water-use efficiency. If the V_c/K_c ratio of the carboxylase was greater, adequate CO_2-fixation rates could be maintained at lower intercellular CO_2 concentrations. This would allow smaller stomatal apertures and, therefore, lower rates of water loss in a plant well-coordinated enough to take advantage of an improved carboxylase. The inherently better water-use efficiency of C_4 plants (Fischer and Turner, 1978), which are able to sustain very low intercellular CO_2 tensions by virtue of their C_4 cycle, illustrates this point.

The amount of nutrient N sequestered in the world's most abundant protein is vast both on a global scale (10 kg/person) and in terms of the N budget of photosynthetic cells (>50% of the soluble protein) (Ellis, 1979). A more catalytically effective Rubisco would improve the N-use efficiency of plants by permitting a smaller investment of N in this component.

III. STRUCTURE AND SYNTHESIS

A. Quaternary Structure

We consider that the available evidence supports the view that there exist in nature only two types of subunit structures for Rubisco. The most common, which occurs in all eukaryotes and the majority of prokaryotes also, is a hexadecameric structure composed of eight copies each of a 50- to 55-kDa large subunit and a 12- to 18-kDa small subunit. The other so far appears to be restricted to the purple, nonsulfur, photosynthetic bacteria. It is an oligomer of large subunits only. The Rubisco from *Rhodospirillum rubrum* is a dimer (Tabita and McFadden, 1974), while in rhodopseudomonads, a higher oligomer, perhaps a hexamer, is favored (Gibson and Tabita, 1977a,c; Shively *et al.*, 1984; H. Sani and C. S. Dow, unpublished, 1984, cited in Ellis and Gatenby, 1984).

There have been many reports of Rubisco subunit structures other than the L8S8 and L2-6 structures mentioned above. L8 and L4 forms lacking small subunits have been reported from cyanobacteria and thiobacilli (Tabita *et al.*, 1974, 1976; Purohit *et al.*, 1976; Codd *et al.*, 1979), but these observations have since proven erroneous (Takabe, 1977; Bowman and Chollet, 1980; Andrews and Abel, 1981; Asami *et al.*, 1983). L6S6 Rubiscos have also been reported from some bacterial sources (Lawlis *et al.*, 1979; Taylor *et al.*, 1980; Taylor and Dow, 1980), but usually these reports were based on estimates of molecular size determined by pore penetration or other hydrodynamic methods. The latter are unreliable with Rubisco because considerable variations in molecular size have been observed between known hexadecameric Rubiscos (Andrews *et al.*, 1981) and even for the same Rubisco under different conditions (Bowien and Gottschalk, 1982; Johal *et al.*, 1985). Apparently, L8S8 Rubiscos are able to undergo changes in conformation in solution that are gross enough to affect the overall size and hydrodynamic behavior of the molecule. The establishment of the subunit stoichiometry of an unknown Rubisco is, therefore, not a simple task.

The three-dimensional arrangement of the subunits in L8S8 Rubiscos has been studied by electron microscopy and x-ray crystallography. All studies agree that the eight large subunits are arranged as two layers of four subunits each in a square array (Baker *et al.*, 1975, 1977a,b; Bowien *et al.*, 1976, 1980; Andrews *et al.*, 1981) (Fig. 4). Electron-microscopic examination of negatively stained molecules showed that the two layers are eclipsed and that there is an obvious channel along the fourfold axis of symmetry of the two layers. There is less information about the positioning of the small subunits. These are too small to be seen in electron micrographs, but images of the complex between *Alcaligenes eutrophus* Rubisco and antibodies to its small subunits showed no antibodies bound to the periphery of the core when viewed down the fourfold axis. Rather, the small-subunit antibodies were bound to the outer layers of the molecule when viewed in side-on projection. Thus it was concluded that four small subunits were bound to each of two faces of the cubic, large-subunit core, that is, the upper and lower faces through which the fourfold axis passes (Bowien and Mayer, 1978). Electron-microscopic images showing views down the fourfold axis reveal pronounced projections (perhaps eight of them) around the periphery of the large-subunit core (Bowien *et al.*, 1976; Andrews *et al.*, 1981). These do not appear to be the small subunits for the reasons mentioned above, and they have led Bowien *et al.* (1976) to postulate that the large subunits are themselves V- or U-shaped. More detailed crystallographic studies presently in

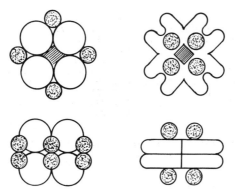

Fig. 4.   Alternative structures proposed for Rubisco. Left: Rubisco from tobacco (Baker *et al.*, 1975, 1977a,b). The positioning of the small subunits is speculative, being only one of several possibilities. Right: Rubisco from *A. eutrophus* (Bowien *et al.*, 1976, 1980). Top: Views down the fourfold axis of symmetry showing the central hole (hatched). Bottom: Side views perpendicular to the fourfold axis. [Reproduced with permission from Miziorko and Lorimer (1983).]

progress will, no doubt, lead to a much better picture of the arrangement of the subunits (Section III,C).

The quaternary structures of Rubiscos from rhodopseudomonads are particularly interesting in that two distinct forms are present. This has been demonstrated so far for *Rhodopseudomonas sphaeroides* (Gibson and Tabita, 1977a), *R. capsulata* (Gibson and Tabita, 1977c; Shively *et al.*, 1984), and *R. blastica* (H. Sani and C. S. Dow, unpublished, 1984, cited in Ellis and Gatenby, 1984). The form I enzyme is a hexadecamer containing both large and small subunits, while the form II enzyme has only large subunits. The large subunits of the two forms do not appear to be identical. They do not cross-react immunochemically (Gibson and Tabita, 1977a,b), they are slightly different in size (Shively *et al.*, 1984), and their peptides maps differ (Gibson and Tabita, 1985). Rather, the single-subunit form II seems to resemble the similar enzyme from *Rhodospirillum rubrum* both antigenically and by partial sequence analysis (Muller *et al.*, 1985). Furthermore, the *R. rubrum* Rubisco gene hybridizes with the gene from the form II enzyme, but not, apparently, with that for the form I enzyme (Muller *et al.*, 1985). Why do rhodopseudomonads need two different kinds of Rubisco? Little is known about their physiological functions. The form I enzyme is apparently only expressed in late logarithmic and stationary phases of cell growth and is more readily repressed by CO_2 than form II (Shively *et al.*, 1984; H. Sani and C. S. Dow, 1984, unpublished, cited in Ellis and Gatenby, 1984). While the s_{rel} value of form I is about sevenfold higher than that of form II (Fig. 3), this should not have much relevance for an anaerobic organism. Nor is the difference in K_c values very large (Jordan and Ogren, 1981). The solution of this mystery awaits further research.

T. John Andrews and George H. Lorimer

```
            10         20         30         40         50         60
1  MSPQTETKAS VEFKAGVKDY KLTYYTPEYE TLDTDILAAF RVSPQPGVPP EEAGAAVAAE
2  .......... .G.....E.. ........Q. .K........ ..T....... ..........
3  ........K. .G........ ......D.Q. .K........ ..T....... ..........
4  .......... .......... ......     .K..F....  ........   .........
5  .......... .G........ .......... .K........ ..T..L.... ..........
6  .V......G  AG........ R.....D.V  VR........ .MT..L.... ..C.......
7  ........TG AG...|.... R.....D.Q  VSE....... .MT......A| .C......
8  .-.K.QS-.A -GY......  ....D.T    PK...L.... P.......A  D....I...
9  MSYA..K.QTK SGY....Q.. R.....D.T PK........ ..T.....F  ..A......

            70         80         90        100        110        120
1  SSTGTWTTVW TDGLTNLDRY KGRCYHIEPV AGEENQYICY VAYPLDLFEE GSVTNMFTSI
2  .......... .....S.... .....R..R. V..KD...A. .......... ..........
3  .......... .....S.... .....E.... P..D..F.A. .......... ..........
4  .......... .          . ...                            .....
5  ..A....... .....S.... .......... P.DPD..... .......... .....
6  .......... .....S.... .....D.... P..D....A. ....I..... ..........
7  .......... .....Q.... ..|..DL... P..S....A. ....I..... ....LL...
8  .......... ..L..DM... ..K...... Q....S.FAF I......... .....IL...
9  .......... ..L..D.... ....D.... P..D..F.A. I......... ..I..VL...

           130        140        150        160        170        180
1  VGNVFGFKAL RALRLEDLRI PVAYVKTFQG PPHGIQVERD KLNKYGRPLL GCTIKPKLGL
2  .......... .......... ...P...... .......... .......... ..........
3  .......... .......... ...Y...... .......... .......... ..........
4  ..........
5  .......... .......... ...P..S... ...R.M.... .......... ..........
6  .......... .......... ...P.....V .......... .......... ...G..
7  .......... .......... ...P..S..W. .......... R......... ..........
8  .........I .S.....I.F ..L....... .......... L......... ....M.....
9  .......... .......... ...I.F     ....I..... .......... ..........

           190        200        210        220        230        240
1  SAKNYGRAVY ECLRGGLDFT KDDENVNSQP FMRWRDRFLF CAEALYKAQA ETGEIKGHYL
2  .......... .......... .......... .......... .......... ..........
3  .......... .......... .......... .......... ...I.S.... ..........
4  .......... .......... .......... ..I...... .......... ..........
5  ........C. .......... .......... .......V.. ...I.S.... ..........
6  .......... .......... .......... .......... V...I..... ....V.....
7  .......... .......... ....|....S .......... ...I...T  ....V.....
8  .......... .......... .....I.... .Q....... V.D.IH.S.. ..........
9  .......... .......... .....I..A. .Q....... V.D.IT....

           250        260        270        280        290        300
1  NATAGTCEDM MKRAVFAREL GVPIVMHDYL TGGFTANTTL SHYCRDNGLL LHIHRAMHAV
2  .......E.  I......... .......... .......... .....S.    A.........
3  .......E.  L......... .......... .......... .......... ..........
4  ..          . I.G......  ...      .......... ......B... ..........
5  ......DE.  I.G.....Q. .......... ....I..... .....S.    AI........
6  .......E.  ....C.K..  .......I... .......... .....S.    AI........
7  .......E.| Y...S..AQI ...I..|... .......... .....S.    AM........
8  .V..P...E. ....E..K.. .M..I..F.  .A....... AKW.....V.
9  .V..P...E. L...EY.K.. KQ..I..... .A....... ARW.......

           310        320        330        340        350        360
1  IDRQKNHGMH FRVLAKALRL SGGDHIHSGT VVGKLEGERD ITLGFVDLLR DDYTEKDRSR
2  ........I. .......M.. .......... .......E   .......... ..FV.Q....
3  .......... .......... .....A..   .......X.E .......... ...IK.....
4  ...B....   M......    .......    .......E   M......... BBFI..B
5  .......... .......M.. .......... .......E   .......... ...FI.....
6  ...R...I.  .......M.. .....L.... .......E   V......M.  ...V......
7  ...R...I.  ...T..M   .....L.... .......E   V......M.  .A.V......
8  ...R...I.  ...C...   .....L.... ...DKA     S.......M. E.HI.R....
9  ........I. .......... .....T..   .........G ..M......  EN.V.Q.K..
```

```
         370        380        390        400        410        420
1  GIYFTQSWVS TPGVLPVASG GIHVWHMPAL TEIFGDDSVL QFGGGTLGHP WGNAPGAVAN
2  .....D... L.....E... .......... .......... .......... ..........
3  .....D... L...I..... .......... .......... .......... ..........
4              M...I..... .......... .......... .......... .......A..
5  ..F...D... M...I..... .......... ...L...... .......... ....H..A..
6  .....D.C. M...M..... .......... V......AC. .......... ....|...A..
7  .....D.CG MG.TM..... .......... .......AC. .......... .......A..
8  .VF...D.A. M......... .......... V......... .......... .......T..
9  .....D.A. L...MA.... .......... V......... .......... ....R..T..

         430        440        450        460        470
1  RVALEACVQA RNEGRDLARE GNTIIREATK WSPELAAACE VWKEIKFE-F PAMDTV
2  ........K. ........Q. ..E.....C. .......... .....V.N.. A.V.VLDK
3  .......... .......... ..A.....C. .......... .......... .....L
4  .......... .......... ......A.C. 
5  .......... .......... VQ-..KA.C. ..A....... I......DG. K....I
6  .......T.. .......... .GDV..S.C. .......... .......... DTI.KL
7  ...S......| .......S.. .GDV....C. .......... .......... ETI.KL
8  .......... .......Y.. .GD.L...G. ........LD L......... ET..KL
9  .......... .....N.... .DV....A. ......V... L......... E.....
```

Fig. 5. Comparison of the amino acid sequences of the large subunits of Rubiscos from nine different species. A dot indicates identity with the residue at that position in sequence 1 (spinach). A dash indicates a deletion arbitrarily inserted to maximize homology. Blanks represent incomplete sequences. Species are (1) spinach (Zurawski *et al.*, 1981), (2) tobacco (Shinozaki and Sugiura (1982), (3) pea (Zurawski *et al.*, 1986), (4) barley partial sequence (Poulsen, 1981), (5) maize (McIntosh *et al.*, 1980), (6) *Chlamydomonas reinhardtii* (Dron *et al.*, 1982), (7) *Euglena gracilis* (Gingrich and Hallick, 1985) with intron positions shown as verticals, (8) *Synechococcus* PCC 6301 (Reichelt and Delaney, 1983), and (9) *Anabaena* PCC 7120 (Curtis and Haselkorn, 1983).

B. Primary Structure

For hexadecameric Rubiscos, the large subunits are very closely related, the small subunits less so. This was first inferred from their amino acid compositions (Takabe and Akazawa, 1975) and confirmed more recently by sequencing.

1. Large Subunit

The amino acid sequences of the large subunits of nine different Rubiscos are compared in Fig. 5. All except that of barley were determined by inference from the nucleotide sequences of the corresponding genes. In addition, the sequence of tobacco was also determined by direct amino acid sequencing of the protein (Amiri *et al.*, 1984), as was the partial sequence of barley (Poulsen, 1981). All of these large subunits were derived from hexadecameric, two-subunit Rubiscos. The *Euglena* gene is unusual in that it is interrupted by nine introns (Gingrich and Hallick, 1985). Their positions are indicated in Fig. 5.

An obvious feature of this sequence comparison is the extraordinary homology between all sequences. The large subunits of all the two-subunit enzymes are more than 80% homologous with each other. Homology between the large subunit of two-subunit Rubiscos and the single subunit of the

```
              10          20          30          40          50
-----------MDQSSRYVNLALKEEDLIAGGEHVLCAY-IMKPKAGYGYVATAAHFAAESSTGTNVEV
MSPQTETKASVEFKAGVKDYK.TYYTPEYETLDTDI.A.FRVSPQPGVPPEE.G..-V........WTT.
    10          20          30          40          50          60

       60          70          80          90         100         110         120
CT---TDDFTRGVDALVYEVDEARELTKIAYPVALFDRNITDGKAMIASFLTLTMGNNQGMGDVEYAKMH
W.DGL.NLDRYKGRCYHI.PVAGE.NQY.C.-..YPLDLFEE.SVT-NM.-.SIV..VF.FKALRALRLE
70          80          90         100         110         120         130

     130         140         150         160         170         180
DFYVPEAYRALFDGPSVNISALW----KVLGRPEVDGGLVVGTIIKPKLGLRPKPFAEACHAFWLGG-DF
.LRI.V..VKT.Q..PHG.QVERDKLN.Y-...------LL.CT.......SA.NYGR.VYECLR..L..
140         150         160              170         180         190

190         200         210         220         230         240         250
IKNDEPQGN-QPFAPLRDTIALV-ADAMRRAQDETGEAKLFSANITADDPFEIIARGEYVLET-FGENAS
T.D..-NV.S...MRW..RF-.FC.E.LYK..A....I.GHYL.A..GTCEDMMK.AVFAR.LGVPIVM-
200         210         220         230         240         250         260

    260         270         280         290         300         310         320
HVALLVDGYVAGAAAITTARRRFPDNFLHYHRAGHGAVTSPQSKR-GYTAFVHCKMARLQGASGIHTGTM
.-DY.TG.FT.NTTLSHYC.DNGLL--..I...M.-..IDR.-.NH.MHFR.LA.AL..S.GDH..S..V
270         280         290         300         310         320         330

    330         340         350         360         370         380
GFGKMEGESS------DRAIAYMLTQDEAQGPFYRQSWGGMKACTPIISGGMNALRMPGFFENLGNANVI
V-..L...RDITLGFV.LLRDDYTEK.RSR.IYFT...VSTPGVL.VA...IHVWH..ALT.IF.DDS.-
     340         350         360         370         380         390

390         400         410         420         430         440         450
LTAGGGAFGHIDGPVAGARSLRQAWQAWRDGVPVL-DYAREHKELAR-AFE-SFPGDADQIYPGWRKALG
.QF...TL..PW.NAP..VAN.V.LE.CVQARNEGR.L...GNTII.E.TKW.-.EL.AACEV-.-.EIK
400         410         420         430         440         450         460

    460
VEDTRSALPA
F.FPAMDTV
470
```

Fig. 6. Comparison of the amino acid sequences of the large subunit of spinach Rubisco (bottom line; Zurawski *et al.*, 1981) with that of the single-subunit Rubisco from *R. rubrum* (top line; Nargang *et al.*, 1984). Dots and dashes have the same significance as in Fig. 5.

R. rubrum enzyme is much less striking, being approximately 25% (Fig. 6). This focuses attention on those regions where the single-subunit enzyme is homologous. Six of the more obvious of these regions are underscored in Fig. 6. With the exception of $C^{L8}459$, which is derivatized by an affinity label only when the enzyme is decarbamylated, all of the residues identified with affinity labels, or otherwise implicated as being near the active site (Section IV,J), are located within these regions. Thus we may infer that the large subunits of all Rubiscos probably have a common heritage.

For the two-subunit Rubiscos, the amino-terminal 13 residues show exceptionally weak homology. Strong homology begins at $K^{L8}14$ (Fig. 5). Even the length of this region varies, ranging from 10 residues in *Anacystis* to 14 residues in *Anabaena*. This nonhomology may indicate posttranslational processing of the large subunit by a trypsin-like protease that cleaves the $L^{L8}14/A^{L8}15$ bond. This is supported by the sequences of barley and tobacco, both determined at the protein level, which begin at $A^{L8}15$. In the case of tobacco, however, there was also some evidence for additional initiation at

$T^{L8}5$, which may indicate that the posttranslational processing occurs in two steps (Amiri *et al.*, 1984). Evidence of posttranslational processing of a precursor of the large subunit was also obtained by Langridge (1981). mRNAs from spinach chloroplasts direct the *in vitro* synthesis, in an *Escherichia coli* cell-free system, of a peptide 1–2 kDa larger than the large subunit obtained from chloroplasts. Treatment of this peptide with chloroplast extracts converted it to the same size as the mature peptide. Interestingly, in *Euglena*, this amino-terminal region occupies the first exon almost entirely. The first intron occurs between $A^{L8}15$ and $G^{L8}16$. If introns do indeed separate functional and structural domains in proteins, as has been proposed (Gilbert, 1978; Blake, 1979), this would be in accordance with a separate function for this amino-terminal region, such as posttranslational removal. The functional role of this processing (perhaps in assembly of the hexadecamer?) is presently completely obscure. A contrary view of the function of this amino-terminal peptide has recently been raised by Gutteridge *et al.* (1986a). Very mild trypsin treatment of several L8S8 Rubiscos caused severe, but not complete, inactivation, concomitant with the release of two peptides whose amino acid compositions indicated that they were derived from positions 1–8 and 9–14. Neither the L8S8 structure nor the ability to bind 2CABP was disturbed. These studies indicate that the amino-terminal 14 residues are present in at least some preparations of purified L8S8 Rubiscos. Further studies are necessary to clarify the function of this amino-terminal region.

2. Small Subunit

Comparison of small-subunit sequences for Rubiscos from nine different species (Fig. 7) reveals a much lower level of homology than that observed with the large subunit. However, two strongly conserved regions are prominent.

The first conserved region, residues 10–21, occurs both in cyanobacterial and in chloroplast small subunits. The second, residues 61–76, is absolutely conserved in the plant enzymes but has an interesting difference in the cyanobacteria. The latter subunits are 12 residues shorter than those from plants, and the deletion occurs in such a position that the first five residues of the second conserved region, residues 61–65, are missing. In order to maximize homology, the other seven missing residues must also be located in the region from positions 45–60. For convenience, they are shown as being contiguous (Fig. 7). Residues 66–76 are identical for all chloroplast small subunits and strongly homologous to the corresponding region of the cyanobacterial small subunits. Secondary structure predictions according to Chou and Fasman (1974) suggest that this region is a short alpha helix (Muller *et al.*, 1983). In some, but not all, plant small-subunit genes, an intron is inserted within the triplet codon of residue 65 (Mazur and Chui, 1985; Dean *et al.*, 1985b), which also suggests that this second highly conserved region

```
            10        20         30         40         50
1   MQVWPPLGLK KFETLSYLPP LTTEQLLAEV NYLLVKGWIP PLEFEVKDGF
2   ......I.K. .......... ..RD...K.. E...R...V. C....LLK..
3   ......I.K. .........D .DDA..AK.. E...R..... C....LEH..
            G                                              H
4   ......YGK. .Y.......D .SQ....L.P D...KD..V. C....TEG..
5            Y.......D ..G....K.. E...D...V. C....L.HK.
6   .....IE.I. .......... .S..A..KQ. D..IRSK.V. C...S-.V..
7   ......E... .......F.L SSV.D.AK.. D...RND.V. CI..S-.E..
8   .SMKTLPKER R...F..... .SDR.IA.QI E.MIEQ.FH. LI..NEHSNP
9   ..--TLPKER RY........ ..DV.IEKQ. Q.I.SQ.Y.. AV..NEVSEP

            60        70         80         90        100
                                                      Y
1   VYREHDKSPG YYDGRYWTMW KLPMFGGTDP AQVVNEVEEV KKAPPDAFVR
2   ..G..N...R .......... ......T... ....K..D.. VA.Y.E....
3   .....NR..X .......... ......C..A S..LK.LQ.A .T.Y.NG.I.
4   ....NN.... .......... ......C..A T..LA..G.A ...Y.E.WI.
5   I...YHA... .......... ......C..A T..LG.LQ.A ...Y.N.GSG
6   .F...NS... .......... ......C..A T..L...... ..EY...Y..
7   ....NNA... .......... ......C..A S..IA....A ...Y.EY...
8   EEF------- -----..... ...L.DCKS. Q..LD..R.C RSEYG.CYI.
9   TEL------- -----...L. ...L..AKTS RE.LA..QSC RSQY.GHYI.

           110       120
1   FIGFNDKREV QCISFIAYKP AGY
2   V....NV.Q. .......HT. ES.
3   I...DNV.Q. .......... P.F
4   I...DNV.Q. .......... E..
5   I...DNV.Q. .......... P.F
        M
6   V...DNL.Q. ..V....FR. P.CEESGKA
7   I...DN..Q. .......... T
8   VA..DNIKQC .TV...VHR. GR.
9   VV..DNIKQC .IL...VH.. SR.
```

Fig. 7. Comparison of the amino acid sequences of the small subunits of Rubiscos from nine different species. Dots, dashes, and blanks have the same significance as in Fig. 5. Heterogeneity at a particular position in a sequence, where known, is shown by inclusion of the symbols for both residues one above the other. Species: (1) spinach (Martin, 1979), (2) pea, *cv.* Feltham First (Bedbrook *et al.,* 1980; Takruri *et al.,* 1981), (3) soybean (Berry-Lowe *et al.,* 1982), (4) tobacco (Muller *et al.,* 1983), (5) petunia (Dunsmuir *et al.,* 1983), (6) wheat (Broglie *et al.,* 1983), (7) *Lemna gibba* (Stiekema *et al.,* 1983), (8) *Synechococcus* PCC 6301 (Shinozaki and Sugiura, 1983), (9) *Anabaena* PCC 7120 (Nierzwicki-Bauer *et al.,* 1984).

may be divided into two functional units, the first of which is lacking in cyanobacterial small subunits.

Toward the carboxyl terminus of the small subunit, there is a third conserved region. Absolute homology is weaker here than in the previous two regions but, between residues 98 and 116, the substitutions that occur are nearly all conservative (Nierzwicki-Bauer *et al.,* 1984).

In addition to the intron at residue 65, which so far has been observed only in tobacco (Mazur and Chui, 1985) and petunia (Dean *et al.,* 1985b), two other intron-insertion positions are known. One occurs between codons 2 and 3, approximately separating the mature peptide from the transit sequence that is cleaved from the small-subunit precursor (Section III,D). An intron occurs in this position in all genomic small-subunit clones from higher plants so far examined. Another intron, which separates residues 47 and 48, is present in the dicotyledons, but absent from wheat. The cyanobacterial small-subunit genes do not, of course, contain introns; nor do they encode amino-terminal transit sequences that are removed posttranslationally.

C. Structure Determination by Crystallography

The structure of Rubisco is currently attracting intense interest among crystallographers. Four independent groups have reported preliminary crystallographic data for the dimeric *R. rubrum* enzyme (Schneider *et al.*, 1984; Branden *et al.*, 1986; Choe *et al.*, 1985; Janson *et al.*, 1984) and for the hexadecamer from tobacco (Baker *et al.*, 1975), spinach (Barcena *et al.*, 1983; Andersson *et al.*, 1983; Andersson and Branden, 1984), and *Alcaligenes eutrophus* (Bowien *et al.*, 1980; Pal *et al.*, 1985).

A similar tetragonal crystal form has been reported by the three groups working on the native *R. rubrum* enzyme (Janson *et al.*, 1984) or the recombinant enzyme that contains an additional 24 amino acids from β-galactosidase (Schneider *et al.*, 1984; Choe *et al.*, 1985; Janson *et al.*, 1984). The crystals of the Swedish group (cell dimensions $a = b = 82$ Å, $c = 324$ Å; space group tetragonal $P4_32_12$ or its enantiomorph $p4_12_12$) were originally believed to be those of the carbamylated quaternary complex, enzyme–$^ACO_2$–Mg^{2+}–2CABP (Schneider *et al.*, 1984). However, these crystals were grown under mildly acidic conditions, and subsequent analysis showed that they contained neither $^ACO_2$ nor 2CABP (Schneider *et al.*, 1986a); that is, the crystals are those of decarbamylated enzyme. The crystals of the other groups are quite similar (cell dimensions $a = b = 82$ Å; $c = 290$ Å; space group, tetragonal $P4_32_12$ or $P4_12_12$). They were grown from solutions of the carbamylated ternary complex E–$^ACO_2$–Mg^{2+} under more alkaline conditions (pH 8.0). However, it is not known if these crystals contain the activator ligands.

An altogether different crystal form of the *R. rubrum* enzyme, one more amenable to crystallographic analysis, has been prepared by the Swedish group (Schneider *et al.*, 1986a; Branden *et al.*, 1986). These are crystals of the decarbamylated enzyme. They are monoclinic with the space group $P2_1$. The cell dimensions are $a = 65.5$ Å, $b = 70.6$ Å, $c = 104.1$ Å, and $\beta = 92.1°$. The assymetric cell contains one dimeric molecule, and the crystals diffract to at least 1.9 Å resolution.

The dimeric molecule has the shape of a distorted ellipsoid with dimensions 45 Å $\times$ 70 Å $\times$ 105 Å, where the 70 Å principal axis of the ellipsoid is the twofold molecular axis (Fig. 8). Each subunit consists of two main domains A and B (Schneider *et al.*, 1986b). Domain B is the N-terminal domain and comprises residues 1–136. The central secondary structural motif is a five-stranded mixed β-sheet with two α-helices on one side and a third α-helix on the other side of the sheet. The larger domain A consists of residues 136–466. This domain has a parallel α/β barrel structure, very similar to that

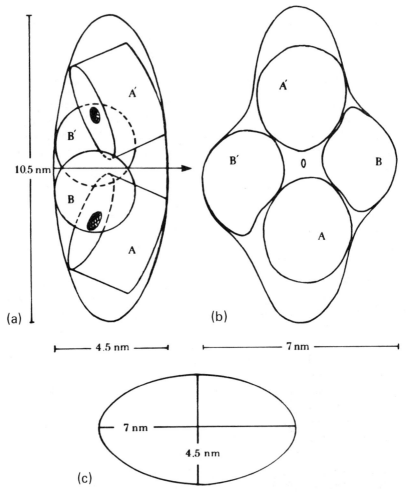

Fig. 8. Three views of the *R. rubrum* Rubisco molecule from different perspectives. A and B are the approximate outlines of the two domains of one subunit. A' and B' are the two domains of the other subunit. Domain A comprises the N-terminal 1–136 residues, while domain B comprises the C-terminal 136–466 residues. (a) View down the *y* axis. The arrow denotes the molecular twofold axis. The active site is in the vicinity of the hatched area. (b) View down the local twofold axis. (c) Cross section through the middle of the molecule in a plane perpendicular to the long axis of the molecule. The local twofold axis is in this plane along the short axis of the ellipsoid. [Reproduced with permission from Branden *et al.* (1986).]

found in triose phosphate isomerase (Banner *et al.*, 1976) and glycolate oxidase (Lindqvist and Branden, 1985). Eight parallel β-strands form the lumen of the barrel, which is surrounded on the outside by eight α-helices. The active site is at the carboxyl end of the strands of the α/β barrel. In both glycolate oxidase and triose phosphate isomerase, the active site is similarly located at the carboxyl end of the α/β barrel. The active site lysines $K^{L2}166$, $K^{L2}191$, and $K^{L2}329$ are found in the loops connecting the carboxyl ends of the β-strands to the α-helices on the exterior of the α/β barrel (Schneider *et. al.*, 1986b). $K^{L2}166$ is located in the loop after strand number 1. The site of carbamylation, $K^{L2}191$, is the first residue of loop number 2., that is, it lies at the bottom of the active site. The substrate RuBP and the reaction intermediate analog, 2CABP, are thought to lie on top of $K^{L2}191$, thus closing off access of the activator ligands $^{A}CO_2$ and Mg^{2+} to the medium and providing an explanation for the nonexchangeability of these ligands from the quaternary complex enzyme–$^{A}CO_2$–Me^{2+}–2CABP (Miziorko, 1979; Miziorko and Sealy, 1980). $K^{L2}329$, another active-site residue identified by affinity-labeling studies (Section IV,J), is part of loop number 6. The crystallographic positioning of $K^{L2}166$ and $K^{L2}329$ within the active site is consistent with the results from Hartman's group. Using carbamylated enzyme, Lee and Hartman (1986) cross-linked $K^{L2}166$ and $K^{L2}329$ with a reagent spanning maximally 12 Å (Section IV,K). Presently, the remaining amino acid side chains are being fitted to the 2.9-Å electron density map, and a complete structure of this, the decarbamylated enzyme, is expected shortly.

The crystallographic analyses of the hexadecameric enzyme have not progressed as far. However, the results are awaited with keen anticipation. For both the spinach and *Alcaligenes* enzymes, the crystals being examined are those of the quaternary complex, enzyme–$^{A}CO_2$–Mg^{2+}–2CABP (Andersson and Branden, 1984; Pal *et al.*, 1985). Since 2CABP is thought to be a very close structural mimic of the hydrated, gem-diol form of the reaction intermediate 3-keto-2CABP, the structure of the quaternary complex that emerges from these studies will be of particular relevance to the mechanism of catalysis. These studies will pinpoint the nature of the interaction of the small subunit(s) with the large subunit(s). A function for the small subunit might also be suggested by the structure.

D. Synthesis and Assembly

1. Prokaryotes

So far as is known, the synthesis and assembly of prokaryotic Rubiscos, whether of the single- or two-subunit types, is unexceptional. For the two-subunit, prokaryotic enzymes, the genes for the large and small subunits occur consecutively on the chromosome and constitute a single operon (Shinozaki and Sugiura, 1983, 195; Nierzwicki-Bauer *et al.*, 1984; Viale *et al.*, 1985; Starnes *et al.*, 1985). In *Alcaligenes eutrophus*, Behki *et al.* (1983) and Andersen and Wilke-Douglas (1984) found evidence for plasmid reiteration of chromosomal Rubisco genes. So far, the pair of genes for the subunits from two cyanobacteria and from *Chromatium vinosum* has been cloned into suitable *E. coli* expression vectors, resulting in the production of fully assembled and catalytically competent Rubisco by the *E. coli* cells (Gatenby *et al.*, 1985; Tabita and Small, 1985; Gurevitz *et al.*, 1985; Christeller *et al.*, 1985; Viale *et al.*, 1985). These observations rule out, for the prokaryotic enzymes at least, any obligate involvement of posttranslational processing or assembly mechanisms, unless these mechanisms also occur in *E. coli.*

Detailed analysis of the expression of the *Anabaena* Rubisco genes in *E. coli* by Gurevitz *et al.* (1985) revealed that the large subunit was produced in much greater quantities than the small subunit. The excess large subunits were insoluble. The soluble enzyme was fully catalytically active and had a full complement of small subunits. From these observations, Gurevitz *et al.* (1985) concluded that assembly occurred via initial formation of L_1S_1 hetero-dimers and that the small subunits were necessary to maintain solubility of the complex. However, other interpretations are possible. For instance, a model for assembly involving preliminary formation of large-subunit octa-mers, followed by progressive binding of small subunits, would be consistent with these observations if subsaturated octamers were less soluble than the fully saturated complex and if precipitation of the subsaturated octamers was accompanied by release of their small subunits. Ultimately, only fully saturated complexes would remain in solution. Release of small subunits during isoelectric precipitation of Rubisco large-subunit octamers is well documented (Section VI,B).

2. Eukaryotes

Little is known about the Rubiscos of nonchlorophytic eukaryotes. However, the chromophyte *Olisthodiscus lteus* has both its large and small sub-unit genes located on the chloroplast genome in the same tandem array as seen in prokaryotes (Reith and Cattolico, 1986).

In chlorophytes, however, the synthesis and assembly of the two-subunit Rubisco has been studied in some detail and is extraordinarily complex. The elements of the process, so far as they are currently understood, are diagra-matically represented in Fig. 9.

Coordinated participation of two genomes is involved. Each chloroplast circular DNA molecule bears a single gene for the large subunit. Therefore, there are hundreds to thousands of presumably identical copies in each photosynthetic cell. The nuclear, small-subunit genes are also multiple and comprise a multigene family of eight or more members (Berry-Lowe *et al.*, 1982; Coruzzi *et al.*, 1983; Broglie *et al.*, 1983; Dean *et al.*, 1985a,b). However, these copies are not identical. Although they differ very little, or not at all, in the amino acid sequence of the mature peptide that they specify, considerable divergence occurs at the nucleotide level. This is particularly apparent in both 5′ and 3′ noncoding regions. Furthermore, the introns vary in length and number, as well as in sequence (Dean *et al.*, 1985b). In pea, Coruzzi *et al.* (1984) found that different members of the gene family were expressed at different levels in different tissues. It is also possible that some

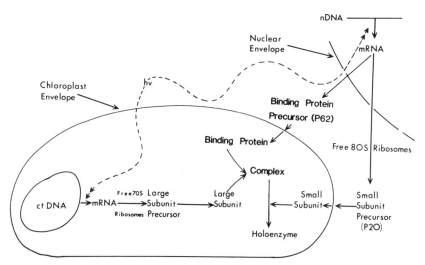

Fig. 9. Synthesis, processing, and assembly of Rubisco in higher plants. [Reproduced with permission from Ellis *et al.* (1984).]

members are not expressed in any tissue and are, therefore, pseudo-genes.

Both subunits have an amino-terminal fragment removed at some stage in the assembly process. This has been discussed earlier (Section III,B,1) for the case of the large subunit. Processing of the small subunit is an integral part of its assimilation by chloroplasts. The precursor enters the plastid via an ATP-dependent transporter on the envelope. A stromal metalloendopeptidase then removes the aminoterminal "transit peptide" (Robinson and Ellis, 1984). This may prevent the mature peptide from leaving the plastid, thus providing the necessary directionality for the uptake process. The manner in which the transit peptide confers transportability on the small subunit and the nature of the transporter are currently being researched. Of great significance is the observation that attachment of the small subunit's transit peptide to an unrelated protein renders the latter transportable into the chloroplast (Van den Broeck *et al.,* 1985). This would appear to rule out the possibility that a conformation acceptable to the transporter is achieved by complementarity between the transit peptide and the protein to be transported. Rather, this observation promotes conjecture that the transit peptide functions with some degree of autonomy, somewhat in the manner of an address label.

The manner of assembly, within the chloroplast, of the two mature subunits to form the functional hexadecamer is currently the subject of considerable conjecture. Another quite abundant chloroplast protein, composed of a pair of nonidentical, but similarly sized (approximately 60-kDa), subunits, tightly binds Rubisco large subunits specifically and reversibly (Hemmingsen and Ellis, 1986; Lennox and Ellis, 1986; Musgrove and Ellis, 1986). This binding protein is of cytoplasmic origin, where it is synthesized as a slightly larger precursor and transported to the chloroplast, perhaps in a manner analogous to the small subunit (Hemmingsen and Ellis, 1986). Labeled large subunits newly synthesized by isolated chloroplasts are predominantly found attached to a 720-kDa oligomeric (dodecameric?) form of the binding protein when the chloroplasts are lysed by extensive dilution. However, addition of MgATP to these extracts to levels similar to those in illuminated chloroplasts causes dissociation of the complex to monomers with the concomitant release of bound large subunits (Bloom *et al.*, 1983; Milos and Roy, 1984; Lennox and Ellis, 1986). Significantly, ATP neither phosphorylates nor adenylates the binding protein during this dissociation (Hemmingsen and Ellis, 1986). The role of these interactions in the assembly of Rubisco is unknown. Certainly, the binding protein sequesters large subunits that would otherwise be insoluble in the absence of small subunits. [Large subunits isolated from higher-plant Rubiscos are insoluble in the absence of detergents or denaturants, as are the large subunits synthesized by *E. coli* harboring large-subunit genes from higher plants (Section VI; Somerville *et al.*, 1984; Gatenby, 1984).] However, the crux of the issue is whether or not this binding is an obligate step in assembly. The kinetics of appearance of bound large subunits and assembled holoenzyme are consistent with, but do not prove, such a precursor–product relationship (Milos and Roy, 1984; Musgrove and Ellis, 1986).

It is attractive to postulate that the oligomeric form of the binding protein facilitates assembly by acting as a kind of molecular scaffold (or "chaperon") that permits only proper interactions between the subunits (Musgrove and Ellis, 1986). This would explain the difficulties with insolubility encountered upon expression of cloned higher-plant Rubisco genes in *E. coli* (Somerville *et al.*, 1984; Gatenby, 1984), where the binding protein is presumably lacking. However, ATP-induced dissociation of the binding protein–large subunit complex occurs at 0°C and may thus be separated from subsequent assembly of the subunits which occurs after warming (Milos and Roy, 1984). Perhaps small subunits are also involved in dissociation, releasing large subunit–small subunit heterodimers that spontaneously assemble to hexadecameric Rubisco. Furthermore, it is at least theoretically possible that the endoproteolytic removal of the small aminoterminal fragment of the large subunit (Langridge, 1981) is catalyzed by the binding protein.

Alternatively, if participation of the binding protein in assembly turns out not be obligatory, it is possible that sequestration of large subunits by the binding protein simply provides a "buffer" against any slight mismatches that may occur in the rate of synthesis of the two subunits. Because of the insolubility of unassembled large subunits, any transient miscoordination causing a temporary insufficiency in the supply of small subunits to the assembly process would result in a disastrous accumulation of insoluble large subunits unless there existed a reservoir capable of accommodating them. Of course, such a circumstance might also be prevented by arranging that the small subunits were always supplied at a faster rate than were the large subunits, with the excess unassembled small subunits eventually being degraded. However, such a wasteful mechanism probably could not be tolerated for a protein as abundant as Rubisco.

How is the close coordination between the rates of synthesis of the two Rubisco subunits achieved? Two separate genomes and protein synthetic machineries are involved. In particular, how are the rates regulated by light, which increases both rates by perhaps as much as two orders of magnitude? It is clear that at least some part of this control is applied at the level of transcription and is mediated by phytochrome and a putative photoreceptor for blue light (Ellis *et al.*, 1984; Fluhr and Chua, 1986). This is indicated by the wavy dashed lines in Fig. 9. At the DNA sequence level, a 33-base-pair sequence immediately preceding the transcriptional initiation site of the small-subunit gene and another upstream, enhancer-like element contained within a 240-base-pair fragment have been implicated in the light regulation of transcription of that gene (Morelli *et al.*, 1985; Timko *et al.*, 1985; Fluhr *et al.*, 1986). For the large subunit, however, transcriptional control cannot be the complete answer. While the level of mRNA for the small subunit closely parallels the level of the mature peptide, accumulation of the message for the large subunit during greening precedes that of the peptide (Ellis *et al.*, 1984). Some other, as yet unknown, posttranscriptional mechanism for coordinating the rate of synthesis of the large subunit to that of the small subunit must exist.

IV. MECHANISMS OF CARBAMYLATION AND CATALYSIS

A. Carbamylation of the Activator Lysine by $^A CO_2$ and Me^{2+}

The catalytically competent form of all Rubiscos, irrespective of their origin, is a ternary complex of enzyme–$^A CO_2$–Me^{2+} (Lorimer *et al.*, 1976; Laing and Christeller, 1976; Badger and Lorimer, 1976; Christeller and Laing, 1978; Whitman *et al.*, 1979; Gibson and Tabita, 1979). The formation of the active ternary complex involves the slow, reversible reaction of a molecule of $^A CO_2$ with the ε-amino group of $K^{L8}201$ (or $K^{L2}191$) to form a carbamate (Fig. 10) (Lorimer *et al.*, 1976; Lorimer and Miziorko, 1980; Lorimer, 1981a; Donnelly *et al.*, 1983). The activator $^A CO_2$ is different from the molecule of substrate $^S CO_2$, which becomes fixed during catalysis (Miziorko, 1979; Lorimer, 1979). Formation of the carbamate is followed by the rapid addition of Me^{2+} to create the active ternary complex.

A recent ^{13}C nuclear magnetic resonance (NMR) analysis of the stable quaternary complex enzyme–$^A CO_2$–Me^{2+}–2CABP (Pierce and Reddy, 1986)

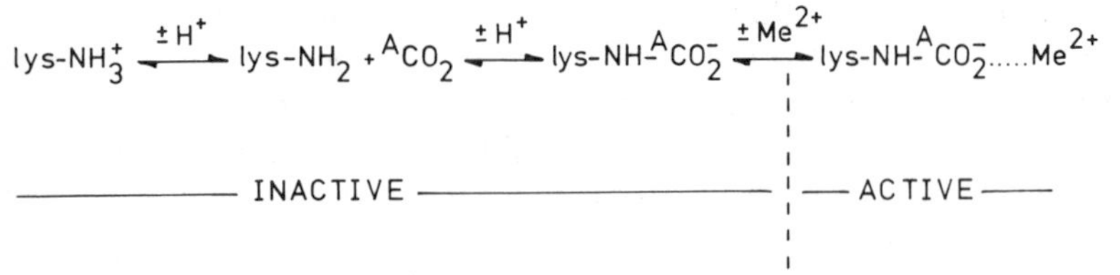

$$\text{lys-NH}_3^+ \xrightarrow{\pm H^+} \text{lys-NH}_2 + {}^A\!CO_2 \xrightarrow{\pm H^+} \text{lys-NH-}{}^A\!CO_2^- \xrightarrow{\pm Me^{2+}} \text{lys-NH-}{}^A\!CO_2^- \cdots Me^{2+}$$

———————— INACTIVE ————————————— ⎸— ACTIVE ——

Fig. 10. The formation of catalytically active enzyme involves the reversible carbamylation of the ε-amino group of $L^{L8}201$ (or $K^{L2}191$). Binding of Me^{2+} to the enzyme carbamate creates the active ternary complex.

has demonstrated that the carbamate, the divalent metal ion, and the carboxyl group of 2CABP are contiguous. Although it is tempting to assume that the carbamate is stabilized by direct coordination to the divalent metal ion, there exists no direct evidence for this. Indeed, attempts to demonstrate direct coordination of ^{17}O-labeled carbamate to Mn^{2+} or Cu^{2+} in quaternary complexes of *R. rubrum* enzyme have failed (Miziorko and Sealy, 1984; Styring and Branden, 1985a,b). However, the activator Me^{2+} is certainly close to the carbamate. Presumably, one or more of the anionic residues surrounding the activator lysine (for example, $D^{L2}193$ or $D^{L8}202$) contribute to the binding of Me^{2+}. An important point emerging from these spectroscopic studies is that the formation of the activator carbamate–Me^{2+} complex occurs within, and, indeed, completes the catalytic site.

The environment of Me^{2+} in the ternary complex (E–$^A\!CO_2$–Me^{2+}) of the L2 and L8S8 Rubiscos has been investigated by electron paramagnetic resonance (EPR) spectroscopy with Co^{2+} (Nilsson *et al.,* 1984; Styring and Branden, 1985a,b) and Mn^{2+} (S. Gutteridge, personal communication). The spectra indicate that the coordination geometry of Me^{2+} is not greatly perturbed upon binding to the catalytic site. The EPR spectrum of the ternary complex E–$^A\!CO_2$–Cu^{2+} displays nitrogen hyperfine structure, and it has been suggested (Branden *et al.,* 1984a,b) that at least one nitrogen atom is coordinated to the Cu^{2+} ion.

The formation of the carbamate–Me^{2+} complex does not require the presence of the small subunit. The *R. rubrum* L2 Rubisco undergoes carbamylation without a small subunit (Whitman *et al.,* 1979). This is also true of the L8S8 enzyme. The L8 core alone forms the carbamate–Me^{2+} complex (Section VI,C) although the equilibrium is less favorable (Andrews and Ballment, 1984b).

The formation of the carbamate is clearly stabilized by the binding of Me^{2+} (Lorimer *et al.,* 1976; Lorimer, 1979; Belknap and Portis, 1986a,b). This role is filled by a wide variety of Me^{2+}. As judged by the criterion of catalytic competency, the L2 Rubisco can accomodate Mg^{2+}, Mn^{2+}, Fe^{2+}, Ca^{2+}, or Cu^{2+} as activating metal ions (Christeller, 1981; Robison *et al.,* 1979; Christeller and Laing, 1978; Nilsson *et al.,* 1984; Styring and Branden, 1985b;

Pierce and Reddy, 1986). Interestingly, *R. rubrum* Rubisco activated with Co^{2+} functions as an oxygenase but not as a carboxylase (Christeller, 1981; Robison *et al.*, 1979). As judged by the less-demanding ability to form a stable $E-^ACO_2-Me^{2+}-2CABP$ complex (i.e., able to withstand passage through a gel-filtration column), this list can be expanded to include Cd^{2+} (Pierce and Reddy, 1986) and Gd^{3+} (G. H. Lorimer, unpublished data). The L8S8 enzyme is even more catholic in its accomodation of various Me^{2+}. Mg^{2+}, Ni^{2+}, Co^{2+}, Fe^{2+}, Mn^{2+}, and Cu^{2+} have all been reported to sustain catalytic activity (Weissbach *et al.*, 1956; Branden *et al.*, 1984a,b; Christeller, 1981; Wildner and Henkel, 1979; Christeller and Laing, 1979). The spinach enzyme can also form a quaternary complex with Cr^{2+} (Miziorko *et al.*, 1982). The data with respect to Ca^{2+} are clear as far as activation is concerned. Barcena (1983), with spinach enzyme, and Parry *et al.* (1983), with wheat enzyme, agree that the carbamate can be stabilized by Ca^{2+}. But there is disagreement about the ability of Ca^{2+} to sustain catalysis. Both Barcena (1983) with spinach Rubisco and Christeller (1981) with soybean Rubisco reported that Ca^{2+} did not sustain catalysis. Parry *et al.* (1983), to the contrary, reported that Ca^{2+} sustained both carboxylation and oxygenation at rates that were about 65% of the rates observed with Mg^{2+}. Substitution of Mg^{2+} with Ca^{2+} did not alter the carboxylase/oxygenase s_{rel} value. This conflict remains to be resolved.

The formation of the carbamylated ternary complex is accompanied by a change in the conformation of the enzyme. A wide variety of physical and chemical evidence supports this conclusion. Carbamylation of both L2 and L8S8 forms of Rubisco is associated with altered reactivity towards a variety of compounds. For example, carbamylated L2 Rubisco reacts with the affinity labels 2-bromoacetylaminopentitol 1,5-bisphosphate and 2-(4-bromoacetamido)-anilino-2-deoxypentitol 1,5-bisphosphate, whereas decarbamylated L2 Rubisco does not (Fraij and Hartman, 1982; Herndon and Hartman, 1984). Group-specific reagents also show differential reactivity toward carbamylated and decarbmylated enzyme. With carbamylated L2 enzyme, $K^{L2}166$ is the major site of arylation by trinitrobenzene sulfonate (TNBS) (Section IV,J). But with decarbamylated L2 enzyme, three additional sites ($K^{L2}300$, $K^{L2}329$, and $M^{L2}1$) become accessible to arylation by TNBS (Hartman *et al.*, 1985). The extent of subunit cross-linking of L8S8 Rubiscos is generally enhanced by carbamylation (Grebanier *et al.*, 1978). For example, the cross-linking of two large subunits of spinach Rubisco with one another by dimethyl suberimidate is almost doubled upon carbamylation. Altered binding of RuBP (Jordan and Chollet, 1983) and of phosphorylated effectors (McCurry *et al.*, 1981; Jordan *et al.*, 1983) to carbamylated and decarbamylated enzyme can also be attributed to the conformational changes accompanying formation of the ternary complex.

Changes in a number of physical properties accompany carbamylation. For example, the circular dichroism spectrum (Grebanier *et al.*, 1978), the fluorescence of added probes such as 1-anilino-8-naphthalene sulfonate (Wildner 1976), and the thermal stability (Tomimatsu and Donovan, 1981) of various L8S8 Rubiscos are all altered upon carbamylation.

A most dramatic physical change has been reported to accompany the carbamylation of the L8S8 Rubisco from *Alcaligenes eutrophus* (Bowien and Gottschalk, 1982). Sedimentation velocity studies showed that the decarbamylated enzyme had a sedimentation coefficient ($s_{20,w}$) of 17.5 S, while the carbamylated form had an $s_{20,w}$ value of 14.3 S. Increases in the radius of gyration from 47.8 to 49.2 nm, in the maximum particle dimension from 13.5 to 15.7 nm, and in the invariant volume from 960 to 979 nm^3 accompanied decarbmylation of the *Alcaligenes* enzyme (Meisenberger *et al.*, 1984). An alteration in the axial ratio during carbamylation/decarbamylation was suggested to account for the drastic change in $s_{20,w}$. However, these properties appear to be unique to the *Alcaligenes* enzyme. Donnelly *et al.* (1984) were unable to detect any significant changes in the $s_{20,w}$ values of *R. rubrum* Rubisco ($s_{20,w} = 5.6–5.9$ S) and spinach Rubisco ($s_{20,w} = 17.4–18.3$ S) upon carbamylation. Additionally, neutron-scattering studies on the carbamylated and decarbamylated forms of the two enzymes failed to reveal any major changes in the distribution of mass. A reason for the discrepancy between the behavior of the *Alcaligenes* L8S8 and spinach L8S8 Rubiscos is not obvious.

Being able to demonstrate that a conformational change occurs is one thing. Defining the conformational change in molecular or atomic terms is quite another. The conformational change accompanying the carbamylation of Rubisco is no exception. This is a problem that is best addressed by the crystallographers.

B.   Binding of RuBP, 2CABP, and other Effector Ligands

The interaction of carbamylated and decarbamylated L8S8 Rubiscos with RuBP and other effectors is very complex. A simple, unifying model has been developed (Badger and Lorimer, 1981) that goes some way toward accounting for the many diverse experimental observations. Results reported since (Jordan and Chollet, 1983; Jordan *et al.*, 1983; Vater *et al.*, 1983; Johal *et al.*, 1985) make it clear, however, that conformational changes and cooperative effects are superimposed on the basic carbamylation/decarbamylation system.

The L8S8 Rubiscos contain eight catalytic sites (one/promoter). Both RuBP and the various effectors interact with the enzyme at a common site, the catalytic site. Two lines of evidence support this conclusion. First, all of the compounds that influence the carbamylation state of the enzyme are linearly competitive inhibitors of catalysis with respect to RuBP (Badger and Lorimer, 1981). Second, the binding of 6-phosphogluconate and NADPH, two of the most powerful effectors, to carbamylated spinach Rubisco is completely eliminated by prior occupancy of the catalytic site with 2CABP (Badger and Lorimer, 1981).

Although catalysis is highly specific for RuBP, the catalytic site can clearly accomodate a wide variety of compounds. McCurry et al. (1981) and Vater et al. (1983) together list 29 compounds that, at one concentration or another, enhance the state of carbamylation. The most effective compounds are characterized by having at least two anionic groups, for example, carboxyl or phosphate residues, linked by maximally six to seven carbon atoms. Although these structural elements are well represented in RuBP itself and in effectors such as 6-phosphogluconate, they can be reduced to its negatively charged terminal groups. Thus Pi alone is an effective enhancer of carbamylation. Binding experiments (Jordan et al., 1983) show that 2Pi/ promoter are bound.

The basic principle underlying the model of Badger and Lorimer (1981) is that effectors enhance carbamylation by stabilizing the ternary $E-^ACO_2-Mg^{2+}$ complex. Kinetic and gel-filtration experiments showed that 6-phosphogluconate and NADPH stabilized the binding of $^ACO_2$ and $^{54}Mn^{2+}$, consistent with the model. McCurry et al. (1981) measured the formation of the $[^{14}C]^ACO_2$-labeled ternary complex in response to the addition of effectors by trapping with 2CABP. Their results are entirely in accord with the basic model.

Hatch and Jensen (1980) classified effectors into two groups. Positive effectors, such as NADPH and 6-phosphogluconate, enhance carbamylation. Negative effectors, ribose 5-phosphate for example, favor the decarbamylated state. Recognizing that a given effector could bind to both decarbamylated and carbamylated enzyme, Badger and Lorimer (1981) proposed that positive effectors bind more tightly to carbamyalted enzyme than they do to decarbamylated enzyme. The reverse was thought to apply to negative effectors. Consequently, it was proposed that the ratio of dissociation constants describing the binding of effector to the carbamylated and decarbamylated forms determines whether a given effector promotes or inhibits carbamylation and the degree to which it does so.

This model was predicated on the assumption (for which there was then no contrary evidence) that each of the eight sites acted independently of the other seven. Subsequently, it has become clear that the eight sites do not behave independently; that is, events at one catalytic site allosterically influence the properties of the others (Jordan et al., 1983; Vater et al., 1983; Johal et al., 1985; Parry et al., 1985; Belknap and Portis, 1986b). For such interactions to occur, changes in the conformation of the enzyme must accompany binding of the ligands to the ternary carbamylated complex. Spectral changes in response to ligand binding have been reported (Vater and Salnikow, 1979). Additionally, EPR studies reveal that the environment of

Me^{2+} changes dramatically upon binding of ligands such as RuBP or 2CABP (Miziorko and Sealy, 1980; Nilsson *et al.*, 1984).

These conformational changes are associated with cooperative interactions between the sites. The binding of 2CABP to carbamylated spinach enzyme has long been known to involve enzyme isomerization (Pierce *et al.*, 1980a). In a series of cleverly conceived experiments, Johal *et al.* (1985) explored the binding of 2CABP in greater detail. Two important points emerged from this study. As revealed by changes in the (ion-exchange) chromatographic and electrophoretic properties of the enzyme upon binding 2CABP, it is clear that ligand binding induces a conformational reorganization that extends to the surface of the protein. Second, when offered an equal concentration of available sites in the form of L8S8(2CABP)$_0$ or L8S-8(2CABP)$_6$, 2CABP binds preferentially to enzyme molecules with the lower level of site occupancy. This is a particularly clear demonstration of negative cooperativity between the catalytic sites.

Another demonstration of cooperativity between the sites is evident in the dilution experiments reported by Jordan *et al.* (1983) and Vater *et al.* (1983). They investigated the ability of 6-phosphogluconate to retard the decarbamylation of the spinach enzyme upon dilution of the ^{A}CO$_2$ and Me^{2+}. Both groups reported that concentrations of 6-phosphogluconate approximately equivalent to one site per octamer significantly retarded decarbamylation. This implies that the approximately seven sites free of ligand were influenced by the one site to which the ligand was bound. Such experiments measure the dissociation constant for the binding of ligand to the first of the eight carbamylated sites available. The values reported (<1 μM) are one to two orders of magnitude less than the bulk dissociation constants measured kinetically in steady-state competition experiments versus RuBP (8.5 μM 6-phosphogluconate; Badger and Lorimer, 1981) or by equilibrium binding experiments (37 μM 6-phosphogluconate; Badger and Lorimer, 1981). Thus, the sites do not behave independently and binding of 6-phosphogulconate to the first site is considerably stronger than subsequent binding events. This finding of cooperativity between the sites has interesting physiological consequences.

It has been pointed out (Lorimer *et al.*, 1978) that the *in vivo* concentration of many of the compounds known to enhance carbamylation is well below the concentration of Rubisco sites. The ability of such effectors to maintain Rubisco in the active carbamylated state *in vivo* was therefore doubted. However, if cooperative effects are taken into consideration, it becomes quite feasible for many of these compounds to play such a role. Of course, the site that these effectors occupy in achieving their cooperative effects is no longer capable of catalysis. The loss of one site might be the price paid to ensure that the other seven sites are carbamylated at concentrations of CO$_2$ and Mg^{2+} previously thought inadequate.

C. Order of Addition of Substrates

Fraom a kineticist's standpoint, Rubisco is a singularly ill-behaved enzyme. Several factors complicate the assays of carboxylase and oxygenase. Among these complications are (1) the requirement of both reactions for activation of the enzyme by preincubation with CO_2 and Me^{2+} (Lorimer *et al.*, 1977); (2) the nonlinear time courses of both carboxylation and oxygenation—for unknown reasons the catalytic capacity of the enzyme declines with a half-time of 5–10 min from the instant RuBP is added, to yield a final "true" steady-state rate that may be as low as 10–20% of the initial rate (Section V,B,4); (3) contamination of RuBP with inhibitors (Paech *et al.*, 1978); (4) the need to avoid the competitive effects of O_2 and CO_2 on carboxylation and oxygenation, respectively; (5) inhibition by HCO_3^- at high $[CO_2]$ and/or high pH; and (6) the low K_m for RuBP, which, when combined with the insensitive response of the O_2 electrode, makes oxygenase assays at [RuBP] $<$ K_m technically demanding. In hindsight, one can now recognise that many of the published steady-state kinetic analyses are flawed by failure to take into account one or more of these complications.

An important mechanistic detail concerns the order in which the substrates add to the enzyme. With many kinetically well-behaved enzymes, the question of kinetic order can be answered by one or other of the steady-state kinetic techniques (Cleland, 1963; Segel, 1975). With Rubisco, such kinetic studies have not yielded unequivocal conclusions.

H_2O_2 is an uncompetitive inhibitor with respect to RuBP in both carboxylase and oxygenase reactions and competitive against O_2 in the oxygenase reaction (Badger *et al.*, 1980). Since H_2O_2 is not a substrate, this result implies that oxygenation is ordered with RuBP binding first. However, in the carboxylase reaction, H_2O_2 is mixed noncompetitive against CO_2, implying a random mechanism for carboxylation. Both interpretations are compromised by the likelihood that H_2O_2 reacts with RuBP (nonenzymatically and/or enzymatically) to form 2-peroxypentitol 1,5-bisphosphate (2PPBP). As an analog of the putative oxygenase reaction intermediate, 3-keto-2-peroxyarabinitol 1,5-bisphosphate, 2PPBP would likely be an effective inhibitor of both carboxylation and oxygenation.

Product inhibition studies (Laing and Christeller, 1980) suggest either random addition or ordered addition with RuBP binding first. A random mechanism for substrate addition was inferred from studies with carbonyl sulfide (COS), an analog of CO_2 (Laing and Christeller, 1980). COS was shown to inhibit carboxylation competitively with respect to CO_2 and noncompetitively with respect to RuBP. An ordered reaction requires that a dead-end inhibitor (which COS was assumed to be) that is competitive for the second substrate be uncompetitive for the first substrate. Therefore, it was concluded that the reaction involved the random addition of substrates.

However, the assumption that COS is a dead-end inhibitor is incorrect. COS is an alternative substrate (Lorimer and Pierce, 1987). The products of thiocarboxylation are 3-phospho-1-thio-D-glycerate and l-3-P-glycerate. The inhibition patterns observed with COS are therefore similar to these observed with the other alternate substrate, O_2. Accordingly, the sequence of substrate binding cannot be inferred from steady-state inhibition studies.

An entirely different approach to the kinetic mechanism of Rubisco was taken by the DuPont group (Pierce et al., 1986a). They reasoned that, if the reaction proceeded with random addition of the substrates or with the addition of the gaseous substrates first, one ought to be able to demonstrate a binary complex of carbamylated enzyme–$^SCO_2$ (or O_2) in the absence of RuBP. Me^{2+} interacts with the carboxyl group of 2CABP, the analog of 3-keto-2CABP (Miziorko and Sealy, 1984; Pierce and Reddy, 1986). For the quaternary complexes of enzyme–$^ACO_2$–Mn^{2+}–2CABP, the relaxation rate of the carboxyl group is so fast that the resonance is undetectable by NMR. Since the carboxyl group of 3-keto-2CABP is derived from $^SCO_2$, Pierce et al. (1986a) sought to demonstrate formation of the binary carbamylated enzyme–$^SCO_2$ complex by observing an enhancement of the relaxation rate of $^{13}CO_2$ by protein-bound Mn^{2+}.

Relaxation rates of $H^{13}CO_3^-$ and $^{13}CO_2$ were determined in the presence of substoichiometric amounts of Mn^{2+} for both spinach and R. rubrum enzymes. The spectral line width of the HCO_3^- resonance was increased by the enzyme-bound Mn^{2+}, but that of the CO_2 resonance was scarcely affected. Additionally, there was a marked enhancement of the longitudinal relaxation rate of HCO_3^- but not of CO_2. The carbon–metal distance for the Mn^{2+}-HCO_3^- interaction was calculated to be 5.7 Å for the spinach enzyme and 5.5 Å for the R. rubrum enzyme. This is in agreement with the value of 5.4 Å previously reported for the spinach enzyme (Miziorko and Midvan, 1974). Pierce and Reddy (1986) found the phosphorus atoms of 2CABP to be within 5–6 Å of the divalent metal ion in the quaternary complex enzyme–$^ACO_2$–Me^{2+}-2CABP. This suggests that HCO_3^- binds to the carbamylated enzyme at or near the phosphate binding site. Consistent with this idea, the enhancement of the relaxation rate of HCO_3^- by enzyme-bound Mn^{2+} was completely eliminated upon the addition of 2CABP.

An attempt was made to create a binding site for $^SCO_2$ by the addition of XuBP (an analog of RuBP). However, even though the relaxation rate of HCO_3^- was much reduced, no increase in the negligible enhancement of the relaxation rate of CO_2 was observed. These observations are inconsistent with the binding of $^SCO_2$ to the carbamylated enzyme, even though HCO_3^-, which is not a substrate, manifestly does bind. Two possible conclusions can be drawn regarding $^SCO_2$ binding in the absence of RuBP: (1) either $^SCO_2$ does not bind to the carbamylated enzyme or (2), if it does so, its rate of dissociation from the enzyme is too slow ($k < 10^2$ sec^{-1}) to permit detection of its binding by the NMR technique.

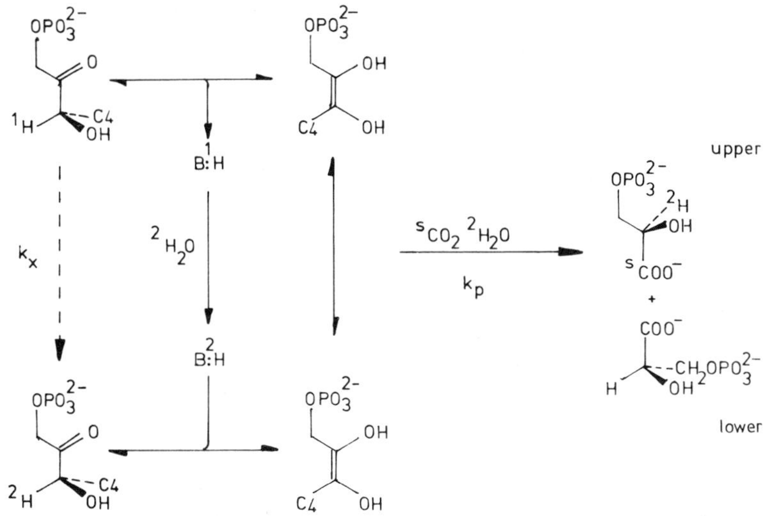

Fig. 11. The loss H-3 of RuBP catalyzed by Rubisco in ^2H_2O occurs via two pathways: via carboxylation (k_p) and exchange (k_x) with the medium. Both pathways involve the intermediacy of the enzyme-bound 2,3-enediol(ate) of RuBP.

With these conclusions in mind, attempts were made to trap the putative carbamylated enzyme–$^SCO_2$ Michaelis complex by the isotope-trapping technique of Rose (1980). No $^SCO_2$ was trapped with either the spinach or the *R. rubrum* Rubisco (Pierce *et al.*, 1986a; Jaworowski and Rose, 1985). Once again, two possible conclusions could be drawn. Either no binary enzyme–$^SCO_2$ complex exists or the dissociation of $^SCO_2$ from the binary and/or ternary complexes is too fast to permit trapping by this method. The rate constant for the release of $^SCO_2$ from the enzyme would have to be greater than 10^2 sec^{-1} in order to escape trapping. Although this is within the range accessible by the NMR method described above, no rapidly exchanging interaction between CO_2 and enzyme was observed. One is therefore forced to conclude that neither spinach nor *R. rubrum* Rubisco forms a Michaelis complex with $^SCO_2$, at least not in the absence of RuBP (Pierce *et al.*, 1986a). A similar conclusion was reached with respect to O_2 on the basis of equilibrium binding experiments.

Obviously RuBP must bind to the enzyme before the binding of $^SCO_2$ or O_2. But does the first step in catalysis *per se*, enolization of RuBP, require the presence of the gaseous substrates?

Consider the reaction of [1-^{1}H, 3-^{1}H]RuBP with Rubisco in ^2H_2O (Fig. 11). A decline in the quantity of H-1 of RuBP occurs only as a result of product formation. H-3 of RuBP, on the other hand, can be lost via two pathways: (a) the exchange pathway (k_x) for the conversion of [1-^{1}H, 3-^{1}H]RuBP to [1-^{1}H, 3-^{2}H]RuBP, which is irreversible in ^2H_2O; (b) the catalytic pathway (k_p),

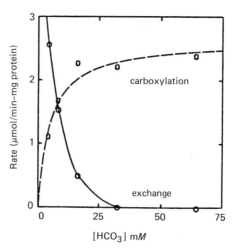

Fig. 12. The effect of [CO_2] on the exchange of H-3 of RuBP with the medium during catalysis. [Reproduced with permission from Pierce *et al.* (1986a).]

which leads to product formation. By allowing Rubisco to interact with [1-^{1}H, 3-^{1}H]RuBP in a ^{2}H$_2$O solution, one can follow, by proton NMR, the change of the resonance intensities associated with H-1 and H-3 of RuBP and H-2 of the product *l*-3-P-glycerate (Gutteridge *et al.*, 1984b). If Rubisco catalyzes the enolization of RuBP and, consequently, the exchange of the H-3 proton with solvent deuterons, faster than it catalyzes product formation, then H-3 of RuBP will be lost faster than H-1. When such an experiment was performed with *R. rubrum* Rubisco at low [CO_2], H-3 of RuBP declined faster than H-1, showing that exchange occurs rapidly. At saturating [CO_2], however, the rates of H-3 and H-1 decline were similar. Under such conditions, exchange is slow relative to product formation. By using this technique over a range of [CO_2], it was found that, whereas the rate of product formation increases hyperbolically with [CO_2] (as expected!), the rate of the exchange reaction is increasingly inhibited as the [CO_2] rises (Fig. 12). When the data are extrapolated to zero [CO_2], it is apparent that exchange (and, by inference, enolization) of RuBP occurs in the absence of CO_2 at rates that exceed the maximum rate of carboxylation. Less extensive measurements with spinach enzyme (Pierce *et al.*, 1986a) and wheat enzyme (Gutteridge *et al.*, 1984b) are consistent with this conclusion. Thus, enolization of RuBP occurs prior to the interaction of the gaseous substrates with the enzyme–2,3-enediolate complex.

As a consequence of the above results, we favor the ordered kinetic mechanism shown in Fig. 13 for the carboxylation and oxygenation of RuBP. Farquhar (1979) has derived the following steady-state rate equations to

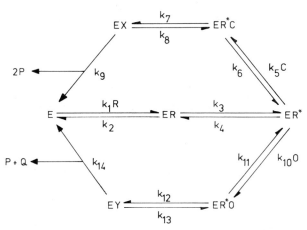

Fig. 13. The ordered kinetic mechanism for the carboxylation and oxygenation of RuBP. E, enzyme; R, RuBP; R*, 2,3-enediolate of RuBP; C, CO_2; O, O_2; X, 3-keto-2CABP; Y, β-keto-2PABP; P, 3-P-glycerate; Q, 2-P-glycolate.

describe the initial rates of carboxylation (v_c) and oxygenation (v_o):

$$v_c = \frac{V_c RC}{K_c R + RC + K_c RO/K_o + K_{Rc}C + K_{Ro}K_c O/K_o + K_{iR}K_c} \tag{I}$$

and

$$v_o = \frac{V_o RO}{K_o R + RO + K_o RC/K_c + K_{Ro}O + K_{Rc}K_o C/K_c + K_{iR}K_o} \tag{II}$$

where the various parameters are defined in the footnote.

For an ordered mechanism, these equations yield two interesting kinetic relationships. First,

$$V_c/V_o = K_{Rc}/K_{Ro} \tag{III}$$

that is, the ratio of the V_{max} values for carboxylation and oxygenation equals the ratio of the RuBP Michaelis constants for these two reactions. Although this relationship was first noted 10 years ago (Badger and Collatz, 1977), it has yet to be confirmed experimentally. The technical problems mentioned earlier make such measurements difficult. Second,

$$V_c/V_o = [V_c/K_c]C/[V_o/K_o]O = s_{rel}C/O \tag{IV}$$

This relationship was discussed in Section II,C. It has been found to be universally true for a wide variety of Rubiscos (Jordan and Ogren, 1981), activated with different metal ions (Jordan and Ogren, 1983), and over an extensive range of $[CO_2]$, $[O_2]$, $[RuBP]$, $[H^+]$, and temperature (Jordan and Ogren, 1984).

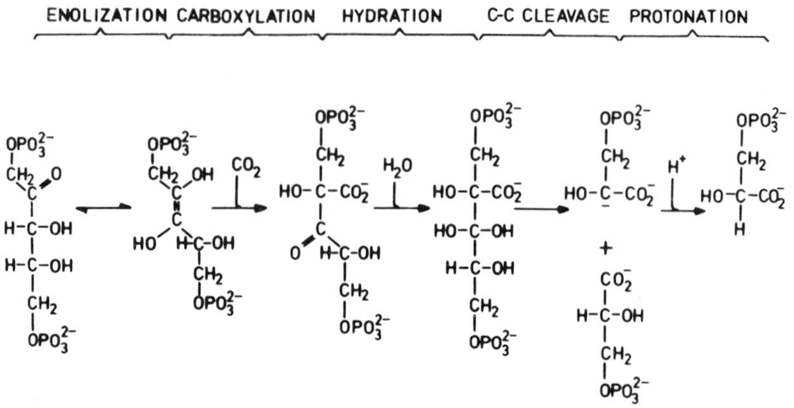

ENOLIZATION   CARBOXYLATION   HYDRATION   C-C CLEAVAGE   PROTONATION

Fig. 14. The mechanism of carboxylation of RuBP.

D. Reaction Intermediates and Their Properties

Following the discovery of RuBP carboxylase (Quayle *et al.*, 1954; Weissbach *et al.*, 1954), Calvin (1956) proposed a mechanism, a more detailed form of which is shown in Fig. 14. The reaction involves several steps and intermediates. *Deprotonation* of C-3 (*enolization*) generates the necessary nucleophilic center at C-2 in the form of the 2,3-enediol(ate). *Carboxylation* of the 2,3-enediol(ate) creates the six-carbon, β-keto acid intermediate, 3-keto-2CABP (ketone form), which undergoes *hydration* to the gem diol form. Deprotonation of the O-3 of 3-keto-2CABP (gem diol) initiates *carbon–carbon cleavage*, yielding a molecule of *l*-3-P-glycerate and the C-2 carbanion (or aci-acid) form of *u*-3-P-glycerate. The stereospecific protonation of this C-2 carbanion to yield *u*-3-P-glycerate completes the reaction.

Isotope labeling experiments support this outline of events. Thus, the substrate CO_2 [not HCO_3^- (Cooper *et al.*, 1969)] becomes attached to C-2 of RuBP and carbon–carbon cleavage occurs between C-2 and C-3 (Mullhofer and Rose, 1965; Pierce *et al.*, 1980b). The proton at C-3 of RuBP is lost to the medium, while the proton at C-2 of *u*-3-P-glycerate is derived ultimately from the solvent (Fiedler *et al.*, 1967). One of the carboxyl oxygen atoms of *l*-3-P-glycerate is derived from the solvent, the other from the C-3 oxygen atom of RuBP. The C-2 oxygen atom of RuBP is also retained to become O-2 of *u*-3-P-glycerate (Lorimer, 1978; Sue and Knowles, 1978). The retention of both C-2 and C-3 oxygen atoms during carboxylation eliminates from consideration mechanisms of carboxylation involving covalent bonds between the enzyme and C-2 or C-3 of the substrate or reaction intermediates.

1. 2,3-Enediol(ate)

The evidence for the intermediacy of the 2,3-enediol(ate) comes from hydrogen isotope exchange experiments and chemical trapping experiments. The intermediacy of the 2,3-enediol(ate) requires the enzyme-catalyzed abstraction of the C-3 proton. Provided that this proton (now bound to an enzyme base) exchanges rapidly with the solvent, an exchange of protons between C-3 of RuBP and the solvent is to be expected. Such exchanges are well precedented in the reactions catalyzed by a number of isomerases, where the intermediacy of enediol(ate)s is not questioned (Rose, 1980). Early attempts to demonstrate this exchange were frustrated by the failure to utilize carbamylated enzyme (Hurwitz et al., 1956; Fiedler et al., 1967). When carbamylated R. rubrum Rubisco was used (Saver and Knowles, 1982; Sue and Knowles, 1982), it was possible to demonstrate both the "wash in" of solvent ^3H into the C-3 of RuBP and the "wash out" of [3-^3H]RuBP into the solvent. Such exchange experiments constitute strong evidence for the intermediacy of the 2,3-enediolate. Additional support for this conclusion comes from the "competition experiments" between exchange and carboxylation referred to in Section IV,C. Exchange is maximal when carboxylation [a competing fate for the 2,3-enediol(ate)] is minimal (Gutteridge et al., 1984b; Pierce et al., 1986a).

The results of chemical trapping experiments also support the intermediacy of the 2,3-enediol(ate). Bhagwat and McFadden (1982) reported that carbamylated spinach Rubisco catalyzed an RuBP-dependent reduction of tetranitromethane to nitroform, indicating the presence of a carbanion intermediate. Since nitroform production was inhibited by high [CO_2], they concluded that the intermediate in question was the 2,3-enediol(ate). However, Mulligan and Tolbert (1983), in comparable experiments using $Fe(CN)_6^{3-}$ and other oxidants, were unable to detect the presence of a carbanion intermediate with spinach Rubisco.

The properties of the 2,3-enediol(ate) of triose phosphate have recently been explored by Richards (1984, 1985). Of particular relevance to the present discussion was the observation that it underwent a very rapid β-elimination of the C-1 phosphate group (Fig. 15). The 2,3-enediol(ate) of RuBP formed nonenzymatically under alkaline conditions from both RuBP and XuBP also undergoes the loss of the C-1 phosphate with the generation of the dicarbonyl compound, 1-deoxy-2,3-pentadiulose 5-phosphate (Paech et al., 1978). Thus, when enzyme-bound 2,3-enediol(ate) is released from the enzyme by acid quenching, the expected fate of the released intermediate is to undergo β-elimination of the C-1 phosphate. When R. rubrum Rubisco was acid quenched during steady-state carboxylation, an intermediate that decomposed to Pi with a half-time of <6 msec was detected (Jaworowski et al., 1984). The release of Pi from this intermediate was prevented by the presence of I_2. The amount of this Pi-producing intermediate increased as the [CO_2] used in its production decreased. These are the properties to be expected of the 2,3-enediol(ate). Similar experiments with spinach Rubisco (Mulligan and Tolbert, 1983) failed to detect this Pi-releasing intermediate, perhaps indicating that the steady-state level of the 2,3-enediol(ate) on the spinach enzyme is much lower than on the R. rubrum enzyme.

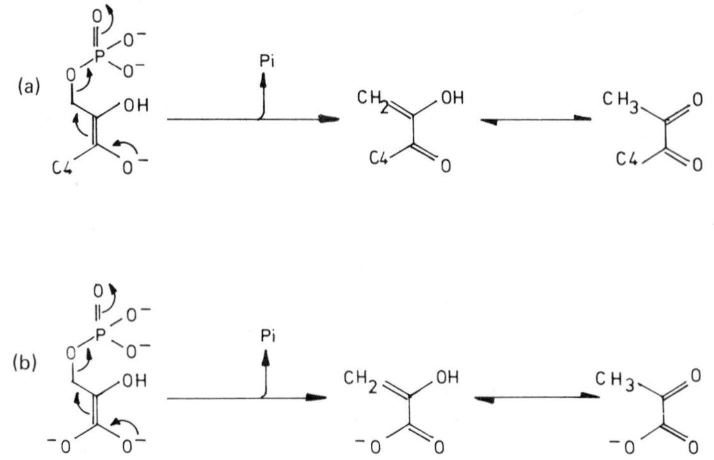

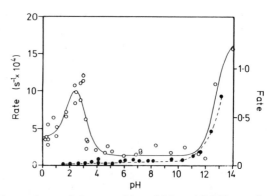

Fig. 15. The β-elimination of the C-1 phosphate group from (a) the 2,3-enediolate of RuBP to yield Pi and 1-deoxy-D-glycero-2,3-pentadiulose 5-phosphate and (b) the aci-acid form of 3-P-glycerate to yield Pi and pyruvate.

2. 2-Carboxy-3-keto-arabinitol 1,5-Bisphosphate

The evidence for the intermediacy of 3-keto-2CABP is compelling. As Calvin (1956) predicted, addition of CO_2 to the C-2 of the 2,3-enediol(ate) of RuBP results in the formation of a β-keto acid. Subsequent attempts to demonstrate its existence and to characterize its properties (Moses and Calvin, 1958; Sjödin and Vestermark, 1973; Siegel and Lane, 1973) were frustrated by its reputed lability. However, the ability of 2-carboxy-pentitol

Fig. 16. The fate and rate of decomposition of 3-keto-2CABP as a function of pH. Fate (closed circles) is numerically represented as the partition coefficient [hydrolysis/(decarboxylation plus hydrolysis)]. Rate is shown by open circles. [Reproduced with permission from Lorimer *et al.* (1986).]

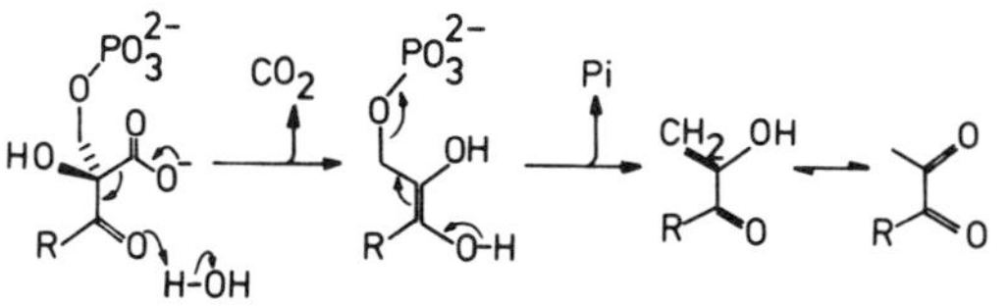

Fig. 17. The nonenzymatic decarboxylation of 3-keto-2CABP and the subsequent β-elimination of the C-1 phosphate group.

1,5-bisphosphate (a mixture of 2CABP and 2CRBP) to act as a potent inhibitor of carboxylation was taken as evidence for the intermediacy of the six-carbon β-keto acid (Siegel and Lane, 1972; Wolfenden, 1972). Borohydride trapping experiments (Schloss and Lorimer, 1982) (Section IV,E) conclusively established the existence of this intermediate and established its stereochemical configuration about C-2.

The ability to trap 3-keto-2CABP with borohydride depends on first denaturing the enzyme with a brief acid quench; that is, the reduction of the C-3 carbonyl group occurs in free solution. The intermediate could not be reduced with borohydride directly on the enzyme, a result that can now be attributed to the fact that 3-keto-2CABP exists on the enzyme predominantly, perhaps solely, as the hydrated C-3 gem diol, which is not amenable to borohydride reduction (Lorimer et al., 1986).

The ability to release 3-keto-2CABP by acid quenching of the reaction in progress led to the discovery that it is much more stable than previously thought (Pierce et al., 1986b; Lorimer et al., 1986). The chemical properties of 3-keto-2CABP in free solution have been explored (Fig. 16). Over a wide range of pH (4–11), 3-keto-2CABP undergoes decarboxylation at a constant, pH-independent rate (2×10^{-4} sec^{-1}). Jaworowski et al. (1984) have shown that this decarboxylation is accompanied by the β-elimination of the C-1 phosphate, presumably via the 2,3-enediolate. A ^{13}C NMR study of the decarboxylation reaction (Pierce et al., 1986b) provided additional evidence for this β-elimination. The product of the decarboxylation (Fig. 17), 1-deoxy-D-glycero-2,3-pentodiulose 5-phosphate, is also formed under alkaline conditions from both RuBP and XuBP (Paech et al., 1978). At pH values below ~4, acid catalysis of the decarboxylation reaction is observed, as is common for β-keto acids (Fig. 18). At pH values above ~11, 3-keto-2CABP is subject to hydrolysis rather than decarboxylation. The products of this nonenzy-

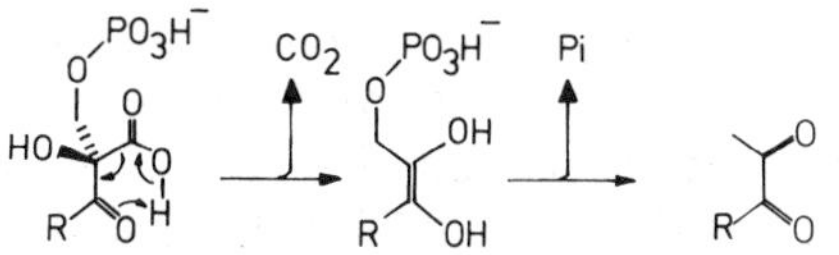

Fig. 18. Acid-catalyzed, nonenzymatic decarboxylation of 3-keto-2CABP.

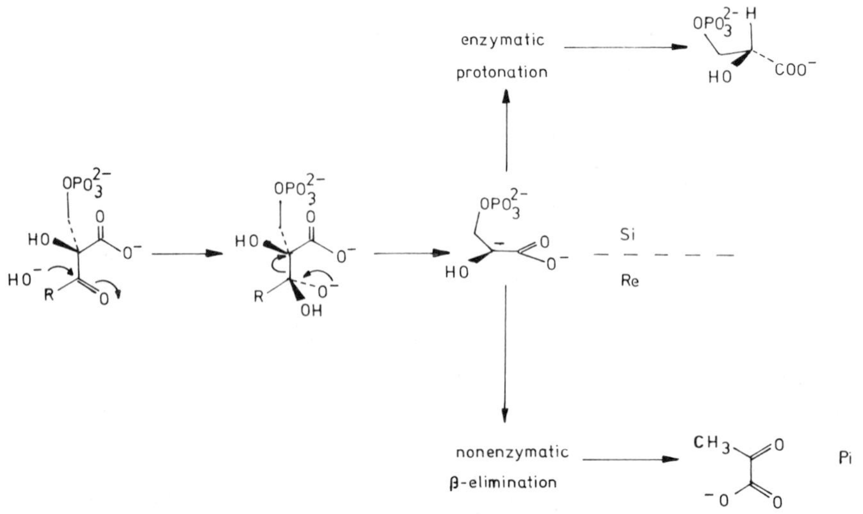

Fig. 19. The hydrolysis of 3-keto-2CABP to yield (bottom) nonenzymatically, *l*-3-P-glycer-ate, pyruvate, and Pi by β-elimination or (top) enzymatically, *u*-3-P-glycerate and *l*-3-P-glycer-ate by stereospecific protonation.

matic hydrolysis are not, however, two molecules of *u*- and *l*-3-P-glycerate but rather pyruvate, Pi, and *l*-3-P-glycerate (Fig. 19). Presumably, the C-2 carbanion of *u*-3-P-glycerate undergoes β-elimination, rather than protona-tion.

In solution, 3-keto-2CABP is capable of existing in two states, the free ketone form and the hydrated carbonyl or gem diol form. ^{13}C NMR analyses of [2,3^{13}C] 3-keto-2CABP established that, in free solution, the free ketone form is favored over the gem diol form by a factor of at least 15 : 1 (Pierce *et al.*, 1986b). The rate of hydration of 3-keto-2CABP in free solution was measured by exchange with ^{18}O-labeled water and determined to be about 2.5×10^{-3} sec^{-1}. Since this rate is about 10^3-fold slower than V_c, it follows that the enzyme must accelerate the hydration of the C-3 carbonyl group of 3-keto-2CABP by at least three orders of magnitude.

Although the predominant form of 3-keto-2CABP in free solution is the free ketone, the situation on the enzyme is reversed. Borohydride trapping experiments using [3-^{18}O]RuBP yielded carboxypentitols that were only 50% enriched (Lorimer *et al.*, 1986). Such a result is to be expected if the car-bonyl group of 3-keto-2CABP is substantially hydrated on the enzyme. The divalent metal ion presumably plays an important role in forming and stabi-lizing the hydrated from of 3-keto-2CABP on the enzyme (Section IV,G).

The O-3 oxygen of RuBP is completely retained during carboxylation (Lorimer, 1978; Sue and Knowles, 1978). Although the enzyme clearly cata-lyzes the hydration step, no wash-out of the carbonyl oxygen of 3-keto-

2CABP occurs. Thus, the hydration step must either be kinetically irreversible and/or sterochemically directed so that the same oxygen atom is removed (as water) during dehydration as was added during hydration.

3. C-2 Carbanion (or Aci-Acid) Form of u-3-P-Glycerate

Abstraction of a proton from the gem diol form of 3-keto-2CABP initiates carbon–carbon bond cleavage between C-2 and C-3 and leads (formally) to the formation of the C-2 carbanion (or aci-acid) form of u-3-P-glycerate (Fig. 19). In free solution this species has the potential to undergo either protonation to give a mixture of D- and L-3-P-glycerates or β-elimination of the C-1 phosphate to yield pyruvate and Pi. The alkaline hydrolysis of 3-keto-2CABP produces pyruvate and Pi, indicating that β-elimination is the predominant fate of the C-2 carbanion in free solution under these conditions.

An attempt to detect this intermediate has been made by acid quenching of the reaction in progress (Jaworowski *et al.*, 1984). The expected products, L-3-P-glycerate or pyruvate, could not be detected. However, when carboxylation was performed in [^{3}H]water, the specific activity of the u-3-P-glycerate was about one-sixth that of the water protons (Saver and Knowles, 1982; Fiedler *et al.*, 1967). This discrimination implies that there is competition between ^{3}H and ^{1}H in the protonation of the putative C-2 carbanion and can be taken as evidence for its intermediacy. The failure to detect it in acid-quenched reactions (Jaworowski *et al.*, 1984) may simply indicate that it does not accumulate to detectable amounts.

E. Stereochemistry

The conversion of D-ribulose 1,5-bisphosphate into two molecules of 3-phospho-D-glycerate (stereoconfiguration $2R$) can potentially occur via two stereochemically distinct pathways (Fig. 20).

1. Re attack/retention: Attack of CO_2 on the 2,3-enediolate occurs on the Re face of C-2 generating a 3-keto-2CRBP with the $2S$ stereoconfiguration. Hydrolysis of 3-keto-2CRBP to u-3-phospho-D-glycerate requires retention of configuration about C-2.

2. Si attack/inversion: Alternatively, attack of CO_2 on the Si face of C-2 generates 3-keto-2CABP ($2R$) as the intermediate. Hydrolysis of 3-keto-2CABP requires inversion of configuration about C-2. The experimental evidence clearly implicates Si attack/inversion as the stereochemical course of carboxylation.

Pierce *et al.* (1980a) were the first to chromatographically resolve the epimeric mixture of 2′-carboxy-D-pentitol bisphosphates ultimately resulting from the addition of cyanide to RuBP. They characterized the purified epimers as 2CABP (2′-phosphohydroxymethyl-D-ribonic acid 5-phosphate) and 2CRBP. Since 2CABP was several orders of magnitude more inhibitory than

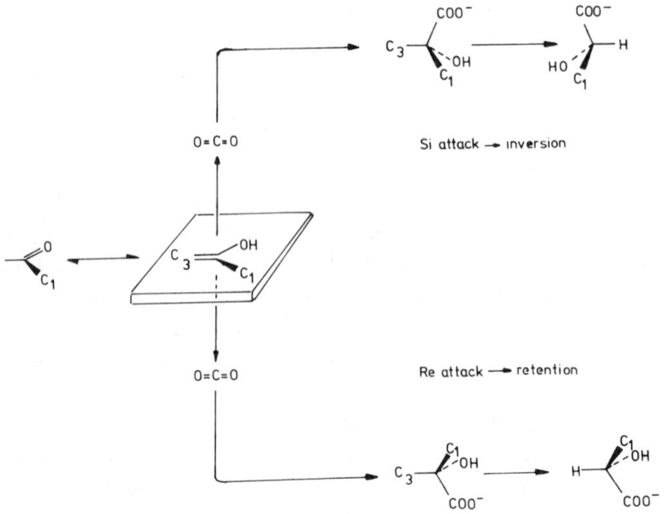

Fig. 20. The two stereochemically possible courses of carboxylation of RuBP to yield u-3-phospho-D-glycerate. Top: Attack of CO_2 on the Si face of C-2 must be coupled with inversion of configuration around C-2 during hydrolysis. Bottom: Attack of CO_2 on the Re face of C-2 must be coupled with retention of configuration about C-2 during hydrolysis.

2CRBP, it was concluded that the authentic reaction intermediate must also have the same stereoconfiguration about C-2 as 2CABP, that is, $2R$. Consequently, the stereochemical course of carboxylation must involve Si attack/inversion.

A more direct approach to the stereochemical course of carboxylation was adopted by the Du Pont group. They have addressed two questions. (1) How is the carbonyl group of the substrate RuBP oriented within the catalytic site? (2) Is this orientation of the carbonyl group of RuBP related to the direction from which CO_2 attacks the 2,3-enediolate? When the carbonyl group of RuBP is reduced by borohydride in free solution, reduction occurs with a very slight stereochemical preference for the Re face (Fig. 21). In contrast, when the RuBP is bound at the active site of carbamylated enzyme, the reduction by borohydride occurs exclusively on the Si face (G. H. Lorimer and S. Gutteridge, unpublished data). Presumably, access to the Re face is blocked by the enzyme. Curiously, when the RuBP is bound at the active site of decarbamylated enzyme, the reduction by borohydride, while showing preference for the Si face, does not occur exclusively on the Si face. Perhaps the divalent metal ion, besides enhancing the reactivity of the carbonyl group through polarization, also limits access to the Re face.

Experiments were designed to trap the 6-carbon intermediate formed upon carboxylation of the 2,3-enediolate and to define the stereochemical course of its formation (Schloss and Lorimer, 1982). These were based on the idea that the labile 6-carbon intermediate could be stabilized by borohydride

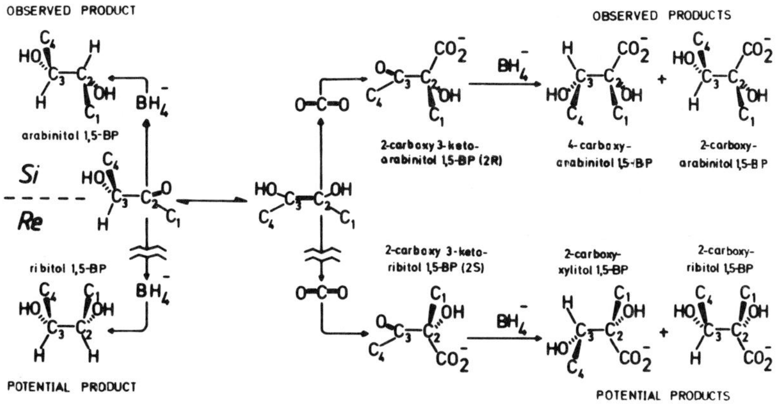

Fig. 21. The stereochemical orientation of the carbonyl group of enzyme-bound RuBP (determined by borohydride reduction) and the stereochemical configuration about C-2 of the six-carbon carboxylated reaction intermediate (determined by borohydride reduction of acid-quenched carboxylase reactions in progress).

reduction to the corresponding 2'-carboxypentitol bisphosphates (Fig. 21). Depending on the stereochemical course of carboxylation, either 3-keto-2CABP (Si attack) or 3-keto-2CRBP (Re attack) are to be expected. Reduction of the former by borohydride yields a stable epimeric mixture of 2CABP and 4CABP (equivalent to 2'-carboxy-D-lyxitol bisphosphate), while reduction of the latter yields a mixture of 2CXBP and 2CRBP. The mixture of 2CABP and 4CABP can be distinguished from the mixture of 2CXBP and 2CRBP both chromatographically and by virtue of their (2CABP and 4CABP) more potent (by a factor of at least 10^5) inhibitory properties. A mixture of 2CABP and 4CABP was recovered following reduction of the 6-carbon intermediate, implying (Fig. 21) that the attack of CO_2 on the 2,3-enediolate proceeds on the Si face of C-2, so as to generate 3-keto-2CABP as an intermediate (Schloss and Lorimer, 1982). This conclusion is consistent with the orientation of RuBP within the active site and also with the evidence that the reaction is ordered with RuBP binding and undergoing enolization before reacting with CO_2 (Pierce *et al.*, 1986a).

One aspect of the stereochemical course of carboxylation which remains to be defined concerns the configuration of the 2,3-enediolate, cis (Z) or trans (E). If the C-3 proton abstracted during enolization is removed by an enzyme group B (Fig. 22), which is located on the Si side with respect to the plane of the C-2 carbonyl, the resultant 2,3-enediolate will have the trans configuration. Alternatively, when the proton-abstracting group is on the Re side, the resultant 2,3-enediolate will have the cis configuration (Fig. 22). Currently, no distinction has been made between these two possibilities, so that the stereochemical orientation about C-3 of the 2,3-enediolate and subsequent intermediates remains undefined.

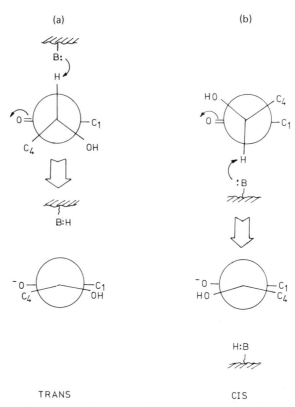

Fig. 22.   Abstraction of H-3 by an enzyme group B: located on the Si side with respect to the plane of the carbonyl group leads to the *trans*-2,3-enediolate (a), while abstraction of H-3 by an enzyme group B: located on the Re side with respect to the plane of the carbonyl group leads to the *cis*-2,3-enediolate (b). Currently, no distinction has been made so that the orientation of the groups at C-3 remains undefined.

A final point of stereochemical interest arises as a consequence of the inversion of configuration about C-2 accompanying the hydrolysis of 3-keto-2CABP. More specifically, it concerns the direction from which the C-2 proton of *u*-3-phospho-D-glycerate arises. Inspection (Fig. 23) shows that this proton is added to the Si face of the carbanion at C-2 of *u*-3-phospho-D-glycerate. It should be noted that the Si face of this carbanion is stereochemically remote from that region of the enzyme that catalyzed the proton-transfer reactions at C-3 and O-3 of the substrate and reaction intermediates, respectively. If an enzyme group is involved in catalyzing this terminal protonation, on stereochemical grounds it must be different from the group(s) catalyzing the other proton-transfer reactions accompanying carboxylation.

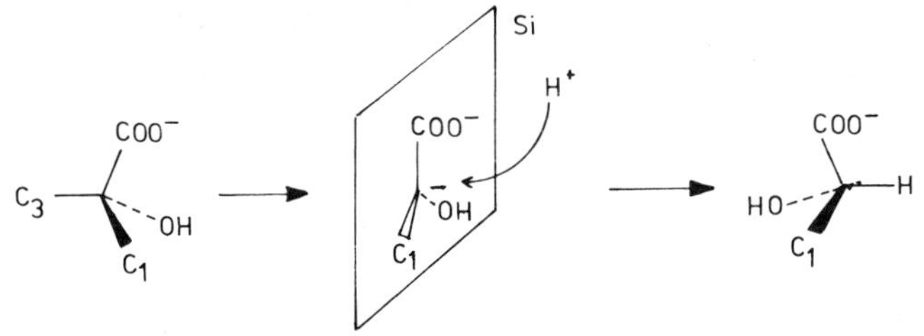

Fig. 23. The stereospecific protonation of the C-2 carbanion form of *u*-3-P-glycerate occurs on the Si face. This face is stereochemically remote from the enzyme group(s) that catalyzed the proton-transfers from C-3 and O-3.

F. Partial Reactions

1. Enolization

Enolization can occur as a partial reaction independently of the reactions involving the gaseous substrates. This can be inferred from the ability of the carbamylated enzyme to catalyze the exchange of the C-3 proton of RuBP with the solvent (Saver and Knowles, 1982; Sue and Knowles, 1982; Gutteridge *et al.*, 1984b; Pierce *et al.*, 1986a) (Section IV,C). Enolization requires that the enzyme be carbamylated, which suggests a possible role for

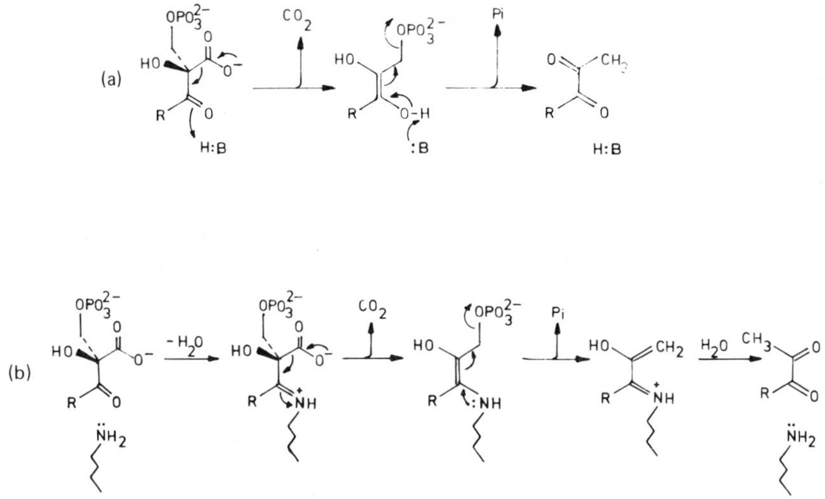

Fig. 24. Two plausible mechanisms for the decarboxylation of 3-keto-2CABP catalyzed by decarbamylated enzyme. (a) General acid catalysis is consistent with the requirement (established in the direction of carboxylation) that the O-3 oxygen atom be retained. (b) Electrophilic catalysis via a Schiff base recognizes the propensity of $K^{L2}166$ ($K^{L8}175$) to form a Schiff base with pyridoxal phosphate.

Me^{2+} in the polarization of the C-2 carbonyl bond (Section IV,G). Additionally, at least for L8S8 Rubiscos, enolization also requires the presence of both L and S subunits (Andrews *et al.*, 1986) (Section VI,C).

2. Decarboxylation

The ability to prepare substrate quantities of [2'-^{14}C] 3-keto-2CABP (Pierce *et al.*, 1986b) led to the discovery of a second partial reaction, the decarboxylation of 3-keto-2CABP. This is specifically catalyzed by decarbamylated enzyme. Since the enzyme-catalyzed decarboxylation is inhibited by 2CABP, it presumably occurs within the catalytic site. Since the decarbamylated enzyme does not catalyze the deprotonation of C-3 of RuBP, not surprisingly, it does not catalyze the reverse reaction, the protonation of C-3 of the 2,3-enediol(ate) product of the decarboxylation of 3-keto-2CABP. Instead, β-elimination of the C-1 phosphate group accompanies the decarboxylation. It is not known, however, if the β-elimination of phosphate occurs on the enzyme or in free solution.

It is not yet known if the decarboxylation of 3-keto-2CABP by decarbamylated enzyme occurs by a mechanism that represents a true reversal of the carboxylation step catalyzed by carbamylated enzyme. A mechanism involving general acid catalysis has been proposed (Fig. 24) (Lorimer *et al.*, 1986). Operating in reverse (Fig. 25) (i.e., in the direction of carboxylation), this mechanism satisfies the requirement that the O-3 oxygen atom of RuBP be retained (Lorimer, 1978; Sue and Knowles, 1978). Additionally, it provides a mechanism whereby a single enzyme group could catalyze the three proton-transfer steps to and from the O-3 oxygen. Alternatively, the enzyme-catalyzed decarboxylation could occur by a mechanism that is not a true reversal of the carboxylation reaction. For example, catalysis could be achieved via Schiff base formation (with L^{L2}166 or K^{L8}175?). Operated in reverse, such a mechanism would not satisfy the criterion of O-3 retention. However, the ability of K^{L2}166 (and K^{L8}175) to form a Schiff base with the carbonyl group of pyridoxal phosphate (Herndon *et al.*, 1982; Spellman *et al.*, 1979) should be mentioned in this context.

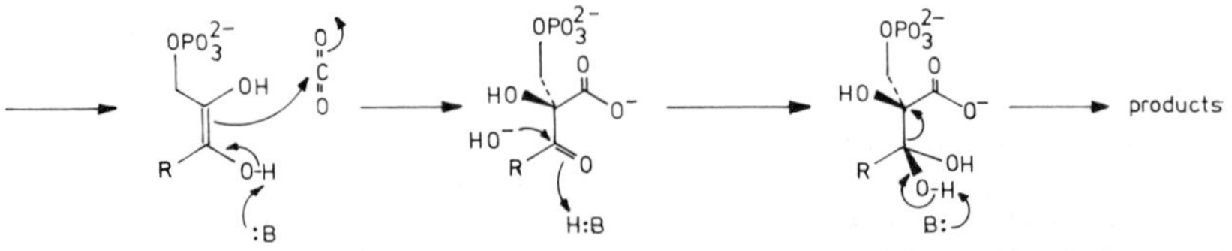

Fig. 25. The proton transfers associated with three steps of the carboxylation reaction (carboxylation, hydration of the carbonyl group of 3-keto-2CABP, and carbon–carbon cleavage) can be facilitated by a single-enzyme group B.

3. *Hydrolysis*

The availability of 3-keto-2CABP enabled a third partial reaction, its hydrolysis to products, to be analyzed independently of the other steps in catalysis (Pierce *et al.*, 1986b). L2 and L8S8 Rubiscos, fully carbamylated with $^ACO_2$ and Mg^{2+}, hydrolyzed 3-keto-2CABP completely. No decarboxylation occurred with Mg^{2+} as the activator Me^{2+}, demonstrating that, with the formation of 3-keto-2CABP, the enzyme is fully committed to the formation of 3-P-glycerate (Section IV,I). This strong committment to product formation is as true for the L2 Rubisco as it is for the cyanobacterial and higher-plant L8S8 Rubiscos, even though these enzymes differ markedly in their V_c/K_c and s_{rel} values.

Some decarboxylation of 3-keto-2CABP was evident when Mn^{2+} and Co^{2+} were used as activator Me^{2+}. But the differences in partitioning of 3-keto-2CABP observed with the different Me^{2+} are insufficient to explain the large effects on s_{rel} that are seen with these Me^{2+}. Even the Co^{2+}-activated *R. rubrum* enzyme, which cannot catalyze the carboxylation of RuBP (Robison *et al.*, 1979; Christeller, 1981), hydrolyzes 30% of the 3-keto-2CABP to 3-P-glycerate.

While various Rubiscos clearly catalyze the hydrolysis of 3-keto-2CABP, they do so at rates which are about 3% of V_c. Nevertheless, these rates exceed by one to two orders of magnitude the nonenzymatic rate of hydration of the carbonyl group. It is therefore the free ketone form of 3-keto-2CABP that is acquired from solution by the various Rubiscos. The low rates of hydrolysis of 3-keto-2CABP are limited by an event that is not on the direct catalytic pathway for the conversion of RuBP to 3-P-glycerate. The inactivation of Rubisco by 2CABP is a two-step process involving a rapid-equilibrium binding step ($K_i = 0.4$ μM) followed by a slow (0.04 sec^{-1}) isomerization of the quaternary complex. The maximal rate of 3-keto-2CABP hydrolysis by spinach Rubisco (0.062 sec^{-1}) and its affinity for the enzyme ($K_m = 1.5$ μM) are similar to the corresponding values for 2CABP inhibition. It is therefore tempting to speculate that both phenomena involve the same rate-limiting enzyme isomerization (Pierce *et al.*, 1986b).

G. Role of the Divalent Metal Ion in Carboxylation

Much circumstantial evidence points to the intimate involvement of Me^{2+} in carboxylation. Two of the three partial reactions (Section IV,F), the exchange of the C-3 proton of RuBP with the solvent and the hydrolysis of 3-keto-2CABP, require the presence of the carbamate–Me^{2+} complex within the catalytic site (Pierce *et al.*, 1986a,b). Involvement of Me^{2+} in the reactions of the gaseous substrates with the 2,3-enediolate of RuBP can be inferred from the differential effects of various Me^{2+} on s_{rel} (Section IV,I).

Useful insights into the environment of Me^{2+} have been provided by spectroscopic studies, particularly with the quaternary complex $E-^ACO_2-Me^{2+}$-2CABP. The structure of this complex is especially interesting since it is likely to resemble quite closely the complex formed with 3-keto-2CABP (gem diol form) during carboxylation.

The EPR spectra of the $E-^ACO_2-Me^{2+}$-2CABP complexes formed with Mn^{2+} (Miziorko and Sealy, 1980, 1984; Gutteridge et al., 1984a), Co^{2+}, and Cu^{2+} (Branden et al., 1984a; Styring and Branden, 1985a,b; Nilsson et al., 1984) have been reported for both L8S8 and L2 Rubiscos. These anisotropic spectra indicate that the coordination sphere of Me^{2+} is highly distorted. This contrasts with the ternary carbamylated $E-^ACO_2-Me^{2+}$ complex, where the coordination geometry of Me^{2+} is not greatly perturbed (Nilsson et al., 1984; Styring and Branden, 1985a). The change in the coordination geometry of Me^{2+} seen upon adding 2CABP or RuBP to the carbamylated enzyme is accompanied in both L2 and L8S8 Rubiscos by the disappearance of the nitrogen hyperfine structure (Branden et al., 1984a; Styring and Branden, 1985a). The narrower hyperfine lines in the $E-^ACO_2-Cu^{2+}$-2CABP spectra are consistent with Cu^{2+} coordination to oxygen atoms only.

The spectral-broadening effects of ^{17}O have been very useful in gaining structural information about the quaternary complex (Miziorko and Sealy, 1984; Styring and Branden, 1985a). This complex was prepared with the L2 enzyme and Mn^{2+} in ^{17}O-enriched water. Spectral braodening of the Mn^{2+} resonances was observed (Miziorko and Sealy, 1984). Replacement of the isotopic water with normal water narrowed the resonances. This indicates that one or more slowly exchanging water molecules are coordinated to the Mn^{2+}. A similar experiment was performed with the L8S8 enzyme and Cu^{2+} (Styring and Branden, 1985a). The narrower hyperfine lines of the Cu^{2+} spectra enabled the conclusion to be drawn that only one water molecule is bound to Cu^{2+} in the $E-^ACO_2-Cu^{2+}$-2CABP complex.

Miziorko and Sealy (1984) prepared $E-^ACO_2-Mn^{2+}$-2CABP complexes with the L2 enzyme labeled with ^{17}O in either the 2'-carboxyl group or in the oxygen atom at C-2. Spectral broadening of the Mn^{2+} resonances was observed, a result that demonstrates the direct coordination of the metal ion to the carboxyl and C-2 oxygen atoms of 2CABP. The Swedish group has made similar observations with quaternary complexes made with Cu^{2+} and the L8S8 enzyme. In addition, they have demonstrated the direct coordination of Cu^{2+} to the carboxyl and C-2 oxygen atoms of 3-P-glycerate (Styring and Branden, 1985a).

These studies have been augmented with ^{13}C and ^{31}P NMR experiments on the quaternary complex of the L2 enzyme (Pierce and Reddy, 1986). The results indicated that the two P atoms of 2CABP were approximately equidistant from Me^{2+} at a distance of <6 Å. This study also showed that Me^{2+} was in close proximity (<6 Å) to the activator carbamate and the carboxyl group of 2CABP.

With the above results in mind, it is reasonable to conclude that Me^{2+} plays a critical role in many of the steps of carboxylation. The deprotonation of C-3 of RuBP and the stabilization of the ensuing 2,3-enediolate require the presence of the carbamate–Me^{2+} complex within the catalytic site. A likely role for Me^{2+} in this process would be to act as an electron sink, polarizing the carbonyl bond and facilitating enolization by increasing the acidity of the C-3 proton. In free solution, 2,3-enediolates of the type formed during the carboxylation of RuBP rapidly undergo β-elimination of the C-1 phosphate group (Richards, 1984; Jaworowski *et al.*, 1984). Yet within the catalytic site of Rubisco, this β-elimination reaction does not occur to a detectable extent. Presumably Me^{2+} stabilizes the 2,3-enediolate so that reaction with CO_2 or O_2 can occur. Neither the manner in which Me^{2+} interacts with the 2,3-enediolate nor the mechanism by which Me^{2+} influences the relative rates with which the gaseous substrates react with the 2,3-enediolate (s_{rel}) is understood. Nevertheless, the addition of $^SCO_2$ yields the free ketone form of 3-keto-2CABP, which is stabilized by coordination to Me^{2+} via the 2' carboxyl and C-2 oxygen atoms. There is evidence (Section IV,D) that the enzyme catalyzes the hydration of the carbonyl group of 3-keto-2CABP and for the presence of one metal-coordinated water molecule in the quaternary complex E–$^ACO_2$-Me^{2+}-2CABP. An important role for Me^{2+} in the hydration of 3-keto-2CABP can be envisaged by considering the role of the Me^{2+} in the hydration of CO_2 catalyzed by carbonic anhydrase. There, a hydroxide anion from the inner coordination sphere of the active site Zn^{2+} is thought to attack the CO_2. A ^{13}C NMR study of the Mn^{2+}-substituted carbonic anhydrase (Led *et al.*, 1982) found that the predominant enzyme-bound species is HCO_3^-, the hydrated substrate/product. The enzyme-bound HCO_3^- is directly coordinated to the Mn^{2+}, whereas CO_2 is only weakly bound to the enzyme, without a direct bond to Mn^{2+}. An analogous mechanism involving the active-site Me^{2+} of Rubisco could account for both the enhanced rate of hydration of enzyme-bound 3-keto-2CABP and the preferential stabilization of the hydrated gem-diol form of 3-keto-2CABP (Section IV,D). Such a mechanism implies coordination of Me^{2+} to one of the C-3 hydroxyl oxygens of the gem diol. ^{17}O-induced spectral broadening of Mn^{2+} or Cu^{2+} resonances in quaternary complexes prepared with [3-^{17}O]2CABP and/or [3-^{17}O]4CABP could test these ideas.

The final steps of the carboxylation reaction could also be facilitated by the presence of Me^{2+} within the active site. In free solution, the C-2 carbanion or aci-acid form of u-3-P-glycerate appears to undergo β-elimination of the C-1 phosphate to form pyruvate and Pi rather than 3-P-glycerate. Yet within the catalytic site this elimination reaction does not occur. Presumably Me^{2+} plays an important role in the stabilization of this intermediate so that it may be stereospecifically protonated to complete the reaction.

H. Mechanism of Oxygenation

Compared with the carboxylase reaction there is a dearth of experimental evidence about the oxygenase reaction. Early ^{18}O-labeling experiments

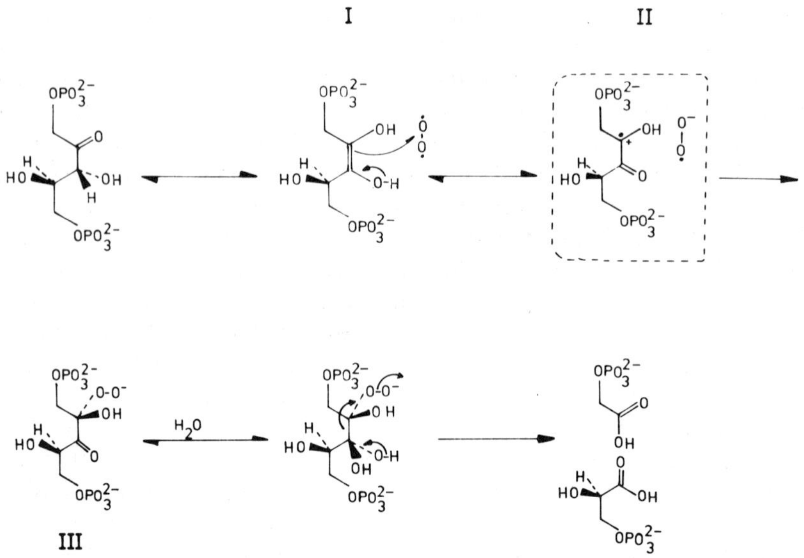

I   II

III

Fig. 26. A mechanism for the oxygenation of RuBP. The stereoconfiguration of the putative hydroperoxide is taken to be the same as for the carboxylation reaction (i.e., arabinitol).

(Lorimer *et al.*, 1973) showed that oxygenation is accompanied by the incorporation of one atom of molecular O_2 into.the carboxyl group of 2-P-glycolate. The other atom of molecular O_2 is lost to the medium. When oxygenation is performed in $H_2{}^{18}O$, an atom of ^{18}O appears in the carboxyl group of 3-P-glycerate. These results were augmented (Pierce *et al.*, 1980b) with the demonstration that the carbonyl oxygen of RuBP is retained during oxygenation and that oxygenation is accompanied by the cleavage of the bond between C-2 and C-3 of RuBP. On the basis of these results, Lorimer *et al.* (1973) proposed a mechanism involving the attack of O_2 on the 2,3-enediol(ate) of RuBP, leading to the formation of an intermediate hydroperoxide at C-2 of RuBP (Fig. 26). The stereochemical configuration about C-2 of this hydroperoxide is presumed to be the same (i.e., arabinitol rather than ribitol) as that of the carboxylated reaction intermediate, 3-keto-2CABP. (Due to the priority rules of stereochemistry, 2'-peroxyarabinitol bisphosphate is assigned the 2*S* stereoconfiguration whereas 2'-carboxyarabinitol bisphosphate is 2*R*). The breakdown of 3-keto-2'-peroxyarabinitol bisphosphate to products clearly involves the addition and elimination of water, since only one of the two atoms of molecular O_2 appears in the carboxyl group of P-glycolate. The alternative mechanism for the breakdown of the intermediate hydroperoxide, via a dioxetane ring, can be eliminated on the grounds that it would lead to the incorporation of the other atom of molecular O_2 into the carboxyl group of 3-P-glycerate.

A comparative view of oxygenation and carboxylation reveals that the proton-transfer steps involving C-3 of the substrate and O-3 of the subsequent intermediates are common to both reactions. Presumably the enzyme group(s) that catalyze these proton-transfer steps are the same in both reactions. However, the terminal step of the carboxylation reaction involves a protonation at C-2 that obviously does not occur during oxygenation. Thus carboxylation may require the participation of an additional enzyme group that is not involved in oxygenation.

Physical evidence supporting the intermediacy of the hydroperoxide in the oxygenase reaction has recently been reported (Branden et al., 1984b). When spinach enzyme, activated with Cu^{2+}, is mixed with RuBP, three different enzyme–Cu^{2+} complexes can be detected by EPR spectroscopy. Like the Mn EPR spectra of the quaternary complexes enzyme–$^ACO_2$–Mn^{2+}–2CABP and enzyme–$^ACO_2$–Mn^{2+}–4CABP (Miziorko and Sealy, 1980, 1984; Gutteridge et al., 1984a), the Cu EPR spectra are distinctly rhombic, indicating that the metal ion is assymetrically coordinated. Two of the Cu^{2+} complexes are formed anaerobically and are attributable to enzyme–Cu^{2+}–RuBP and enzyme–Cu^{2+}–3-P-glycerate. The third complex, however, displays some of the properties to be expected of an enzyme–Cu^{2+}–hydroperoxide intermediate. It is rapidly but transiently formed. Its formation is oxygen-dependent and precedes the appearance of the product, P-glycolate. When [^{17}O] oxygen is used, distinct broadening of the ^{63}Cu EPR hyperfine lines of this complex occurs, establishing that one atom of oxygen, derived from molecular O_2, must be directly coordinated to the Cu^{2+} ion of this complex. Additional spectroscopic evidence suggests that this putative enzyme–Cu^{2+}–hydroperoxide complex contains at least one molecule of water coordinated to the Cu^{2+} ion. It is not known if this water molecule is involved in the decomposition of the hydroperoxide. The enzyme–Cu^{2+}–hydroperoxide complex has also been prepared using [2-^{17}O]RuBP (Styring and Branden, 1985a). The labeled carbonyl oxygen is expected to become the C-2 hydroxyl oxygen of the hydroperoxide. Splitting of the hyperfine lines of this complex is observed, indicating that the C-2 hydroxyl oxygen of the hydroperoxide intermediate is coordinated to Cu^{2+} (together with one of the peroxide oxygen atoms). The structure of the enzyme–$^ACO_2$–Me^{2+}–hydroperoxide complex that emerges from these and other studies bears a close resemblance (as might be expected) to the structure proposed for the enzyme–$^ACO_2$–Me^{2+}–3-keto-2CABP complex (Fig. 27).

Kinetic evidence (Badger et al., 1980; Pierce et al., 1986a) supports the notion that the oxygenation, like that carboxylation, is an ordered reaction. Steady-state kinetic experiments showed that H_2O_2 inhibits oxygenation in a manner that is reversible, competitive with respect to O_2, and uncompetitive with respect to RuBP. H_2O_2 is not an alternative substrate. This pattern of inhibition is generally indicative of an ordered reaction, with the uncompetitive substrate (RuBP in this instance) binding first. An interesting complication to the straightforward interpretation of the above result arises from the susceptibility of the carbonyl group of RuBP to nucleophilic attack by H_2O_2. The addition products, an epimeric mixture of 2′-peroxy-pentitol bisphosphates, would probably be quite effective inhibitors of both carboxylation

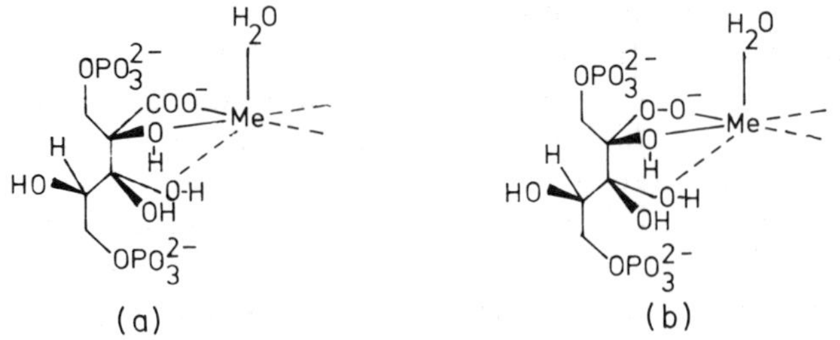

Fig. 27. The coordination of Me^{2+} to the reaction intermediates (a) of the carboxylase reaction and (b) of the oxygenase reaction results in very similar structures. Solid bonds to Me^{2+} are drawn where there is experimental evidence for their existence. Broken bonds to Me^{2+} are conjectural.

and oxygenation. If the inhibition of H_2O_2 is indeed due to the formation of 2'-peroxypentitol bisphosphate, in free solution or within the catalytic site in competition with CO_2 and O_2, one may regard this as additional evidence for the intermediacy of the hydroperoxide in the oxygenase reaction. (The inhibition of carboxylation by 2'-carboxypentitol bisphosphates was long considered as evidence for the intermediacy of 3-keto-2CABP in the carboxylation reaction).

Equilibrium binding experiments have been performed with the carbamylated enzyme from *R. rubrum* in an attempt to demonstrate binding of O_2 (Pierce *et al.*, 1986a). No evidence was obtained for O_2 binding to activated enzyme either in the presence or absence of the RuBP analog xylulose bisphosphate. The conditions were such that a binding constant of less than 10 mM (i.e., 25 times K_m for O_2) could have been measured. While this result is subject to the usual caveats concerning negative results, it is clearly consistent with the view that the interaction of O_2 with the enzyme occurs after the binding of RuBP. Since enolization can occur in the absence of CO_2 or O_2 (Pierce *et al.*, 1986a), it seems reasonable to conclude that the alternative substrates, CO_2 and O_2, compete with one another for reaction with the 2,3-enediol(ate).

Since ground-state O_2 is a triplet molecule, oxygenation reactions where a singlet substrate (such as RuBP) reacts with O_2 to yield singlet product(s) (P-glycolate and 3-P-glycerate) involve spin inversion. This is often kinetically and/or thermodynamically unfavorable. Yet systems capable of oxygenating a wide variety of relatively inert substrates have evolved. Almost without exception, these systems contain spin-delocalizing devices, transition-metal ions such as iron or copper and/or organic cofactors (flavin, heme, etc.) capable of resonance stabilization. In this respect Rubisco is exceptional, for it contains neither organic cofactors nor transition-metal ions (Lorimer *et al.*, 1973; Chollet *et al.*, 1975; Johal *et al.*, 1980). RuBP oxygenase differs from other oxygenation reactions in an important biological respect. Other oxygenations are presumed to serve a biological purpose, that is, to confer some selective advantage upon the organism. The evolution of mechanisms (transition-metal ions, organic cofactors) that assist these reactions is to be expected. The oxygenation of RuBP, however, seems to serve no useful biological purpose. Rather, it is disadvantageous. One would therefore expect evolution to suppress mechanisms that enhance the oxygenation of RuBP. Seen in this light, the absence of transition-metal ions and organic cofactors from Rubisco is understandable.

The two most outstanding problems concerning oxygenation relate to the mechanism of the oxygenation step and the manner in which the various divalent metal ions influence partitioning between carboxylation and oxygenation. A plausible, but speculative, mechanism for the oxygenation step is shown in Fig. 26. Carbanions have the propensity to react with O_2, usually via free radical mechanisms (Russell *et al.,* 1967; Kochi, 1973). Similarly, the oxygenation step may be initiated by an electron transfer from the 2,3-enediolate (**I**) to O_2 so as to create a radical pair, the radical cation at C-2 and a superoxide radical anion (**II**). In free solution the lifetime of such a radical pair is usually too short for spin inversion to occur (Kochi, 1973; Hamilton, 1974). Consequently they either revert to reactants or dissociate to become independent radical species. However, the longevity of a radical pair and the probability of spin inversion can be enhanced by cage effects (Nelson and Bartlett, 1966; Koenig and Fischer, 1973). Cage effects usually refer to the restraining influence that the solvent molecules encaging the radical pair impart to the motion of the latter. The catalytic sites of enzymes are "cages" par excellence. In order for oxygenation to proceed it becomes necessary to hold the radical pair in place within the catalytic site or "cage" for sufficient time to allow spin inversion to occur. The sugar radical cation is structurally similar to the 2,3-enediol(ate) and is thus unlikely to escape from the catalytic site. The superoxide radical anion has been reported to form coordination complexes in aprotic media with Li^+ and a variety of divalent metal ions (Sawyer and Gibian, 1979). It is therefore not unreasonable to suppose that the radical pair might be stabilized within the catalytic site. Perhaps the different divalent metal ions influence partitioning between carboxylation and oxygenation through their effect on the stability of the radical pair. In any event, the 3-keto-2-peroxyarabinitol bisphosphate (**III**) formed as a result of the recombination of the radical pair undergoes conversion to product by the addition and elimination of water. Some experiments to test these speculations would clearly be in order.

I. Relative Specificity

As previously mentioned (Section IV,C), the ordered kinetic mechanism of carboxylation and oxygenation and the linear, mutual competition between the gaseous substrates for the 2,3-enediolate of RuBP require that the relative rates of carboxylation and oxygenation (v_c/v_o) vary in proportion to the ratio of $[CO_2]/[O_2]$. The partitioning coefficient or relative specificity, s_{rel}, is simply the ratio of the V_{max}/K_m values for CO_2 and O_2. These kinetic parameters can be further resolved into elementary rate constants (Pierce *et al.*, 1986b) (Fig. 13). Thus

$$V_c/K_c = k_5[k_3/(k_3 + k_4)]\{k_7/[k_6(1 + k_8/k_9) + k_7]\} \tag{V}$$

and

$$V_o/K_o = k_{10}[k_3/(k_3 + k_4)]\{k_{12}/[k_{11}(1 + k_{13}/k_{14}) + k_{12}]\} \tag{VI}$$

Since dissociation of the gaseous substrates is a faster process than chemical bond formation, (i.e., $k_6 \gg k_7$ and $k_{11} \gg k_{12}$), these kinetic equations simplify to

$$V_c/K_c = k_c[k_3/(k_3 + k_4)][k_9/(k_8 + k_9)] \tag{VII}$$

and

$$V_o/K_o = k_o[k_3/(k_3 + k_4)][k_{14}/(k_{13} + k_{14})] \tag{VIII}$$

Hence,

$$\begin{aligned} s_{rel} &= [V_c/K_c]/[V_o/K_o] \\ &= [k_c/k_o]\{[k_9/(k_8 + k_9)]/[k_{14}/(k_{13} + k_{14})]\} \end{aligned} \tag{IX}$$

where $k_c = (k_5k_7/k_6)$ and $k_o = (k_{10}k_{12}/k_{11})$ are the rates of catalytically effective, gaseous substrate collision with the enzyme. In the absence of the ternary Michaelis complexes, ER*C and ER*O, $k_c = k_5$ and $k_o = k_{10}$, that is, the intrinsic second-order rate constants for interaction of the enzyme-bound 2,3-enediolate with the gaseous substrates to yield 3-keto-2CABP and 3-keto-2PABP.

According to Eq. (IX), s_{rel} is simply the ratio of the rates of gaseous substrate addition (k_c/k_o) times the ratio of the partition coefficients for 3-keto-2CABP and 3-keto-2PABP, $[k_9/(k_8 + k_9)]/[k_{14}/(k_{13} + k_{14})]$.

The DuPont group has recently measured the partitioning of 3-keto-2CABP $[k_9/(k_8 + k_9)]$ with various Rubiscos and with various Me^{2+} (Pierce *et al.*, 1986b). For Mg^{2+}, the values were near unity, even for Rubiscos with different V_c/K_c values; that is, 3-keto-2CABP is strongly committed to product formation. Even the Co^{2+} activated *R. rubrum* enzyme that is unable to caboxylate RuBP (Robison *et al.*, 1979) hydrolyzed much of the exogenously added 3-keto-2CABP to products.

Since the partitioning of 3-keto-2CABP is almost totally toward products $([k_9/(k_8 + k_9)] \approx 1)$, Eq. (VII) reduces to

$$V_c/K_c = k_c[k_3/(k_3 + k_4)] \tag{X}$$

It follows that for changes in V_c/K_c to occur, either the rate of reaction of CO_2 with the enzyme-bound 2,3-enediolate (k_c) must increase or the relative proportion of RuBP that is present on the enzyme as the 2,3-enediolate $[k_3/(k_3 + k_4)]$ must increase.

Similar experiments have yet to be performed concerning the partitioning of the oxygenase reaction intermediate, 3-keto-2PABP. However, by drawing analogy with the decomposition of similar hydroperoxides, which invariably partition to form two acids by hydrolytic carbon–carbon bond cleavage, a similar conclusion can be reached for the partitioning of 3-keto-2PABP. Thus, the partitioning term $[k_{14}/(k_{13} + k_{14})]$ of Eq. (VIII) is assumed to be unity and Eq. (VIII) simplifies to

$$V_o/K_o = k_o[k_3/(k_3 + k_4)] \tag{XI}$$

Combining Eqs. (X) and (XI) one obtains the following simplified expression for the relative substrate specificity s_{rel}:

$$s_{rel} = [V_c/K_c]/[V_o/K_o]$$
$$= (k_c/k_o) = [k_5k_7/k_6]/[k_{10}k_{12}/k_{11}] \tag{XII}$$

We have noted earlier (Fig. 3) that different Ribiscos display different s_{rel} values. According to Eq. (XII) this is because they differentially enhance the reactivities of CO_2 and O_2 toward the 2,3-enediolate of RuBP. Quite how this is achieved is the very heart of the problem. As the structural details of the catalytic site emerge and responses to specific site-directed mutations are assessed, some clues to the solution of this central problem are sure to be uncovered.

J. Identification of Groups within the Catalytic Site

The goal of many studies is to identify the specific amino acids that comprise the catalytic site. Many of these studies employ group-specific reagents that react more or less specifically with a particular amino acid residue. These reagents generally have no intrinsic affinity for the catalytic site, and therefore can, and frequently do, react at sites that are remote from the catalytic site. A more rigorous approach to the identification of the amino acids comprising the catalytic site employs affinity labeling. In the case of Rubisco this involves the use of reactive analogs of RuBP. The technique has been applied to L8S8 Rubiscos and to the L2 form of the enzyme in an especially imaginative and rigorous manner by Hartman and his colleagues. The structures of five of these affinity labels are shown in Fig. 28. These compounds have been shown to satisfy all or most of the following criteria of affinity labeling: (1) In short-term experiments they behave as competitive inhibitors of catalysis (both carboxylation and oxygenation) with respect to RuBP. (2) Inactivation displays pseudo-first-order kinetics and proceeds to completion. (3) Compounds known to bind to the catalytic site (e.g., RuBP or 2CRBP) strongly retard the progress of inactivation. (4) Their interaction

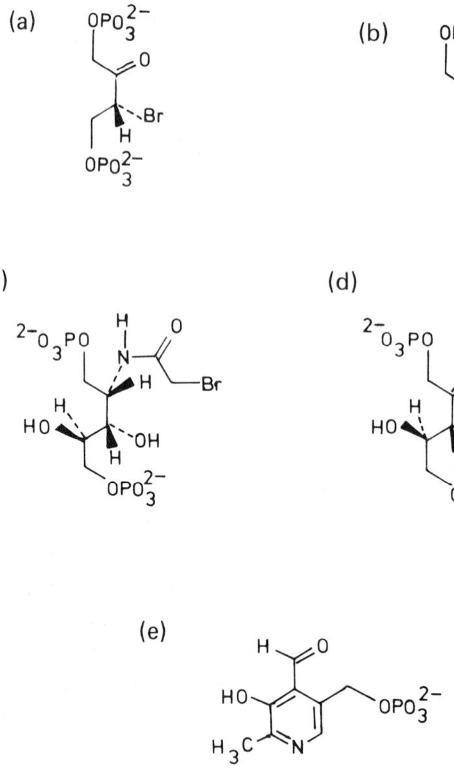

Fig. 28. Affinity labels used by Hartman's group to identify groups within the domain of the active site of L2 and L8S8 Riboscos. (a) 3-Bromo-1,4-dihydroxy-2-butanone-1,4-bisphosphate; (b) N-bromoacetylethanolamine phosphate; (c) 2-[(bromoacetyl)-amino]pentitol 1,5-bisphosphate; (d) 2-(4-bromoacetamido)anilino-2-deoxy-arabinitol 1,5-bisphosphate; (e) pyridoxal phosphate.

with the enzyme resembles the interaction of substrate with the enzyme; that is, they display saturation kinetics rather than bimolecualr kinetics. (5) Inactivation is associated with the incorporation of 1 mol reagent per mol protomer inactivated. (6) Inactivation is correlated with the modification of a limited number of residues (one or two) per protomer. (7) Site-specific modification by these affinity labels requires the catalytic site to be in the native state.

In assessing whether or not a targeted residue is essential for catalysis or merely within the domain of the catalytic site, Hartman and his colleagues made comparisons between the evolutionarily distant and structurally diverse spinach and *R. rubrum* Rubiscos. For example, N-bromoacetylethanolamine phosphate labels $C^{L8}172$ and $C^{L8}459$ of the decarbamylated spinach enzyme (Schloss *et al.*, 1978). Examination of the corresponding *R.*

rubrum sequences (Fig. 6) shows that neither residue is conserved. Thus neither residue is likely to participate in catalysis. Similarly, 2-(4-bromoacetamido)anilino-2-deoxyarabinitol, 1,5-bisphosphate labels both $H^{L2}44$ and $C^{L2}58$ of the *R. rubrum* enzyme (Herndon and Hartman, 1984). While neither residue is conserved in the L8S8 enzyme, comparison of the two sequences (Fig. 6) shows that $H^{L2}44$ and $C^{L2}58$ bracket the strongly conserved octapeptide AAESSTGT. Conservation of structure is usually taken as an indication of functional importance. This reagent has another interesting property that is perhaps helpful in defining the structural changes accompanying carbamylation. The maximal distance from C-2 of the pentitol bisphosphate moiety to the alpha-C of the reactive group is 10 Å. Yet this reagent only reacts to a significant extent with carbamylated enzyme. This suggests that carbamylation is accompanied by structural movements that bring the target residues to within 10 Å of the position within the catalytic site that is occupied by C-2 of RuBP. Recent experiments on the carbamylated L2 enzyme with cross-linking reagents spanning a maximal distance of 7 Å have linked $K^{L2}166$ with $C^{L2}58$, in confirmation of the result with the affinity label (F. C. Hartman, personal communication).

2-[(Bromoacetyl)amino]pentitol 1,5-bisphosphate modifies $M^{L2}330$ only when the enzyme is carbamylated (Fraij and Hartman, 1982, 1983). While this methionine is not conserved in the L8S8 enzyme, it is immediately adjacent to the conserved $K^{L2}329$, which is itself the target for modification by 3-bromo-1,4-dihydroxy-2-butanone 1,4-bisphosphate (Stringer and Hartman, 1978). These two results clearly place this region of the protein within the domain of the catalytic site.

The most consistent target for the affinity labels is $K^{L8}175$ or its L2 counterpart, $K^{L2}166$. This residue, part of the strongly conserved heptapeptide IKPKLGL, is interesting for a several reasons. Using the cross-linker 4,4′-dithiocyano 2,2′-disulfonate stilbene, Lee and Hartman (1986) have linked $K^{L2}166$ with $K^{L2}329$. Both residues have been modified by affinity labels, and the cross-linking reaction itself is retarded by 2CRBP. This particular reagent has a maximal span of 12 Å between its reactive groups, so that $K^{L2}166$ and $K^{L2}329$ can be brought to that distance of one another.

The properties of $K^{L2}166$ and of $K^{L8}334$ have been explored by Hartman *et al.* (1985). Trinitrobenzene sulfonate (TNBS), a group-selective reagent, specific for reaction with unprotonated amines, was shown to inactivate both enzymes. Inactivation of the spinach enzyme by TNBS was shown to result from the specific arylation of $K^{L8}334$. The specificity of alkylation, but not its rate, was independent of pH and the carbamylation state of the enzyme. Thus, by following the reaction of TNBS with the enzyme, Hartman *et al.* (1985) determined the pK_a of $K^{L8}334$, its intrinsic reactivity (k_o) toward TNBS, and the consequences of carbamylation/decarbamylation upon these two parameters. Carbamylation of the spinach enzyme lowered the pK_a of $K^{L8}334$ from 9.8 to 9.0, an effect that is perhaps attributable to the proximity of the divalent metal ion to $K^{L8}334$. In the case of the *R. rubrum* enzyme, the specificity of TNBS was dependent on the carbamylation state of the enzyme. Only when the enzyme was carbamylated at $K^{L2}191$ did TNBS react specifically at $K^{L2}166$. The decarbamylated enzyme reacted with TNBS at several sites, including $K^{L2}166$ and $K^{L2}329$ (the L2 equivalent of $K^{L8}334$), and thus lacked the specificity for further kinetic measurements. With a pK_a of 7.9, $K^{L2}166$ of the carbamylated L2 enzyme is unusually acidic. Proximal positive charges, on the activator Mg^{2+} and on the ε-amino group of $K^{L2}329$, might contribute to the very low pK_a of $K^{L2}166$.

The clear identification of $K^{L2}166$ and $K^{L2}329$ and their counterparts $K^{L8}175$ and $K^{L8}334$ as active-site residues raises the thorny question of functionality. The arguments here are necessarily more circumstantial than definitive. Several lines of evidence suggest that $K^{L8}334$ is not essential for binding of RuBP. The competitive inhibitor 2CRBP affords only a modest degree of protection against TNBS inactivation. Second, L8S8 enzyme, arylated at $K^{L8}334$, can nevertheless undergo carbamylation and form a moderately stable quaternary complex with 2CABP (Hartman *et al.*, 1985). The binding of carbohydrates by proteins involves many contact residues (Quiocho, 1986). The exceptionally tight binding of 2CABP by Rubisco must also involve multiple contact residues. It is conceivable that the loss of only one of many contact residues, such as might accompany the inactivation by TNBS, might merely weaken rather than eliminate substrate binding (Hartman *et al.*, 1985). The pK_a of $K^{L2}334$ (9.0) is compatible with binding of one of the two phosphate moieties. However, that role for $K^{L2}166$ is excluded by its strongly acidic nature ($pK_a = 7.9$). Alternatively, rather than passively binding the substrate, $K^{L2}166$ and $K^{L2}329$ (or $K^{L8}175$ and $K^{L8}334$) may actively participate in catalysis. We have earlier noted that the overall carboxylation reaction involves several proton-transfer steps. Obviously, $K^{L2}166$ and $K^{L2}329$ must be considered as the residues catalyzing these proton transfers.

Kinetic data (Van Dyk and Schloss, 1986; Christeller, 1982) point to the existence of a catalytically essential base with a pK of about 7.5. Based on the primary deuterium isotope effect with [3-^{2}H]RuBP, this essential base is likely to be the one that abstracts the C-3 proton of RuBP to initiate catalysis (Van Dyk and Schloss, 1986). The similarity of this (kinetic) pK value with the (intrinsic) pK value determined for $K^{L2}166$, the species invariance of $K^{L2}166$, together with its enhanced nucleophilicity, constitutes strong circumstantial evidence that $K^{L2}166$ is indeed the active-site residue that initiates catalysis by abstracting the C-3 proton of RuBP. Site-directed mutagenesis (Section IV,K) further strengthens this conclusion.

Kinetic analyses (Van Dyk and Schloss, 1986) also point to the existence of an essential acid with a pK in the range 7.8–8.3. Catalysis involves several steps that would be facilitated by the presence of a protonated group. For example, the hydration of carbonyl groups is acid-catalyzed. Thus one would expect protonation of the C-3 carbonyl oxygen of 3-keto-2CABP to facilitate its hydration. Additionally, an active-site group is probably involved in the stereochemically directed protonation of the C-2 carbanion (or aci-acid) form of the u-3-glycerate. Given the difficulties in comparing kinetic and intrinsic pK values, the protonated form of $K^{L8}334$ or $K^{L2}329$ could correspond to the essential acid that has been revealed kinetically.

K. Site-Directed Mutagenesis

Site-directed mutagenesis promises to add considerably to our understanding of enzymatic mechanisms. It is most powerfully practiced in conjunction with crystallographic information, on chemically and kinetically well-defined systems (Fersht *et al.*, 1986). Our knowledge of Rubisco is not yet so advanced. Nevertheless, the technique has been applied in a limited way to probe the function of amino acid residues thought to be within the domain of the active site of the *R. rubrum* enzyme.

1. The Activator Carbamate–Me²⁺ Site

This is an interesting site on at least two counts. Formation of the activator carbamate–Me^{2+} complex on $L^{L2}191$ creates a catalyst capable of carboxylating or oxygenating RuBP. Additionally, some as yet undefined property of Me^{2+} strongly influences the relative specificity toward CO_2 and O_2 (Jordan and Ogren, 1983). Two mutations have been created at this site, $E^{L2}188$ (Gutteridge *et al.*, 1984a) and $E^{L2}191$ (Estelle *et al.*, 1985). The first mutation effectively inserts a methylene group and was made in an attempt to perturb the environment of Me^{2+}. Electron spin resonance (ESR) studies of the mutant quaternary complexes $E-^ACO_2-Mn^{2+}-2CABP$ and $E-^ACO_2-Mn^{2+}-4CABP$ revealed that the environment of Me^{2+} was perturbed in the mutant. Perhaps this perturbation accounts for the slightly altered kinetic properties associated with the mutant $E^{L2}188$ ($V_c = 4.8\ sec^{-1}$ versus $5.9\ sec^{-1}$ for $D^{L2}188$; $V_c/K_c = 3.4 \times 10^4\ M^{-1}\ sec^{-1}$ versus $7.0 \times 10^4\ M^{-1}\ sec^{-1}$ for $D^{L2}188$). However, the s_{rel} value was unaltered (Gutteridge *et al.*, 1984a). With the wisdom conferred by hindsight, we can see why this mutation had such little effect. Crystallographic analysis (Schneider *et al.*, 1986b) reveals that $D^{L2}188$ is not in the catalytic site at all. Rather, it lies at the bottom of the α/β barrel!

The $E^{L2}191$ mutation was created to test the proposition that the γ-carboxyl group of glutamate might substitute for the lysine carbamate as a site for binding Me^{2+} (Estelle *et al.*, 1985). $E^{L2}191$ proved completely incapable of catalyzing the carboxylation or oxygenation of RuBP or of forming a tight complex with 2CABP. Nor was it able to catalyze the hydrolysis of 3-keto-2CABP in the presence of Mg^{2+}. However, $E^{L2}191$ did catalyze the decarboxylation of 3-keto-2CABP (G. H. Lorimer and C. R. Somerville, unpublished data), implying that the mutant was not otherwise structurally impaired. The simplest interpretation of these results is that $E^{L2}191$ is unable to bind Me^{2+}. The oxygen atoms of the γ-carboxyl group of $E^{L2}191$ would be about 3.6 Å closer to the polypeptide backbone than the ε-carbamate of $K^{L2}191$. This distance may simply be too large to permit proper coordination of Me^{2+} with other protein ligands and/or the ligands provided by substrate or reaction intermediates (Estelle *et al.*, 1985).

2. Lysine-166

The evidence that $K^{L2}166$ functions as a proton-transfer group has been summarized (Section IV,J). Hartman's group have created three site-directed mutants at this position (Hartman *et al.*, 1986). The mutant $R^{L2}166$ was created so as to substitute a cationic residue in place of $K^{L2}166$. $R^{L2}166$ is less than 0.02% as active as $K^{L2}166$, suggesting that electrostatic interactions between $K^{L2}166$ and the substrate or reaction intermediates are not important. In the mutants $S^{L2}166$ (relative activity 0.2%) and $H^{L2}166$ (relative activity 2%), the nucleophilic nature of $K^{L2}166$ has been maintained. The substantial activity of $H^{L2}166$ compared with $S^{L2}166$ is consistent with $K^{L2}166$ playing a catalytic role in one or more of the proton-transfer reactions accompanying the carboxylation of RuBP.

3. Histidine 291

Following studies of the reactivity of L8S8 Rubisco with the group-specific reagent diethyl pyrocarbonate (Igarashi *et al.*, 1985; Paech, 1985), the suggestion was made that $H^{L8}298$ might be the essential base initiating catalysis by abstraction of the C-3 proton of RuBP. Hartman and his colleagues (Niyogi *et al.*, 1986) have therefore created the mutant $A^{L2}291$ to test this hypothesis. $A^{L2}291$ displays altered kinetic properties ($V_c = 1.5$ sec^{-1} versus 3.6 sec^{-1} for $H^{L2}291$; $V_c/K_{Rc} = 1.7 \times 10^4\ M^{-1}$ sec^{-1} versus $6 \times 10^5\ M^{-1}$ sec^{-1} for $H^{L2}291$), which might possibly reflect a secondary role for $H^{L2}291$ in the maintainance of conformation. However, the substantial activity of $A^{L2}291$ clearly excludes $H^{L2}291$ (and by inferrence $H^{L8}298$) as an essential base catalyzing any of the proton-transfer steps.

4. Leucine 330

The New Zealand group (Terzaghi *et al.*, 1986) have converted $M^{L2}330$ to $L^{L2}330$. This residue is adjacent to the conserved $K^{L2}329$ ($=K^{L8}334$) and is a target of affinity-labeling studies (Fraij and Hartman, 1983). In the L8S8 enzyme, $M^{L2}330$ is replaced with $L^{L8}335$. Thus the mutation was designed to confer some L8-like structures upon the L2 catalytic site. The value for V_c was reduced fivefold in the mutant; that for V_c/K_{Rc} was reduced 125-fold in the mutant; and that for V_c/K_c was reduced 73-fold in the mutant while the value for K_{iR} was increased fivefold. Nevertheless, the s_{rel} values for the mutant and wild-type enzymes were the same. The authors therefore concluded that the mutation has a decreased ability to stabilize the 2,3-enediolate.

5. *Future Prospects*

When coupled to the use of the partial reactions (Section IV,F) and guided by the results of the crsytallographic analysis (Section III,C), site-directed mutagenesis may enable us to define the catalytic roles of specific residues. Can mutant enzymes be created that, though lacking the ability to catalyze the overall reaction, are nevertheless able to catalyze one or more of the partial reactions? For example, a mutant enzyme that was unable to catalyze the exchange of the C-3 proton of RuBP with the solvent, but that was able to catalyze *both* the hydrolysis of 3-keto-2CABP by carbamylated enzyme and the decarboxylation of 3-keto-2CABP by decarbamylated enzyme, would be of considerable mechanistic interest. Its existence would support the idea that not one, but two, discrete groups are responsible for the transfer of the proton from the C-3 of RuBP and for the transfer of the protons to and from the O-3 of the 2,3-enediol and of 3-keto-2CABP. And would a mutant enzyme, unable to stereospecifically protonate the C-2 carbanion of u-3-P-glycerate, resolve its dilemma by eliminating the C-1 phosphate to produce pyruvate and Pi in place of u-3-P-glycerate?

V. MECHANISMS FOR MODULATING ACTIVITY

A. Does Rubisco Need to Be Regulated?

Rubisco's substrate, RuBP, is produced by the phosphoribulokinase reaction, which requires ATP. Since the production of ATP requires light, one might expect that the supply of RuBP would be fairly closely regulated by light and that this, alone, should provide a sufficient mechanism for modulating Rubisco's activity according to the prevailing level of chloroplast metabolism. This concept may be put another way in the form of the question: "Why would it ever be beneficial to have a supply of RuBP, but not to carboxylate it?" Given the instability of RuBP [it degrades at a rate of approximately 1% per hour in solution at pH 8.3 and 30°C to form a variety of compounds that may inhibit Rubisco (Paech *et al.*, 1978], it would seem, at least superficially, safer for Rubisco to operate in an unregulated manner, thus keeping the pool size of RuBP as small as possible under all conditions.

Evidence from a variety of sources contradicts such simple notions. Rubisco's activity definitely is modulated other than by the supply of its substrate, and proably by several different mechanisms. There is no direct evidence bearing on the reasons for such regulation, but consideration of the stoichiometry of the carboxylase reaction (Fig. 1) may provide a clue (J. A. Berry, personal communication). As well as two molecules of phosphoglycerate, two protons are also produced for each molecule of RuBP consumed. Of course, these protons are consumed in the subsequent reactions that reduce phosphoglycerate to triose phosphate, making the C_3 cycle electroneutral overall, but let us consider what would happen if some blockage in the utilization of phosphoglycerate occurred. Such a bottleneck might be expected at limiting light levels or in any other condition that limited the flow of ATP and reductant from the light-harvesting machinery. For maximum utilization of Rubisco at high rates of photosynthesis in saturating light, the steady-state concentration of RuBP in the stroma must exceed the concentration of Rubisco active sites (approximately 4 mM; Jensen and Bahr, 1977). Indeed, direct measurements indicate higher concentrations (6–15 mM) (Perchorowicz *et al.*, 1981; Perchorowicz and Jensen, 1983; Badger *et al.*, 1984; Mott *et al.*, 1984; Dietz and Heber, 1984, 1986; von Caemmerer and Edmondson, 1986). If all of this pool were converted to phosphoglycerate, as it might during a sudden reduction in light level if Rubisco were unregulated, then 12–30 mM H^+ would be released into the stroma from this source alone. In the absence of buffering by other stromal components, this would result in a stromal pH below 2 and the buffer capacity of the stroma would have to be very large, indeed, to absorb these protons without a significant drop in pH. Such acidification would be prevented if Rubisco's activity was regulated so that the supply of protons generated during carboxylation was matched by the demand for them by the subsequent reduction steps. A result of such regulation would be the maintenance of a reasonably constant ratio between the pool sizes of RuBP and phosphoglycerate, regardless of the rate of metabolism through the C_3 cycle. Such a constant ratio between these pools has been observed experimentally when light variation was the source of variation in metabolic flux (von Caemmerer and Edmondson, 1986). The transient reduction in the RuBP pool size observed when illumination is suddenly reduced, followed by a slower return to the RuBP level prevailing before the reduction in illumination (Mott *et al.*, 1984), is consistent with the operation of a feedback mechanism that regulates Rubisco's activity in accordance with the supply of RuBP (which reflects the demand for phosphoglycerate).

B. Possible Regulatory Mechanisms

Given the necessity of a means of matching Rubisco's activity *in vivo* to the demand for phosphoglycerate, what mechanisms might be involved in achieving such regulation? A variety has been documented. All of these mechanisms would, of course, modulate both carboxylase and oxygenase activities together.

1. Simple Product Inhibition

Phosphoglycerate inhibits carboxylation competitively with respect to RuBP (Badger and Lorimer, 1981). This inhibition would be, by itself, a sufficient means of matching the rates of supply and demand for phosphoglycerate if the ratio between the K_i for phosphoglycerate and the K_m for RuBP were sufficiently small. However, this ratio has been measured to be approximately 40 (Badger and Lorimer, 1981), which is much too large for phosphoglycerate to have a significant inhibitory effect at the phosphoglycerate/RuBP ratios observed *in vivo* (approximately 2; von Caemmerer and Edmondson, 1986).

2. Carbamylation/Decarbamylation

One might imagine that stromal acidification promoted by excessively high phosphoglycerate concentrations would inhibit Rubisco by driving the equilibrium of the CO_2/Mg^{2+}-induced carbamylation reactions (Section IV,A) back toward inactive (decarbamylated) enzyme. This is the result of acidification *in vitro* (Lorimer et al., 1976), and it would provide a very direct means of regulating Rubisco's activity according to the demand for phosphoglycerate. However, the stroma is well supplied with phosphorylated compounds, including RuBP itself, that bind to Rubisco and perturb the equilibrium between the carbamylated (ECM) and decarbamylated (E) forms (Badger and Lorimer, 1981; McCurry et al., 1981; Jordan et al., 1983; Jordan and Chollet, 1983). Therefore, it is difficult to predict the degree of carbamylation *in vivo* under various conditions. However, in theory at least, this kind of carbamylation/decarbamylation equilibrium has the potential for providing the necessary modulation mechanism.

Experimental evidence from intact leaves supports a role for the carbamylation reactions in modulating Rubisco's activity *in vivo*, particularly in response to variations in light intensity. Rubisco appears to be nearly fully carbamylated in saturating light, and a substantial degree of decarbamylation is seen at low light levels (Perchorowicz et al., 1981; Perchorowicz and Jensen, 1983; Taylor and Terry, 1984; Servaites et al., 1984; Salvucci et al., 1986; von Caemmerer and Edmondson, 1986). Some decarbamylation is also seen at both very high and very low CO_2 partial pressures (Perchorowicz and Jensen, 1983; von Caemmerer and Edmondson, 1986). Decarbamylation at high CO_2 accords with the present model because high CO_2 would favor phosphoglycerate overproduction and the concomitant stromal acidification would reduce the carbamylation level. Decarbamylation at low CO_2 may simply reflect the CO_2 requirement of the carbamylation reactions.

In darkness, the carbamylation level is also reduced, but sometimes not as much as at very low light levels (Perchorowicz et al., 1981; Salvucci et al., 1986; von Caemmerer and Edmondson, 1986). Of course, the level of carbamylation in darkness is essentially irrelevant because of the low concentrations of RuBP that prevail (Perchorowicz et al., 1981; Badger et al., 1984; Dietz and Heber, 1984, 1986; von Caemmerer and Edmondson, 1986). However, such carbamylation levels are not in accord with the equilibrium of carbamylation observed *in vitro* (Lorimer et al., 1976) and the low pH of the stroma of isolated chloroplasts in darkness (approximately pH 7, Heldt *et al.*, 1973; Boag and Portis, 1984). If the pH of the chloroplast stroma in leaves is as low, almost complete decarbamylation would be expected, whereas carbamylation levels of approximately 50% or higher are observed. Perhaps other factors, such as binding of metabolites preferentially to the carbamylated form, may be at work.

It has also been suggested (Lorimer, 1981b; Ogren *et al.*, 1986) that full carbamylation of Rubisco at high light intensities is inconsistent with the simple carbamylation model and the conditions thought to prevail in the stroma in the light (10 μM CO_2, 5–10 mM Mg^{2+}, pH about 8). However, the estimates of stromal pH were obtained with isolated chloroplasts (Heldt *et al.*, 1973; Boag and Portis, 1984), whose stromal pH is affected by the pH of the medium (Heldt *et al.*, 1973) and may differ from the stromal pH prevailing in chloroplasts in illuminated leaves. No means of estimating stromal pH in intact leaves currently exists, but if the pH in strong light was half a pH unit higher than the above estimates, then the problem would largely disappear (Mott and Berry, 1986). Alternatively, preferential binding of RuBP to the carbamylated form of Rubisco would promote higher carbamylation levels (Badger and Lorimer, 1981; von Caemmerer and Edmondson, 1986). However, demonstration of apparent extremely tight binding of RuBP to the decarbamylated form [K_d^{inact} perhaps two to three orders of magnitude less than K_m^{act} (Jordan and Chollet, 1983)] negates this possibility. Indeed, such tight binding to the decarbamylated form presents a conundrum as to how Rubisco could be carbamylated at all in the presence of RuBP under any circumstances.

In Section IV,B, we suggested that cooperative interactions between the eight active sites of the L8S8 enzyme might provide another explanation for carbamylation levels *in vivo* being higher than predicted from the characteristics of the isolated enzyme. If binding of an effector to one site promotes carbamylation at the remaining seven free sites, then higher carbamylation levels would be attained in the presence of such an effector. Furthermore, modulation of the concentrations of such effectors at levels much lower than the concentration of Rubisco sites would provide an efficient means for regulating Rubisco's carbamylation state. Further experimentation is necessary to test this hypothesis.

3. Tight-Binding Inhibitors

We have already mentioned the tendency of some phosphorylated compounds to bind preferentially to the carbamylated, that is, ECM, form of Rubisco, thus promoting the level of carbamylation at equilibrium. 6-Phosphogluconate is a good example of such an effector (Section IV,B). However, while the enzyme may be carbamylated, it is nevertheless catalytically incompetent because the effector blocks the active site. If the availability of such an effector were regulated by the light level, then this would constitute another, independent means of modulating Rubisco's catalytic throughput.

Recently, observations that Rubisco from many, but by no means all, higher-plant species was partially inhibited when isolated from darkened leaves, and that the inhibition could not be relieved by incubation with CO_2 and Mg^{2+} (Vu *et al.*, 1983, 1984; Servaites *et al.*, 1984), led to the demonstration that the darkened-leaf enzyme was sequestered by a tight-binding inhibitor with a K_d of approximately 100 nM (Seemann *et al.*, 1985). This inhibitor could only be displaced by high concentrations of SO_4^{2-} (Servaites, 1985) or, most easily, by the reaction-intermediate analog, 2CABP (Seemann *et al.*, 1985). Inhibited preparations could also be reactivated by phosphatase treatment (Seemann *et al.*, 1985). Eventually, the inhibitor was isolated in quantity from darkened-leaf Rubisco and shown to be identical to 2CABP, except for the lack of the C-5 phosphate group, that is, it was 2′-carboxyarabinitol 1-phosphate (Berry *et al.*, 1987; Gutteridge *et al.*, 1986b). Only a few minutes of strong illumination of darkened leaves is sufficient to remove the inhibitor and, at lower light intensities, the level of inhibition reflects the light level (Seemann *et al.*, 1985). Clearly, this system displays all the necessary elements of a light-modulation mechanism for Rubisco. The pathways of synthesis and degradation of this inhibitor and their regulation by light remain to be elucidated.

Why is not Rubisco's activity modulated by light in this manner in all plants? For the majority, of which spinach is an example, fully active Rubisco may be extracted from darkened leaves. This is a question for further inquiry. Perhaps species that lack the tight-binding inhibitor simply rely on other regulatory mechanisms. However, we note that this kind of light-modulation mechanism would function just as efficiently with an inhibitor that bound considerably less tightly than 2′-carboxyarabinitol 1-phosphate. It is only necessary that the inhibitor bind tightly enough so that it is not easily displaced by RuBP. 6-Phosphogluconate, for example, which has a K_i of 10 μM (Badger and Lorimer, 1981), might serve this purpose. However, an inhibitor with this kind of K_i would not remain bound to Rubisco during extraction and purification. Therefore, it is possible that light regulation of Rubisco by inhibitors might be a general phenomenon, but only in some cases does the inhibitor involved bind tightly enough to be recovered with the enzyme after extraction. As mentioned earlier, binding of such an inhibitor may be an explanation for the higher-than-expected carbamylation levels observed in darkness.

4. *Time-Dependent Kinetics*

A curious phenomenon, which is manifested by all higher-plant Rubiscos, but infrequently reported, may also be relevant to Rubisco's activity *in vivo*. When assayed *in vitro*, fully carbamylated Rubisco does not show a linear rate of conversion of RuBP to phosphoglycerate characteristic of a typical well-behaved enzyme. There is a decline in activity following exposure to RuBP, which produces curvature in the time course (Andrews and Hatch, 1971; Laing and Christeller, 1976; Sicher *et al.*, 1981; McCurry *et al.*, 1981;

Mott and Berry, 1986). While this might be expected when CO_2 or Mg^{2+} are subsaturating (resulting from decarbamylation following exposure of fully carbamylated enzyme to the less than fully carbamylating conditions of the assay medium), the phenomenon still occurs when CO_2 and Mg^{2+} are saturating during assay, or are at the same concentration in the assay medium as in the preactivation solution. Moreover, the curvature does not continue until no activity remains. Eventually a linear rate, which is substantially smaller than the initial rate, is obtained after many minutes. The presence of a slow, tight-binding inhibitor, such as xylulose 1,5-bisphosphate, in RuBP preparations (Paech *et al.,* 1978) would be consistent with such a time-dependent inhibition. However, inhibition is still seen with very pure RuBP preparations, prepared without exposure to above-neutral pH, which have little or no detectable xylulose 1,5-bisphosphate, and when an enzymatic system for removing xylulose 1,5-bisphosphate is present (D. L. Edmondson, T. J. Andrews, M. R. Badger, and S. von Caemmerer, unpublished). An explanation for this intriguing phenomenon is not presently available, although there have been suggestions that it is caused by RuBP binding to the decarbamylated form, as well as to the carbamylated, catalytic form, thus decreasing the level of carbamylation in the steady state (McCurry *et al.,* 1981; Mott and Berry, 1986). Slow, unproductive binding of RuBP to the carbamylated form is also a possibility. This time-dependent decline in activity is relevant to the present discussion for two reasons. First, since it is induced by exposure to RuBP, it may well occur *in vivo* and the final, rather than the initial, rate may be more relevant to the physiological situation. Second, there is evidence that the extent of the decline is reduced as pH increases, being eliminated at pH values above 8.6 (Mott and Berry, 1986). Therefore, falling pH may cause a steeper decline in Rubisco's *in vivo* activity than the rather small dependency of the initial rate on pH might suggest (Andrews *et al.,* 1975; Mott and Berry, 1986). Such a pH dependence obviously would assist in limiting overproduction of phosphoglycerate and its accompanying protons.

5. Rubisco Activase

Evidence for yet another means of regulating Rubisco's activity, which may or may not be related to any of the above mechanisms, has been obtained using a mutant of the C_3 crucifer *Arabidopsis thaliana* (Somerville *et al.,* 1982). This mutant, which grows well at elevated CO_2 but is severely impaired in air, has a Rubisco that is indistinguishable from the wild type. However, the carbamylation level in the light is abnormally low (Somerville *et al.,* 1982; Salvucci *et al.,* 1986). Studies with this mutant revealed that it lacks two previously unstudied stromal polypeptides with molecular masses of 50 and 47 kDa (Salvucci *et al.,* 1985). Lack of the 50-kDa species (and presumably also the 47-kDa species) cosegregated with the carbamylation deficiency. Furthermore, rapid light-induced carbamylation of Rubisco in the presence of a high concentration of RuBP [which binds tightly to the decarbamylated form of isolated Rubisco and greatly slows its rate of carbamylation (Laing and Christeller, 1976; Jordan and Chollet, 1983)] was demonstrated both in lysed chloroplasts and in reconstituted systems containing thylakoid membranes, isolated Rubisco, and a macromolecular stromal fraction containing the two polypeptides. If the stromal fraction was omitted, or if it was derived from the mutant, no light-induced carbamylation occurred. Thus Salvucci *et al.* (1985) suggested that the two polypeptides are the subunits of a "Rubisco activase" that catalyzes light-induced carbamylation of Rubisco *in vivo*. However, it remains to be established whether or not these polypeptides have a truly catalytic function.

In the complete reconstituted system, the $K_{0.5}(CO_2)$ for carbamylation also appeared to be substantially lower than that expected from measurements with the isolated enzyme (Ogren *et al.*, 1986; Lorimer *et al.*, 1976). This suggests that activase perturbs the equilibrium of the carbamylation reactions toward the activated (EM) form. However, enzymes, like all catalysts, cannot alter the equilibria of the reactions they catalyze. Therefore, the system must be more complex than simple catalysis by activase of the previously well-studied CO_2/Mg^{2+}-induced carbamylation reactions. Perhaps activase effects some change in Rubisco itself that alters its carbamylation equilibrium. If so, such a change cannot be induced by stoichiometric binding of activase to Rubisco, because the activase subunits are very much less abundant than Rubisco (Salvucci *et al.*, 1985). Alternatively, activase might cause a change in the concentration of some RuBP-derived effector molecule that binds to Rubisco and alters its activity or carbamylation level. These details and other aspects of this interesting system, such as the manner in which activase is coupled to the light reactions, remain to be elucidated.

VI. SUBUNIT INTERACTIONS

A. Why Two Subunits?

Two-subunit, hexadecameric Rubiscos occur in all aerobic, and even the majority of anaerobic, lithotrophs. However, the existence of functional Rubiscos in purple, nonsulfur bacteria composed solely of large subunits raises intriguing questions about the role of the small subunit in the predominant class of Rubiscos which have it. Why are two subunits usually required when clearly one will do, at least in some circumstances? Inspection of the s_{rel} values of Rubiscos from a wide range of organisms (Fig. 3) indicates one possible explanation. There is an obvious discontinuity. Single subunit Rubiscos have the lowest s_{rel} values (<20) and are well separated from the two-subunit class, which all have values in excess of 50. One infers that possession of the small subunit enables more effective discrimination against O_2. No other kinetic parameter, such as K_m or V_{max}, for either carboxylase or oxygenase activities shows any particular correlation with presence or absence of small subunits.

Since the small subunit appears to affect this most crucial aspect of Rubisco's catalysis, study of the role of the small subunit in Rubisco's catalytic mechanism is clearly warranted.

B. Reversible Dissociation of Small Subunits from the Large-Subunit Octamer

Studies of the role of the small subunit were long hampered by inability to dismantle Rubisco's quaternary structure and separate the large and small subunits without denaturing them irreversibly. Dissociation into monomers and separation are easily effected by detergents, high concentrations of chaotropic salts, or strongly alkaline conditions, but the separated subunits will not renature or reassociate when returned to milder conditions. Furthermore, the separated large subunits precipitate when the denaturant is removed (Rutner and Lane, 1967; Kawashima and Wildman, 1971; Voordouw *et al.,* 1984; Jordan and Chollet, 1985).

The first departure from this unhappy situation was achieved by exploiting the precipitation of Rubisco that occurs when the pH is lowered toward the isoelectric point. When a particularly robust cyanobacterial Rubisco from a *Synechococcus* species was precipitated in this manner, it was discovered that some of the small subunits did not precipitate and were released into solution (Andrews and Abel, 1981; Andrews and Ballment, 1983). At high ionic strength, the fraction of small subunits removed in this way exceeded 80%. Furthermore, the precipitated large subunits remained assembled as octamers and could be redissolved readily in neutral buffer. The process could be repeated until the large-subunit octamers were stripped of small subunits to the point of nondetectability. Furthermore, by varying the ionic strength during precipitation, it was possible to obtain preparations with any required percentage of residual small subunits. These preparations were used to show that catalytic activity was directly proportional to the degree of saturation with small subunits (Fig. 29), leading to the inescapable inference that small subunits were required for catalysis—that is, only those large subunits that had a small subunit attached to them were catalytically competent (Andrews and Ballment, 1983).

Both the isolated large-subunit octamers and the free small-subunit monomers remained in good condition after separation. This was evidenced by their ability to recombine to reform the native quaternary structure with full return of catalytic ability (Andrews and Abel, 1981; Andrews and Ballment, 1983). Reassembly proceeded very rapidly when the separated subunit preparations were mixed at neutral pH, showing an approximately first-order

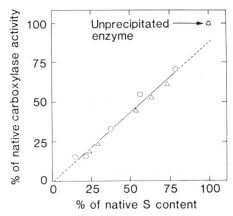

Fig. 29. Relationship between small-subunit content and carboxylase activity of small-subunit-depleted *Synechococcus* Rubisco preparations. The variation in small-subunit content was induced by near-isoelectric precipitation at varying NaCl concentrations. [Data from Andrews and Ballment (1983).]

dependence on the concentration of each subunit (Andrews and Ballment, 1984b). Titration curves obtained by adding increasing concentrations of small subunits to a fixed amount of large-subunit octamers (Fig. 30) approximated the binding constant of the small subunits to the nanomolar range (Andrews and Ballment, 1983). Addition of excess small subunits over and above full saturation did not promote further activity, nor did addition of excess small subunits to the undissociated enzyme (Andrews and Ballment, 1983).

The release of small subunits during near-isoelectric precipitation is curious. It has not been possible to separate small-subunit dissociation from precipitation. Gel filtration of *Synechococcus* Rubisco at acidic pH values only slightly (0.2–0.3 pH units) higher than the pH required to induce precipitation did not resolve the subunits (T. J. Andrews and B. Ballment, unpublished). Furthermore, at low ionic strength, precipitation still occurred even though most of the small subunits remained attached (Andrews and Ballment, 1983). Therefore, precipitation cannot be considered as being a consequence of the loss of small subunits. Rather, small-subunit release must be intimately associated with the precipitation process.

Subsequently, other laboratories, using other bacterial hexadecameric Rubiscos, devised different means of separating small subunits from the large-subunit octamer. Rubisco from the hypersaline cyanobacterium *Aphanothece halophytica* is unusually prone to losing its small subunits when exposed to low ionic strength, even at neutral pH. Sedimentation at low ionic strength removed the majority of the small subunits (Asami *et al.*, 1983; Takabe *et al.*, 1984a; Incharoensakdi *et al.*, 1985a). In this respect it appears diametrically different from the *Synechococcus* enzyme, where high

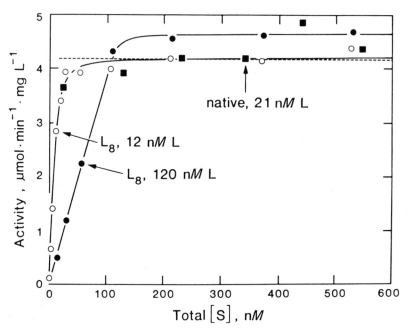

Fig. 30. Reactivation of *Synchococcus* large-subunit octamers by titration of fixed concentrations of large subunits [L] with increasing concentrations of isolated *Synechococcus* small subunits [S]. The lack of an effect of additional small subunits on the activity of the native holoenzyme is also shown. [Data from Andrews and Ballment (1983).]

ionic strength promotes dissociation, indicating that application of these techniques to an unknown Rubisco must be approached with caution. *Chromatium vinosum* Rubisco releases its small subunits from the large-subunit octamer at pH 9.3–9.6, allowing separation by sedimentation (Jordan and Chollet, 1985; Incharoensakdi *et al.*, 1985b). These studies all confirmed that catalytic activity of small-subunit-depleted preparations was linearly proportional to small-subunit content and that reconstitution was accompanied by return of catalytic competence.

Both oxygenase and carboxylase activities were eliminated by removal of the small subunits, and partially stripped preparations had the same oxygenase/carboxylase activity ratio as the undissociated enzyme (Andrews and Ballment, 1983). Clearly, undersaturation with small subunits does not alter the s_{rel} parameter.

There is very little evidence for cooperativity in the binding of small subunits to the large subunit octamer. The shape of the curves in the subsaturating range of small subunit content (Figs. 29 and 30) is quite linear. However, subsaturated preparations form a heterogeneous population of molecules varying in their degree of saturation with small subunits (Andrews and Abel, 1981). These observations are most simply explained by a model where only those large subunits that have a small subunit attached to them are active, and fully so in both carboxylase and oxygenase functions, regardless of whether or not other large subunits in the same octamer have small subunits bound.

Reversible separation of the small subunits from the large-subunit octamer has, so far, only been achieved with Rubiscos from prokaryotes. Application of such techniques to higher-plant Rubiscos results either in no dissociation or in irreversible denaturation of the large subunits, or both. However, near-isoelectric precipitation of spinach Rubisco does cause release of a fraction of the small subunits and these are able to bind to the cyanobacterial large-subunit octamer. The remaining undissociated small subunits and all of the large subunits precipitate and cannot be redissolved without the aid of detergents (Andrews and Lorimer, 1985).

Torres-Ruiz and McFadden (1985) have reported the isolation of large-subunit octamers from *Chromatium vinosum* Rubisco by high-speed centrifugation in solutions containing polyethylene glycol 6000. The small subunits were reported to be pelleted by this procedure, leaving large-subunit octamers in the supernatant. In marked contradiction to all other data for large-subunit octamers from this and other species, these octamers apparently retained full catalytic activity. The conflict between this observation and those from other laboratories remains to be resolved.

C. Capabilities of the Large-Subunit Octamer

What is the condition of the catalytic site after removal of the small subunit? Is its architecture grossly changed or is the influence of the small subunit quite subtle? The evidence seems to be in favor of the latter and comes from a variety of experiments.

1. Binding of 2CABP

The reaction-intermediate analog, 2CABP, binds extraordinarily tightly to the native catalytic site and, in so doing, renders the activator carbamate and the metal ion, as well as the bound 2CABP molecule itself, nonexchangeable with the surrounding solution (Section IV). Similar very tight binding, indicated by the nonexchangeability of the activator carbamate, labeled with $^{14}CO_2$, occurs with all large-subunit octamers so far studied (Andrews and Ballment, 1984b; Takabe *et al.*, 1984b; Jordan and Chollet, 1985). This is strong evidence that the active site is not radically altered by removal of the small subunits. In the absence of direct measurements of the binding constant for 2CABP to the large-subunit octamer, we cannot be sure that this binding is quite as tight as that exhibited by the native enzyme, but it is certainly not reversible on the time scale of gel-filtration experiments.

In this respect, the large-subunit octamer resembles a mutant Rubisco from *Chlamydomonas* ($G^{L8}171 \rightarrow D^{L8}171$), which has a full complement of small subunits, but whose large subunit has a single amino acid substitution (Dron *et al.*, 1983) which renders it inactive. This inactive Rubisco also still binds 2CABP tightly (Spreitzer *et al.*, 1982).

2. CO_2/Mg^{2+} *Carbamylation*

The 2CABP-binding data mentioned above also indicate that the carbamylation of $K^{L8}201$ and the binding of the divalent metal ion (Section IV,A) still proceed without the small subunits. In the absence of 2CABP, it was possible to measure the kinetics and equilibrium of these reactions when the small subunits were missing by exploiting the virtually instantaneous reassembly of active enzyme that occurs on addition of small subunits. Using the *Synechococcus* enzyme, it was shown that, although the initial rate of carbamylation was not altered by removal of the small subunits, the equilibrium was clearly less in favor of carbamylation when the small subunits were missing (Andrews and Ballment, 1984b). This might indicate some subtle change in the environment of $K^{L8}201$, perhaps slightly raising its pK_a or affecting chelation of the metal.

3. *Incapable of Partial Reactions*

Not only does removal of the small subunits render Rubisco incapable of catalyzing both of its overall reactions, but three partial reactions, which may be dissected from the carboxylase reaction sequence, are also inactivated (Andrews *et al.*, 1986). Thus it has not been possible to localize the small subunit's influence to a specific subsection of the catalytic sequence.

The three partial reactions investigated were (1) hydrolysis of the six-carbon intermediate, 3-keto-2CABP, which is catalyzed by the carbamylated enzyme; (2) decarboxylation of this intermediate, which is catalyzed by the decarbamylated enzyme; and (3) enolization of RuBP to its enediol form, which may be measured by following the loss of proton at C-3 of RuBP to solvent ^2H_2O by proton NMR techniques. These partial reactions are described in greater detail in Section IV,F. None of them was catalyzed by preparations of *Synechococcus* large-subunit octamers to extents greater than the trace of overall reaction capability still retained by such preparations and attributable to traces of small subunits remaining (Andrews *et al.*, 1986).

4. *Incapable of Binding 3-Keto-2CABP Tightly*

Since, as mentioned above, the large-subunit octamer is not capable of catalyzing conversion of 3-keto-2CABP, either to products or to reactants, one might imagine that 3-keto-2CABP should bind to the catalytic site of the large subunit at least as tightly as does its analog, 2CABP. However, gel-filtration experiments showed that this was not so (Andrews *et al.*, 1986). Binding of the two ligands was directly compared in the same experiment by mixing ^3H-labeled 2CABP and ^{14}C-labeled 3-keto-2CABP with the large-subunit octamer such that the molar concentration of active sites was in excess of the combined ligand concentrations. After gel filtration to separate the unbound ligands, it was found that, while the expected amount of ^{3}H was recovered bound to the protein, no ^{14}C at all was retained. The conclusion that the catalytic sites on the large-subunit octamer do not bind 3-keto-2CABP nearly as tightly as its analog, 2CABP, seems inescapable.

How might this unexpected inability to bind 3-keto-2CABP tightly be explained? At least two hypotheses may be entertained. First, it is possible that 2CABP is a better analog of the form of the intermediate actually bound by the enzyme than is the form of the intermediate which predominates in solution. In solution, 3-keto-2CABP exists predominantly as the unhydrated ketone (Section IV,D). However, the intermediate tightly bound on the enzyme is predominantly in the hydrated, gem diol form. This form has tetrahedral stereochemistry at C-3, thus resembling 2CABP but not 3-keto-2CABP. If the large-subunit octamer is not able to catalyze the hydration of 3-keto-2CABP, the latter may not bind particularly tightly. However, the nonenzymatic rate of hydration of 3-keto-2CABP has been measured (Lorimer *et al.*, 1986) and it appears to be fast enough ($t_{1/2}$ = 4.6 min at 25°C) to produce detectable quantities of the hydrate during the course of the above gel-filtration experiment if the hydrate were efficiently sequestered by the large-subunit octamer. If this hypothesis is thus ruled out, we are then forced to consider an alternative possibility that has far-reaching ramifications. Perhaps the isolated large subunits are not entirely catalytically blocked.

5. Very Slow Catalysis?

The variety of subunit-separation techniques employed all achieve the same result: namely, the removal of the small subunits from the large-subunit octamer. However, none of them seems to be able to remove the last traces of catalytic activity. A residual activity of 1–3% of that of the undissociated enzyme remains in even the most exhaustively stripped preparations. The position is further complicated by difficulties encountered in accurately measuring residual traces of small subunits by direct means. The most quantitatively reliable method is based on high-performance gel filtration of the subunits dissociated to monomers with Na dodecyl sulfate (Andrews and Ballment, 1984a). However, its accuracy deteriorates when the molar ratio of small subunits to large subunits is very low (<0.05) because of baseline irregularities. Immunochemical quantitation would probably be more accurate at these low levels, but the question would still remain as to whether any last traces of small subunits detected by these means were really in good condition and capable of promoting catalytic competence. It is possible that these traces of small subunits cannot be removed simply because they are denatured and thus precipitate with the large-subunit octamer. Therefore an intriguing, but so far unproven, possibility exists. Perhaps this last trace of catalytic activity, which is exhibited in all three partial reactions mentioned above as well as in the overall carboxylase reaction, is not due to vestiges of undissociated enzyme. Perhaps it represents an intrinsic, but very slow, catalytic potential of the large subunits by themselves.

Such a capacity for very slow catalysis would provide a convenient explanation for the large-subunit octamer's apparent inability to bind 3-keto-2CABP tightly (Andrews *et al.*, 1986). If there remained any intrinsic capacity to convert the intermediate to products, even if this capacity was two orders of magnitude less than that of the intact enzyme, any bound intermediate would be discharged as products during the course of gel filtration and thus not recovered with the protein.

D. Hybridization of Rubisco Subunits Derived from Different Species

The large-subunit octamer of Rubisco binds small subunits and reassembles to form a species indistinguishable from the native, undissociated enzyme. Furthermore, heterologous small subunits derived from a different Rubisco from the one that provided the large-subunit octamer also bind to assemble hybrid enzymes. These hybrids are catalytically active to a greater or lesser extent. Sometimes they are almost as active, in terms of the V_{max} for the carboxylase reaction, as the undissociated parent Rubiscos from which the large-subunit octamer was obtained.

The first hybrids were constructed between the Rubiscos of *Synechococcus* and the unusual, chlorophyll *b*-containing prokaryote, *Prochloron*, which grows symbiotically with a coral-reef ascidian (Andrews *et al.*, 1984). Hybrids were constructed in both senses, that is, *Synechococcus* large subunits with *Prochloron* small subunits, and vice versa. The heterologous small subunits bound as much as an order of magnitude less tightly than the homologous small subunits, but, when fully saturated with foreign small subunits, the hybrids were only slightly less proficient in catalyzing the carboxylase reaction in the presence of saturating substrates than when homologous small subunits were used. Subsequently, subunit hybrids were constructed between Rubiscos from *Synechococcus, Aphanothece,* and *Chromatium* in all six possible senses (Incharoensakdi *et al.*, 1985b). These confirmed the earlier observations. Foreign small subunits bound less tightly and promoted lower V_{max} activities. The situation with *Chromatium* was particularly interesting. Its small subunits were able to bind and activate either of the cyanobacterial large-subunit octamers, but its large-subunit octamer was not activated at all by either of the cyanobacterial small subunits.

Perhaps the most informative Rubisco subunit hybrid constructed so far is that between *Synechococcus* large subunits and spinach small subunits (Andrews and Lorimer, 1985). As mentioned previously, near-isoelectric precipitation of spinach Rubisco (and probably most other higher-plant Rubiscos) irreversibly denatures the large subunits, but some of the small subunits are

TABLE I

**Kinetic Parameters of the Carboxylase Activities at
25°C of Spinach and *Synechococcus* Rubiscos and the
Hybrid Enzyme Formed from *Synechococcus* Large
Subunits and Spinach Small Subunits.**

	K_c (μM)	V_c (sec^{-1})
Spinach[a]	8.5	1.6
Synechococcus[b]	185	11.5
Synechococcus L/spinach S[b]	332	5.2

[a] Data from Jordan and Ogren (1984) recalculated as
described in the legend to Fig. 3.
[b] Data from Andrews and Lorimer (1985).

released in good condition. These were hybridized with the *Synechococcus* large subunit octamer. Once again, they bind more than an order of magnitude less tightly than the homologous small subunits and they promote only about half of the V_{max} of the homologously reassembled *Synechococcus* enzyme. Kinetic parameters of the carboxylase activity of the hybrid are compared with those of the homologously reassembled *Synechococcus* and native spinach enzymes in Table 1. Most notable is the observation that the K_c of the hybrid is nearly twice as high as the K_c of the *Synechococcus* enzyme, which is itself more than an order of magnitude higher than the K_c of the spinach carboxylase. As a result, the V_c/K_c ratio (an index of the catalytic efficiency of an enzyme in processing a limiting concentration of its substrate) of the hybrid is fourfold lower than that of the *Synechococcus* enzyme and 12-fold lower than that of the spinach enzyme. Clearly, substitution of spinach small subunits does not confer the catalytic properties of the spinach enzyme upon the *Synechococcus* large subunits.

Of great significance is the observation that the hybrid's s_{rel} parameter is the same as that of the native *Synechococcus* enzyme (Fig. 3) (Andrews and Lorimer, 1985). Partitioning between carboxylase and oxygenase functions must, therefore, be specified entirely by the large subunit. Any improvements in this parameter, if indeed any are possible, will require engineering of the chloroplast genome.

Construction of a functional hybrid across such a large phylogenetic divergence as that existing between cyanobacteria and higher plants underscores the conservation of the tertiary and quaternary structure of Rubiscos. In this context, the deletion of 12 amino acid residues from the cyanobacterial small subunits, as compared to higher-plant small subunits, that occurs adjacent to the most conserved region of the peptide (Fig. 7) is of particular interest. This deletion also aligns with an exon boundary in some eukaryotic small subunits (Mazur and Chui, 1985; Dean *et al.*, 1985b). Therefore, the deleted fragment may constitute a discrete functional unit and it may lie at the periphery of the protein, as exon junctions often do (Craik *et al.*, 1982). Whatever its function is, it does not interfere with binding to, and activation of, the cyanobacterial large-subunit octamer (Andrews and Lorimer, 1985). Lack of hybridization of the cyanobacterial small subunits with the *Chromatium* large-subunit octamer (Incharoensakdi *et al.*, 1985b) may be indicative of wider divergence in the tertiary structure of Rubisco between *Chromatium* and cyanobacteria than between the latter and higher plants.

E. Possible Roles for the Small Subunit

1. Solubility

The small subunit may help to maintain the solubility of the hexadeca-meric complex. Even for those prokaryotic Rubiscos whose isolated large-subunit octamers remain soluble after removal of the small subunits, this solubility is probably very limited. At high protein concentrations, such as those experienced in the cell or plastid or when Rubisco is expressed in *E. coli*, even these kinds of large subunits may need to be associated with small subunits to remain in solution. However, it is clear from the foregoing that the small subunit also has a profound influence on the catalytic mechanism. Two alternative models for this kind of involvement may be imagined. These are diagramatically represented in Fig. 31.

2. Part of the Catalytic Site?

The simplest hypothesis is that the catalytic site lies at the interface be-tween the large and small subunits with residues of both participating in substrate binding and catalysis (Fig. 31a). Two categories of evidence miti-gate against acceptance of such a simple model. First, no affinity labels have

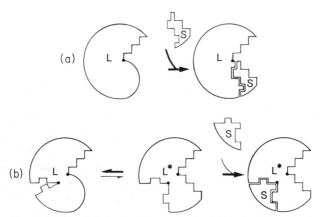

Fig. 31. Alternative possibilities for the small subunit's influence on catalysis. (a) Obliga-tory involvement; (b) remote control.

yet been found that label residues of the small subunit. Such labeling might be expected if reactive small-subunit residues were near the catalytic center. Second, the retention of even the merest trace of catalytic activity by the large subunit octamer after total removal of small subunits would eliminate this model from consideration. As mentioned earlier, lack of binding of 3-keto-2CABP to the large-subunit octamer can be explained most satisfyingly by such a trace of catalytic competence (Section VI,C), although more direct is proof is needed.

3. A Conformational Lock?

A more plausible explanation is that binding of the small subunit, at a site remote from the catalytic center, alters the tertiary structure of the large subunit to produce an active conformation at the catalytic site (Fig. 31b). This model allows introduction of a refinement permitting a trace of intrinsic activity for the large subunit by itself. Perhaps the large subunit is able to achieve the active conformation without assistance from the small subunit but this active conformation is so thermodynamically unfavorable that only a trace of it is present at equilibrium in the absence of small subunits. If the small subunits bound only to the large subunits when they were in the active conformation, the catalytic conformation would be locked in place by such binding. These concepts require further research and, in particular, a reliable means of producing large-subunit octamers totally, and verifiably, free of small subunits.

VII. EVOLUTION OF RUBISCO

A. A Single Phylogeny

In Section II we stressed that Rubisco is a unique catalyst on which all life depends. It is also clear that Rubisco evolved uniquely, all present-day large subunits being descended from a common ancestral protein. Early suspicions that the single-subunit Rubiscos in the purple bacteria might have an independent phylogeny (Robison *et al.*, 1980; Lorimer, 1981b) have been allayed following the determination of the complete sequence of the *R. rubrum* enzyme (Fig. 6). Although this sequence differs markedly from that of the large subunit of two-subunit Rubiscos, several regions of strong homology may be identified and these occur in the same order, and with approximately the same spacing, in both kinds of subunit. Moreover, the overall length of the *R. rubrum* sequence is very similar to that of the large subunits of the two-subunit enzymes. All of the residues of the *R. rubrum* sequence identified by affinity labeling as being close to the active site are contained within these conserved regions.

The origins of the small subunit are less clear because its sequence is more

```
                  456                                              466
R. rubrum         gly val glu asp thr arg ser ala leu pro alaCOO‾

                  10                                               20
Small Subunit     lys lys phe glu thr leu ser tyr leu pro pro leu
                  arg arg tyr         phe         phe     asp ser
                                                          leu
```

Fig. 32. Comparison of the sequence at the carboxy terminus of *R. rubrum* Rubisco with one of the conserved regions of the small subunit. For the small-subunit sequence, the upper line represents the commonest sequence. Less common substitutions are shown underneath.

divergent and its function more mysterious. Nevertheless, several regions of its sequence are highly conserved (Fig. 7), as is its overall size, and it also may have a monophyletic origin. An important question is: "Did the original Rubisco have small subunits?" Two scenarios may be imagined, and we cannot choose between them. Perhaps Rubisco first appeared as a single-subunit protein, like that found in present-day purple bacteria. The latter's poor selectivity for CO_2, as opposed to O_2, would not have been a problem because the primitive atmosphere contained little, if any O_2, and the CO_2 concentration was probably much higher than at present (Broda, 1975). As oxygenic photosynthesis rendered the CO_2/O_2 ratio increasingly unfavorable, Rubiscos with higher s_{rel} values would have been selected. As mentioned in Section VI,A, appearance of small subunits may have enabled such an increase in s_{rel}. Alternatively, the first Rubisco may have had both subunits, perhaps arising from the combination and modification of two preexisting proteins. The single-subunit variant might then have arisen from a fusion and reduction of the two peptides such that the functionality of small subunit became incorporated into the large subunit. A faint homology between one of the conserved regions of the small subunit and a short sequence at the carboxyl terminus of *R. rubrum* Rubisco (Fig. 32) may be relevant in this context. It remains to be seen if this short sequence serves some essential purpose that requires its conservation, or if it is simply a statistical accident.

Therefore, it seems safe to assume that Rubisco, one of the cornerstones of autotrophy, must have arisen only once in the evolution of life, presumably appearing with the first chemolithotrophs, which arose from primitive fermentative bacteria more than 3.5×10^9 years ago, before the genesis of even the oldest rocks on this planet (Broda, 1975).

B. Adaptive Responses to the Declining CO_2/O_2 Ratio

What changes have occurred in Rubisco during its long evolution? In particular, how has it adapted to the declining CO_2/O_2 ratio that it, itself, has largely caused? The original Rubisco's inherent oxygenase activity would not have been expressed in the primitive anaerobic environment. However, with the advent of cyanobacteria and oxygenic photosynthesis, the CO_2/O_2 ratio would have become progressively less favorable for carboxylation, to the point that the positive carbon balance of photosynthesis would have been threatened if early Rubiscos had s_{rel} values much lower than those of present-day, higher-plant Rubiscos. In general, it seems that Rubisco has been remarkably unresponsive to this extreme selective pressure. Rather, adaptations appear to have occurred in Rubisco's metabolic environment instead.

1. Increases in CO_2/O_2 Relative Specificity?

As stated above, an obvious response to the worsening atmospheric CO_2/O_2 ratio would have been the selection of Rubiscos with increased s_{rel} values. This may very well have occurred, but we have no evidence either for or against such adaptation. Inspection of the s_{rel} values for various Rubiscos (Fig. 3) indicates that they range over about an order of magnitude. It is tempting to speculate that this variation represents an evolutionary spectrum. However, this is unwise and flawed fundamentally by the erroneous assumption that "primitive" prokaryotes are less highly evolved than the more "advanced" eukaryotes. Because of the difference in generation times, probably the reverse is nearer the truth. It is probably safer to assume that all Rubiscos are, to a greater or lesser extent, adapted to the differing CO_2/O_2 ratios in which they function. Nevertheless, it is clear that there is at least some limited plasticity in s_{rel}, with the lowest values being found in Rubiscos from anaerobic bacteria and the highest in C_3 plants that have no means of elevating the CO_2/O_2 ratio at the carboxylation sites. Organisms possessing CO_2-concentrating mechanisms, such as algae and C_4 plants, have Rubiscos with intermediate s_{rel} values.

2. Photorespiration

In the presence of O_2, and the absence of extraneous means for manipulating the CO_2/O_2 ratio in the environment of Rubisco, oxygenation of RuBP and production of phosphoglycolate is unavoidable. In these circumstances, the photorespiratory pathway functions to remove and recycle as much of the phosphoglycolate carbon as possible (Lorimer and Andrews, 1981). No other generally applicable rationale for photorespiration has been found, and therefore this process, with its inherent CO_2 loss and energy costs, must be regarded as an adaptive response to RuBP oxygenation in current atmospheric conditions (Lorimer and Andrews, 1981).

3. CO_2-Accumulating Mechanisms

Another way that nature has found to circumvent RuBP oxygenation involves the energy-driven accumulation of CO_2 so that its concentration at the site of Rubisco is manyfold higher than that in equilibrium with air. Two general categories of mechanisms for achieving this result exist. These are the C_4 pathway found predominantly in plants from warm climates and the inorganic-carbon-concentrating mechanism found in unicellular algae. Both of these mechanisms are the subjects of other chapters in this volume.

Regardless of the details of the CO_2-accumulation mechanism employed, the end result is the virtually complete suppression of RuBP oxygenation and photorespiration under normal conditions. Why, then, have these mechanisms not found universal acceptance in nature? The answer lies in their energy costs and the temperature dependence of s_{rel}. The energy costs of the C_4 pathway are better understood than are those of the algal inorganic-carbon uptake system. Additional ATP, over and above that consumed by the C_3 cycle, is required to drive the C_4-acid cycle, and leakage of some of the accumulated CO_2 from the bundle-sheath cells constitutes a further drain on the energetics of the overall system. These additional costs cause the quantity of light required to energize a given amount of CO_2 fixation by the C_4 mechanism to be 30–50% greater than that required by the C_3 pathway operating in the absence of photorespiration (e.g., at low O_2 or high CO_2) (Ehleringer and Bjorkman, 1976; Ku and Edwards, 1978; Ehleringer and Pearcy, 1983). These costs are not dependent on temperature. On the other hand, the energy costs of photorespiration, which derive from the ATP and NADH consumed by the glycolate pathway, are temperature-dependent for two reasons. First, Rubisco's s_{rel} parameter decreases with increasing temperature. Second, the temperature coefficients of the solubilities of CO_2 and O_2 differ in such a manner that the CO_2/O_2 ratio of a solution in equilibrium with air decreases as temperature is increased (Ku and Edwards, 1978; Jordan and Ogren, 1984). Both of these phenomena cause the fraction of RuBP that is oxygenated to increase with temperature. As a result, photorespiration consumes less energy than CO_2 accumulation at low temperatures, but at higher temperatures the reverse is true (Ehleringer and Bjorkman, 1976; Ku and Edwards, 1978; Ehleringer and Pearcy, 1983).

4. Carboxysomes?

Another way of dealing with the inefficiencies arising from Rubisco's oxygenase activity would be to eliminate oxygenation completely by preventing access of O_2 to the active site. Such a strategy might not have the energy costs associated with the CO_2-accumulation mechanisms mentioned above. In theory, access of O_2 might be prevented by surrounding Rubisco with a membrane that was impermeable to gases but that did not impede the movement of ionic species, such as HCO_3^-, protons, RuBP, and products. Such a mechanism may, in fact, be the rationale for carboxysomes.

Carboxysomes, also called polyhedral bodies, are subcellular inclusions found in many, but not all, CO_2-fixing prokaryotes. These structures contain a large fraction of the cell's Rubisco in a paracrystalline array surrounded by a single proteinaceous membrane. Carboxysomal Rubisco is of the two-subunit, hexadecameric type and is fully active and assayable *in situ* in carboxysomes isolated by centrifugation from cell extracts. The occurrence, properties, and possible functions of carboxysomes have been comprehensively reviewed recently (Codd and Marsden, 1984).

Evidence concerning the degree of O_2 sensitivity of carboxysomal Rubisco is conflicting. Coleman *et al.* (1982) observed that Rubisco from cyanobacteria was immune to O_2 inhibition while in the carboxysomes, but normally sensitive after release into solution. However, these observations have not yet been repeatable in other laboratories (A. Kaplan, personal communication; M. Badger, personal communication). Greater abundance of carboxysomes in cyanobacteria grown in limiting-CO_2 conditions (Turpin *et al.*, 1984) is consistent with an O_2-protective role. Further investigations of this interesting possibility are clearly warranted.

VIII. PROSPECTS FOR A BETTER RUBISCO

A. Is a Better Rubisco Possible?

Given that, for more than 3.5×10^9 years, nature has been conducting a selection experiment with Rubisco where increases in s_{rel} and V_c/K_c would be very advantageous, is Rubisco in higher plants already perfect in the sense that no further increases in these parameters are possible? This may, indeed, be true but we have no direct evidence. Most importantly, despite detailed probing of the catalytic mechanism, no *theoretical* limit to s_{rel} based on mechanistic understanding has emerged (Section IV,I). Nor is there reason for believing that s_{rel} and V_c/K_c are inversely related, which might lead to suspicion that the optimal compromise between these two parameters has already been reached. Indeed, those Rubiscos with the highest s_{rel} values also have the highest V_c/K_c ratios, although we must qualify the latter statement with the observation that it is difficult to be sure of V_c estimates because of the possibility of contamination of all enzyme preparations with unknown amounts of catalytically incompetent, degraded species. However, if we restrict our attention to intensively studied, highly purified preparations showing the highest specific activities, we see that the higher-plant Rubisco has the highest V_c/K_c ratio, as well as the highest s_{rel} (Table II). Therefore, in the absence of known theoretical limits to the advancement of Rubisco, attempts to improve it are, at least, not hopeless.

However, since Rubisco has been under such intense selection for so long, it is not likely that improvements may be achieved by simple means, such as single amino acid changes, which would, presumably, have evolved naturally and frequently. Any beneficial changes arising in this manner will already have occurred. The gulf between present-day Rubiscos and any more efficient Rubiscos that may exist in principle must be too wide to have been bridged by natural evolutionary means in the available time. Therefore, it

TABLE II

Comparison of the CO_2/O_2 Relative Specificities (s_{rel}) of Various Rubiscos with the Catalytic Efficiency of Their Carboxylase Activities at Limiting CO_2 Concentration (V_c/K_c)

	s_{rel}[a]	V_c/K_c $(sec^{-1}M^{-1})$
Spinach	105	19×10^{4}[b]
Synechococcus	68	6.2×10^{4}[a]
R. rubrum	15	7.0×10^{4}[c]

[a] Data from Andrews and Lorimer (1985).
[b] Data from Jordan and Ogren (1984) recalculated as described in the legend to Fig. 3.
[c] Data from Gutteridge *et al.* (1984a).

seems probable that improvements will require more extensive changes involving multiple residues, perhaps in both subunits. If coordinated multiple changes are necessary, with any changes occurring singly being deleterious, then it is easy to imagine that not enough time has yet elapsed for them to appear.

B.   How Might a Better Rubisco Be Achieved?

1.   Molecular Engineering

If we knew the structure of a more efficient Rubisco, it is reasonable to imagine, given the power of modern molecular genetics, that it could be engineered. Incorporation of any new genetic information for large subunits into the plastid genome might be more difficult at the present stage of development of this technology, but we do not consider that this is, by any means, the major obstacle hindering this kind of progress. The major obstacle is lack of understanding of the changes necessary to achieve a better Rubisco. While, as far as mechanism is concerned, Rubisco may be the best understood plant enzyme, such understanding does not approach the level of detail required to predict what might be beneficial changes at the active site. More, and more detailed, studies of Rubisco's complex mechanism are necessary and they must address the role of enzymatic reactive groups in the chemistry.

The lack of x-ray crystallographic information about the topography of the active site and the interaction of enzymatic groups with substrate or reaction-intermediate analogs is a severe hindrance to the kind of mechanistic and directed-mutagenic studies that will be necessary. However, crystallographic studies are proceeding (Section III,C) and this roadblock may be removed in the near future.

2. *Selective Mutagenesis* in Vitro

In theory, another method for accelerating the evolution of greater efficiency in Rubisco may be imagined (J. Pierce, personal communication). It does not require any knowledge of the reaction mechanism or of active-site structure, but other requirements are rather stringent. An aerobic organism would be required that depended on Rubisco for autotrophic CO_2 fixation and that was amenable to genetic manipulation. The organism should also be able to be maintained heterotrophically during genetic manipulations, and it should lack, genetically, the ability to accumulate CO_2 internally. The Rubisco genes would be deleted and the excised genes (or Rubisco genes from another source) subjected to intense mutagenesis under extreme conditions *in vitro*. The mutagenized products would then be reinserted into the genome, if possible, and selection applied under autotrophic conditions by poising the CO_2/O_2 ratio so that growth was possible only if a mutant Rubisco with an s_{rel} higher than that of the wild type appeared. This extreme mutagenesis might be expected to achieve the desired multiple changes simultaneously, thus bridging the gap, which natural evolution may not be able to cross, between existing Rubiscos and more efficient Rubiscos that may exist in principle. Difficulties in finding an organism meeting these stringent requirements have, so far, obstructed this elegant approach.

REFERENCES

Amiri, I., Salnikow, J., and Vater, J. (1984). *Biochim. Biophys. Acta* **784**, 116–123.

Andersen, K., and Wilke-Douglas, M. (1984). *J. Bacteriol.* **159**, 973–978.

Andersson, I., and Branden, C. I. (1984). *J. Mol. Biol.* **172**, 363–366.

Andersson, I., Tjader, A. C., Cedergren-Zeppezauer, E., and Branden, C. I. (1983). *J. Biol. Chem.* **258**, 14088-14090.

Andrews, T. J., and Abel, K. M. (1981). *J. Biol. Chem.* **256**, 8445–8451.

Andrews, T. J., and Ballment, B. (1983). *J. Biol. Chem.* **258**, 7514–7518.

Andrews, T. J., and Ballment, B. (1984a). *Plant Physiol.* **75**, 503–510.

Andrews, T. J., and Ballment, B. (1984b). *Proc. Natl. Acad. Sci. U. S. A.* **81**, 3660-3664.

Andrews, T. J., and Hatch, M. D. (1971). *Phytochemistry* **10**, 9–15.

Andrews, T. J., and Lorimer, G. H. (1978). *FEBS Lett.* **90**, 1–9.

Andrews, T. J., and Lorimer, G. H. (1985). *J. Biol. Chem.* **260**, 4632–4636.

Andrews, T. J., Lorimer, G. H., and Tolbert, N. E. (1973). *Biochemistry* **12**, 11–18.

Andrews, T. J., Badger, M. R., and Lorimer, G. H. (1975). *Arch. Biochem. Biophys.* **171**, 93–103.

Andrews, T. J., Abel, K. M., Menzel, D., and Badger, M. R. (1981). *Arch. Microbiol.* **130**, 344–348.

Andrews, T. J., Greenwood, D. M., and Yellowlees, D. (1984). *Arch. Biochem. Biophys.* **234**, 313–317.

Andrews, T. J., Lorimer, G. H., and Pierce, J. (1986). *J. Biol. Chem.* **261**, 12184–12188.

Asami, S., Takabe, T., Akazawa, T., and Codd, G. A. (1983). *Arch. Biochem. Biophys.* **225**, 713–721.

Badger, M. R., and Andrews, T. J. (1974). *Biochem. Biophys. Res. Commun.* **60,** 204–210.
Badger, M. R., and Collatz, G. J. (1977). *Year Book—Carnegie Inst. Washington* **76,** 355–361.
Badger, M. R., and Lorimer, G. H. (1976). *Arch. Biochem. Biophys.* **175,** 723–729.
Badger, M. R., and Lorimer, G. H. (1981). *Biochemistry* **20,** 2219–2225.
Badger, M. R., Andrews, T. J., Canvin, D. T., and Lorimer, G. H. (1980). *J. Biol. Chem.* **255,** 7870–7875.
Badger, M. R., Sharkey, T. D., and von Caemmerer, S. (1984). *Planta* **160,** 305–313.
Baker, T. S., Eisenberg, D., Eiserling, F. A., and Weisman, L. (1975). *J. Mol. Biol.* **91,** 391–399.
Baker, T. S., Eisenberg, D., and Eiserling, F. (1977a). *Science* **196,** 293–295.
Baker, T. S., Suh, S. W., and Eisenberg, D. (1977b). *Proc. Natl. Acad. Sci. U. S. A.* **74,** 1037–1041.
Banner, D. W., Bloomer, H. C., Petsko, G. A., Phillips, D. C., Pogson, C. I., and Wilson, A. I. (1976). *Biochem. Biophys. Res. Commun.* **72,** 146–155.
Barcena, J. A. (1983). *Biochem. Int.* **7,** 755–760.
Barcena, J. A., Pickersgill, R. W., Adams, M. J., Phillips, D. C., and Whatley, F. R. (1983). *EMBO J.* **2,** 2363–2367.
Bassham, J. A., Benson, A. A., and Calvin, M. (1950). *J. Biol. Chem.* **185,** 781–787.
Bedbrook, J. R., Smith, S. M., and Ellis, R. J. (1980). *Nature (London)* **287,** 692–697.
Behki, R. M., Selvaraj, G., and Iyer, V. N. (1983). *Can. J. Microbiol.* **29,** 767–774.
Belknap, W. R., and Portis, A. R. (1986a). *Plant Physiol.* **80,** 707–710.
Belknap, W. R., and Portis, A. R. (1986b). *Biochemistry* **25,** 1864–1869.
Benson, A. A., Bassham, J. A., Calvin, M., Goodale, T. C., Haas, V. A., and Stepka, W. (1950). *J. Am. Chem. Soc.* **72,** 1710–1718.
Berry, J. A., Seemann, J. R., Lorimer, G. H., Pierce, J., Meeks, J., and Freas, S. (1987). *Proc. Natl. Acad. Sci. U. S. A.* **84,** 734–738.
Berry-Lowe, S. L., McKnight, T. D., Shah, D. M., and Meagher, R. B. (1982). *J. Mol. Appl. Genet.* **1,** 483–498.
Bhagwat, A. S., and McFadden, B. A. (1982). *FEBS Lett.* **145,** 313–316.
Bjorkman, O., and Demmig, B. (1987). *Planta* **170,** 489–504.
Blake, C. C. F. (1979). *Nature (London)* **277,** 598.
Bloom, M. V., Milos, P., and Roy, H. (1983). *Proc. Natl. Acad. Sci. U. S. A.* **80,** 1013–1017.
Boag, S., and Portis, A. R. (1984). *Planta* **160,** 33–40.
Bowes, G., Ogren, W. L., and Hageman, R. H. (1971). *Biochem. Biophys. Res. Commun.* **45,** 716–722.
Bowien, B., and Gottschalk, E. M. (1982). *J. Biol. Chem.* **257,** 11845–11847.
Bowien, B., and Mayer, F. (1978). *Eur. J. Biochem.* **88,** 97–107.
Bowien, B., Mayer, F., Codd, G. A., and Schlegel, H. G. (1976). *Arch. Microbiol.* **110,** 157–166.
Bowien, B., Mayer, F., Spiess, E., Pahler, A., Englisch, U., and Saenger, W. (1980). *Eur. J. Biochem.* **106,** 405–410.
Bowman, L. H., and Chollet, R. (1980). *J. Bacteriol.* **141,** 652–657.
Branden, C. I., Schneider, G., Lindqvist, Y., Andersson, I., Knight, S., and Lorimer, G. H. (1986). *Philos. Trans. R. Soc. London, Ser. B* **313,** 359–365.
Branden, R., Nilsson, T., and Styring, S. (1984a). *Biochemistry* **23,** 4373–4377.
Branden, R., Nilsson, T., and Styring, S. (1984b). *Biochemistry* **23,** 4378–4382.
Broda, E. (1975). "The Evolution of the Bioenergetic Processes." Pergamon, Oxford.
Broglie, R., Coruzzi, G., Lamppa, G., Keith, B., and Chua, N.-H. (1983). *Bio Technology* **1,** 55–61.
Calvin, M. (1956). *J. Chem. Soc.,* pp. 1895–1915.
Calvin, M., and Benson, A. A. (1948). *Science* **107,** 476–480.
Choe, H. W., Jakob, R., Hahn, U., and Pal, G. P. (1985). *J. Mol. Biol.* **185,** 781–783.

Chollet, R., Anderson, L. L., and Hovsepian, L. C. (1975). *Biochem. Biophys. Res. Commun.* **64,** 97–107.

Chou, P. Y., and Fasman, G. D. (1974). *Biochemistry* **13,** 222–245.

Christeller, J. T. (1981). *Biochem. J.* **193,** 839–844.

Christeller, J. T. (1982). *Arch. Biochem. Biophys.* **217,** 485–490.

Christeller, J. T., and Laing, W. A. (1978). *Biochem. J.* **173,** 467–473.

Christeller, J. T., and Laing, W. A. (1979). *Biochem. J.* **183,** 747–750.

Christeller, J. T., Terzaghi, B. E., Hill, D. F., and Laing, W. A. (1985). *Plant Mol. Biol.* **5,** 257–263.

Cleland, W. W. (1963). *Biochim. Biophys. Acta* **67,** 104–137.

Codd, G. A., and Marsden, W. J. N. (1984). *Biol. Rev. Cambridge Philos. Soc.* **59,** 389–422.

Codd, G. A., Cook, C. M., and Stewart, W. D. P. (1979). *FEMS Microbiol. Lett.* **6,** 81–86.

Coleman, J. R., Seemann, J. R., and Berry J. A. (1982). *Year Book—Carnegie Inst. Washington* **81,** 83–87.

Cooper, T. G., Filmer, D. L., Wishnick, M., and Lane, M. D. (1969). *J. Biol. Chem.* **244,** 1081–1083.

Coruzzi, G., Broglie, R., Cashmore, A., and Chua, N.-H. (1983). *J. Biol. Chem.* **258,** 1399–1402.

Coruzzi, G., Broglie, R., Edwards, R., and Chua, N.-H. (1984). *EMBO J.* **3,** 1671–1679.

Craik, C. S., Sprang, S., Flettrick, R., and Rutter, W. J. (1982). *Nature (London)* **299,** 180–182.

Curtis, S. E., and Haselkorn, R., (1983). *Proc. Natl. Acad. Sci. U. S. A.* **80,** 1835–1839.

Dean, C., Van den Elzen, P., Tamaki, S., Dunsmuir, P., and Bedbrook, J. (1985a). *Proc. Natl. Acad. Sci. U. S. A.* **82,** 4964–4968.

Dean, C., Van den Elzen, P., Tamaki, S., Dunsmuir, P., and Bedbrook, J. (1985b). *EMBO J.* **4,** 3055–3061.

Dietz, K.-J., and Heber, U. (1984). *Biochim. Biophys. Acta* **767,** 432–443.

Dietz, K.-J., and Heber, U. (1986). *Biochim. Biophys. Acta* **848,** 392–401.

Donnelly, M. I., Stringer, C. D., and Hartman, F. C. (1983). *Biochemistry* **21,** 4346–4352.

Donnelly, M. I., Hartman, F. C., and Ramakrishnan, V. (1984). *J. Biol. Chem.* **259,** 406–411.

Dron, M., Rahire, M., and Rochaix, J.-D. (1982). *J. Mol. Biol.* **162,** 775–793.

Dron, M., Rahire, M., Rochaix, J.-D., and Mets, L. (1983). *Plasmid* **9,** 321–324.

Dunsmuir, P., Smith, S., and Bedbrook, J. (1983). *Nucleic Acids Res.* **11,** 4177–4183.

Ehleringer, J., and Bjorkman, O. (1976). *Year Book—Carnegie Inst. Washington* **75,** 418–421.

Ehleringer, J., and Pearcy, R. W. (1983). *Plant Physiol.* **73,** 555–559.

Ellis, R. J. (1979). *Trends Biochem. Sci.* **4,** 241–244.

Ellis, R. J., and Gatenby, A. A. (1984). *Annu. Proc. Phytochem. Soc. Eur.* **23,** 41–60.

Ellis, R. J., Gallagher, T. F., Jenkins, G. I., and Lennox, C. R. (1984). *J. Embryol. Exp. Morphol.* **83,** 163–178.

Estelle, M., Hanks, J., McIntosh, L., and Somerville, C. (1985). *J. Biol. Chem.* **260,** 9523–9526.

Evans, M. C. W., Buchanan, B. B., and Arnon, D. I. (1966). *Proc. Natl. Acad. Sci. U.S.A.* **55,** 928–934.

Farquhar, G. (1979). *Arch. Biochem. Biophys.* **193,** 456–465.

Fersht, A. (1985). "Enzyme Structure and Function," 2nd ed. Freeman, New York.

Ferhst, A. R., Shi, J.-P., Knill-Jones, J., Lowe, D. M., Wilkinson, A. J., Blow, P., Carter, P., Waye, M. M. Y., and Winter, G. (1985). *Nature (London)* **314,** 235–238.

Fersht, A. R., Leatherbarrow, R. J., and Wells, T. N. C. (1986). *Trends Biochem. Sci.* **11,** 321–325.

Fiedler, F., Mullhofer, G., Trebst, A., and Rose, I. A. (1967). *Eur. J. Biochem.* **1,** 395–399.

Fischer, R. A., and Turner, N. C. (1978). *Annu. Rev. Plant Physiol.* **29,** 277–317.

Fluhr, R., and Chua, N.-H. (1986). *Proc. Natl. Acad. Sci. U. S. A.* **83,** 2358–2362.

Fluhr, R., Kuhlemeier, C., Nagy, F., and Chua, N.-H. (1986). *Science* **232,** 1106–1112.

Fraij, B., and Hartman, F. C. (1982). *J. Biol. Chem.* **257,** 3501–3505.

Fraij, B., and Hartman, F. C. (1983). *Biochemistry* **22**, 1515–1520.
Gatenby, A. A. (1984). *Eur. J. Biochem.* **144**, 361–366.
Gatenby, A. A., Van der Vies, S. M., and Bradley, D. (1985). *Nature (London)* **314**, 617–620.
Gibson, J. L., and Tabita, F. R. (1977a). *J. Biol. Chem.* **252**, 943–949.
Gibson, J. L., and Tabita, F. R. (1977b). *J. Bacteriol.* **131**, 1020–1022.
Gibson, J. L., and Tabita, F. R. (1977c). *J. Bacteriol.* **132**, 818–823.
Gibson, J. L., and Tabita, F. R. (1979). *J. Bacteriol.* **140**, 1023–1027.
Gibson, J. L., and Tabita, F. R. (1985). *J. Bacteriol.* **164**, 1188–1193.
Gilbert, W. (1978). *Nature (London)* **271**, 501.
Gould, S. J., and Lewontin, R. C. (1979). *Proc. R. Soc. London, Ser. B* **205**, 581–598.
Grebanier, A. E., Champagne, D., and Roy, H. (1978). *Biochemistry* **17**, 5150–5155.
Gringrich, J. C., and Hallick, R. B. (1985). *J. Biol. Chem.* **260**, 16162–16168.
Gurevitz, M., Somerville, C. R., and McIntosh, L. (1985). *Proc. Natl. Acad. Sci. U. S. A.* **82**, 6456–6550.
Gutteridge, S., Sigal, I., Thomas, B., Arentzen, R., Cordova, A., and Lorimer, G. H. (1984a). *EMBO J.* **3**, 2737–2743.
Gutteridge, S., Parry, M. A. J., Schmidt, C. N. G., and Feeney, J. (1984b). *FEBS Lett.* **170**, 355–359.
Gutteridge, S., Millard, B. N., and Parry, M. A. J. (1986a), *FEBS Lett.* **196**, 263–268.
Gutteridge, S., Parry, M. A. J., Burton, S., Keys, A. J., Mudd, A., Feeney, J., Servaites, J., and Pierce, J. (1986b). *Nature (London)* **324**, 274–276.
Hamilton, G. A. (1974). *In* "Molecular Mechanisms of Oxygen Activation" (O. Hayaishi, ed.), pp. 405–448. Academic Press, New York.
Hartman, F. C., Milanez, S., and Lee, E. H. (1985). *J. Biol. Chem.* **260**, 13968–13975.
Hartman, F. C., Mural, R. J., Mitra, S., Soper, T. S., Niyogi, S. K., Foote, R. S., Machanoff, R., Lee, E. H., and Larimer, F. W. (1986). *Fed. Proc., Fed. Am. Soc. Exp. Biol.* **45**, Abstr. 2519.
Hatch, A. L., and Jensen, R. G. (1980). *Arch. Biochem. Biophys.* **205**, 587–594.
Heldt, H. W., Werdan, K., Milovancev, M., and Geller, G. (1973). *Biochim. Biophys. Acta* **314**, 224–241.
Hemmingsen, S. M., and Ellis, R. J. (1986). *Plant Physiol.* **80**, 269–276.
Herndon, C. S., and Hartman, F. C. (1984). *J. Biol. Chem.* **259**, 3102–3110.
Herndon, C. S., Norton, I. L., and Hartman, F. C. (1982). *Biochemsitry* **21**, 1380–1385.
Hurwitz, J., Jakoby, W. B., and Horecker, B. L. (1956). *Biochim. Biophys. Acta* **22**, 194–195.
Igarashi, Y., McFadden, B. A., and El-Gul, T. (1985). *Biochemistry* **24**, 3957–3962.
Incharoensakdi, A., Takabe, T., and Akazawa, T. (1985a). *Arch. Biochem. Biophys.* **237**, 445–453.
Incharoensakdi, A., Takabe, T., and Akazawa, T. (1985b). *Biochem. Biophys. Res. Commun.* **126**, 698–704.
Janson, C. A., Smith, W. W., Eisenberg, D., and Hartman, F. C. (1984). *J. Biol. Chem.* **259**, 11594–11596.
Jaworowski, A., and Rose, I. A. (1985). *J. Biol. Chem.* **260**, 944–948.
Jaworowski, A., Hartman, F. C., and Rose, I. A. (1984). *J. Biol. Chem.* **259**, 6783–6789.
Jensen, R. G., and Bahr, J. T. (1977). *Annu. Rev. Plant Physiol.* **28**, 379–400.
Johal, S., Bourque, D. P., Smith, W. W., Suh, S. W., and Eisenberg, D. (1980). *J. Biol. Chem.* **255**, 8873–8880.
Johal, S., Partridge, B. E., and Chollet, R. (1985). *J. Biol. Chem.* **260**, 9894–9904.
Jordan, D. B., and Chollet, R. (1983). *J. Biol. Chem.* **258**, 13752–13758.
Jordan, D. B., and Chollet, R. (1985). *Arch. Biochem. Biophys.* **236**, 487–496.
Jordan, D. B., and Ogren, W. L. (1981). *Nature (London)* **291**, 513–515.
Jordan, D. B., and Ogren, W. L. (1983). *Arch. Biochem. Biophys.* **227**, 425–433.
Jordan, D. B., and Ogren, W. L. (1984). *Planta* **161**, 308–313.

Jordan, D. B., Chollet, R., and Ogren, W. L. (1983). *Biochemistry* **22,** 3410–3418.
Kawashima, N., and Wildman, S. G. (1971). *Biochim. Biophys. Acta* **229,** 749–760.
Kochi, J. K., ed. (1973). "Free Radicals." Wiley, New York.
Koenig, T., and Fischer, H. (1973). *In* "Free Radicals" (J. K. Kochi, ed.), pp. 157–190. Wiley, New York.
Ku, S.-B., and Edwards, G. E. (1978). *Planta* **140,** 1–6.
Laing, W. A., and Christeller, J. T. (1976). *Biochem. J.* **159,** 563–570.
Laing, W. A., and Christeller, J. T. (1980). *Arch. Biochem. Biophys.* **202,** 592–600.
Laing, W. A., Ogren, W. L., and Hageman, R. H. (1974). *Plant Physiol.* **54,** 678–685.
Langridge, P. (1981). *FEBS Lett.* **123,** 85–89.
Lawlis, V. B., Gordon, G. L. R., and McFadden, B. A. (1979). *J. Bacteriol.* **139,** 287–298.
Led, J. J., Neesgaard, E., and Johansen, J. T. (1982). *FEBS Lett.* **147,** 74–80.
Lee, E. H., and Hartman, F. C. (1986). *Fed. Proc., Fed. Am. Soc. Exp. Biol.* **45,** Abstr. 772.
Lennox, C. R., and Ellis, R. J. (1986). *Biochem. Soc. Trans.* **14,** 9–11.
Lindqvist, Y., and Branden, C. I. (1985). *Proc. Natl. Acad. Sci. U. S. A.* **82,** 6855–6859.
Lorimer, G. H. (1978). *Eur. J. Biochem.* **89,** 43–50.
Lorimer, G. H. (1979). *J. Biol. Chem.* **254,** 5599–5601.
Lorimer, G. H. (1981a). *Biochemistry* **20,** 1236–1240.
Lorimer, G. H. (1981b). *Annu. Rev. Plant Physiol.* **32,** 349–383.
Lorimer, G. H., and Andrews, T. J. (1973). *Nature (London)* **243,** 359–360.
Lorimer, G. H., and Andrews, T. J. (1981). *In* "The Biochemistry of Plants" (M. D. Hatch and N. K. Boardman, eds.), Vol. 8, pp. 329–374. Academic Press, New York.
Lorimer, G. H., and Miziorko, H. M. (1980). *Biochemistry* **19,** 5321–5328.
Lorimer, G. H., and Pierce, J. (1987). *J. Biol. Chem.* (in press).
Lorimer, G. H., Andrews, T. J., and Tolbert, N. E. (1973). *Biochemistry* **15,** 529–536.
Lorimer, G. H., Badger, M. R., and Andrews, T. J. (1976). *Biochemistry* **15,** 529–536.
Lorimer, G. H., Badger, M. R., and Andrews, T. J. (1977). *Anal. Biochem.* **78,** 66–75.
Lorimer, G. H., Badger, M. R., and Heldt, H. W. (1978). *In* "Photosynthetic Carbon Assimilation" (H. W. Siegleman and G. Hind, eds.), pp. 283–306. Plenum, New York.
Lorimer G. H., Andrews, T. J., Pierce, J., and Schloss, J. V. (1986). *Philos. Trans. R. Soc. London, Ser. B* **313,** 397–407.
McCurry, S. D., Pierce, J., Tolbert, N. E., and Orme-Johnson, W. H. (1981). *J. Biol. Chem.* **256,** 6623–6628.
McFadden, B. A. (1973). *Bacteriol. Rev.* **37,** 289–319.
McIntosh, L., Poulsen, C., and Bogorad, L. (1980). *Nature (London)* **288,** 556–560.
Martin, P. G. (1979). *Aust. J. Plant Physiol.* **6,** 401–408.
Mazur, B. J., and Chui, C.-F. (1985). *Nucleic Acids Res.* **13,** 2373–2386.
Meisenberger, O., Pilz, I., Bowien, B., Pal, G. P., and Saenger, W. (1984). *J. Biol. Chem.* **259,** 4463–4465.
Milos, P., and Roy, H. (1984). *J. Cell. Biochem.* **24,** 153–162.
Miziorko, H. M. (1979). *J. Biol. Chem.* **254,** 270–272.
Miziorko, H. M., and Lorimer, G. H. (1983). *Annu. Rev. Biochem.* **52,** 507–535.
Miziorko, H. M., and Mildvan, A. (1974). *J. Biol. Chem.* **249,** 2743–2750.
Miziorko, H. M., and Sealy, R. C. (1980). *Biochemistry* **19,** 1167–1171.
Miziorko, H. M., and Sealy, R. C. (1984). *Biochemistry* **23,** 479–485.
Miziorko, H. M., Behnke, C. E., and Houkom, E. C. (1982). *Biochemistry* **21,** 6669–6674.
Morelli, G., Nagy, F., Frayley, R. T., Rogers, S. G., and Chua, N.-H. (1985). *Nature (London)* **315,** 200–204.
Moses, V., and Calvin, M. (1958). *Proc. Natl. Acad. Sci. U. S. A.* **44,** 260–277.
Mott, K. A., and Berry, J. A. (1986). *Plant Physiol.* **82,** 77–82.
Mott, K. A., Jensen, R. G., O'Leary, J. W., and Berry, J. A. (1984). *Plant Physiol.* **76,** 968–971.
Muller, E. D., Chory, J., and Kaplan, S. (1985). *J. Bacteriol.* **161,** 469–472.

Muller, K. D., Salnikow, J., and Vater, J. (1983). *Biochim. Biophys. Acta* **742**, 78–83.

Mullhofer, G., and Rose, I. A. (1965). *J. Biol. Chem.* **240**, 1341–1347.

Mulligan, R. M., and Tolbert, N. E. (1983). *Arch. Biochem. Biophys.* **225**, 610–620.

Musgrove, J. E., and Ellis, R. J. (1986). *Philos. Trans. R. Soc. London, Ser. B* **313**, 419–428.

Nargang, F., McIntosh, L., and Somerville, C. (1984). *Mol. Gen. Genet.* **193**, 220–224.

Nelson, S. F., and Bartlett, P. D. (1966). *J. Amer. Chem. Soc.* **88**, 143–149.

Nierzwicki-Bauer, S. A., Curtis, S. E., and Haselkorn, R. (1984). *Proc. Natl. Acad. Sci. U. S. A.* **81**, 5961–5965.

Nilsson, T., Branden, R., and Styring, S. (1984). *Biochim. Biophys. Acta* **788**, 274–280.

Niyogi, S. K., Foote, R. S., Mural, R. J., Larimer, F. W., Mitra, S., Soper, T. S., Machanoff, R., and Hartman, F. C. (1986). *J. Biol. Chem.* **261**, 10087–10092.

Ogren, W. L., Salvucci, M. E., and Portis, A. R. (1986). *Philos. Trans. R. Soc. London, Ser. B* **313**, 337–344.

Osmond, C. B. (1981). *Biochim. Biophys. Acta* **639**, 77–98.

Peach, C. (1985). *Biochemistry* **24**, 3194–3199.

Paech, C., Pierce, J., McCurry, S. D., and Tolbert, N. E. (1978). *Biochem. Biophys. Res. Commun.* **83**, 1084–1092.

Pal, G. P., Jakob, R., Hahn, U., Bowien, B., and Saenger, W. (1985). *J. Biol. Chem.* **260**, 10768–10770.

Parry, M. A. J., Schmidt, C. N. G., Keys, A. J., and Gutteridge, S. (1983). *FEBS Lett.* **159**, 107–111.

Parry, M. A. J., Schmidt, C. N. G., Cornelius, M. J., Keys, A. J., Millard, B. N., and Gutteridge, S. (1985). *J. Exp. Bot.* **36**, 1396–1404.

Perchorowicz, J. T., and Jensen, R. G. (1983). *Plant Physiol.* **71**, 955–960.

Perchorowicz, J. T., Raynes, D. A., and Jensen, R. G. (1981). Proc. Natl. Acad. Sci. U. S. A. **78**, *2985–2989*.

Pierce, J., and Reddy, G. S. (1986). *Arch. Biochem. Biophys.* **245**, 483–493.

Pierce, J., Tolbert, N. E., and Barker, R. (1980a). *Biochemistry* **19**, 934–942.

Pierce, J., Tolbert, N. E., and Barker, R. (1980b). *J. Biol. Chem.* **255**, 509–511.

Pierce, J., Lorimer, G. H., and Reddy, G. S. (1986a). *Biochemistry* **25**, 1636–1644.

Pierce, J., Andrews, T. J., and Lorimer, G. H. (1986b). *J. Biol. Chem.* **261**, 10248–10256.

Poulsen, C. (1981). *Carlsberg Res. Commun.* **46**, 259–278.

Purohit, K., McFadden, B. A., and Cohen, A. L. (1976). *J. Bacteriol.* **127**, 505–515.

Quayle, J. R., Fuller, R. C., Benson, A. A., and Calvin, M. (1954). *J. Am. Chem. Soc.* **76**, 3610–3611.

Quiocho, F. (1986). *Annu. Rev. Biochem.* **55**, 287–316.

Reichelt, B. Y., and Delaney, S. F. (1983). *DNA* **2**, 121–129.

Reith, M., and Cattolico, R. A. (1986). *Proc. Natl. Acad. Sci. USA* **83**, 8599–8603.

Richards, J. P. (1984). *J. Am. Chem. Soc.* **106**, 4926–4936.

Richards, J. P. (1985). *Biochemistry* **24**, 949–953.

Robinson, C., and Ellis, R. J. (1984). *Eur. J. Biochem.* **142**, 337–342.

Robison, P. D., Martin, M. N., and Tabita, F. R. (1979). *Biochemistry* **18**, 4453–4458.

Robison, P. D., Whitman, W. B., Wadill, F., Riggs, A. F., and Tabita, F. R. (1980). *Biochemistry* **21**, 4848–4853.

Rose, I. A. (1980). *In* "Methods in Enzymology" (D. Purich, ed.), Vol. 64, pp. 47–59. Academic Press, New York.

Russell, G. A., Bemis, A. G., Geels, E. J., Janzen, E. G., and Moye, A. J. (1967). *Adv. Chem. Ser.* **75**, 174–201.

Rutner, A. C., and Lane, M. D. (1967). *Biochem. Biophys. Res. Commun.* **28**, 531–537.

Salvucci, M. E., Portis, A. R., and Ogren, W. L. (1985). *Photosynth. Res.* **7**, 193–201.

Salvucci, M. E., Portis, A. R., and Ogren, W. L. (1986). *Plant Physiol.* **80**, 655–659.

Saver, B. G., and Knowles, J. R. (1982). *Biochemistry* **21**, 5398–5403.

Sawyer, D. T., and Gibian, M. J. (1979). *Tetrahedron* **35**, 1471–1481.

Schloss, J. V., and Lorimer, G. H. (1982). *J. Biol. Chem.* **257**, 4691–4694.

Schloss, J. V., Stringer, C. D., and Hartman, F. C. (1978). *J. Biol. Chem.* **253**, 5707–5711.

Schneider, G., Branden, C. I., and Lorimer, G. H. (1984). *J. Mol. Biol.* **175**, 99–102.

Schneider, G., Branden, C. I., and Lorimer, G. H. (1986a). *J. Mol. Biol.* **187**, 141–143.

Schneider, G., Lindqvist, Y., Branden, C. I., and Lorimer, G. H. (1986b). *EMBO J.* **5**, 3409–3415.

Seemann, J. R., Berry, J. A., Freas, S. M., and Krump, M. A. (1985). *Proc. Natl. Acad. Sci. U. S. A.* **82**, 8024–8028.

Segel, I. H. (1975). "Enzyme Kinetics." Wiley, New York.

Servaites, J. C. (1985). *Plant Physiol.* **78**, 839–843.

Servaites, J. C., Torisky, R. S., and Chao, S. F. (1984). *Plant Sci. Lett.* **35**, 115–121.

Shinozaki, K., and Sugiura, M. (1982). *Gene* **20**, 91–102.

Shinozaki, K., and Sugiura, M. (1983). *Nucl. Acids Res.* **11**, 6957–6964.

Shinozaki, K., and Sugiura, M. (1985). *Mol. Gen. Genet.* **200**, 27–32.

Shively, J. M., Davidson, E., and Marrs, B. L. (1984). *Arch. Microbiol.* **138**, 233–236.

Sicher, R. C., Hatch, A. L., Stumpf, D. K., and Jensen, R. G. (1981). *Plant Physiol.* **68**, 252–255.

Siegel, M. I., and Lane, M. D. (1972). *Biochem. Biophys. Res. Commun.* **48**, 508–516.

Siegel, M. I., and Lane, M. D. (1973). *J. Biol. Chem.* **248**, 5486–5498.

Sjödin, B., and Vestermark, A. (1973). *Biochim. Biophys. Acta* **297**, 165–173.

Somerville, C. R., and Ogren, W. L. (1982). *Trends Biochem. Sci.* **7**, 171–174.

Somerville, C. R., Portis, A. R., and Ogren, W. L. (1982). *Plant Physiol.* **70**, 381–387.

Somerville, C. R., Fitchen, J., Somerville, S., McIntosh, L., and Nargang, F. (1948). *In* "Advances in Gene Technology: Molecular Genetics of Plants and Animals" (K. Downey, R. W. Voellmy, J. Schultz, and F. Ahmad, eds.), pp. 295–309. Academic Press, New York.

Spellman, M., Tolbert, N. E., and Hartman, F. C. (1979). *Abstr. Pap., 178th Natl. Meet., Am. Chem. Soc.,* BIOL 3.

Spreitzer, R. J., Jordan, D. B., and Ogren, W. L. (1982). *FEBS Lett.* **148**, 117–121.

Starnes, S. M., Lambert, D. H., Maxwell, E. S., Stevens, S. E., Porter, R. D., and Shively, J. M. (1985). *FEMS Microbiol. Lett.* **28**, 165–169.

Stiekema, W. J., Wimpee, C. F., and Tobin, E. M. (1983). *Nucleic Acids Res.* **11**, 8051–8061.

Stringer, C. D., and Hartman, C. D. (1978). *Biochim. Biophys. Res. Commun.* **80**, 1043–1048.

Styring, S., and Branden, R. (1985a). *Biochemistry* **24**, 6011–6019.

Styring, S., and Branden, R. (1985b). *Biochim. Biophys. Acta* **832**, 113–118.

Sue, J. R., and Knowles, J. R. (1978). *Biochemistry* **17**, 4041–4044.

Sue, J. R., and Knowles, J. R. (1982). *Biochemistry* **21**, 5404–5410.

Tabita, F. R., and McFadden, B. A. (1974). *J. Biol. Chem.* **249**, 3459–3464.

Tabita, F. R., and Small, C. L. (1985). *Proc. Natl. Acad. Sci. U. S. A.* **82**, 6100–6103.

Tabita, F. R., Stevens, S. E., and Quijano, R. (1974). *Biochem. Biophys. Res. Commun.* **61**, 45–52.

Tabita, F. R., Stevens, S. E., and Gibson, J. L. (1976). *J. Bacteriol.* **125**, 531–539.

Takabe, T. (1977). *Agric. Biol. Chem.* **41**, 2255–2260.

Takabe, T., and Akazawa, T. (1975). *Plant Cell Physiol.* **16**, 1049–1060.

Takabe, T., and Akazawa, T. (1977). *Plant Cell Physiol.* **18**, 753–765.

Takabe, T., Incharoensakdi, A., and Akazawa, T. (1984a). *Biochem. Biophys. Res. Commun.* **122**, 763–769.

Takabe, T., Rai, A. K., and Akazawa, T. (1984b). *Arch. Biochem. Biophys.* **229**, 202–211.

Takruri, I. A. H., Boulter, D., and Ellis, R. J. (1981). *Phytochemistry* **20**, 413–415.

Taylor, S. C., and Dow, C. S. (1980). *J. Gen. Microbiol.* **116**, 81–87.

Taylor, S. C., Dalton, H., and Dow, C. S. (1980). *FEMS Microbiol. Lett.* **8**, 157–160.

Taylor, S. E., and Terry, N. (1984). *Plant Physiol.* **75**, 82–86.

Terzaghi, B. E., Laing, W. A., Christeller, J. T., Petersen, G. B., and Hill, D. F. (1986). *Biochem. J.* **235,** 839–846.

Timko, M. P., Kausch, A. P., Castresana, C., Fassler, J., Herrera-Estrella, L., Van den Broeck, G., van Montagu, M., Schell, J.,and Cashmore, A. R. (1985). *Nature (London)* **318,** 579–582.

Tolbert, N. E. (1971). *In* "Photosynthesis and Photorespiration" (M. D. Hatch, C. B. Osmond, and R. O. Slatyer, eds.), pp. 458–471. Wiley (Interscience), New York.

Tomimatsu, Y., and Donovan, J. W. (1981). *Plant Physiol.* **68,** 808–813.

Torres-Ruiz, J. A., and McFadden, B. A. (1985). *Arch. Microbiol.* **142,** 55–60.

Turpin, D. H., Miller, A. G., and Canvin, D. T. (1984). *J. Phycol.* **20,** 249–253.

Van den Broeck, G., Timko, M., Kausch, A. P., Cashmore, A. R., van Montagu, M., and Herrera-Estrella, L. (1985). *Nature (London)* **313,** 358–363.

Van Dyk, D., and Schloss, J. V. (1986). *Biochemistry* **25,** 5145–5156.

Vater, J., and Salnikow, J. (1979). *Arch. Biochem. Biophys.* **194,** 190–197.

Vater, J., Gaudszun, T., Lange, B., Erdin, N., and Salnikow, J. (1983). *Z. Naturforch., C: Biosci.* **38C,** 418–427.

Viale, A. M., Kobayashi, H., Takabe, T., and Akazawa, T. (1985). *FEBS Lett.* **192,** 283–288.

von Caemmerer, S., and Edmondson, D. L. (1986). *Aust. J. Plant Physiol.* **13,** 669–688.

Voordouw, G., Van der Vies, S. M., and Bouwmeister, P. P. (1984). *Eur. J. Biochem.* **141,** 313–318.

Vu, C. V., Allen, L. H., and Bowes, G. (1983). *Plant Physiol.* **73,** 729–734.

Vu, J. C. V., Allen, L. H., and Bowes, G. (1984). *Plant Physiol.* **76,** 843–845.

Weissbach, A., Smyrniotis, P. Z., and Horecker, B. L. (1954). *J. Am. Chem. Soc.* **76,** 3611–3612.

Weissbach, A., Horecker, B. L., and Hurwitz, J. (1956). *J. Biol. Chem.* **218,** 795–810.

Whitman, W. B., Martin, M. N., and Tabita, F. R. (1979). *J. Biol. Chem.* **254,** 10184–10189.

Wildner, G. F. (1976). *Z. Naturforsch., C: Biosci.* **31C,** 267–271.

Wildner, G. F., and Henkel, J. (1979). *Planta* **169,** 223–228.

Wolfenden, R. (1972). *Acc. Chem. Res.* **5,** 10–18.

Yokota, A., and Kitaoka, S. (1985). *Biochem. Biophys. Res. Commun.* **131,** 1075–1079.

Zurawski, G., Perrot, B., Bottomley, W., and Whitfeld, P. R., (1981). *Nucleic Acids Res.* **9,** 3251–3270.

Zurawski, G., Whitfeld, P. R., and Bottomley, W. (1986). *Nucleic Acids Res.* **14,** 3975–3992.

The CO_2-Concentrating Mechanism in Aquatic Phototrophs

4

MURRAY R. BADGER

I. Introduction
II. Historical Perspectives
III. The Significance of the CO_2-Concentrating Mechanism
IV. The Essential Components of the CO_2-Concentrating Mechanism
V. The CO_2-Concentrating Mechanism in Cyanobacteria
   and Green Microalgae
   A. Cyanobacteria
   B. Green Microalgae
VI. Induction of the CO_2-Concentrating Mechanism
   A. Protein Synthesis
   B. Factors Affecting Induction
   C. Adaptation to High CO_2
VII. Evidence for the CO_2-Concentrating Mechanism
   in Other Aquatic Phototrophs
   A. Freshwater Angiosperms and Giant Algae
   B. Marine Algae and Angiosperms
VIII. The Ecological Importance of the CO_2-Concentrating Mechanism
   A. Environmental Variation in CO_2 Supply
   B. Nitrogen Limitation
   C. Organisms without the CO_2-Concentrating Mechanism
   References

The Biochemistry of Plants, Vol. 10

I. INTRODUCTION*

Aquatic phototrophs are presented with a number of unique problems with regard to the efficient operation of photosynthesis. Among these, the acquisition of CO_2 from the external medium and its supply as a substrate for the primary CO_2-fixing enzyme, ribulose bisphosphate carboxylase/oxygenase (Rubisco; EC 4.1.1.39), has been of great significance. The evolution of terrestrial plants represents a major step in overcoming this problem, as the effective aqueous diffusion path has been reduced to a few micrometers (see Raven, 1970, 1984). Since the diffusion coefficient for CO_2 is some 10^4 higher in air than water, this greatly reduces the problem of the supply of CO_2 to Rubisco.

The acquisition of CO_2 from an aquatic environment presents problems, largely as a result of the physical chemistry relating to inorganic carbon (Ci) species in solution. As mentioned above, the diffusion rate of CO_2 is slow in aqueous medium, and this also applies to the other major forms of Ci, namely HCO_3^- and CO_3^-. These three forms of Ci are interconvertible according to the following series of reactions:

$$H_2O + CO_2 \underset{k_{-1}}{\overset{k_{+1}}{\rightleftarrows}} H_2CO_3 \underset{k_{-2}}{\overset{k_{+2}}{\rightleftarrows}} HCO_3^- + H^+ \underset{k_{-3}}{\overset{k_{+3}}{\rightleftarrows}} CO_3^{2-} + H^+$$

$$OH^- + CO_2 \underset{k_{-4}}{\overset{k_{+4}}{\rightleftarrows}} HCO_3^-$$

The ionic reactions with rate constants k_{+2}, k_{-2}, k_{+3}, and k_{-3}, are rapid (diffusion-limited). At neutral pH the uncatalyzed hydration/dehydration of CO_2 is a slow process with the rate constants having the following approximate values at 25°C: $k_{+1} = 0.037$ sec^{-1} (Edsall, 1969); $k_{-1} = 14$ sec^{-1}

---

* Abbreviations: ADP, adenosine diphosphate; ATP, adenosine triphosphate; air-grown cells, cells grown with air bubbling through the culture medium; CA, carbonic anhydrase; Ci, inorganic carbon; C_3, plants that use Rubisco as the primary CO_2-fixing enzyme; C_4, plants with the C_4-dicarboxylic acid pathway of photosynthesis; CAM, crassulacean acid metabolism; DCMU, 3-(3,4-dichlorophenyl)-1,1-dimethyl urea; DMSO, dimethyl sulfoxide; $\delta^{13}C$, $^{13}C/^{12}C$ ratio, relative to a standard; $\Delta\mu X$, the electrochemical potential gradient of the chemical species X; F1F0-ATPase, the proton translocating ATPase in the cell membrane of bacteria; high-CO_2 grown cells, cells grown at high levels of CO_2, displaying a low affinity for CO_2 and Ci in photosynthesis; high-Ci cells, cells grown at high levels of Ci, showing a low affinity for CO_2 and Ci in photosynthesis; H^+-ATPase, proton-translocating ATPase; K_m, the Michaelis–Menten constant for an enzyme reaction; $K_{0.5}(X)$, the substrate concentration of X that gives half maximal rate for a process; kDa, kilodalton; low-CO_2 grown cells, cells grown at low levels of CO_2, displaying a high affinity for CO_2 and Ci in photosynthesis; low-Ci cells, cells grown at low levels of Ci, showing a high affinity for CO_2 and Ci in photosynthesis; NADP, nicotinamide adenine dinucleotide phosphate; PCO cycle, photosynthetic carbon oxidation cycle; PCR cycle, photosynthetic carbon reduction cycle; Pi, inorganic phosphate; P_{CO_2}, permeability coefficient for CO_2; P_{O_2}, permeability coefficient for O_2; PEP, phosphoenolpyruvate; Rubisco, ribulose bisphosphate carboxylase/oxygenase (EC 4.1.1.39); V_{max}, substrate-saturated rate for a process.

(Gutknecht *et al.*, 1977); $k_{+4} = 8.5$ m³ mol⁻¹ sec⁻¹ (Edsall, 1969); $k_{-4} = 19 \times 10^{-4}$ sec⁻¹ (Walker *et al.*, 1980).

As can be seen from the above equations, the chemical equilibrium between the species is influenced by the pH of the medium. Above about pH 6, there is an increasing predominance of the HCO_3^- ion, and then CO_3^- above about pH 9. If the aqueous medium is in equilibrium with the gas phase, then the CO_2 in solution will remain constant regardless of pH; however, the total Ci will increase roughly 10-fold with each unit of pH increase above about 7.

Supply of Ci to the surface of an aquatic cell is not limited to the diffusion of CO_2, as the diffusion of HCO_3^- and CO_3^- and their conversion to CO_2 can also contribute. As the concentrations of HCO_3^- may be up to several orders of magnitude higher than CO_2, this represents a potentially major pathway for CO_2 supply. The relatively slow conversion of HCO_3^- to CO_2 (especially at pH levels above 7) means, however, that this may limit CO_2 supply, particularly if photosynthesis is rapid. The kinetic limitations imposed on CO_2 supply via diffusion and interconversion pathways have been dealt with extensively, and readers are referred elsewhere for a more detailed treatment (Lucas, 1975; Lehman, 1978; Miller, 1985).

These physical and chemical factors complicate the approaches that aquatic phototrophs must adopt to obtain CO_2 efficiently. The major strategy that has evolved in all levels of aquatic organisms (cyanobacteria, microalgae, macroalgae, and aquatic angiosperms) appears to be a mechanism for the active transport of Ci into cells. This transport generally results in the accumulation of Ci within the cell and the establishment of an elevated internal CO_2 concentration. The term "CO_2-concentrating mechanism" has thus arisen. This mechanism utilizes both CO_2 and HCO_3^- as substrates from the external medium, and its operation as a "front end" to photosynthesis considerably modifies the characteristics of many of the organisms in which it has been identified. It generally results in a high affinity for CO_2 in photosynthesis, a reduced photorespiration and sensitivity to O_2, and a low CO_2 compensation point, thus achieving functional similarities to C_4 photosynthesis in higher terrestrial plants. This review details what is known about the operation of this CO_2-concentrating mechanism in the groups of aquatic phototrophs where it has been studied and highlights those areas where further research is necessary.

II. HISTORICAL PERSPECTIVES

The discovery of a CO_2 concentrating mechanism in aquatic phototrophs came from a lengthy interest of many researchers in the ability of aquatic organisms (both macrophytes and microphytes) to use the HCO_3^- ion as a direct carbon source for photosynthesis (see reviews by Steemann-Nielsen, 1947, 1960; Lucas, 1983). There was considerable discussion that species could be divided into those that showed the ability to use HCO_3^-, those that utilized CO_2, and those that exhibited a mixture of both capabilities. Speculations about the basis for the direct use of HCO_3^- have inevitably proposed an HCO_3^- pumping mechanism linked in some way to the input of photosynthetic energy, either directly as ATP or through energy gradients produced by ion fluxes between the cell and its aqueous environment. Despite this fascination with the subject of utilization, it has only been relatively recently in the evolution of this field that HCO_3^- utilization has been examined more closely, to assess the impact it has on the photosynthetic biochemistry and physiology of the organism.

With an expansion of the interest in the photosynthetic properties of microalgae, it was clearly established that the process of HCO_3^- utilization could act as a photosynthetic CO_2 concentrating mechanism, in a similar fashion to that already established for C_4 photosynthesis in higher plants. In 1976, investigations by Berry *et al.* with the green alga *Chlamydomonas rheinhardtii* showed that the photosynthetic response of cells to external CO_2 was dramatically affected by the previous growth conditions of the algae. Cells grown with air levels of CO_2 (330 $\mu l/l$) had a much higher affinity for external CO_2 in photosynthesis than did cells grown with the high levels of CO_2 (50 ml/l) commonly used in laboratory culture of many algal strains. This increase in affinity was correlated with a reduction in the O_2 sensitivity of photosynthesis, as measured by growth, glycolate excretion, and inhibition of net photosynthesis. This adaptation produced the terminology "low-CO_2" and "high-CO_2" grown cells. The affinity of photosynthesis for CO_2 of low-CO_2 grown cells was greater than that of C_3 higher plants [$K_m(CO_2) < 1$ μM], but there was no evidence that this was achieved through alteration of the properties of the primary CO_2-fixing enzyme, Rubisco, or the acquisition of a C_4-type carboxylation pathway. At this same time there were other reports emerging that the growth of the green algae *Scenedesmus obliquus* (Findenegg, 1976) and *Chlorella vulgaris* (Hogetsu and Miyachi, 1977) at high and low CO_2 caused similar changes in the response of photosynthesis to CO_2.

Carbonic anhydrase activity had been previously shown to dramatically increase when high-CO_2 grown cells of the green algae *Chlorella* (Graham and Reed, 1971), *Chlamydomonas* (Nelson *et al.*, 1969), and *Scenedesmus* (Findenegg, 1976) and cyanobacteria (Dohler, 1974; Ingle and Colman, 1976) were transferred to growth with air levels of CO_2. Considering this biochemical change, it was suggested that this increase in carbonic anhydrase may by itself be responsible for the increased affinity of the cells for external CO_2. Whilst this enzyme will promote the equilibration of CO_2 and HCO_3^- and reduce the drop in CO_2 concentration between the external medium and the site of CO_2 fixation, the overall affinity of the cells would still be limited by the intrinsic properties of primary CO_2 fixing enzyme. Realizing that low-CO_2 grown cells of *Chlamydomonas* had an affinity for CO_2 that was at least an order of magnitude higher than the measured *in vitro* properties of their

Rubisco, Berry *et al.* (1976) suggested that these cells may actively accumulate HCO_3^- by a metabolic influx pump, and thus elevate the internal CO_2 above external concentrations. Following on from this initial suggestion, it was subsequently shown, using silicon-oil centrifugation techniques, that it was indeed possible for low-CO_2 grown cells to concentrate CO_2 internally up to 40-fold higher than the external concentrations (Badger *et al.*, 1980). These studies were extended to include the cyanobacterium *Anabaena variabilis,* which showed an even greater capacity to concentrate Ci, achieving a ratio between internal and external Ci concentrations of greater than 1000 (Kaplan *et al.*, 1980). Using similar techniques, the ability to concentrate CO_2 and Ci internally has been shown in a number of other air-grown cyanobacteria (Miller and Colman, 1980a; Coleman and Colman, 1981; Kaplan, 1981; Badger and Andrews, 1982) and green algae (Beardall, 1981; Beardall and Raven, 1981; Zenvirth and Kaplan, 1981; Spalding and Ogren, 1983; Beardall and Entwisle, 1984; Tsuzuki *et al.*, 1985).

III. THE SIGNIFICANCE OF THE CO₂-CONCENTRATING MECHANISM

The CO_2-concentrating mechanism can be seen to have one major role in improving the photosynthetic performance of the aquatic organisms in which it is found. Simply stated, it modifies the limitations placed on the cell by the properties of the primary carboxylation enzyme, Rubisco. These limitations can be specifically identified as follows:

1. The competitive inhibition effects of O_2 on the carboxylase reaction (see Andrews and Lorimer, this volume) are reduced, if not eliminated, by the elevation of CO_2 at the active site of the enzyme. This reduces the deleterious effects of photorespiratory processes (see Lorimer and Andrews, Volume 8 of this series) on the rate of net CO_2 fixation.

2. The effective specific activity of Rubisco is increased to the intrinsic CO_2 and RuP_2 saturated rate measured for the purified enzyme. This has the effect that less Rubisco protein is needed to achieve a given rate of photosynthesis.

3. The affinity of the cells for total external Ci is raised far above the affinity that could be achieved if the cells were forced to rely on the relatively low affinity for CO_2 of Rubisco. As will be discussed later, this is the result of the active uptake of both CO_2 and HCO_3^- from the external medium via a transport system that has a relatively high affinity for both species. This means that aquatic organisms, with this increased affinity, can photosynthesize at the relatively low levels of both CO_2 and total Ci that occur in aquatic ecosystems, especially when the potential rate of photosynthesis exceeds the rate of CO_2 supply from the surrounding medium (see Spence and Maberly, 1985; Talling, 1985). This is particularly important at alkaline pH, where a decreasing proportion of the external Ci is present as CO_2. It is here

that an organism with a relatively low affinity for CO_2 and an inability to directly access the HCO_3^- pool, will be severely limited.

It has been proposed that an aquatic organism, using the CO_2-concentrating mechanism, may increase the utilization efficiency of two other major limiting resources in aquatic environments, namely, light and nitrogen (Raven, 1984, 1985). Nitrogen may be used more efficiently for two reasons: (1) because there is a smaller proportion of total cell nitrogen tied up in the form of Rubisco protein and perhaps components of the enzymatic machinery necessary for photorespiratory metabolism of glycolate (Beardall *et al.*, 1982; Raven *et al.*, 1985). This proposal assumes that the molecular components of the CO_2-concentrating machinery require a smaller investment of nitrogen than the savings made in the above components. (2) The reduction in photorespiratory metabolism decreases the potential losses of NH_3 that might otherwise occur through the metabolism of glycine to serine in the PCO cycle (Beardall *et al.*, 1982).

The savings that may be made with respect to light energy are much less certain. The energy costs of photosynthesis in the presence of the CO_2-concentrating mechanism must be compared to the situation that occurs in its absence. The savings made through the reduction in photorespiration and its energy-draining metabolism must be costed against the extra energy necessary to run the CO_2-concentrating mechanism. While estimates have been made of this (Raven and Lucas, 1985), the assumptions that have had to supplement the scarce supply of facts mean that the photon cost efficiency of the CO_2-concentrating mechanism is unknown.

IV. THE ESSENTIAL COMPONENTS OF THE
CO₂-CONCENTRATING MECHANISM

At present, four major physical or biochemical elements of the CO_2-concentrating mechanism can be identified, and these are shown in a simple model in Fig. 1. The elements depicted are:

1. A pumping mechanism, to actively transport inorganic carbon into the cell or chloroplast.

2. An energy supply, to drive the pumping mechanism.

3. A leak rate control device, to reduce the efflux of CO_2 out of the cell and into the surrounding medium.

4. A mechanism to provide rapid interconversion between CO_2 and HCO_3^-, both inside and outside the cell, so that the slow chemical interconversion between these species is speeded up and both forms of Ci may serve as substrates for the transport and fixation processes.

I will discuss these four elements in broad terms before progressing to a detailed analysis of the evidence for the mechanistic basis of each element in the various phototrophs that have been studied.

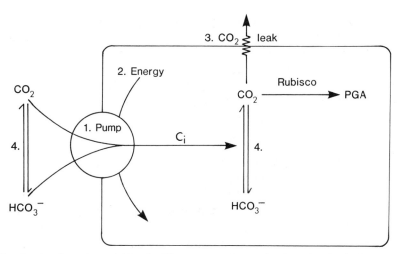

Fig. 1. A schematic model for the CO_2-concentrating mechanism, showing its four essential components: (1) a pumping mechanism to transport Ci into the cell compartment; (2) a means of supplying energy to enable the pump to transport against the electrochemical gradient for Ci; (3) a diffusion barrier to minimize the backflux of CO_2 out of the internal compartment; and (4) mechanisms for interconverting Ci species both outside and inside the cell compartment. See text for further discussion.

First, the accumulation of Ci and CO_2 has an obvious requirement for a mechanism to transport Ci between the external medium and the compartment of the cell where the final accumulation occurs. This is generally viewed as a biophysical pump, which may be linked to cellular energy supplies.

Second, as the accumulation of CO_2 is an active process, it requires the input of some form of physical or biochemical energy. As we shall see, the energy for pumping is derived from the photosynthetic reactions, with little accumulation occurring in the dark.

Third, considering that CO_2 is relatively permeable to biological membranes (Gutknekt *et al.*, 1977), a CO_2-concentrating mechanism can be an energetically wasteful process if a significant proportion of the Ci that is being pumped into the cell is effluxing into the surrounding medium. Thus, it is important to reduce this wasteful pathway by the development of a diffusion barrier, similar to the hypothesized role for the suberized layer around the bundle sheath in C_4 plants.

Finally, as the Ci in solution around the cell can exist as various proportions of CO_2, HCO_3^-, and CO_3^-, depending on the pH, it is logical that mechanisms should have evolved to enable the Ci pump to have access to the dominant form in each aquatic environment. Thus at alkaline pH the pump will be able to directly utilize HCO_3^-, while in neutral to acidic environments, an ability to use CO_2 will be advantageous. Arguments supporting these proposals consider that at either extreme of pH, the absolute concentrations of either CO_2 or HCO_3^- are too low to support the operation of a pumping mechanism if only one form of Ci were used. In addition to external interconversions, the Ci will be delivered to the concentrating compartment as either CO_2 or HCO_3^-. As the carboxylating enzyme uses CO_2 as a substrate (Cooper et al., 1969), there may be a requirement for a rapid interconversion mechanism within this region. This will be particularly so if HCO_3^- is arriving internally.

V. THE CO_2-CONCENTRATING MECHANISM IN CYANOBACTERIA AND GREEN MICROALGAE

Almost all the experimental work designed to describe and elucidate the operation of the CO_2-concentrating mechanism has been carried out with microphytes of the Chlorophyceae and Cynaophyceae. This has been because of their traditional roles as laboratory material and the ease with which elements of the mechanism can be measured in them. In this section I will assess the evidence that exists for the operation of the four elements of the CO_2-concentrating mechanism, described in Section III, as it has been gathered for cyanobacteria and green algae.

A. Cyanobacteria

The manifestations of the CO_2-concentrating mechanism are clearly evident in air-grown cyanobacteria. The affinity of photosynthesis for CO_2 (0.1-0.5 μM; Kaplan et al., 1980; Miller and Colman, 1980a,b; Kaplan, 1981; Badger and Andrews, 1982) can be several orders of magnitude lower than that measured for the primary carboxylating enzyme, Rubisco (100–170 μM; Badger, 1980; Andrews and Abel, 1981), although the primary CO_2-fixing pathway is the PCR cycle of C_3 photosynthesis (Coleman and Colman, 1980). The CO_2 compensation point is very low (around 0.03 μM at alkaline pH; Birmingham and Colman, 1979) and insensitive to O_2, and there is no evidence for the occurrence of an O_2 effect on net photosynthesis or the presence of photorespiration (Lloyd et al., 1977; Birmingham et al., 1982). As well as these physiological properties, there is an easily demonstrable accumulation of Ci within the cell in the light, with accumulation ratios between inside and outside reaching values in excess of 1000 (Kaplan et al., 1980; Badger and Andrews, 1982; Miller and Colman, 1980a).

1. Pumping

a. Location of the Pump. As cyanobacteria are prokaryotes, the study of many of the aspects of the CO_2-concentrating mechanism is greatly simplified. There is only one major internal compartment in which Ci accumulation can be proposed to be occurring (the cytosol), and only the inner and outer cell membranes separate this from the external medium. Thus it is assumed that the pumping mechanism is associated with the double membrane surrounding the cell.

The structure of the inner and outer cell membrane complex is very similar to that described for gram-negative bacteria (see Drews and Weckesser, 1982; Nikaido and Nakae, 1979). The outer membrane is composed of lipopolysaccharide and phospholipid, and contains a considerable amount of a limited number of proteins (Murata *et al.,* 1981; Resch and Gibson, 1983). As with the enteric bacteria (Nikaido and Vaara, 1985), much of this protein may be associated with protein channels (porins), which may control the diffusion of hydrophilic and charged substrates to the periplasmic space, while the lipopolysaccharide layer prevents the entry of many hydrophobic compounds. The outer membrane is connected on the inside to the peptidoglycan layer, which serves a largely structural function. The inner membrane is analogous to the plasmamembrane and serves as the major permeability barrier for the cell. This membrane contains numerous protein components (Murata *et al.,* 1981; Omata and Ogawa, 1985, 1986), some of which seem to be involved in such processes as ion symport and antiport systems (such as H$^+$/Na$^+$ antiport; Miller *et al.,* 1984a; Kaplan *et al.,* 1984), H$^+$-translocating electron-transport chains (Nitschmann and Peschek, 1985), and possibly the pumping of Ci into the cell (Omata and Ogawa, 1985, 1986). The inner membrane is thus the only logical structure that can be considered to "house" the components of the Ci pumping system. Recently a 42-kDa polypeptide has been identified, in the cytoplasmic membrane, that is induced when high-CO$_2$ grown cells are transferred to air, correlating with the induction of a high affinity for Ci in photosynthesis (Omata and Ogawa, 1985, 1986).

b. Specificity for Substrate. Much of the initial work aimed at characterizing the CO$_2$-concentrating mechanism assumed that the mechanism of Ci accumulation utilized HCO$_3^-$ as the sole species for transport. This hypothesis has since been shown to be incorrect, as it appears that the pump is able to directly use both CO$_2$ and HCO$_3^-$ as species for transport (Badger and Andrews, 1982; Volokita *et al.,* 1984). In air-grown cells of *Anabaena variabilis,* it appers that the CO$_2$ uptake process has a higher apparent affinity for its substrate than does HCO$_3^-$ uptake ($K_{0.5}$ = 17 μM versus 60 μM), although the V_{max} values for transport were similar (Volokita *et al.,* 1984). The situation is similar for *Synechococcus* PCC6301, with high-affinity cells showing a $K_{1/2}$ for CO$_2$ and HCO$_3^-$ of about 5 and 50 μM, respectively (Badger and Gallagher, 1987).

c.   Rate of Pumping. The capacity of the pump to transport Ci into the cell can be estimated from experiments conducted under both non-steady-state and steady-state conditions. For non-steady-state conditions, time courses of Ci accumulation have been conducted under conditions where the internal Ci pool has been depleted. The initial slope of Ci accumulation for these experiments represents an estimate of the maximum rate at which the transport system can function, without the inhibitory effects of an elevated internal pool. Initial experiments with *Anabaena variabilis* (Kaplan *et al.*, 1980) estimated that low-CO_2 grown cells had a V_{max} for Ci transport (3800 μmol mg^{-1} Chl h^{-1}) that was about 10-fold higher than the V_{max} of photosynthesis. These experiments did not consider, however, that CO_2 could act as a species for uptake and that a time course initiated with HCO_3^- will represent an initial uptake of HCO_3^- followed by a longer-term accumulation of CO_2. When the initial rate of accumulation was measured with either HCO_3^- or CO_2 as substrate (Volokita *et al.*, 1984), the V_{max} estimated for each species was similar although some sixfold lower than the previous value (600 μmol mg^{-1} Chl h^{-1}). Experiments conducted with species of *Synechococcus* (Badger and Andrews, 1982; Badger and Gallagher, 1987) were not able to arrive at estimates of V_{max}, as the rate of accumulation of CO_2 was so rapid as to be kinetically unresolvable. When CO_2 is presented to these cells, the internal pool reaches near maximum levels at the shortest time point (5 sec) and only when HCO_3^- is given is there a time course that shows a relatively slow accumulation over the following 90 sec. In this latter case it is almost certain that the slow rise in pool is due to the subsequent increase in CO_2 after the addition of HCO_3^-, rather than slow kinetics of HCO_3^- transport. The only conclusion that can be reached is that accumulation of the internal pool is very rapid when both CO_2 and HCO_3^- are given. Analysis of the size of the internal pool after 10 sec of exposure to varying concentrations of CO_2 and HCO_3^- indicated that, with *Synechococcus* PCC6301, saturating concentrations of CO_2 may support higher internal pool levels than HCO_3^- (Badger and Gallagher, 1987).

The fluxes of Ci during steady-state photosynthesis are quite different to the initial influx measurements. Estimates of gross Ci uptake, gross photosynthesis, and Ci leakage in a marine *Synechococcus* species (Badger *et al.*, 1985) indicate that under saturating external Ci concentrations the gross uptake of Ci by high-affinity cells may only exceed the rate of photosynthesis by less than 50%. Similarly, in air-grown cells of *Anabaena variabilis* (Marcus *et al.*, 1982), the leakage of CO_2 during steady-state photosynthesis was estimated to be around 40% of net photosynthesis. Thus, although the potential rate of transport when the internal Ci pool is depleted may be very high, the actual rate during steady-state photosynthesis is regulated so it is only sufficiently in excess of photosynthesis so as to maintain internal CO_2 levels that are probably just saturating for Rubisco.

d. Acclimation to External Ci. The characteristics of the pump are changed when cyanobacteria are grown at different concentrations of external Ci (Kaplan *et al.*, 1980; Badger and Gallagher, 1987). Experiments with *Anabaena variabilis* (Kaplan *et al.*, 1980) indicated that the capacity of the pump to transport Ci was reduced in low-affinity cells (grown on 5% CO_2 in air), being about 10% of the V_{max} of high-affinity cells (grown on air). The affinity of the transport system for external Ci was not altered. Similar studies with *Synechococcus* PCC6301 (Badger and Gallagher, 1987) give a somewhat different picture when the utilization of both CO_2 and HCO_3^- was examined. Low-affinity cells (grown at 1000 μl CO_2/l) showed a dramatic reduction in the capacity for HCO_3^- transport to the extent that it may have been totally absent, compared to high-affinity cells (grown at 30 μl CO_2/l). However, the ability to transport CO_2 was retained, although the affinity was reduced some fivefold to 10-fold and the V_{max} decreased 30–40%.

e. Species of Ci Entering the Cytosol. It has been generally assumed that the Ci that the pump delivers to the inside of the cell is in the form of HCO_3^- ; however, since the realisation that CO_2 can act as a species for transport, this must be questioned. Kaplan *et al.* (1980) argued that the high levels of internal Ci in cyanobacteria were the result of HCO_3^- transport into the cytosol and its subsequently slow conversion to CO_2 in the absence of carbonic anhydrase. This would require high levels of HCO_3^- to accumulate before the rate of uncatalyzed conversion to CO_2 was sufficient to support photosynthesis. This hypothesis has subsequently been shown to be invalid, as the levels of internal Ci have been calculated to be only just saturating for Rubisco, assuming rapid equilibrium between CO_2 and HCO_3^- , and there is evidence that rapid exchange between these species does occur (Badger *et al.*, 1985). In a series of experiments measuring the internal Ci pool and photosynthesis when either CO_2 or HCO_3^- was given to *Anabaena* cells, Volokita *et al.* (1984) argued that there is evidence to suggest that HCO_3^- is arriving internally, regardless of the species taken up. This is not convincing, considering the evidence that there is rapid equilibrium between the species inside the cell. At present it is impossible to determine which species arrives internally or whether this differs with CO_2 or HCO_3^- uptake.

f. Mechanistic Hypotheses. By assessing what is known about the pump, some tentative hypotheses can be developed to describe its mechanism of operation. One model has been proposed by Volokita *et al.* (1984), in which the pump is envisaged to be primarily a HCO_3^- transporter to which a carbonic anhydrase-like moiety is attached (Fig. 2a). This carbonic anhydrase (CA) "front end" would allow CO_2 to act as a source of HCO_3^- for the pump and explain how both HCO_3^- and CO_2 uptake can be achieved. A key experiment on which this proposal is based is the observation that ethoxyzolamide (an inhibitor of CA) inhibits CO_2 transport preferentially over HCO_3^- uptake in *Anabaena variablis* (Volokita *et al.*, 1984). Two lines of evidence make this proposal unlikely. First, a repeat of the experiments with *Anabaena* shows that, with high-affinity cells, ethoxyzolamide preferentially decreases HCO_3^- utilization for photosynthesis (G. D. Price and M. R. Badger, unpub-

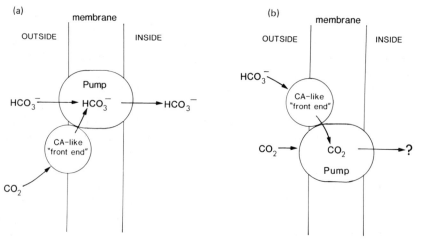

Fig. 2. Two models showing possible interactions of CO_2 and HCO_3^- with the pumping mechanism in cyanobacteria, which allows both species to act as substrates. Both models show a carbonic anhydrase-like "front end" to the pump, but in one case (a) an HCO_3^- pump is proposed, while in the other (b) a CO_2 pump is hypothesized.

lished results). Second, the observation that high- and low-affinity cells of *Synechococcus* (Badger and Gallagher, 1987) differ most significantly in their abilities to transport HCO_3^- suggests that the induction of a high-affinity cell is achieved primarily through the acquisition of an HCO_3^--utilizing "front end" for a CO_2 transporter. Thus, it seems most reasonable to suggest that the pump itself is primarily a CO_2 transport system and that HCO_3^- serves as a substrate via its conversion to CO_2 by a CA-like attachment. This hypothesis is depicted in Fig. 2b.

    g. The Role of Carboxysomes. Outside the considerations of the pumping mechanism mentioned above, it has become apparent that the polyhedral bodies within cyanobacteria, called carboxysomes, may have an important function in the CO_2-concentrating mechanism. These bodies appear to contain most of the active Rubisco in cyanobacteria (Coleman *et al.*, 1982) with few other enzymatic inclusions, especially carbonic anhydrase (Marsden *et al.*, 1984; Lanaras *et al.*, 1985).

    The carboxysome content of *Synechococcus leopoliensis* (measured by the cross-sectional area in electron micrographs) has been shown to increase when the cells are adapted to low CO_2 conditions (Turpin *et al.*, 1984). This could not be confirmed, however, when measuring the percent pellatibility of Rubisco activity (with *Synechococcus* PCC6301; M. R. Badger, unpublished results). More interestingly, the Rubisco activity of carboxysomes appears to be insensitive to competitive O_2 inhibition, unlike the free enzyme (Coleman *et al.*, 1982), although other workers have been unable to duplicate these results (J. R. Seemann, M. R. Badger, and G. D. Price, unpublished results). These observations have raised the question of whether the carboxysome is of central importance to the CO_2-concentrating mechanism, or whether it may be an adaptation to reduce the possible effects of elevated O_2, which may occur during photosynthesis (see Section V,A,3,d). If carboxysomes do have a special role to play, then the protein shell of these bodies (Holthuijzen et al., 1986) may be of central importance to controlling the fluxes of gases and charged substrates between the inside and outside. Presumably this coat will have distinctly different permeability properties from those of lipid membranes with charged species showing higher permeabilities relative to uncharged nonpolar molecules.

2. Energetics

In cyanobacteria, the accumulation ratio ($[Ci]_{inside}/[Ci]_{outside}$) has been observed to reach values in excess of 1000, particularly at low external [Ci]. If we assume that CO_2 and HCO_3^- are near chemical equilibrium within the cell, then this represents the ratio for both species. For a pumping system transporting either CO_2 or HCO_3^- across the inner membrane, the accumulation of Ci will proceed against a considerable energy barrier. For HCO_3^- accumulation this will be equal to $\Delta\mu HCO_3^-$, while for CO_2 this will be $\Delta\mu CO_2$ ($= \Delta\mu HCO_3^- - \Delta\mu H^+$).

a. Coupling to Ion Fluxes. All proposals for coupling of Ci uptake to ion fluxes across the inner membrane have assumed that it is the HCO_3^- ion that is taken up by the pump and delivered to the cytoplasm. No proposal has dealt with the possibility that the pump may in fact transport CO_2; however, this must also be considered.

Three models have been proposed to account for the transport of HCO_3^- into the cell: (1) a primary electrogenic HCO_3^- pump, operating together with a secondary Na^+/H^+ antiporter to control internal pH (Kaplan et al., 1982; Kaplan, 1985); (2) H^+/HCO_3^- symport or OH^-/HCO_3^- antiport, secondary to an H^+ extrusion pump (Kaplan et al., 1982; Reinhold et al., 1984); (3) Na^+/HCO_3^- symport, secondary to an Na^+/H^+ antiporter (Kaplan et al., 1984). The H^+/HCO_3^- symport mechanism has been rejected in Anabaena variabilis, where the stoichiometry of H^+ to HCO_3^- transport would have to range from 5:3 (pH 8) to 17:1 (pH9.5) to account for the observed Ci accumulation ratios (Zenvirth et al., 1984; Raven and Lucas, 1985). The position with regard to the Na^+-linked schemes [Fig. 3(a) and Fig. 3(b)] is unclear. Both can be manipulated to explain the internal hyperpolarization observed when HCO_3^- is added to Ci depleted cells (Kaplan et al., 1982; Kaplan, 1985); however, neither of them can explain the apparent Na^+ requirement for HCO_3^- but not CO_2 uptake (Miller and Canvin, 1985). This failure, combined with the strong possibility that the pump may use CO_2 rather than HCO_3^- (see Section V,A,1,b), casts doubts on both models.

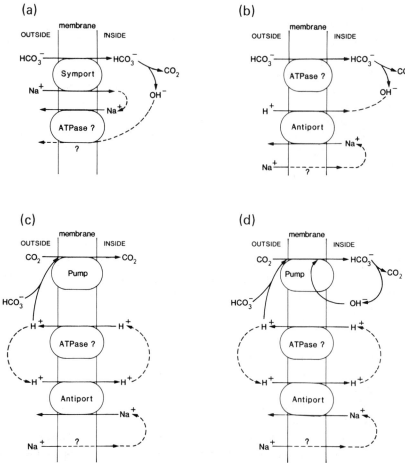

Fig. 3. Four possible models to explain the ways in which Ci uptake may be coupled to ion fluxes across the plasma membrane in cyanobacteria. (a) An Na^+/HCO_3^- symport pumping mechanism that is energized by the operation of an Na^+/OH^- symport (or H^+ antiport) pump, possibly utilizing ATP as an energy source. (b) A primary electrogenic HCO_3^- pump, utilizing ATP as an energy source, acting together with an Na^+/H^+ antiport system to control internal pH. (c) A primary CO_2 pump utilizing an unknown energy source and delivering CO_2 to the inside of the cell. This would act together with an H^+-effluxing ATPase, which enables HCO_3^- to be converted to CO_2, and an Na^+/H^+ antiport system to control internal pH when HCO_3^- is utilized by the pump. (d) A primary CO_2 pump energised by a localized OH^- gradient and delivering HCO_3^- to the interior of the cell. As for (c), this would act together with an H^+-effluxing ATPase and an Na^+/H^+ antiport mechanism.

Another scheme can be proposed that satisfies all current observations [Fig. 3(c)]. This model is based on the mechanism proposed to operate in the giant alga *Chara corallina* (Walker *et al.*, 1980; Lucas, 1983), where HCO_3^- utilization occurs in discrete acid bands along the cell. It is difficult to physically prove that microenvironments of high proton concentration can form on the outside of the inner membrane, yet there is strong evidence for such regions in alkalophilic bacteria operating at pH > 9 (Krulwich, 1986), where the respiratory chain may deliver some portion of effluxing protons to the F1F0-ATPase and the Na^+/H^+ antiporter by some localized pathway. The model of Fig. 3(c) is a direct result of such a consideration with respect to localized delivery of protons to a region close to a CO_2 transporter, so that protons can facilitate the conversion of HCO_3^- to CO_2. Cell hyperpolarization would be due to HCO_3^- stimulation of proton efflux, and the Na^+/H^+ antiporter would be required to regulate internal pH. CO_2 uptake would not require Na^+, as its transport would be electroneutral with no net proton exchange. The major unknown with this scheme is how the pump operates to transport CO_2 and how it is energised. The model of Fig. 3(c) proposes that CO_2 (or $H^+ + HCO_3^-$) is delivered to the inside of the cell, but it is possible that HCO_3^- may enter alone. This would require a mechanism for the combination of CO_2 with OH^- during transport, resulting in base-catalyzed HCO_3^- production [Fig. 3(d)]. CO_2 uptake would still result in no net proton exchange, while HCO_3^- uptake would release an OH^- internally. The vectorial consumption of OH^- near the outside of the membrane and the internal release of HCO_3^- would drive CO_2 accumulation with the maximum energy available equal to the localized $\Delta\mu OH^- - \Delta\mu HCO_3^-$.

b. Photosynthesis as an Energy Supply. The accumulation of Ci by the cyanobacteria *Anabaena variabilis* and *Anacystis nidulans* (*Synechococcus* sp.) is dependent on the supply of photosynthetic energy, and is greatly reduced by the whole-chain electron-transport inhibitor DCMU (Miller and Colman, 1980a; Badger and Andrews, 1982; Kaplan *et al.*, 1982; Ogawa *et al.*, 1985a) as well as darkness (Miller and Colman, 1980a; Badger and Andrews, 1982; Ogawa and Inou, 1983; Ogawa *et al.*, 1985a).

It was initially hypothesized that the photosynthetic reactions were linked to Ci transport through the synthesis of ATP. This was supported by the effects of proton ionophores on the ability of cells to accumulate Ci (Miller and Colman, 1980a). The inhibition of accumulation by DCMU initially suggested that ATP was synthesized largely as a result of linear electron transport. Contradictory results were obtained, however, with high-CO₂ grown cells of *Anabaena variabilis* (Ogawa and Inoue, 1983; Ogawa *et al.*, 1984), where accumulation was insensitive to DCMU and showed an action spectrum similar to that of PS1. Further work with air-grown cells of *Anacystis nidulans* (Ogawa *et al.*, 1985a,b) has subsequently shown that Ci transport appears to require both PS1-driven cyclic electron flow and ATP synthesis. Interestingly, growth of *Anacystis nidulans* at low CO₂ leads to an increase in the relative amount of PS1 and the potential for cyclic electron flow (Manodori and Melis, 1984).

In the light of possible models for Ci transport (see Section V,A,1,f), several schemes for linking both electron transport and ATP to pump energization are feasible. Ogawa *et al.* (1985a) have proposed that Ci transport is directly linked to an electron transport chain driving H^+ influx in the inner membrane, and ATP would be required to energize Na^+ efflux. Considering that CO_2 uptake does not require Na^+ (Miller and Canvin, 1985), this would predict that CO_2 accumulation by the pump may not require ATP. If the pump primarily utilizes CO_2 [Fig. 3(c) and 3(d)], then it is also possible that a cytoplasmic electron transport chain may drive proton efflux to catalyze the localized conversion of HCO_3^- to CO_2. The transfer of electrons between the thyalkoids and the cytoplasmic membrane would probably occur via a reduced mobile carrier such as ferrodoxin.

c. **Energy Requirement for Pumping.** The minimum free energy required to transport a mole of Ci from the external medium to the cytosol of the cell, where CO_2 fixation by Rubisco occurs, can be derived from the electrochemical differences for the transported species. These energy requirements have been considered by Raven and Lucas (1985) for cyanobacteria and may be used to set limits on the photosynthetic energy expenditure that is necessary to run the CO_2-concentrating mechanism. If the actively transported species is CO_2, then it is likely that its movement into the cell will be electroneutral and the energy required can be calculated from the activity difference of CO_2 across the inner membrane ($\Delta\mu CO_2$). For HCO_3^- transport, however, all mechanisms predict that its movement will be associated with the appearance of one net negative charge internally. In this case the energy required will be calculated from the activity difference of the HCO_3^- ion plus the electrical potential difference between the cell and the medium. As pointed out by Raven and Lucas (1985), the free-energy difference for CO_2 across the inner membrane is less than that for HCO_3^- by approximately the free-energy difference for H^+, due to the fact that the accumulation of one mole of HCO_3^- is equivalent to the separate accumulation of one mole of CO_2 and the expulsion of one mole of H^+.

Using the data of Zenvirth *et al.* (1984), for *Anabaena variabilis*, Raven and Lucas (1985) have calculated that for a cell achieving an accumulation ratio for Ci of about 2000 (at pH 8, external Ci = 20 μM, electrical potential difference = 90 mV), the free-energy differences between cytosol and medium for the relevent species are CO_2, +19.5 kJ mol^{-1}; HCO_3^-, +27.5 kJ mol^{-1}; and H^+, +7.5 kJ mol^{-1}. Considering that the hydrolysis of one mole of ATP to ADP + Pi will potentially yield a free energy of 50–55 kJ mol^{-1}, they suggest that the highest computed free-energy difference for the transport of the HCO_3^- ion is consistent with the movement of two moles of HCO_3^- for each mole of ATP. Ignoring the possible involvement of cyclic electron transport, as well as ATP consumption (see Section V,A,2,b), one mole of ATP could be produced via cyclic photophosphorylation with the expenditure of three moles of photons (assuming $3H^+$/ATP, and $1H^+$ separated across the thylakoid membrane per electron). Thus the transport of one mole of HCO_3^- may require 1.5 moles of photons.

The above calculations assume that there is no escape of transported Ci back into the bulk medium via a leakage pathway. As we shall see in the next section, the proportion of Ci that leaks back to the medium will dramatically affect the energetic efficiency of the pumping process, because for every mole of leakage that occurs, another 1.5 mole of photons may have to be expended to maintain the steady-state internal CO_2 concentration. Measurement of light-response curves at saturating external Ci concentrations, with a marine *Synechococcus* species (Badger and Andrews, 1982), has shown that high-CO_2 grown cells appear to have larger apparent quantum yields for net O_2 evolution than do low-CO_2 grown cells. The difference is about a twofold change of initial slope, and may give an indication of the extra costs incurred in running the CO_2-concentrating mechanism. If the cost of fixing a CO_2 molecule in high-CO_2 cells is close to the value for C_3 photosynthesis in the absence of photorespiration, then a cost of eight photons will be incurred. A doubling of the quantum yield would indicate an extra eight are required to concentrate CO_2 in low-CO_2 cells. If the cost of transporting a Ci species into the cell is about 1.5 photons, then the extra energy cost is presumably associated with the leakge of Ci back out of the cell, with around five Ci species being transported for every one CO_2 fixed in photosynthesis.

3. Leakage of CO₂

For higher plants operating in a gaseous environment, the supply of CO_2 to the cytosol occurs via diffusion through the water and lipid layers surrounding the cell. This is largely made possible by the relatively high permeability of polar lipid bilayers to CO_2 (permeability coefficient $P_{CO_2} = 3.5 \times 10^{-3}$ m sec^{-1} for high-cholesterol membranes *in vitro* at 25°C; Gutknecht *et al.*, 1977). If these properties applied to the membranes surrounding the cyanobacterial cell, then it can be shown that the back-diffusion of CO_2 from the concentrated internal pool would pose a major energetic problem for cells operating the CO_2-concentrating mechanism (Badger *et al.*, 1985; Lucas and Raven, 1985).

a. Effect of CO₂ Leakage on Energy Costs. The ratio of Ci transport to CO_2 efflux that could be maintained during steady-state photosynthesis is dependent on both the absolute concentration gradient for CO_2 that is established between the cell and the bulk medium and the permeability of the cell to CO_2 efflux. This assumes of course that leakage of HCO_3^- is not significant. These relationships have been calculated for a marine *Synechococcus* species (Badger *et al.*, 1985), with the assumption that the internal CO_2 concentration in cyanobacteria using the CO_2-concentrating mechanism must be close to saturating for Rubisco (1000 μM). With these cells, it was calculated that to maintain a ratio of Ci uptake to CO_2 fixation of <10 the cell would have to show a P_{CO_2} of $<10^{-6}$ m sec^{-1}. Indeed, to achieve a ratio of 5:1, as suggested by the quantum yield measurements (see Section V,A,2,c), a value of 10^{-7} would be more reasonable. If the cells had permeability coefficients similar to those of lipid bilayers, then ratios in excess of 1000 would occur.

b. Measured Permeability to CO_2. The permeability of cyanobacterial cells to CO_2 has been measured by preloading the cells with Ci during steady-state photosynthesis and monitoring the subsequent efflux of Ci following a change to unlabeled substrate or darkness (Marcus *et al.*, 1982; Badger *et al.*, 1985). For the marine *Synechococcus* species, P_{CO_2} estimates for air-grown cells ranged from 5.7×10^{-7} to 6×10^{-9} m sec^{-1}, assuming internal pH ranging from 7.5 to 8.5.

c. Ratio between Transport and Fixation. An estimate of the magnitude of net photosynthesis, gross Ci transport, and CO_2 leakage has also been made for air-grown cells of the marine *Synechococcus* species (Badger *et al.*, 1985). At levels of external Ci that were just saturating for photosynthesis, Ci transport did not exceed net photosynthesis by more than 50%, and at supersaturating Ci levels this figure approched 100%. To obtain a CO_2 leakage rate of around 50% of net photosynthesis, the cell permeability to CO_2 was calculated to be about 10^{-8} m sec^{-1}, a figure well in accord with the theoretical estimates calculated by the authors for the same cells.

d. Oxygen Buildup Problems. If the cell has an effective permeability coefficient for CO_2 that is of the order of 10^{-8} m sec^{-1}, then it seems reasonable to predict that the permeability for O_2 will be similar. If this is the case, then the problem of internal O_2 buildup during photosynthesis cannot be ignored. For the marine *Synechococcus* species, Badger *et al.* (1985) have calculated that if the P_{O_2} is $<5 \times 10^{-8}$ m sec^{-1}, then O_2 inside a photosynthesising cell will reach levels in excess of 1 atm partial pressure. It is likely that these levels, if they occurred, would pose severe toxicity problems for the cell. The questions of whether these levels actually exist and how the cyanobacterial cell may deal with the potential O_2 problem remain to be answered.

e. The Barrier to CO_2 Leakage. The very low permeability to CO_2 displayed by cyanobacterial cells is far below that measured for any other biological membrane system. A value of $P_{CO_2} < 10^{-7}$ m sec^{-1} is considerably less than that measured for condensed monolayers of long-chain fatty acids or alcohols (10^{-5} m sec^{-1}; Gutknecht *et al.*, 1977). How this barrier to CO_2 diffusion is achieved remains an area for speculation, but it is possible that use of hydrophilic protein costs, with low permeability to hydrophobic molecules, may be involved. Evidence has been presented for *Anabaena variabilis* (Marcus *et al.*, 1982), to suggest that growth of cells at low CO_2 leads to a thickening of the inner and outer membrane structure and the possible production of a new layer, interior to the inner membrane. This correlated with low-CO_2 grown cells also showing a greater resistance to lysozyme treatment. It is unlikely that high-CO_2 grown cells lack the CO_2 diffusion barrier, as they still show the potential to concentrate CO_2; however, there may well be a reduction in its absolute effectiveness.

4. *Interconversion of Ci Species*

The catalysis of the interconversion of Ci species, both outside and inside the cell, has a major role to play in enabling the CO_2-concentrating mechanism to operate efficiently. The most likely factor that can influence the rate of this interconversion is the presence or absence of the enzyme carbonic anhydrase (CA). However, it must also be considered that regions of both acidity and alkalinity can also have an effect.

a. Evidence for Regions of Rapid Interconversion. There is no evidence for the presence of carbonic anhydrase activity when suspensions of intact cyanobacteria are assayed in the dark (Yagawa *et al.*, 1984; Badger *et al.*, 1985), indicating that it is unlikely that an exterior form of carbonic anhydrase exists in a form that is freely accessible to external substrates. When cells are assayed in the light, however, a different picture is found. Using the marine *Synechococcus* species, Badger *et al.* (1985) were able to show that the CO_2 that was effluxing from the cell during steady-state photosynthesis had passed through a region of rapid interconversion, where the [18]O from labeled Ci had been lost to water. This clearly indicated that a region of rapid interconversion was associated with the cells; however, the access of Ci species to it was dependenton the operation of the pumping mechanism. This could be inferred to support the existence of internal CA.

A vital theoretical role exists for internal CA if the CO_2-concentrating mechanism is proposed to operate via the transport of HCO_3^- into the cytosol. For *Synechococcus* species, Badger *et al.* (1985) have calculated that, in the absence of internal CA activity, the internal Ci pool would have to build up to concentrations in excess of 2 M to support CO_2-saturated photosynthesis. This would clearly be unfeasible and is not observed. The maximum pool size for Ci, of 30–60 mM, appears to be only just sufficient to provide saturating CO_2 to Rubisco, if it is assumed that rapid interconversion between Ci species occurs in the cytosol (Badger *et al.*, 1985).

Finally, inhibitors of carbonic anhydrase have been shown to reduce the affinity of low-CO_2 grown cells for external Ci, without reducing the V_{max} of photosynthesis (Kaplan *et al.*, 1980; Shiraiwa and Miyachi, 1985a; Badger, 1985). This would futher indicate a central role for CA in the CO_2-concentrating mechanism; however, it is not universal for all cyanobacteria. Inhibitor effects have only been shown conclusively with *Anabaena variabilis* cells, while *Synechococcus* species appear to be insensitive (Badger *et al.*, 1985; Badger, 1985; M. R. Badger, unpublished results).

b. Internal Interconversion. Homogenates of cyanobacteria generally show levels of carbonic anhydrase activity that are low and sometimes undetectable (Ingle and Colman, 1976; Kaplan *et al.*, 1980; Yagawa *et al.*, 1984; Shiraiwa and Miyachi, 1985a; Badger *et al.*, 1985; Lanaras *et al.*, 1985). This activity is elevated in low-CO_2 grown cells of *Coccochloris peniocystis* (Ingle and Colman, 1976) and *Anabaena variabilis* (Yagawa *et al.*, 1984; Shiraiwa and Miyachi, 1985a) and is thus implicated in the functioning of the CO_2-concentrating mechanism. In *Anabaena variabilis* cells, both soluble and membrane-bound forms of this activity have been detected, with strains differing in whether they possess both or just the soluble form (Yagawa *et al.*, 1984). In *Chlorogloepsis fritschii,* however, the predominant form is particulate, and presumably membrane-bound in nature (Lanaras *et al.*, 1985). The soluble form is presumably active in the cytosol, and is sufficient to maintain internal Ci species close to chemical equilibrium (Badger *et al.*, 1985). The role of the membrane form remains to be determined; however, it is possible that it functions on the inner membrane in association with the pumping mechanism.

c. External Interconversion. There is no clear evidence that any form of rapid interconversion occurs externally to the inner membrane in cyanobacteria; however, one is tempted to speculate. With *Anabaena variabilis* cells, Volokita *et al.* (1984) have found that the CA inhibitor ethoxyzolamide inhibited Ci accumulation, with CO_2 uptake being preferentially reduced compared to HCO_3^-. From this they speculated that the pump may have a CA-like moiety associated with it, its role being to convert CO_2 to HCO_3^- for transport by the pump. Other evidence suggests that in fact the reverse may occur, with HCO_3^- being converted to CO_2 prior to uptake (see Section V,A,1,f). In either case, a membrane-bound form of CA may be associated with the pump to fulfill this role.

The postulation of a pumping mechanism that may generate microenvironments of high proton concentration [see Figs. 3(c) and 3(d)] provides another way in which the conversion of HCO_3^- to CO_2 may be increased. Likewise, the postulation of a region of high OH^- concentration associated with CO_2 transport [see Fig. 3(d)] provides a mechanism of rapid conversion of CO_2 to HCO_3^-. Whether either of these occurs remains to be seen.

B. Green Microalgae

Eukaryotic algae of the Chlorophyceae family were the first aquatic organisms in which the CO_2-concentrating mechanism was identified. As for cyanobacteria, air-grown cells of green algae clearly show physiological responses that indicate the presence of this mechanism and its ability to modify the basic response of C_3 photosynthesis. The affinity of photosynthesis for CO_2 is considerably lower than that measured for Rubisco in these organisms, but the size of this difference is dependent on the species. *Chlamydomonas reinhardtii* has been recorded to have $K_{0.5}(CO_2)$ values in the range of 0.2–2.8 μM, with the higher values occurring with cells measured at pH below 6 (Berry *et al.*, 1976; Moroney and Tolbert, 1985). This compares with a $K_m(CO_2)$ for Rubisco, from the same species, of about 25 μM (Berry *et al.*, 1976). *Chlorella vulgaris* shows a similar range of $K_{0.5}(CO_2)$ for photosynthesis, and the properties of extracted Rubisco are similar to those of *Chlamydomonas* (Tsuzuki *et al.*, 1985). Properties of Rubisco have not been determined from other green microalgae; however, the $K_{0.5}(CO_2)$ values have been in the 1–5 μM range for *Chlorella emersonii* (Beardall and Raven, 1981), *Chlorella saccharophila* (Beardall, 1981), *Chlorella pyrenoidosa* (Shelp and Canvin, 1980), *Dunaliela salina* (Zenvirth and Kaplan, 1981), *Botrydiopsis intercedans* (Beardall and Entwisle, 1984), and *Scenedesmus odliquus* (Findenegg, 1976). The CO₂ compensation point is low for species that have been measured, ranging from around 0.03 μM at alkaline pH (>8) to 0.3 μM in acidic medium (<6.5) (Lloyd *et al.*, 1977; Birmingham and Colman, 1979; Beardall *et al.*, 1982), and it is insensitive to O_2 from 0 to 21%. Additionally, there is no effect of O_2 on the rate of net photosynthesis or its involvement in the production of photorespiratory metabolites (Lloyd *et al.*, 1977).

Biochemical analysis of the photosynthetic pathway in *Chlorella* was the initial source for the discovery of the C_3 pathway of photosynthesis; however, this was achieved with high-CO₂ grown cells that presumably did not show the physiological characteristics mentioned above. Subsequent comparison of high- and low-CO₂ grown cells of *Chlorella* have shown no change in the basic nature of this pathway that can be correlated with the change in the physiology of the cells (Reed and Graham, 1977; Hogetsu and Miyachi, 1979b). It can be shown, however, that low-CO₂ grown cells do have the ability to elevate intracellular Ci and CO₂ during photosynthesis, up to 50-fold higher than the expected passive distribution between the cell and the external medium (Badger *et al.*, 1980; Beardall and Raven, 1981; Beardall, 1981; Beardall *et al.*, 1982; Spalding *et al.*, 1983a,b; Shelp and Canvin, 1984; Tsuzuki *et al.*, 1985). Modification of the basic process of C_3 photosynthesis is thus achieved by the elevation of internal CO₂.

1. Pumping

The existence of a Ci pumping mechanism has been firmly established in *Chlamydomonas reinhardtii*, where a mutant defective in the ability to transport Ci into the cell has been isolated (Spalding *et al.*, 1983b). This mutant has been shown to be the result of a single recessive nuclear mutation probably associated with a component of the Ci transporter (Spalding *et al.*, 1983c, 1985). Although the actual physical nature of these mutants has not been established, this is strong evidence to suggest the physical existence of a Ci pumping mechanism in this and other green microalgae.

a. Location of the Pump. For the eukaryotic green microalgae, it is possible for the inorganic pumping mechanism to be located either on the cytoplasmic membrane or the inner membrane of the chloroplast envelop. If pumping was to occur across the cytoplasmic membrane, then Ci would be accumulated in the cytosol. Assuming that the chloroplast envelope is impermeable to HCO_3^- ions, then elevated CO_2 in the cytosol would diffuse passively into the chloroplast, where it would be fixed by Rubisco. If, however, the pump were located on the chloroplast envelope, then accumulation would occur primarily within the stroma, and Ci would gain entry to the cell primarily by the passive diffusion of CO_2 across the plasma membrane.

At present, there is no clear indication of where the pump is located. Experiments with *Chlamydomonas reinhardtii* (Marcus *et al.*, 1984), in which the cytoplasmic membrane was selectively permeabilized by treatment with DMSO and polylysine, have been interpreted to indicate that pumping occurs across the plasma membrane. These results are not unequivocal, however, as accumulation may have been decreased by an increase in the leakage of CO_2 from the cells. Other workers have examined the species of Ci used by both *Chlorella vulgaris* (Miyachi *et al.*, 1983) and *Chlamydomonas reinhardtii* (Tsuzuki, 1983; Imamura *et al.*, 1983; Moroney and Tolbert, 1985) and concluded that it is CO_2 that primarily enters the cell. This has then been used as evidence to imply that active transport of HCO_3^- occurs at the chloroplast envelope. However, this ignores the possibility that the pump may primarily use CO_2 as a species (see Section V,B,1,b) and could thus be located on the plasma membrane.

b. Specificity for Substrate. The question of which species of Ci may be the substrate for the pumping mechanism is difficult to answer without a knowledge of where the pump is located. Active species experiments with *Chlorella vulgaris* (Miyachi *et al.*, 1983), *Chlamydomonas reinhardtii* (Tsuzuki, 1983; Imamura *et al.*, 1983; Marcus *et al.*, 1984; Moroney and Tolbert, 1985), and *Dunaliella tertiolecta* (Aizawa *et al.*, 1986) have all established that in air-grown cells of thse species, CO_2 is the primary form of Ci that enters the cell. The ability of HCO_3^- to support photosynthesis arises because of the presence of carbonic anhydrase in the periplasmic space, which allows HCO_3^- to be rapidly converted to CO_2 close to the cell surface (see Section V,B,4,a). In the case of *Chlamydomonas* (Marcus *et al.*, 1984), location of the pump has been presumed to occur on the plasma membrane and the pump is thus presumed to use CO_2 as the primary substrate. In all other cases, CO_2 is assumed to passively diffuse into the cell and be converted internally to HCO_3^-, which can then act as a substrate for the pump located on the chloroplast envelope. Further work on determining the location of the pump will be necessary before this problem is resolved.

c. **Species of Ci Arriving Internally.** As for cyanobacteria (see Section V,A,1,e), it has generally been assumed that HCO_3^- is actively pumped into either the cytoplasm or the stroma. This assumption has ignored the evidence accumulated in cyanobacteria (see Section V,A,1,b), that CO_2 may be the pump's substrate. The only real evidence which can be used to shed light on the question comes from experiments with inhibitors of internal carbonic anhydrase and the study of a mutant of *Chlamydomonas* depleted in this enzyme. When *Chlamydomonas reinhardtii* is treated with ethoxyzolamide, an inhibitor of both internal and external carbonic anhydrase, the affinity of photosynthesis for external Ci is greatly reduced. In spite of this, the internal Ci pool is considerably higher in the presence of the inhibitor (Badger *et al.,* 1980; Spalding *et al.,* 1983a,b; Moroney *et al.,* 1985). This can only be interpreted to mean that HCO_3^- is being actively accumulated, but its rate of conversion to CO_2 is drastically reduced, thus leading to a disequilibrium where HCO_3^- is unusually high and CO_2 is low. However, it must be considered that ethoxyzolamide may have effects on the rate of pumping, either through a direct effect on the pump or through the simultaneous inhibition of external carbonic anhydrase. Similar results are seen with the CA-deficient *Chlamydomonas* mutant, where photosynthesis can be much reduced compared to wild type even though internal Ci pools are fivefold or more higher (Spalding *et al.,* 1983a). In this case, it can be assumed that the properties of the pump are not affected.

The only contradictory evidence, to suggest that both CO_2 and HCO_3^- may be accumulated, comes from work with *Chlorella vulgaris* (Tsuzuki *et al.,* 1985). With this species, supplying cells with either CO_2 or HCO_3^- during isotope disequilibrium experiments resulted in greatly different relationships between internal Ci and photosynthetic rate. At rate-limiting concentrations of external Ci, internal Ci pools were higher when HCO_3^- was given, particularly at alkaline pH, but the rates of photosynthesis were less. This suggests that HCO_3^- and CO_2 may not be in rapid equilibrium within the cell and that HCO_3^- may be accumulated internally when it is presented as the external substrate, but when the cells are given CO_2, it is CO_2 that is transported into the cell and elevated.

d. **Rate of Pumping.** The accumulation of Ci by air-grown cells of *Chlamydomonas reinhardtii,* previously depleted of internal Ci, is rapid. Internal Ci pools reach maximum levels after 5–10 sec of incubation and remain constant until the external pool is depleted (Badger *et al.,* 1980; Spalding and Ogren, 1983; Marcus *et al.,* 1984; Moroney *et al.,* 1985). An attempt to resolve the kinetics of this process, using a modified silicon-oil centrifugation technique, has estimated a V_{max} for Ci uptake of around 1400 μmol (mg Chl)$^{-1}$ h^{-1} and a K_m(Ci) of 160 μM (Spalding and Ogren, 1983). Thus the potential V_{max} for accumulation exceeds the V_{max} of photosynthesis by at least fivefold. The flux rate during steady-state photosynthesis has not been estimated; however, as in cyanobacteria (see Section V,A,1,c), it is likely that the rate of pumping is considerably below this maximum level. Modeling of photosynthesis in *Chlamydomonas* (Spalding and Portis, 1985) has shown that under conditions of steady-state photosynthesis and with reasonable estimates of cell leakage to CO_2, the rate of pumping is not required to exceed CO_2 fixation by more than twofold in order to support internal CO_2 concentrations which are saturating for Rubisco.

In all these cases, the periplasmic levels of carbonic anhydrase associated with the cells would have been sufficient to promote rapid equilibrium between the Ci species, so no distinction can be made between the rates at which CO_2 or HCO_3^- are accumulated. The activity of external carbonic anhydrase has been reduced by the use the CA inhibitors which do not penetrate the cell (Marcus *et al.*, 1984; Moroney *et al.*, 1985). When this is done, active species experiments show that CO_2 is a better substrate for accumulation by the cell than HCO_3^-. Similar results are also obtained with a cell wall-less mutant, in which the external CA is greatly reduced (Marcus *et al.*, 1984).

2. Energetics

In the green microalgae, the accumulation ratio for both total Ci and CO_2 may reach values in excess of 40, at rate-limiting concentrations of external Ci. From measurements of intracellular pH and membrane potential, it can be concluded that this accumulation proceeds against the electrochemical gradient for both CO_2 and HCO_3^- and is therefore an active process (Badger *et al.*, 1980; Beardall and Raven, 1981; Zenvirth and Kaplan, 1981).

    a. Coupling to Ion Fluxes. While it is appropriate to speculate about the way in which Ci transport may be linked to transmembrane energy fluxes, or to the hydrolysis of ATP itself, no evidence exists to allow any hypothesis to be confirmed. As discussed for the cyanobacteria (see Section V,A,1,f), all proposals have assumed that the pumping mechanism transports HCO_3^- across the appropriate membrane and no consideration has been given to a mechanism that postulates CO_2 as the primary species for the pump.

Two models for the transport of HCO_3^- have been suggested, both having direct analogies with the mechanisms discussed for cyanobacteria. The first is a primary electrogenic HCO_3^-, which would operate together with a secondary OH^- efflux or H^+ influx mechanism (Beardall and Raven, 1981). This scheme seems to provide the only reasonable explanation for the observed hyperpolarization of cell membrane potential when Ci is given to cells in the light (Beardall and Raven, 1981). The second possibility is that of an H^+/HCO_3^- symport mechanism. This would require a $H^+ : HCO_3^-$ transport ratio of greater than 1; however, this scheme could not easily account for the hyperpolarization of membrane potential (Beardall and Raven, 1981). Both schemes for transport could operate at either the plasmalemma or chloroplast envelope.

No requirement for the flux of another counterion has been obtained. In the short term, photosynthesis in *Chlamydomonas reinhardtii* is unaffected by the removal of both K^+ and Na^+ from the medium (M. R. Badger unpublished results). In *Scenedesmus obliquus,* some evidence has been presented for the uptake of Cl^- ions that may accompany photosynthesis (Findenegg, 1977); however, it is difficult to see how this relates to a mechanism directly related to active Ci uptake.

b. Photosynthesis as an Energy Supply. As for cyanobacteria (Section V,A,2,b), the accumulation of Ci in *Chlamydomonas reinhardtii* and *Chlorella pyrenoidosa* is dependent on the supply of photosynthetic energy. Internal pools are reduced to near passive equilibrium levels in the presence of DCMU and darkness (Badger *et al.,* 1980; Spalding *et al.,* 1983a; Shelp and Canvin, 1984). As accumulation is also sensitive to the use of proton ionophores, it has been proposed that the supply of ATP by the photosynthetic reactions provides the energy for transport. If the transport system is located on the plasmalemma, then a means of exporting ATP from the chloroplast would have to be envisaged.

c. Energy Requirement for Pumping. The transport costs for the accumulation of Ci in green microalgae are lower than those calculated for cyanobacteria (see Section V,A,2,c), due to the 20- to 50-fold lower accumulation ratios for Ci species during steady-state photosynthesis. For HCO_3^- accumulation, this difference may be offset by a more negative membrane potential in green algae, compared to cyanobacteria (Beardall and Raven, 1981). The free-energy difference for CO_2 accumulation, however, will be less than half that for cyanobacteria (<10 kJ mol^{-1}), making it possible for one mole of ATP to energize the accumulation of up five moles of CO_2, and a somewhat lower amount of HCO_3^-. Thus where it was estimated that the transport of one mole of HCO_3^- in cyanobacteria may require 1.5 moles of photons (Section V,A,2,c), this figure will be reduced in green algae, according to the reduction in the requirement of ATP for the movement of Ci species.

Measurements of quantum yields in *Chlamydomonas reinhardtii* have shown that the energy cost to fix a CO_2 molecule may be only about 20% more expensive in cells with the CO_2-concentrating mechanism, compared to those without it (Spalding and Portis, 1985). This energy cost is about 20% of that estimated from cyanobacterial light response curves (Badger and Andrews, 1982). Assuming a photon requirement for pumping of one per Ci species transported, a 20% increase in costs will be incurred if it is assumed that there is a 60% leakage of Ci back out of the cell as CO_2 (assuming a quantum requirement of eight for cells without the CO_2-concentrating mechanism).

3. Leakage of CO_2

The barrier restricting the leakage of CO_2 does not have to be as "CO_2-tight" in the green algae, compared with cyanobacteria, due to the lower gradients of CO_2 established between the cell and the external medium, the lower energy costs for the transport of Ci species, and the lower ratio of surface area to cell volume in the larger eukaryotic cells. A relationship between Ci transport (V_t, mol sec^{-1} cell^{-1}), CO_2 fixation (V_c, mol sec^{-1} cell^{-1}), and the permeability of the cell to CO_2 efflux (P_{CO_2}, m sec^{-1}) can be formulated for conditions where CO_2 fixation is just saturated with external Ci and internal CO_2 concentration is saturating for Rubisco [for these calculations I will assume that $K_m(CO_2)$ is 25 μM (Berry et al., 1976) and that internal CO_2 is 100 μM]. Under conditions of steady-state photosynthesis, then,

$$V_t = V_c + (P_{CO_2})(\Delta CO_2)A \qquad (1)$$

where A is the surface area of a cell (m^2 cell^{-1}) and ΔCO_2 is the CO_2 concentration gradient between the cytosol and the external medium. This relationship assumes that, for a eukaryotic cell, the barrier to CO_2 efflux is associated with the plamalemma and cell wall and not the chloroplast envelope. Using data for air-grown cells of *Chlamydomonas reinhardtii* (Badger et al., 1978), this relationship can be quantified ($V_c = 7.37 \times 10^{-17}$, $A = 1.23 \times 10^{-6}$, and $\Delta CO_2 = 10^{-1}$ mol m^{-3}) for various ratios of V_t/V_c and assumed P_{CO_2} values. If it were assumed that the CO_2 leakage rate is 50% of the CO_2 fixation rate (i.e., $V_t/V_c = 1.5$), the cell P_{CO_2} would be 2.8 $\times$ 10^{-6} m sec^{-1}. This is about two orders of magnitude higher than the value calculated for cyanobacteria, assuming the same V_t/V_c ratio (see Section V,A,3,a). Taking the P_{O_2} value to be the same as for CO_2, the O_2 gradient developed in these cells during photosynthesis will be 200 μM. This is about twice the atmospheric concentration of O_2, but is much less than the concentration that could be potentially accumulated in cyanobacteria (see Section V,A,3,d), and should not pose the toxicity problems that cyanobacteria may encounter. Similar results have been obtained from a more detailed photosynthetic model for *Chlamydomonas* (Spaling and Portis, 1985), which considers the effects of varying properties of the transport system, the permeability of the cell to CO_2, and the internal activity of carbonic anhydrase.

The permeability of green algal cells to CO_2 has only been measured in *Dunaliella salina* (Zenvirth and Kaplan, 1981), where a P_{CO_2} value of 10^{-6}–10^{-7} m sec^{-1} was calculated. This would put the permeability of *Dunaliella* some one to two orders of magnitude higher than that measured for cyanobacteria, and would establish a V_t/V_c ratio similar to that calculated above, for *Chlamydomonas*. Although there is an estimate of the permeability to CO_2, there are no direct measurements of the amount of CO_2 leakage from a green algal cell during steady-state photosynthesis.

The assumption that the barrier to CO_2 efflux is associated with the plasmalemma has no experimental support and is largely based on the premise that the Ci transport system is located in this membrane. If the transporter were on the chloroplast envelope, then it is likely that the diffusion barrier would be associated with the envelope as well, as the plasmalemma would have to be reasonably permeable to CO_2 to allow the supply of substrate to the pump. As discussed for cyanobacteria (Section V,A,3,e), there is little evidence to explain how a diffusive barrier of such surprising magnitude can be developed, although the barrier in green microalgae may be two orders of magnitude more leaky than that in cyanobacteria.

4. Interconversion of Ci Species

Inorganic carbon accumulated by the eukaryotic cell must pass through three separate compartments before fixation by Rubisco: the periplasmic space, the cytosol, and the stroma. Within each of these zones there is evidence for the presence of carbonic anhydrase and its functional role in the operation of the CO_2-concentrating mechanism. The postulated roles of carbonic anhydrase in the overall mechanism depends, however, on the view adopted on the location and substrate specificity of the Ci transporter, and these are depicted in Fig. 4.

a. Periplasmic Carbonic Anhydrase. A form of periplasmic carbonic anhydrase, either free in the periplasmic space or attached to the cell wall or plasma membrane, has been shown to be present in air-grown cells of *Chlamydomonas reinhardtii* (Kimpel *et al.*, 1983; Coleman *et al.*, 1984), some species of *Chlorella* (Pronina *et al.*, 1981; Miyachi *et al.*, 1983), *Dunaliella tertiolecta* (Aizawa *et al.*, 1986) and *Scenedesmus* species (Pronina *et al.*, 1981; Findenegg, 1976). This activity can be detected by the ability of whole cells to catalyze the interconversion of external Ci species, in both the dark and light, and its presence correlates exactly with the ability of cells to utilize HCO_3^- for photosynthesis (Tsuzuki, 1983; Miyachi *et al.*, 1983; Imamura *et al.*, 1983; Marcus *et al.*, 1984; Moroney *et al.*, 1985; Aizawa *et al.*, 1986). Thus, species of *Chlorella* that lack a periplasmic carbonic anhydrase have been shown to utilize only CO_2 for photosynthesis (Miyachi *et al.*, 1983). This has been considered in Fig. 4, and periplasmic carbonic anhydrase is proposed to participate in Ci acquisition by allowing HCO_3^- to be converted to CO_2 close to the cell surface for subsequent passive diffusion or active transport into the cell.

b. Internal Carbonic Anhydrase. All air-grown cells of green algae studied to date have carbonic anhydrase activity that is located within the cell (Hogetsu and Miyachi, 1979a; Pronina *et al.*, 1981; Miyachi *et al.*, 1983; Kimpel *et al.*, 1983; Spalding *et al.*, 1985). This activity may be present in both a soluble and a membrane-bound form, depending on the species exam-

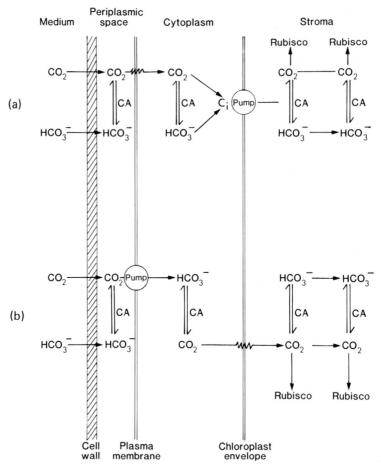

Fig. 4. The possible locations of the Ci pump in green microalgae, and its functional relationship with external and internal carbonic anhydrase (CA). (a) The pump, using an unknown form of Ci, located on the chloroplast envelope, and CO_2 entering the cell by passive diffusion across the plasma membrane. (b) A CO_2 pump located on the plasma membrane, and CO_2 entering the chloroplast by passive diffusion across the chloroplast envelope.

ined (Pronina *et al.*, 1981; Miyachi *et al.*, 1983; Tu *et al.*, 1986); however, the soluble form appears to be the activity that shows a dramatic increase when high-CO_2 grown cells are adapted to growth on air (Pronina *et al.*, 1981; Miyachi *et al.*, 1985).

Several lines of evidence suggest that the soluble internal carbonic anhydrase functions primarily to convert HCO_3^-, delivered by the Ci pump to either the cytosol or the chloroplast, to CO_2 for fixation by Rubisco (see Fig. 4). When the internal CA of *Chlamydomonas reinhardtii* is reduced, either by the use of CA inhibitors (Badger *et al.*, 1980; Spalding *et al.*, 1983a,b; Moroney *et al.*, 1985) or the study of a mutant defective in internal CA

(Spalding *et al.*, 1983a; Spalding *et al.*, 1985), the internal Ci pool accumulates to levels in excess of the wild-type cells but photosynthesis is reduced. This can only be interpreted to mean that HCO_3^-, accumulated by the pump, is being elevated due to the reduced rate of conversion to CO_2 and that CO_2 is subsequently reduced. A model analyzing the theoretical relationship between the internal Ci pools, photosynthesis, and carbonic anhydrase activity shows that if HCO_3^- is transported into the cell, then internal CA is necessary to maintain internal Ci pools at the levels measured in air-grown cells (Spalding and Portis, 1985). If CA activity is reduced to zero in this model, then the internal HCO_3^- pool sizes necessary to produce a CO_2 conversion rate capable of supporting photosynthesis reach levels in excess of those measured in the CA mutant. A minor role for the soluble internal CA can also be postulated in promoting the facilitated diffusion of Ci species within both the cytosol and the chloroplast, regardless of whether the Ci transport system is operating at the plasmamembrane or the chloroplast envelope.

Little work has been done on localizing the internal CA. Evidence with *Chlorella* cells (Hogetsu and Miyachi, 1979a) suggests that much of the soluble internal activity is associated with the chloroplasts. If this were the general case, then an argument could be made for the Ci transporting system being located on the chloroplast envelope. More evidence is necessary, however, before any such conclusions can be drawn. The location and function of an internal membrane-bound CA remain even more elusive. The only speculation that can be made at this stage is that it may be associated with the Ci pump located either on the plasmamembrane or the chloroplast envelope.

c. **Acid and Alkaline Zones.** The possible involvement of localized acid and alkaline environments in the utilization of Ci species by eukaryotic microalgae has not been extensively considered, generally because of the limitations that the physical size of the organisms would impose on the maintenance of such zones. However, if it is possible to create microenvironments with such properties, as discussed for cyanobacteria (see Section V,A,2,a), then they will also act as regions in which the interconversion of Ci species is speeded up. Whether an acid region may be associated with production of CO_2 for its subsequent transport by a CO_2 pump remains an area for speculation.

VI. INDUCTION OF THE CO₂-CONCENTRATING MECHANISM

The CO_2-concentrating mechanism is induced when either cyanobacteria and green microalgae are transferred from conditions of high-CO_2 (or non-limiting Ci supply) to growth under low-CO_2 (or limiting Ci supply). This experimental change in Ci supply has generally been imposed by changing the CO_2 concentration in the gas with which the growth medium is bubbled, decreasing it from 1–5% CO_2 in air down to air levels of CO_2 (0.03–0.04% CO_2). The biochemical, physiological, and structural changes that occur during this induction have been discussed in the previous section, when describing the operation of the concentrating mechanism in air-adapted cells. These changes include an increase in one or more forms of carbonic anhydrase (see Miyachi et al., 1985); an increase in Ci transport capacity (Badger et al., 1980; Kaplan et al., 1980; Marcus et al., 1982); an increase in glycolate dehydrogenase activity in green algae (Bruin et al., 1970); and an increase in the permeability barrier, preventing excessive CO_2 loss from the cell (Marcus et al., 1982; Zenvirth et al., 1984; Badger et al., 1985). The complete adaptation to low Ci conditions takes from 4–8 h in cyanobacteria (Marcus et al., 1982; Omata and Ogawa, 1986) and green algae (Graham et al., 1971; Shiraiwa et al., 1981; Coleman et al., 1984; Shiraiwa and Miyachi, 1985b).

A. Protein Synthesis

The adaptation process depends on the synthesis of new proteins, some of which have been specifically identified. In the cyanobacteria Anabaena variablis and Anacystis nidulans the adaptation to low CO_2 is inhibited by the protein synthesis inhibitor spectinomycin, as well as the 70 S ribosome inhibitor chloramphenicol and the RNA synthesis inhibitor rifampicin (Marcus et al., 1982; Omata and Ogawa, 1986). In green algal species of Chlorella vulgaris and Chlamydomonas reinhardtii, the induction of carbonic anhydrase is prevented by the cytosolic ribosome inhibitor cycloheximide, but not the chloroplast ribosome inhibitor chloramphenicol (Shiraiwa et al., 1981; Spencer et al., 1983), and in Chlorella vulgaris, cycloheximide prevents the increase in the affinity of photosynthesis for Ci (Shiraiwa et al., 1981).

In Chlamydomonas reinhardtii, the induction of periplasmic carbonic anhydrase has been studied in some detail. It is synthesized on cytoplasmic ribosomes and is glycosylated to form a 37-kDa polypeptide, prior to export into the perplasmic space (Coleman and Grossman, 1984; Toguri et al., 1984; Coleman et al., 1985). Translationally active RNA is only observed in air-adapted cells, so it has been suggested that the induction of this RNA species and its translation involves the activation of gene(s) encoding for carbonic anhydrase, or a posttranscriptional modification that converts translationally inactive messenger to active messenger (Toguri et al., 1984; Coleman et al., 1985).

Other changes also occur in protein synthesis during the adaptation of C. reinhardtii to low CO_2. There is a transient decrease in the synthesis of several proteins, including Rubisco, and a transient increase in the synthesis of two polypeptides during the initial stages of adaptation (Coleman et al.,

1985). Similarly, in *Chlamydomonas segnis,* three polypeptides have been identified as being specific to low-CO$_2$ adapted cells, while a number of others may only be present in high-CO$_2$ adapted cells (Badour and Kim, 1986). The significance of these other protein changes remains to be determined.

The adaptation process in *Chlamydomonas reinhardtii* is linked to the cell cycle in synchronous cultures kept under a light and dark regime (Marcus *et al.,* 1986). The activity of the CO$_2$-concentrating mechanism, including the ability to accumulate Ci and the carbonic anhydrase levels, peaked after 6–9 hr in the light and reached a minimum after 6–9 in the dark. Thus, at the end of the dark period, the cells had the properties of high-CO$_2$ cells even though they were still bubbled with air. Whether these oscillations occur in continuously illuminated cultures, kept under steady-state conditions of either synchronous or nonsynchronous growth, is unknown. However, as the periodic changes of other proteins are eliminated under growth with constant environmental conditions (Rollins *et al.,* 1983), it is likely that these oscillations are a phenomenon associated with the periodic stress of light changes.

In the cyanobacterium *Anacystis nidulans* the changes induced by low CO$_2$ in the polypeptide composition of cell walls, cytoplasmic membranes, thylakoid membranes, and cytosol fractions have been extensively studied (Omata and Ogawa, 1985, 1986). The major alteration was a marked increase of a 42-kDa polypeptide in the cytoplasmic membrane, with a linear relationship being found between the amount of this polypeptide and Ci-accumulating ability of the cells. During the process of adaptation, the synthesis of most proteins, other than this polypeptide, was greatly reduced, and no significant changes were observed in the amount of other polypeptides in the cytoplasmic membranes or other cell fractions. These results suggest that this 42-kDa polypeptide is involved in Ci transport.

B. Factors Affecting Induction

1. Inorganic Carbon

The relationship between the induction process and the concentration of the dissolved species of Ci during growth has been poorly studied. The general notion has existed that high-affinity cells are produced when cells are grown with air bubbling, while low-affinity cells are obtained when the culture is aerated with 1–5% CO$_2$ in air. Recently, however, it has become apparent that the induction of the CO$_2$-concentrating mechanism may occur at much lower Ci concentrations than previously realized.

a. Cyanobacteria. For the cyanobacterium *Anacystis nidulans* (*Synechococcus* PCC6301), it has been shown that what was previously termed a high-CO$_2$ cell [$K_{0.5}$(Ci) = 0.2–1.0 m*M*] is produced when there is as little as

2–4 mM Ci species in the culture medium (Miller *et al.*, 1984b; Mayo *et al.*, 1986; Badger and Gallagher, 1987). The highest affinity or low-CO_2 cells [$K_{0.5}$(Ci) = 2–10 μM] are produced when Ci levels are below 50 μM. This response is independent of pH between pH 6 and 10 (Mayo *et al.*, 1986) and indicates that the cells are responding to the total Ci concentration rather than the CO_2 concentration *per se*. If the Ci concentration during growth is maintained between these extremes, then cells showing intermediate $K_{0.5}$(Ci) values are obtained (Mayo *et al.*, 1986; Badger and Gallagher, 1987). The response of cyanobacteria seen to the total Ci concentration over a wide range of pH, rather than the concentration of CO_2, suggests that there is an ability to induce variable levels of HCO_3^- useage. This would be consistent with the Ci transport system being primarily a CO_2 pump with an inducible "front end" enabling HCO_3^- to be converted to CO_2.

The levels of Ci that cause adaptation of the cells to occur are concentrations that would be experienced under natural growth conditions. Thus, in dilute cell suspensions bubbled with air, where the potential rate of Ci supply to the cell exceeds the rate of photosynthesis, high Ci conditions will prevail. Conversely, when cell densities increase and the potential photosynthesis exceeds Ci supply rates, then low-Ci conditions will exist (see Turpin *et al.*, 1985). Thus the induction of the CO_2-concentrating mechanism in cyanobacteria is of great ecological importance. The production of the low-CO_2 cell type in the past by growth on bubbling air has been highly fortuitous, with low Ci concentrations arising as a conseuence of poor gas equilibration and high cell densities. In the future, more attention will have to be given to growth of cultures at Ci concentrations that are known to be adequate to produce the appropriate cell type.

b. Green Microalgae. Uncertainty exists as to the concentrations of Ci that are required to produce the extremes of cell adaptation, largely because no measurements have been made of the Ci concentrations in the medium during growth. For *Chlorella vulgaris,* it was established that the $K_m(CO_2)$ of photosynthesis of cells showed a decrease when the CO_2 concentration in the bubbling gas supply was lowered stepwise from 3% to 0.04% (Shiraiwa and Miyachi, 1985b). Values of $K_m(CO_2)$ intermediate between the extremes were obtained; however, it was unclear what the exact concentration of Ci in the medium was. For instance, when cells were grown on 0.04% CO_2 for 12 h, the Ci in the medium was reported to be 2.5 μM and the pH above 8. This indicates the problem of severe disequilibrium between the gas phase and the liquid, experienced in most batch cultures, but generally not quantified.

When careful measurements are made of the response of cell physiology to known, steady-state, concentrations of CO_2 and Ci, it is likely that a similar situation will be discovered for cyanobacteria. High-CO_2 physiology will prevail in cells adapted to growth at Ci levels close to equilibrium with air, while low-CO_2 physiology will be produced when Ci supply is limiting and Ci decreases to low concentrations. It is also likely that cells showing intermediate physiology will also exist if Ci is maintained at levels between those that produce the adaptive extremes.

2. *Oxygen*

The rate and extent of the induction to low Ci concentrations is accentuated in 21% O$_2$ when compared with 1–2% O$_2$ with the cyanobacterium *Anabaena variabilis* (Marcus *et al.*, 1983) and the green algae *Chlorella* sp. (Ramazanov *et al.*, 1984) and *Chlamydomonas reinhardtii* (Spalding and Ogren, 1982). This has led to the suggestion by each of these groups that the signal triggering the induction of the CO$_2$-concentrating mechanism may be associated with the production of photorespiratory metabolites. Indeed, Marcus *et al.* (1983) and Kaplan (1985) have specifically suggested that phosphoglycolate may be the triggering metabolite sensed by the cell, noting that this compound builds up to high levels following transfer of high-Ci cell to low-Ci conditions. This theory, involving photorespiration as a trigger for induction, would predict in the extreme that if oxygen effects on Rubisco were eliminated completely, through removal of oxygen, induction might not occur at all. This experiment is impossible to do, as the oxygen evolution from photosystem 2 would maintain a finite level of internal oxygen. However, it would seem for cyanobacteria that adaptation to low Ci in the absence of O$_2$ would still have advantages, due to the low affinity of Rubisco for CO$_2$ in these organisms (Badger, 1980; Andrews and Abel, 1981). More work, carefully measuring the interaction between the concentration of Ci and O$_2$, needs to be carried out before more definitive conclusions can be drawn.

3. *The Requirement for Photosynthesis*

The adaptation to low Ci concentrations appears to occur only when some or all of the photosynthetic reactions are functioning. If cells are maintained in the dark while the Ci is lowered, by changing the bubbling gas, then little change from the high-Ci state is observed in *Anabaena variabilis* (Marcus *et al.*, 1982), *Chlorella* sp. (Semenenko *et al.*, 1979; Shiraiwa *et al.*, 1981), *Chlamydomonas reinhardtii* (Spalding and Ogren, 1982; Spencer *et al.*, 1983), and *Scenedesmus* sp. (Semenenko *et al.*, 1979). While light is thus required for the adaptation to occur, there seems to be species variability of the absolute response to light intensity. A low light intensity, around the light compensation point, was insufficient to initiate the adaptation to low Ci in *Chlamydomonas reinhardtii* and some species of *Chlorella* and *Scenedesmus* (Spalding and Ogren, 1982; Semenenko *et al.*, 1979), but was adequate to cause this to occur in *Chlorella vulgaris* (Shiraiwa and Miyachi, 1983). A similar species difference was observed in the effects of DCMU on the induction process, with *Chlorella vulgaris* adapting normally while *Chlamydomonas reinhardtii* was inhibited (Shiraiwa and Miyach, 1983; Spalding and

Ogren, 1982; Spencer *et al.*, 1983). Adaptation of carbonic anhydrase activity has been studied in a number of *Chlamydomonas* mutants defective in various photosynthetic components ranging from Rubisco to photophosphorylation (Spalding and Ogren, 1982; Spencer *et al.*, 1983). All mutants showed an inability to induce CA activity when the Ci concentration was reduced, which suggests, in this species, that the regulation of the induction process is coordinated with the photosynthetic process. The suggestion that adaptation may be triggered by the increase of photorespiratory metabolites, such as phosphoglycolate (see Section VI,B,2), would provide a mechanism for this coordination.

A possible artifact in the above experiments is the fact that the Ci level during the induction will vary with the potential rate of photosynthesis of the cell culture. Thus cells in the dark or at the light compensation point will experience a Ci concentration that is roughly in equilibrium with the gas being bubbled through the solution, while cultures at high light will establish a considerable disequilibrium between the solution and the gas phase. If adaptation does occur at Ci concentrations ranging from air equilibrium levels downward, as is the case for cyanobacteria (see Section VI,B,1,a), then in the dark, at low light, or when photosynthesis is inhibited, the Ci concentrations in air bubbled media may not fall sufficiently low to cause the adaptation to occur.

4. Nitrogen Limitation

It has been observed with *Chlorella emersonii* (Beardall *et al.*, 1982), that the CO_2-concentrating mechanism can be induced by a nitrogen nutrition limitation under conditions of high CO_2, as well as by Ci limitation. The mechanism by which this induction occurs is difficult to envisage given the hypotheses presented above; however, the significance of such a change has been discussed by the authors. The induction is suggested to be a response to minimize the interference between the reassimilation of NH_3, produced in photorespiratory metabolism, and the primary assimilation of NH_4, from the external medium. This assumes that both processes make use of a rate-limiting level of glutamine synthetase within the cell. As limiting nitrogen caused the induction in the presence of 5% CO_2, a level sufficient to suppress all photorespiration, it is hard to see how any photorespiratory release of nitrogen would have occurred at all. In fact, it is hard to see how the induction of the CO_2-concentrating mechanism would have produced any benefits not already conferred by the external high CO_2.

C. Adaptation to High CO_2

A high-Ci adapted cell is generally viewed as deficient in most of the components of the CO_2-concentrating mechanism. Whether they are completely absent, however, is probably different between cyanobacteria and green algae, primarily due to the properties of their Rubiscos. High-Ci green microalgae show $K_{0.5}(CO_2)$ values for photosynthesis that are not much less than the *in vitro* $K_m(CO_2)$ of their Rubisco. Thus it is reasonable to suggest that these cells do not have a significant ability to concentrate CO_2. For cyanobacteria, however, the situation is different, with high-Ci cells showing a $K_{0.5}(CO_2)$ at least one-tenth that of their Rubisco and an ability to concentrate Ci internally (Kaplan *et al.*, 1980; Badger and Gallagher, 1987). This suggests that high-Ci grown cyanobacteria retains some ability to concentrate CO_2. This can be rationalized if the high-Ci cell state is seen as the form that the cell adopts when the Ci concentration is near equilibrium with air (see Section VI,B,1,a). Under these conditions when CO_2 is about 10 μM, and because of the lower affinity of cyanobacterial Rubisco for CO_2 [$K_m(CO_2)$ = 100–150 μM versus 20–30 μM for green algae] (Badger, 1980; Andrews and Abel, 1981; Berry *et al.*, 1976; Tsuzuki *et al.*, 1985), cyanobacteria would still be severely CO_2-limited without some CO_2-concentrating capacity.

The process of adaptation to high Ci concentrations has been studied much less frequently than the reverse transition; however, where it has been reported, the change from a low-Ci cell to a high-Ci cell takes considerably longer. For example, in *Chlamydomonas reinhardtii,* the carbonic anhydrase activity falls over a period of 24 h following the switch from air to 4% CO_2 (Yang *et al.*, 1985), whereas induction to low CO_2 took only 4–6 h. It is reasonable to suspect, then, that the process of deadaptation may occur through the lack of any signal to renew or sustain the changes that were necessary to produce the low-Ci cell, rather than being due to a specific signal to cause the reverse transition to be initiated. Indeed, Kaplan (1985) has suggested that it would be difficult for a low-Ci cell to sense a shift to high-Ci conditions, as the internal Ci would already be high.

The change to a high Ci cell may be viewed as being due to the loss of those components that allowed the CO_2-concentrating mechanism to operate effectively. This would include the loss of carbonic anhydrase and Ci pumping proteins, photorespiratory enzymes, and perhaps components of the CO_2 diffusion barrier. As we have no idea of the turnover rate of these components in the cell, it is hard to estimate the rate at which these components would be broken down once their synthesis was stopped. It is possible to envisage, however, how this process of protein turnover, combined with the dilution that will occur when components are divided between daughter cells upon cell division, would lead to a gradual decline in the components of the CO_2-concentating mechanism over a period of 24 h.

It has been suggested that the synthesis of components of the CO_2-concentrating mechanism may cease at some particular stage of the cell cycle and may not be renewed unless the continuation of low-Ci conditions is experienced (Kaplan, 1985). Consistent with this hypothesis is the observation that synchronous cultures of *Chlamydomonas reinhardtii* grown continuously on air show oscillations between low-Ci properties in the light and high-Ci properties at the end of the dark period and that these oscillations persist if illumination is continued into the scheduled dark period (Marcus *et al.*, 1986). Studies on *Chlamydomonas segnis* (Tan and Badour, 1983) also suggest that adaptation to low Ci depends on the stage of the cell cycle. Whether there is a specific relationship between the cell cycle and the ability to renew or cease synthesis of components of the CO_2-concentrating mechanism remains to be confirmed and to be extended to other species.

VII. EVIDENCE FOR THE CO_2-CONCENTRATING MECHANISM IN OTHER AQUATIC PHOTOTROPHS

The cyanobacteria and the microalgae are the only aquatic phototrophs that have been shown to unequivocally possess a functional CO_2-concentrating mechanism. This stems solely from the fact that only in these organisms is it possible to directly measure the accumulation of Ci and CO_2 within the cell, by a technique such as silicon-oil centrifugation. Despite this lack of hard physical evidence for the mechanism, other classes of aquatic photosynthetic organisms also show physiological features that would predict that a form of CO_2-concentrating mechanism was functionally active. A universal property of these organisms is the ability to develop the capacity to directly utilize HCO_3^- as a source of carbon for photosynthesis (see Lucas, 1983) and achieve an apparent $K_m(CO_2)$ that is considerably lower than that predicted by the kinetics of their Rubisco and the resistance to diffusion of CO_2 into the cell. Reduction in the characteristics associated with photorespiration, such as the oxygen sensitivity of photosynthesis and the CO_2 compensation point, are also consistent with the notion of an elevated internal CO_2 concentration.

A. Freshwater Angiosperms and Giant Algae

It has long been recognised that the leaves of aquatic angiosperms, such as *Potamogeton, Elodea* (Eigeria), and *Valisneria* and the giant algal cells of the Characeae, are able to assimilate HCO_3^- during photosynthesis (see Steemann-Nielsen, 1960). This process of HCO_3^- utilization has been extensively studied in these organisms, and there has been a significant development of the models relating this process to the fluxes of H^+ and OH^- ions between the external cell surface and the medium.

1. *OH^- and H^+ Fluxes*

In the light, when photosynthesis is occurring, the leaves of many aquatic angiosperms and the internodal cells of the giant algae develop discrete acid and alkaline zones, which have been intimately linked to the utilization of

HCO$_3^-$. For the angiosperms, this is manifested by the development of polar leaf surfaces, with the lower side being acid with respect to the external medium, while the upper is alkaline. In the Characeae, the phenomenon is manifest as alternate acid and alkaline bands along the internodal cell.

If CO$_2$ is consumed directly from the external medium, then there is no net exchange of OH$^-$ and H$^+$ ions, either internally or externally. However, HCO$_3^-$ utilization during photosynthesis must inevitably lead to an approximately stoichiometric net production of an OH$^-$ ion for each CO$_2$ consumed by Rubisco. This has generally been seen as the result of HCO$_3^-$ entering the cell, where it subsequently consumes a proton in its conversion to CO$_2$. The cell must deal with the net production of OH$^-$ internally, to maintain internal pH control, and this could be achieved by either OH$^-$ efflux or H$^+$ influx. As bicarbonate utilisation generally occurs at pH > 8, where the external proton concentration is low, an OH$^-$ efflux mechanism is favored (see Lucas, 1983). This OH$^-$ efflux in the light would thus be responsible for the production of the alkaline regions. An alternative mechanism for OH$^-$ generation involves the operation of an H$^+$-extruding ATPase, which enables HCO$_3^-$ to be taken up by either a H$^+$/HCO$_3^-$ symport mechanism or through an acid-catalyzed conversion of HCO$_3^-$ to CO$_2$, external to the cell. In either case, the H$^+$ efflux leads to the net production of OH$^-$ ions internally.

2. Models of HCO$_3$ Utilization

a. Aquatic Angiosperms. Currently, the latest model to gain widest acceptance in explaining the mechanism of HCO$_3^-$ utilization is that developed by Prins *et al.* (1982) (see Fig. 5). This model clearly rejects the previously held notions that HCO$_3^-$/H$^+$ symport is responsible for HCO$_3^-$ transport into the cell. Instead, it argues that HCO$_3^-$ does not enter the cell itself, but is converted to CO$_2$ externally by means of the acid environment at the lower leaf surface. This CO$_2$ then diffuses passively across the plasma membrane and chloroplast envelope to the site of carboxylation. The acidification on the lower surface of the leaf is proposed to be the result of an H$^+$-translocating ATPase, while alkalization of the upper surface is due to an OH$^-$ exretion mechanism. The associated K$^+$ transport through the leaf occurs predominantly through the cell walls (left-hand side of Fig. 5) and is driven by the transleaf electrical potential difference. A symplastic pathway for K$^+$ movement is also envisaged (right-hand side of Fig. 5), energized by the proton motive force at the upper and lower surfaces. At present, the available experimental data from *Elodea* and *Potamogeton* are consistent with the model presented in Fig. 5. Whether other aquatic angiosperms have adopted the same stategy remains to be seen.

b. Characeae. For *Chara* species, there are two models of HCO$_3^-$ utilization that have a degree of acceptance. A hypothesis similar to that already described for aquatic angiosperms was proposed by Ferrier (1980) and

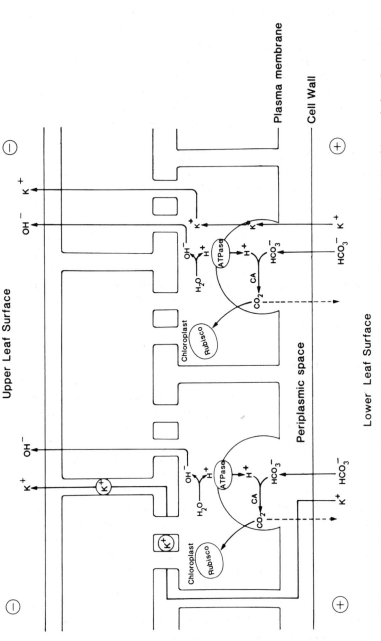

Fig. 5. A model for photosynthetic bicarbonate utilization in *Elodea* and *Potamogeton* (after Prins *et al.*, 1982, with permission). Protons are pumped into the periplasmic space, probably by an ATPase, and these are used to convert HCO_3^- to CO_2. This conversion process may be aided by the presence of extracellular carbonic anhydrase (CA) activity. The CO_2 thus produced then diffuses into the cell and chloroplast, and some proportion of it will efflux back into the medium. The possibility that this CO_2 production is confined to a localized region such as plasma membrane invaginations is also indicated. The internal production of OH^- ions, resulting from H^+ pumping, is dealt with by a transleaf efflux mechanism that releases them on the opposite surface. Two possible pathways are given for K^+ transport: in the left side through the cell wall, and on the right side via the symplast. Further explanation is given in the text.

Walker *et al.* (1980). This postulated that active H^+ efflux (via an H^+-ATPase in the acid bands) could drive the external conversion of HCO_3^- to CO_2 [Fig. 6(a)], while the internally generated OH^- is excreted in the alkaline band. The CO_2 that is formed next to the plasma membrane is envisaged to diffuse passively into the cell.

An alternative to this hypothesis has been proposed by Lucas (1985). Central to this model is the notion that direct HCO_3^- uptake occurs as the result of H^+/HCO_3^- symport mechanism, located in the plasma membrane of the acid regions [Fig. 6(b)]. Other aspects of this scheme relating to the H^+-ATPase and OH^- excretion are the same as in the first model.

It has been suggested that carbonic anhydrase in the periplasmic space has a role in enabling the CO_2 production process to operate effectively at the pH of the acid band (Price *et al.*, 1985), by allowing the CO_2 production rate to be greatly enhanced. Both models have incorporated the presence of periplasmic CA as being beneficial (Fig. 6); however, it must be questioned whether this would be so for the H^+/HCO_3^- symport model. In this case, rather than acting as a proposed savenging mechanism for CO_2 leaking out of the cell (Lucas, 1985), it is much more likely that the presence of CA would allow HCO_3^- conversion to CO_2, and its subsequent leakage back into the external solution, to uncouple H^+ efflux from HCO_3^- uptake.

It has been difficult to design experiments to distinguish between the two mechanisms. A crucial parameter that needs to be known is the pH of the acid zone next to the plasmamembrane. If there is a sufficient pH drop from the external medium to the cell surface, then the CO_2 generation model can be justified (see Walker, 1983, 1985). At present, the pH measured next to the exterior of the cell wall with microelectrodes seems insufficient to fully support this hypothesis, even with the presence of carbonic anhydrase to speed up the interconversion of HCO_3^- to CO_2 (Walker, 1985). However, there is no positive evidence to support the H^+/HCO_3^- symport mechanism either.

3. Spatial Organization of Pumping

The spatial separation of acid and alkaline zones seems to be important to allow efficient HCO_3^- utilization in both the aquatic angiosperms and the giant algae. This is presumably because the physical separation of these zones would allow the generation of an acid region without the possibility that OH^- efflux will inefficiently uncouple H^+ efflux from HCO_3^- usage. Within the acid zones there may also be futher morphological development that will allow the maintenance of higher H^+ concentrations. The cells of the lower epidermis of *Elodea* and *Potamogeton* leaves are reported to be transfer cells, in which the plasmalemma surface area is greatly increased by the presence of cell wall invaginations (Falk and Sitte, 1963; Pate and Gunning, 1972). It has been proposed that the H^+-ATPase pumps are located in these plasmalemma invaginations and produce localized acid environments within

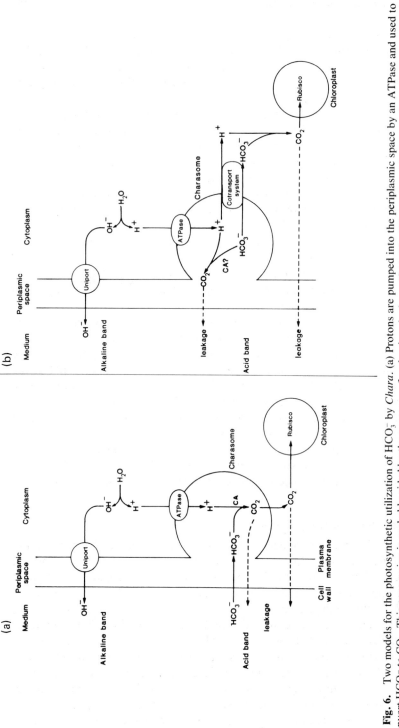

Fig. 6. Two models for the photosynthetic utilization of HCO_3^- by *Chara*. (a) Protons are pumped into the periplasmic space by an ATPase and used to convert HCO_3^- to CO_2. This conversion is probably aided by the presence of carbonic anhydrase (CA). The CO_2 produced is proposed to diffuse into the cell and chloroplast, as well as leak back into the external medium. The OH^- ions produced internally are removed from the cell by an efflux pump located in the alkaline band. (b) This model is the same as (a) except that HCO_3^- rather than CO_2 primarily enters the cell by means of an H^+/HCO_3^- cotransport system. Carbonic anhydrase is depicted as being deleterious to this process by promoting inefficient leakage of CO_2 back to the external medium and thus uncoupling H^+ efflux from HCO_3^- uptake. In both models the H^+-ATPase is shown to be located in the invaginations of the charasome.

them (Prins *et al.*, 1982). In the Characeae, comparable membrane invaginations have been observed to be specifically associated with the acid regions and correlate in density with the ability to utilize HCO$_3^-$ (Price *et al.*, 1985). These invaginations have been termed charasomes or plasmalemmasomes, and have been cytochemically stained to show the presence of ATPase activity (Price and Whitecross, 1983). The charasome has been further implicated as a structure to localize carbonic anhydrase activity so that its participation in HCO$_3^-$ utilization may be optimized (Price *et al.*, 1985; Walker, 1985). The role of invaginations in both the Characeae and aquatic angiosperms is incorporated into Figs. 5 and 6.

While the development of invaginations within the acid region may increase the efficiency of HCO$_3^-$ utilization, it appears not to be absolutely essential. It has been reported that membrane foldings are not always present in leaves of *Potamogeton* and *Elodea* capable of HCO$_3^-$ utilization (Prins *et al.*, 1982). Similarly, while the charasome develops in members of the genus *Chara,* certain species of the genus *Nitella* appear to lack such invaginations, even though they exhibit similar acid banding and show at least some ability to use HCO$_3^-$ for photosynthesis (Lucas, 1985).

4. The Ability to Concentrate CO$_2$

While it is obvious that the mechanisms proposed for HCO$_3^-$ utilization in freshwater aquatic angiosperms and the Characeae can confer the ability to concentrate CO$_2$ internally, there is little evidence to indicate the extent to which this is achieved. Several properties of photosynthesis suggest that under some conditions a considerable degree of CO$_2$ accumulation must occur. Aquatic angiosperms are often less inhibited by oxygen than are terrestrial C$_3$ plants under equivalent conditions (Chollet and Ogren, 1975; Bowes, 1985), and their CO$_2$ compensation points may also be considerably reduced (Bowes, 1985). Indeed, the Rubisco from members of the Characeae and aquatic angiosperms also differs from that of C$_3$ land plants, in that the K_m(CO$_2$) is almost twice as high (Yeoh *et al.*, 1981) and in this respect resembles the enzyme from green microalgae and C$_4$ plants. This would strongly suggest that the ability to concentrate CO$_2$, to levels similar to those found in the microalgae and C$_4$ plants, exists in these aquatic macrophytes. A model of photosynthesis has been developed for the aquatic angiosperm *Eigeria densa,* which incorportes an analysis of the effects of an HCO$_3^-$ pumping mechanism on photosynthesis and the internal CO$_2$ concentration (Laing and Browse, 1985). This theoretical analysis demonstrates that internal CO$_2$ elevation can be achieved by a limited amount of HCO$_3^-$ pumping, although it additionally suggests that the internal release of photorespiratory CO$_2$ can also achieve this, particularly when the unstirred layer diffusive resistance surrounding the cell is high.

It is possible that several aspects of the HCO$_3^-$ utilization mechanism in freshwater aquatic macrophytes remain to be discovered, and that these aid the effectiveness of this system in concentrating CO_2. The current schemes outlined in Figs. 5 and 6 acknowledge that CO_2 may only enter the cell by passive diffusion through the plasmamembrane. If an active CO_2 transport system were to exist in the acid zones, similar to that discovered in cyanobacteria (see Section V,A,1), then the effectiveness of the CO_2 generation hypothesis would be greatly enhanced. A similar increase in efficiency of concentrating ability would also be achieved if a barrier existed to prevent the free diffusion of CO_2 out of the cell. In their model of photosynthesis, Laing and Browse (1985) have shown that unstirred layer resistance of up to 6×10^{-6} m sec^{-1} may have this effect; however, it also reduces the rate of supply of substrate to the pumping mechanism. A barrier that preferentially reduced CO_2 leakage compared to Ci supply would be greatly beneficial.

5. Induction of HCO$_3^-$ Utilization

Some aquatic angiosperms and members of the Characeae show fluctuations in their capacities to utilize HCO$_3^-$ for photosynthesis, depending on the environmental conditions in which they are currently growing. For the aquatic angiosperms *Myriophyllum* and *Lymnophila sessifolia,* high- and low-photorespiration forms have been found to occur in the winter and summer growth months, respectively. The change to the low-photorespiration state is correlated with an increase in the ability to use HCO$_3^-$ and a doubling of the carbonic anhydrase activity associated with the plants (Salvucci and Bowes, 1981, 1983a,b; Bowes, 1985). These changes have been suggested to be associated with the induction of a HCO$_3^-$ utilizing mechanism that can elevate internal CO_2. The induction of this ability during the summer months correlates with periods of rapid photosynthesis and elevated temperatures; when the Ci supply becomes limiting, the O_2 levels are elevated, and the potential for photorespiration is maximal. Much more dramatic changes in HCO$_3^-$-utilizing ability have been reported for *Chara corallina* (Price *et al.,* 1985). Laboratory cultures grown at pH > 8 showed much greater potential to utilize HCO$_3^-$, compared with those grown at pH 6.8. This suggested that the availabiltiy of HCO$_3^-$ in the medium had a great affect on the induction of a HCO$_3^-$ utilizing mechanism. At pH8, increasing the total Ci 20-fold during growth almost completely eliminated the capacity for HCO$_3^-$ to support photosynthesis. This indicates that if CO_2 supply to the cell can be maintained at a rate sufficient to sustain a relatively high internal CO_2 concentration, then HCO$_3^-$ utilization may not be induced. Similar adaptation has also been shown for *Elodea, Vallisneria,* and *Ceratophyllum* species (Price, 1985).

6. The Addition of an Internal C$_4$ Acid Cycle

While in general it has been shown that the central aspects of C_3 photosynthesis are operational in aquatic macrophytes, including the PCR and PCO cycles, it has become apparent that certain aquatic plants have mechanisms that resemble the C$_4$ and crassulacean acid metabolism (CAM) pathways of terrestrial plants (see Bowes, 1985; Raven *et al.,* 1985).

A type of C$_4$ acid metabolism has been established for the freshwater aquatic angiosperm *Hydrilla verticillata* (Ascencio and Bowes, 1983; Salvucci and Bowes, 1983a) and in the marine Ulvophycean *Udotea* sp. (Reiskind *et al.,* 1984; Bowes, 1985). In both species the internal production of the C$_4$ acid malate appears to be the result of elevated phosphoenolpyruvate carboxylase activity. This C$_4$ acid production may be used to elevate the internal CO$_2$ concentration by the mechanism proposed for *Hydrilla* (Bowes, 1985). Here the malate would pass to the chloroplast for decarboxylation by NADP-malic enzyme, and the CO$_2$ released would be refixed by Rubisco. The pyruvate, formed from this decarboxylation, would be reconverted to phosphoenolpyruvate by pyruvate Pi dikinase in the chloroplast and this would then return to the cytoplasm to participate in C$_4$ carboxylation. This CO$_2$ release within the cell, if linked to a mechanism to reduce back-diffusion of CO$_2$ from the chloroplast or the cell to the external medium, could readily result in the elevation of stromal CO$_2$. *Hydrilla* has been shown to utilize HCO$_3^-$, particularly at alkaline pH where the external Ci is reduced (Bowes, 1985; Spence and Maberly, 1985). Thus, it seems reasonable to suggest that this modified C$_4$ acid cycle may be supplied with substrate by an HCO$_3^-$-utilizing mechanism, located on the plasmalemma. In this scheme, the internal phosphoenolpyruvate (PEP) carboxylase may serve to reduce the cytosolic level of HCO$_3^-$ and CO$_2$, to allow the entry of Ci to proceed more efficiently.

A significant level of CAM-like metabolism (contributing more than 25% of total fixed carbon) has been reported from a number of aquatic macrophytes, predominantly species of the Isoetid family (Raven *et al.,* 1985). The designation of CAM is largely based on a large diurnal fluctuation in titratable acidity and malic acid of up to 200 μeq (g fr wt)$^{-1}$, which is of the order of those observed in terrestrial CAM plants. The most extensively investigated aquatic CAM plant, *Isoetes howellii,* employs PEP carboxylase for malate production and seems to use PEP carboxykinase as its decarboxylase (Keely and Busch, 1984). Decarboxylation of the malate occurs during the daytime, presumable resulting in elevated internal CO$_2$. It has been suggested that *Isoetes* is similar to CAM plants in that it fixes a large proportion of external CO$_2$ via the PCR cycle, after the internal acid has been exhausted (Keeley, 1981). However, no O$_2$ inhibition of daytime photosynthesis can be detected, even when the leaves are split open to eliminate the possibility of lacunal gas refixation (Keeley and Bowes, 1982). Thus it has been suggested that substantial C$_4$ acid metabolism in the light, similar to that found in *Hydrilla,* cannot be ruled out (Bowes, 1985), especially as malate production in the light does occur (Keeley, 1983). More detailed biochemical ellucidation of this mechanism is necessary before a more complete assessment of its functioning can be made. It is interesting that *Isoetes howellii* (lacustris) seems to be restricted to growth in regions of low alkalinity and is not able to use HCO_3^- (Spence and Maberly, 1985). Thus, in this case, the C_4 metabolism may not be linked to an HCO_3^--utilizing mechanism.

B. Marine Algae and Angiosperms

The ability to utilize HCO_3^- for photosynthesis is widespread among the marine algae and angiosperms (Beer *et al.*, 1977; Beer and Eshel, 1983; Sand-Jensen and Gordon, 1984; Kerby and Raven, 1985; Bidwell and McLachlan, 1985). It is not clear, however, whether this is always correlated with physiological characteristics that would suggest that an internal elevation of CO_2 was achieved by this process.

1. Marine Microalgae

Two species of marine microphototrophs, the cyanobacteria *Synechococcus* sp. (Badger and Andrews, 1982) and the green alga *Dunaliella salina* (Zenvirth and Kaplan, 1981), have been conclusively shown to be able to accumulate CO_2 internally as a result of active Ci transport. These species, when grown on air, show photosynthetic characteristics typical of the green algae and cyanobacteria discussed previously (see Section V), having a high photosynthetic affinity for HCO_3^- and CO_2, and a very low CO_2 compensation point. A few other marine microalgae (*Dunaliella tertiolecta, Thalassiosira pseudonana,* and *Porhyridium* sp.) have been tested for their CO_2 compensation points and the effect of O_2 on photosynthesis, and these parameters have been found to be much reduced compared to what would be expected in the absence of a CO_2 concentrating mechanism (Lloyd *et al.*, 1977; Kerby and Raven, 1985). This limited information would suggest that many marine microalgae may have the potential to accumulate CO_2 by a mechanism that is similar to that already discussed for the cyanobacteria and green algae.

2. Marine Macroalgae

A significant number of marine macroalgae (seaweeds) have been shown to possess a photosynthetic physiology that would suggest that CO_2 accumulation was possible by these organisms, although the CO_2-fixation pathway appears to resemble that of C_3 plants rather than C_4 metabolism (Kremer, 1980, 1981). The great majority tested have been shown to have CO_2 compensation concentrations that are lower than would be predicted based on the diffusive entry of CO_2 and Rubisco activity (see Kerby and Raven, 1985; Brown and Tregunna, 1967; Coughlan and Tattersfield, 1977; Raven *et al.*, 1982; Bidwell and McLachlan, 1985; Colman and Cook, 1985). Furthermore, oxygen inhibition of photosynthesis has also been shown to be greatly reduced or absent in the majority of species tested, providing that the Ci concentration is the one that is normally in equilibrium with air (see Kerby and Raven, 1985; Bidwell and McLachlan, 1985). However, there are a number of notable exceptions such as *Codium* and *Sargassum* (Bowes, 1985).

In a number of these macroalgae showing reduced CO$_2$ compensation points, photosynthesis has been found to be saturated with Ci at concentrations around 2 mM, the concentration that exists in the ocean (Beer and Eshel, 1983; Sand-Jensen and Gordon, 1984; Bidwell and McLachlan, 1985). Under these conditions, the CO$_2$ concentration is only 10 μM and would be far from sufficient to support CO$_2$-saturated photosynthesis, considering the total diffusive resistance into the chloroplast (Kerby and Raven, 1985) and the kinetic properties of Rubisco, which in at least one red algal macrophyte shows a $K_m(CO_2)$ in the range of C$_4$ plants (Colman and Cook, 1985). Indeed, in the red alga *Chondrus crispus*, it has been unequivocally shown that over 90% of photosynthesis is supported by HCO$_3^-$ uptake (Brechignac *et al.*, 1986). While these physiological characteristics are consistent with the operation of a CO$_2$-concentrating mechanism based on HCO$_3^-$ utilization, the magnitude of this concentration is not clear. Studies on the production of photorespiratory metabolites during photosynthesis generally show evidence for the operation of some RuBP oxygenase activity (see Burris, 1980). This would suggest that intracellular CO$_2$ may not be high enough to completely inhibit this activity.

3. Marine Angiosperms

In the aquatic angiosperms (seagrasses), the studies of photosynthetic properties have been less extensive. Many of the HCO$_3^-$ users show photosynthetic responses that are saturated by air-equilibrium values of Ci (Beer *et al.*, 1977; Beer and Waisel, 1979; Sand-Jensen and Gordon, 1984), although photosynthesis can be increased at lower pH where the CO$_2$ concentration may be raised to significantly higher levels. The PCR cycle seems to be the major pathway for CO$_2$ fixation (Andrews and Abel, 1979; Benedict *et al.*, 1980; Beer *et al.*, 1980), although there have been some reports of the formation of C$_4$ acids during photosynthesis (Benedict and Scott, 1976). There is also evidence for the operation of photorespiration during photosynthesis in seawater, from the detection of photorespiratory metabolites during pulse labeling experiments (Andrews and Abel, 1979). Seagrasses show $\delta^{13}C$ isotope composition values similar to those found in C$_4$ terrestrial plants, and the reasons for this appear to include (1) the diffusional limitations placed on photosynthesis, (2) the existence of an internal Ci pool, which may be recycled, and (3) the supply of Ci to the internal pool by the transport of Ci (Andrews and Abel, 1979; Abel, 1984). All these observations support the concept that HCO$_3^-$ utilization in seagrasses may lead to elevated internal CO$_2$ concentrations; however, as with seaweeds, the concentration achieved may not be sufficient to completely eliminate the photorespiratory effects of O_2.

4. Mechanisms of HCO$_3^-$ Utilization

Little is known specifically about the mechanism of HCO_3^- utilization by marine phototrophs. As discussed in Section VII,B,1, the microalgae and cyanobacteria probably possess mechanisms similar to those discussed for cyanobacteria and green microalgae in general. Further investigations of a wider range of marine microorganisms are necessary before these generalizations can be verified. For the marine macrophytes, postulations about the mechanism can only come from the models developed for freshwater macrophytes (see Section VII,A,2). Whether acid efflux models form a basic part of this mechanism remains to be determined, but it is likely that fluxes of other ions such as Na^+ and Cl^- may be intimately linked to active Ci utilization. It also seems apparent that carbonic anhydrase does participate in HCO_3^- utilization, as CA inhibitors significantly inhibit photosynthesis and increase the level of O_2 inhibition (Bidwell and McLachlan, 1985; Bowes, 1985).

VIII. THE ECOLOGICAL IMPORTANCE OF THE CO$_2$-CONCENTRATING MECHANISM

The physiological significance of the CO_2-concentrating mechanism was discussed in Section II, and it is clear that the primary benefit of this mechanism is to increase the affinity of photosynthesis for external Ci. A secondary benefit would be to open up the possibility of utilizing limiting resources of nitrogen and light more efficiently (see Section III). When we examine the Ci environments of the aquatic organisms discussed in the previous sections, it can be concluded that, generally, the possession of a CO_2-concentrating mechanism (or the ability to utilize HCO_3^-) would allow the organisms to grow more competitively in situations where the supply of CO_2 for photosynthesis, by passive diffusion from the medium, would otherwise severely limit their growth.

A. Environmental Variation in CO$_2$ Supply

For aquatic organisms, the supply of CO_2 to the cell surface is determined by the concentration of CO_2 in the external solution, the boundary-layer conductance, and the extent to which HCO_3^- diffusion may also contribute to CO_2 supply. Each one of these supply factors can vary, depending on the physicochemical nature of the aquatic environment and the morphological and physiological adaptation displayed by each particular organism. An examination of the natural variation found in these parameters strengthens the conclusion that the CO_2-concentrating mechanism is of greatest ecological significance in environments where the CO_2 supply is most limiting.

1. Variation in CO_2 and HCO_3^- Concentration

   a. Freshwater Habitats. In freshwater environments, the levels of CO_2, HCO_3^-, total Ci, and pH can vary dramatically, depending largely on the biological activity of the photosynthetic organisms that inhabit them (Spence and Maberly, 1985; Talling, 1985; Adams, 1985). In water bodies of low photosynthetic activity, the rates of CO_2 fixation are probably always less than the supply from the atmosphere, so neither carbon depletion nor biologically induced increases in pH are likely to occur. These waters will be around neutral pH, with CO_2 and HCO_3^- concentrations being of similar magnitude. In productive lakes, however, the photosynthetic activity is high enough to exceed the rate of supply of CO_2 from the atmosphere on both a daily and seasonal basis. This leads to a depletion in the level of Ci and an increase in the pH of the water, often to values in excess of 9. Under these conditions, the levels of free CO_2 are well below air equilibrium value, and HCO_3^- and sometimes CO^{2-} are the predominant Ci species.

   The CO_2-concentrating mechanism will be most important as an adaptive mechanism for species growing in productive waters that commonly experience Ci depletion and pH is consequently high. It is here that the ability to photosynthesize at low Ci concentrations and to access the HCO_3^- pool will be of vital significance for positive growth. For microalgae and cyanobacteria in natural situations, there appears to be a gradation of abilities to grow successfully under increasing levels of CO_2 depletion and alkalinity. The cyanobacteria are most successful under conditions of maximum Ci depletion, while in the green algae, species of *Scenedesmus* appear to do better than *Chlorella* (Talling, 1985). Whether this gradation of abilities is solely related to the operation of the CO_2-concentrating mechanism remains open to question, but it is significant that cyanobacteria are able to concentrate CO_2 to higher levels than the green algae and to display higher affinities for external Ci in photosynthesis at alkaline pH. Similarly, many species of *Chlorella* seem to lack external carbonic anhydrase activity, which prevents them from utilizing HCO_3^- (see Section V,B,4).

   The high-CO_2 adapted state of the microphytes probably represents the physiological state under non-Ci-depleted conditions, when the Ci is at air-equilibrium levels. This is certainly so for the cyanobacteria, and is probably so for the green algae, although this remains to be determined (see Section VI,B,1,b). Thus the induction to the low-Ci state will occur as the Ci levels fall in the lake and the pH rises. It could be proposed that the CO_2-concentrating mechanism only has a role to play when HCO_3^- is a significant component of the Ci, as this would certainly occur at high alkalinity. But this ignores the possibility that the accumulation mechanism may use CO_2 as a substrate, which is the case for the cyanobacteria (Sectin V,A,1,b) and is probably so for green algae. This should mean that CO_2 accumulation will be able to operate under conditions of neutral pH, where the HCO_3^- concentration is much lower. It is tempting to speculate that cells growing under air equilibrium concentrations of Ci may have a reduced activity of the CO_2-concentrating mechanism, which uses only CO_2 as a substrate. As the Ci is depleted and the alkalinity rises, the total activity of the CO_2 concentrating mechanism should increase and the ability to utilize HCO_3^- will be induced. There is strong evidence for this in cyanobacteria and to a lesser extent in green algae (Section VI,B,1). Those species of green algae that are unable to grow under the extremes of Ci depletion may not possess the ability to induce HCO_3^- utilization, but they may still be able to concentrate CO_2 under conditions when CO_2 is more readily available.

It has long been realized that certain macrophytes are characteristic of lakes with high alkalinity and low Ci, while others are found only on lakes of low alkalinity and high Ci (see Spence and Maberly, 1985; Adams, 1985). These latter species show a little ability to photosynthesis at low Ci concentrations at alkaline pH and are apparently unable to utilize HCO_3^- (Spence and Maberly, 1985). In contrast, those species found at high alkalinity all show a high affinity for Ci and the ability to access the HCO_3^- pool. These observations support the assertion that the CO_2-concentrating mechanism and the ability to utilize HCO_3^- are important to survival in waters of high pH and low Ci. Whether the CO_2-concentrating mechanism in freshwater macrophytes uses HCO_3^- as its only substrate remains to be determined. However, the $K_{0.5}(CO_2)$ for photosynthesis in many freshwater macrophytes growing at neutral to acid pH appears to be equal to or greater than the *in vitro* $K_m(CO_2)$ for Rubisco, suggesting that little CO_2 accumulation is occurring under these conditions.

As for the microalgae, it appears that many macrophytes vary their ability to utilize HCO_3^- depending on the conditions in which they are grown. Thus, two species of *Elodea* and *Chara* and *Vallisneria spiralis* and *Ceratophyllum demersum* showed an ability to utilize HCO_3^- when cultured in the laboratory at pH 9.5 but lacked this capacity when grown at pH 6.8 or bubbled with 5% CO_2 in air (Price, 1985; Price *et al.*, 1985). Thus it is possible that many macrophytes growing in waters of low alkalinity and air-equilibrium [Ci] do not possess the ability to utilize HCO_3^-. However, as the water body becomes more productive, perhaps with the onset of summer, and the pH rises and Ci falls, then HCO_3^- utilization and CO_2-concentrating ability will be induced.

b.   The Marine Habitat. Unlike the feshwater environment, the CO_2, Ci, and pH do not show large fluctuations in the ocean. Seawater has a high and constant HCO_3^- concentration (about 2 mM), a pH close to 8.2, and a CO_2 concentration in equlibrium with air (14 μM at 15°C). Thus the HCO_3^-/CO_2 ratio is constant at about 150 (Stumm and Morgan, 1970). Under these conditions, the supply of CO_2 to a photosynthetic organism will be similar to that in a low-alkalinity lake with little Ci depletion; however, the HCO_3^- concentration is equivalent to that found under alkaline conditions. The potential CO_2 supply rate for photosynthesis will in general be greater than that found in freshwater situations, as CO_2 supply from the atmosphere is not the major source of dissolved Ci, and localized depletions of Ci are much less likely to be experienced.

Given the above view of Ci in marine situations, it seems reasonable to propose that the CO_2-concentrating mechanism should be less important as a means of utilizing limited Ci supplies, compared to the freshwater habitat. For cyanobacteria, if the adaptation of freshwater species is the same as marine relatives (see Section VI,B,1,a), then the physiological form that would exist in the open ocean will be high-Ci in nature, with a capacity for some CO_2 concentration but probably utilizing CO_2 rather than HCO_3^-. The position for the other microphytes is unclear, but as the CO_2 supply rate almost always exceeds the photosynthetic rate, it is reasonable to suggest that these organisms should also display a high-Ci physiology. It is likely that the high-Ci form will be capable of concentrating CO_2 and suppressing aspects of photorespiration, but not to the extent of freshwater microalgae grown at low Ci. In addition, the high levels of HCO_3^- in seawater may mean that the high-Ci form will display some ability to use this as a carbon species. It is hard to see a role for the induction of variable CO_2-concentrating states in marine microphytes, given the high and constant Ci supply, but at least one marine cyanobacteria displays this ability (Badger and Andrews, 1982). Under bloom conditions in inshore waters, localized depletions of Ci may occur. It is also possible that the CO_2-concentrating mechanism may play an important role in utilizing the limiting nitrogen supplies found in seawater (see Section VIII,B).

For marine macrophytes, a similar comparison with the freshwater environment can be made. They too should experience a CO_2 supply condition that never approaches the limitation found in productive freshwater lakes. Given this view, it is interesting that many marine macrophytes show a higher affinity for HCO_3^- than their freshwater counterparts (Sand-Jensen and Gordon, 1984). This would suggest that a CO_2-concentrating mechanism based on HCO_3^- utilization is operational in many species at a level that is at least equivalent to that in freshwater macrophytes. This may be an adaptive feature for growth in an environment where HCO_3^- is always the predominant form of Ci. The problems of large boundary-layer resistances in macrophytes may mean that even at the relatively high levels of Ci in the marine environment, it is mostly advantageous to utilize HCO_3^- and concentrate CO_2 to reduce the limitations of diffusion. In dense macrophyte beds it is possible to experience high pH values and reduced Ci (Sand-Jensen and Gordon, 1984), and under these conditions an increased ability to utilize HCO_3^- may occur. In general, with little variation in the Ci supply there should not be changes in the HCO_3^--utilizing capacity of the macrophyte populations.

2. Boundary-Layer Conditions

The CO_2 supply also depends on the magnitude of the boundary-layer conductance (see Raven *et al.*, 1985). Generally the boundary-layer conductance is higher for a microphyte than a macrophyte species. In stagnant solution, the unstirred layer around a macrophyte may be in excess of 1 mm thickness, reducing to 50–100 μm under rapidly stirred conditions (Walker, 1985). The boundary layer for microphytes is probably always less than 30 μm and is relatively unaffected by stirring. Variation in boundary-layer conductance can only be achieved by macrophytes either as a result of the turbulence of the water surrounding the organism or through morphological adaptation, such as increasing leaf dissection.

The problem of limitation of CO_2 supply by boundary-layer conductance will be greatest for macrophytes in relatively stagnant environments. The capacity to have a high affinity for Ci will be of most value to macrophyte species that normally experience such situations, and this will occur for productive species in weed beds. Conversely, species that normally exist in fast-flowing water will experience greater CO_2 supply and a high affinity for Ci will be less important. In general, then, the CO_2-concentrating mechanism and the ability to utilize HCO_3^- will have greater relevance where the boundary layer conductance is low.

B. Nitrogen Limitation

Although many aquatic environments are limited in nitrogen supply, it remains to be shown to what extent the CO_2-concentrating mechanism can serve to increase the efficiency of nitrogen utilization. The observation that limiting nitrogen may induce the CO_2-concentrating mechanism in *Chlorella* (Beardall *et al.*, 1982) has not been repeated with any other micro- or macroalgae species. Until this is done, it is only possible to speculate about the interaction of nitrogen and Ci limitations with the CO_2-concentrating physiology of the cell. In both lake and ocean environments, the distribution of light and nitrogen in the water column is generally inversely linked. Light is high in the nutrient-depleted surface waters, while it is low in the deeper nutrient-enriched layers. It can be readily envisaged that a strategy could thus exist whereby those species adapted to surface waters invest excess light energy in a CO_2-concentrating mechanism in return for the ability to become nitrogen-use efficient. At depth, however, the reverse would apply, whereby species would benefit by being light-use efficient. This may result in organisms lacking the CO_2-concentrating mechanism but choosing instead to invest the more plentiful supplies of nitrogen directly in photosynthetic machinery, particularly Rubisco.

C. Organisms without the CO_2-Concentrating Mechanism

1. Aquatic Phototrophs

Based on the foregoing discussion, the absence of any ability to acquire the CO_2-concentrating mechanism is most likely in aquatic species that normally experience no limitation of CO_2 supply. In freshwater environments, this may occur in species such as *Lobelia dortmanna* and *Littorella uniflora*, which have been shown to use CO_2 from the sediment by taking it up through the roots (see Spence and Maberly, 1985). This makes them relatively independent of the external Ci supply. It has been suggested (Steemann-Nielsen, 1960; Spence, 1967) that plants with their shoots near the sediment also have a better supply of CO_2 because of the localized elevation in CO_2 due to organic matter decomposition. Prostrate-growing species such as the aquatic moss *Fontinalis antipyretica* and the giant alga *Nitella flexilis* have been suggested as such examples (Spence and Maberly, 1985). Aquatic species with floating leaves that directly access the air, such as *Nuphar lutea*, may also overcome the problems of CO_2 supply. In addition, the possession of biochemically based CO_2-concentrating mechanisms, such as those discusesd in Section VII,A,6, may eliminate the necessity for a CO_2-concentrating mechanism based on active Ci transport.

2. Terrestrial Plants

The cells and chloroplasts of higher plants do not show any significant ability to concentrate CO_2 or utilize HCO_3^- for photosynthesis (see Colman and Espie, 1985). In this case it must be concluded that the increase in boundary-layer conductance associated with leaf structure, combined with the development of a higher-affinity Rubisco (Yeoh *et al.*, 1981), has been sufficient to assure adequate CO_2 supply. The fact that terrestrial C_3 leaves show photorespiration and O_2 inhibition would suggest, however, that their photosynthesis is only about half saturated with CO_2. This would mean that they would benefit from a CO_2-concentrating mechanism, as evidenced by the development of C_4 photosynthesis. There seems to be no reason *a priori* to suggest that a CO_2-concentrating mechanism, such as that found in green algae, could not function in a C_3 leaf, especially if transport were across the chloroplast envelope. At present we are left to speculate either that there is no significant advantage for a C_3 leaf to have retained such a mechanism or that there is a mechanistic problem that cannot be surmounted by a terrestrial leaf in a gaseous medium. The development of C_4 and CAM CO_2-concentrating mechanisms suggests that other mechanisms have had to be developed to cope with the remaining problems of CO_2 limitation and water loss in a terrestrial environment.

REFERENCES

Abel, K. M. (1984). *Plant Physiol.* **76**, 776–781.

Adams, M. S. (1985). *In* "Inorganic Carbon Uptake by Aquatic Photosynthetic Organisms" (W. J. Lucas and J. A. Berry, eds.), pp. 421–435. Am. Soc. Plant Physiol., Rockville, Maryland.

Aizawa, K., Tusuzuki, M., and Miyachi, S. (1986). *Plant Cell Physiol.* **27**, 37–47.

Andrews, T. J., and Abel, K. M. (1979). *Plant Physiol.* **63**, 650–656.

Andrews, T. J., and Abel, K. M. (1981). *J. Biol. Chem.* **256**, 8445–8451.

Ascencio, J., and Bowes, G. (1983). *Photosynth. Res.* **4**, 151–170.

Badger, M. R. (1980). *Arch. Biochem. Biophys.* **201**, 247–254.

Badger, M. R. (1985). *In* "Inorganic Carbon Uptake by Aquatic Photosynthetic Organisms" (W. J. Lucas and J. A. Berry eds.), pp. 39–52. Am. Soc. Plant Physiol., Rockville, Maryland.

Badger, M. R., and Andrews, T. J. (1982). *Plant Physiol.* **70**, 517–523.

Badger, M. R., and Gallagher, A. (1987). *Aust. J. Plant Physiol.* **14**, 189–202.

Badger, M. R., Kaplan, A., and Berry, J. A. (1978). *Year Book—Carnegie Inst. Washington* **77**, 251–261.

Badger, M. R., Kaplan, A., and Berry, J. A. (1980). *Plant Physiol.* **66**, 407–413.

Badger, M. R., Bassett, M., and Comins, H. N. (1985). *Plant Physiol.* **77**, 465–471.

Badour, S. S., and Kim, W. K. (1986). *Biochem. Physiol. Pflanz.* **181**, 9–16.

Beardall, J. (1981). *J. Phycol.* **17**, 371–373.

Beardall, J., and Entwisle, L. (1984). *Phycologia* **23**, 511–513.

Beardall, J., and Raven, J. A. (1981). *J. Phycol.* **17**, 131–141.

Beardall, J., Griffiths, J. A., and Raven, J. A. (1982). *J. Exp. Bot.* **33**, 729–737.

Beer, S., and Eshel, A. (1983). *J. Exp. Mar. Biol. Ecol.* **70**, 99–106.

Beer, S., and Waisel, Y. (1979). *Aquat. Bot.* **7**, 129–138.

Beer, S., Eshel, A., and Waisel, Y. (1977). *J. Exp. Bot.* **28**, 1180–1189.

Beer, S., Shomer-Ilan, A., and Waisel, Y. (1980). *J. Exp. Bot.* **31**, 1019–1026.

Benedict, C. R., and Scott, J. R. (1976). *Plant Physiol.* **57**, 876–880.

Benedict, C. R., Wong, W. L., and Wong, J. H. H. (1980). *Plant Physiol.* **65**, 512–517.

Berry, J. A., Boynton, J., Kaplan, A., and Badger, M. R. (1976). *Year Book—Carnegie Inst. Washington* **75**, 423–432.

Bidwell, R. G. S., and McLachlan, J. (1985). *J. Exp. Mar. Biol. Ecol.* **86**, 15–46.

Birmingham, B. C., and Colman, B. (1979). *Plant Physiol.* **64**, 892–895.

Birmingham, B. C., Coleman, J. R., and Colman, B. (1982). *Plant Physiol.* **69**, 259–262.

Bowes, G. (1985). *In* "Inorganic Carbon Uptake by Aquatic Photosynthetic Organisms" (W. J. Lucas and J. A. Berry eds.), pp. 187–210. Am. Soc. Plant Physiol., Rockville, Maryland.

Brechignac, F., Andre, M., and Gerbaud, A. (1986). *Plant Physiol.* **80**, 1059–1062.

Brown, D. C., and Tregunna, E. B. (1967). *Can. J. Bot.* **45**, 1135–1143.

Bruin, W. J., Nelson, E. B., and Tolbert, N. E. (1970). *Plant Physiol.* **46**, 386–391.

Burris, J. E. (1980). *In* "Primary Productivity of the Sea" (P. G. Falkowski, ed.), pp. 411-432. Plenum, New York.

Chollet, R., and Ogren, W. L. (1975). *Bot. Rev.* **41**, 137–179.

Coleman, J. R., and Colman, B. (1980). *Planta* **149**, 318–320.

Coleman, J. R., and Colman, B. (1981). *Plant Physiol.* **67**, 917–921.

Coleman, J. R., and Grossman, A. R. (1984). *Proc. Natl. Acad. Sci. U.S.A.* **81**, 6049–6053.

Coleman, J. R., Seemann, J. R., and Berry, J. A. (1982). *Year Book—Carnegie Inst. Washington* **81**, 83–87.

Coleman, J. R., Berry, J. A., Togasaki, R. T., and Grossman, A. R. (1984). *Plant Physiol.* **76**, 472–477.

Coleman, J. R., Green, L. S., Berry, J. A., Togasaki, R. K., and Grossman, A. R. (1985). *In*

"Inorganic Carbon Uptake by Aquatic Photosynthetic Organisms" (W. J. Lucas and J. A. Berry eds.), pp. 339–359. Am. Soc. Plant Physiol., Rockville, Maryland.

Colman, B., and Cook, C. M. (1985). *In* "Inorganic Carbon Uptake by Aquatic Photosynthetic Organisms" (W. J. Lucas and J. A. Berry, eds.), pp. 97–110. Am. Soc. Plant Physiol., Rockville, Maryland.

Colman, B., and Espie, G. S. (1985). *Plant Cell Environ.* **8,** 449–457.

Cooper, T. G., Filmer, D. L., Wishnick, M., and Lane, M. D. (1969). *J. Biol. Chem.* **244,** 1081–1083.

Coughlan, S., and Tattersfield, D. (1977). *Bot. Mar.* **20,** 265–266.

Dohler, G. (1974). *Planta* **117,** 97–99.

Drews, G., and Weckesser, J. (1982). *In* "The Biology of Cyanobacteria" (N. G. Carr and B. A. Whitton, eds.), pp. 125–141. Blackwell, Oxford.

Edsall, J. T. (1969). *NASA [Spec. Publ.] SP* **NASA SP-188,** 15–27.

Falk, H., and Sitte, P. (1963). *Protoplasma* **57,** 290–303.

Ferrier, J. M. (1980). *Plant Physiol.* **66,** 1198–1199.

Findenegg, G. R. (1976). *Z. Pflanzenphysiol.* **79,** 428–437.

Findenegg, G. R. (1977). *Planta* **135,** 33–38.

Graham, D., and Reed, M. L. (1971). *Nature (London)* **231,** 81–83.

Graham, D., Atkins, C. A., Reed, M. L., Patterson, B. D., and Smillie, R. M. (1971). *In* "Photosynthesis and Photorespiration" (M. D. Hatch, C. B. Osmond, and R. O. Slatyer, eds.), pp. 267–274. Wiley (Interscience), New York.

Gutknecht, J., Bisson, M. A., and Tosteson, D. C. (1977). *J. Gen. Physiol.* **69,** 779–794.

Hogetsu, D., and Miyachi, S. (1977). *Plant Cell Physiol.* **18,** 347–352.

Hogetsu, D., and Miyachi, S. (1979a). *Plant Cell Physiol.* **20,** 747–756.

Hogetsu, D., and Miyachi, S. (1979b). *Plant Cell Physiol.* **20,** 1427–1432.

Holthuijzen, Y. A., van Breemen, J. F. L., Kuenen, J. G., and Konings, W. N. (1986). *Arch. Microbiol.* **144,** 398–404.

Imamura, M., Tsuzuki, M., Shiraiwa, Y., and Miyachi, S. (1983). *Plant Cell Physiol.* **24,** 533–540.

Ingle, R. K., and Colman, B. (1976). *Planta* **128,** 217–223.

Kaplan, A. (1981). *J. Exp. Bot.* **32,** 669–677.

Kaplan, A. (1985). *In* "Inorganic Carbon Uptake by Aquatic Photosynthetic Organisms" (W. J. Lucas and J. A. Berry, eds.), pp. 325–338. Am. Soc. Plant Physiol., Rockville, Maryland.

Kaplan, A., Badger, M. R., and Berry, J. A. (1980). *Planta* **149,** 219–226.

Kaplan, A., Zenvirth, D., Reinhold, L., and Berry, J. A. (1982). *Plant Physiol.* **69,** 978–982.

Kaplan, A., Volokita, M., Zenvirth, D., and Reinhold, L. (1984). *FEBS Lett.* **176,** 166–168.

Keeley, J. E. (1981). *Am. J. Bot.* **68,** 420–424.

Keeley, J. E. (1983). *Oecologia* **58,** 57–62.

Keeley, J. E., and Bowes, G. (1982). *Plant Physiol.* **70,** 1455–1458.

Keeley, J. E., and Busch, G. (1984). *Plant Physiol.* **76,** 525–530.

Kerby, N. W., and Raven, J. A. (1985). *Adv. Bot. Res.* **11,** 71–123.

Kimpel, D. L., Togasaki, R. T., and Miyachi, S. (1983). *Plant Cell Physiol.* **24,** 255–259.

Kremer, B. P. (1980). *Planta* **150,** 189–190.

Kremer, B. P. (1981). *Oceanogr. Mar. Biol.* **19,** 41–94.

Krulwich, T. A. (1986). *J. Membr. Biol.* **89,** 113–125.

Laing, W. A., and Browse, J. (1985). *Plant Cell Environ.* **8,** 639–649.

Lanaras, T., Hawthornthwaite, A. M., and Codd, G. A. (1985). *FEMS Microbiol. Lett.* **26,** 285–288.

Lehman, J. T. (1978). *J. Phycol.* **14,** 33–42.

Lloyd, N. D. H., Canvin, D. T., and Culver, D. A. (1977). *Plant Physiol.* **59,** 936–940.

Lucas, W. J. (1975). *J. Exp. Bot.* **26,** 331–346.

Lucas, W. J. (1983). *Annu. Rev. Plant Physiol.* **34,** 71–104.

Lucas, W. J. (1985). *In* "Inorganic Carbon Uptake by Aquatic Photosynthetic Organisms" (W.

J. Lucas and J. A. Berry, eds.), pp. 229–254. Am. Soc. Plant Physiol., Rockville, Maryland.

Manodori, A., and Melis, A. (1984). *Plant Physiol.* **74,** 67–71.

Marcus, Y., Zenvirth, D., Harel, E., and Kaplan, A. (1982). *Plant Physiol.* **69,** 1008–1012.

Marcus, Y., Harel, E., and Kaplan, A. (1983). *Plant Physiol.* **71,** 208–210.

Marcus, Y., Volokita, M., and Kaplan, A. (1984). *J. Exp. Bot.* **35,** 1136–1144.

Marcus, Y., Schuster, G., Michaels, A., and Kaplan, A. (1986). *Plant Physiol.* **80,** 604–607.

Marsden, W. J. N., Lanaras, T., and Codd, G. A. (1984). *J. Gen. Microbiol.* **130,** 2089–2093.

Mayo, W. P., Williams, T. G., Birch, D. G., and Turpin, D. H. (1986). *Plant PHysiol.* **80,** 1038–1040.

Miller, A. G. (1985). *In* "Inorganic Carbon Uptake by Aquatic Photosynthetic Organisms" (W. J. Lucas and J. A. Berry, eds.), pp. 17–29. Am. Soc. Plant Physiol., Rockville, Maryland.

Miller, A. G., and Canvin, D. T. (1985). *FEBS Lett.* **187,** 29–32.

Miller, A. G., and Colman, B. (1980a). *J. Bacteriol.* **143,** 1253–1259.

Miller, A. G., and Colman, B. (1980b). *Plant Physiol.* **65,** 397–402.

Miller, A. G., Turpin, D. H., and Canvin, D. T. (1984a). *J. Bacteriol.* **159,** 100–106.

Miller, A. G., Turpin, D. H., and Canvin, D. T. (1984b). *Plant Physiol.* **75,** 1064–1070.

Miyachi, S., Tsuzuki, M., and Avramova, T. (1983). *Plant Cell Physiol.* **24,** 441–451.

Miyachi, S., Tsuzuki, M., and Yagawa, Y. (1985). *In* "Inorganic Carbon Uptake by Aquatic Photosynthetic Organisms" (W. J. Lucas and J. A. Berry, eds.), pp. 145–154. Am. Soc. Plant Physiol., Rockville, Maryland.

Moroney, J. V., and Tolbert, N. E. (1985). *Plant Physiol.* **77,** 253–258.

Moroney, J. V., Husic, D. H., and Tolbert, N. E. (1985). *Plant Physiol.* **79,** 177–183.

Murata, N., Sato, N., Omata, T., and Kuwabara, T. (1981). *Plant Cell Physiol.* **22,** 855–866.

Nelson, E. B., Cenedella, A., and Tolbert, N. E. (1969). *Phytochemistry* **8,** 2305–2306.

Nikaido, H., and Nakae, T. (1979). *Adv. Microb. Physiol.* **20,** 163–250.

Nikaido, H., and Vaara, M. (1985). *Microbiol. Rev.* **49,** 1–33.

Nitschmann, W. H., and Peschek, G. A. (1985). *Arch. Microbiol.* **141,** 330–336.

Ogawa, T., and Inoue, Y. (1983). *Biochim. Biophys. Acta* **724,** 490–493.

Ogawa, T., Inoue, Y., Lilley, R. McC., and Ogren, W. L. (1984). *In* "Advances in Photosynthesis Research" (C. Sybesma ed.), Vol. 2, pp. 723–726. Martinuus Nijhoff/Dr. W. Junk, The Hague, The Netherlands.

Ogawa, T., Miyano, A., and Inoue, Y. (1985a). *Biochim. Biophys. Acta* **808,** 77–84.

Ogawa, T., Omata, T., Miyano, A., and Inoue, Y. (1985b). *In* "Inorganic Carbon Uptake by Aquatic Photosynthetic Organisms" (W. J. Lucas and J. A. Berry, eds.), pp. 287–304. Am. Soc. Plant Pysiol., Rockville, Maryland.

Omata, T., and Ogawa, T. (1985). *Plant Cell Physiol.* **26,** 1075–1081.

Omata, T., and Ogawa, T. (1986). *Plant Physiol.* **80,** 525–530.

Pate, J. S., and Gunning, B. E. S. (1972). *Annu. Rev. Plant Physiol.* **23,** 173–196.

Price, G. D. (1985). Ph.D. Thesis, Australian National University, Canberra.

Price, G. D., and Whitecross, M. I. (1983). *Protoplasma* **116,** 65–74.

Price, G. D., Badger, M. R., Bassett, M. E., and Whitecross, M. I. (1985). *Aust. J. Plant Physiol.* **12,** 241–256.

Prins, H. B. A., Snel, J. F. H., Zanstra, P. E., and Helder, R. J. (1982). *Plant Cell Environ.* **5,** 207–214.

Pronina, N. A., Avramova, S., Georgiev, D., and Semenenko, V. E. (1981). *Fiziol. Rast.* **28,** 43–52.

Ramazanov, Z. M., Pronina, N. A., and Semenenko, V. E. (1984). *Fiziol. Rast.* (*Moscow*) **31,** 448–455.

Raven, J. A. (1970). *Biol. Rev. Cambridge Philos. Soc.* **45,** 167–221.

Raven, J. A. (1984). "Energetics and Transport in Aquatic Plants." Alan R. Liss, Inc., New York.

Raven, J. A. (1985). *In* "Inorganic Carbon Uptake by Aquatic Photosynthetic Organisms (W. J. Lucas and J. A. Berry, eds.), pp. 67–82. Am. Soc. Plant Physiol., Rockville, Maryland.

Raven, J. A., and Lucas, W. J. (1985). *In* "Inorganic Carbon Uptake by Aquatic Photosynthetic Organisms" (W. J. Lucas and J. A. Berry, eds.), pp. 305–324. Am. Soc. Plant Physiol., Rockville, Maryland.

Raven, J. A., Beardall, J., and Griffiths, H. (1982). *Oecologia* **53**, 68–78.

Raven, J. A., Osborne, B. A., and Johnston, A. M. (1985). *Plant Cell Environ.* **8**, 417–425.

Reed, M. L., and Graham, D. (1977). *Aust. J. Plant Physiol.* **4**, 87–98.

Reinhold, L., Volokita, M., Zenvirth, D., and Kaplan, A. (1984). *Plant Physiol.* **76**, 1090–1092.

Reiskind, J. B., Seamen, P. T., and Bowes, G. (1984). *Plant Physiol.* **75**, 65.

Resch, C. M., and Gibson, J. (1983). *J. Bacteriol.* **155**, 345–350.

Rollins, M. J., Harper, D. I., and John, P. C. L. (1983). *J. Gen. Microbiol.* **129**, 1899–1919.

Salvucci, M. E., and Bowes, G. (1981). *Plant Physiol.* **67**, 335–340.

Salvucci, M. E., and Bowes, G. (1983a). *Planta* **158**, 27–34.

Salvucci, M. E., and Bowes, G. (1983b). *Plant Physiol.* **73**, 488–496.

Sand-Jensen, K., and Gordon, D. M. (1984). *Mar. Biol. (Berlin)* **80**, 247–253.

Semenenko, V. E., Avramova, S., Georgiev, D., and Pronina, N. A. (1979). *Fiziol. Rast. (Moscow)* **26**, 1069–1075.

Shelp, B. J., and Canvin, D. T. (1980). *Plant Physiol.* **65**, 774–779.

Shelp, B. J., and Canvin, D. T. (1984). *Can. J. Bot.* **63**, 1249–1254.

Shiraiwa, Y., and Miyachi, S. (1983). *Plant and Cell Physiol.* **24**, 919–924.

Shiraiwa, Y., and Miyachi, S. (1985a). *Plant Cell Physiol.* **26**, 109–116.

Shiraiwa, Y., and Miyachi, S. (1985b). *Plant Cell Physiol.* **26**, 543–549.

Shiraiwa, Y., Fakler, J., and Miyachi, S. (1981). *In* "Photosynthesis IV. Regulation of Carbon Metabolism" (G. Akoyunoglou, ed.), pp. 493–499. Balaban Int. Sci. Serv., Philadelphia, Pennsylvania.

Spalding, M. H., and Ogren, W. L. (1982). *FEBS Lett.* **145**, 41–44.

Spalding, M. H., and Ogren, W. L. (1983). *FEBS Lett.* **154**, 335–338.

Spalding, M. H., and Portis, A. R. (1985). *Planta* **164**, 308–320.

Spalding, M. H., Spreitzer, R. J., and Ogren, W. L. (1983a). *Plant Physiol.* **73**, 268–272.

Spalding, M. H., Spreitzer, R. J., and Ogren, W. L. (1983b). *Plant Physiol.* **73**, 273–276.

Spalding, M. H., Spreitzer, R. J., and Ogren, W. L. (1983c). *Planta* **159**, 261–266.

Spalding, M. H., Sprietzer, R. J., and Ogren, W. L. (1985). *In* "Inorganic Carbon Uptake by Aquatic Photosynthetic Organisms" (W. J. Lucas and J. A. Berry, eds.), pp. 361–387. Am. Soc. Plant Physiol., Rockville, Maryland.

Spence, D. H. N. (1967). *J. Ecol.* **55**, 147–170.

Spence, D. H. N., and Maberly, S. C. (1985). *In* "Inorganic Carbon Uptake by Aquatic Photosynthetic Organisms" (W. J. Lucas and J. A. Berry, eds.), pp. 125–143. Am. Soc. Plant Physiol., Rockville, Maryland.

Spencer, K. G., Kimpel, D. L., Fisher, M. L., Togasaki, R. K., and Miyachi, S. (1983). *Plant Cell Physiol.* **24**, 301–304.

Steemann-Nielsen, E. (1947). *Dan. Bot. Ark.* **12**, 1–71.

Steemann-Nielsen, E. (1960). *In* "Encyclopedia of Plant Physiology" (W. Ruhland, ed.), Vol. 5, Part 1, pp. 70–84. Springer-Verlag, Berlin and New York.

Stumm, W., and Morgan, J. J. (1970. "Aquatic Chemistry: An Introduction Emphasising Chemical Equilibria in Natural Water." Wiley (Interscience), New York.

Talling, J. F. (1985). *In* "Inorganic Carbon Uptake by Aquatic Photosynthetic Organisms" (W. J. Lucas and J. A. Berry, eds.), pp. 403–420. Am. Soc. Plant Physiol., Rockville, Maryland.

Tan, G. K., and Badour, S. S. (1983). *Z. Pflanzenphysiol.* **109**, 113–125.

Toguri, T., Yang, S. Y., Okabe, K., and Miyachi, S. (1984). *FEBS Lett.* **170**, 117–120.

Tsuzuki, M. (1983). *Z. Pflanzenphysiol.* **110**, 29–37.

Tsuzuki, M., Miyachi, S., and Berry, J. A. (1985). *In* "Inorganic Carbon Uptake by Aquatic Photosynthetic Organisms" (W. J. Lucas and J. A. Berry, eds.), pp. 53–66. Am. Soc. Plant Physiol., Rockville, Maryland.

Tu, C. K., Acevedo-Duncan, M., Wynns, G. C., and Silverman, D. N. (1986). *Plant Physiol.* **80,** 997–1001.

Turpin, D. H., Miller, A. G., and Canvin, D. T. (1984). *J. Phycol.* **20,** 249–253.

Turpin, D. H., Miller, A. G., Parslow, J. S., Elrifi, I. R., and Canvin, D. T. (1985). *In* "Inorganic Carbon Uptake by Aquatic Photosynthetic Organisms" (W. J. Lucas and J. A. Berry, eds.), pp. 449–458. Am. Soc. Plant Physiol., Rockville, Maryland.

Volokita, M., Zenvirth, D., Kaplan, A., and Reinhold, L. (1984). *Plant Physiol.* **76,** 599–602.

Walker, N. A. (1983). *Plant Cell Environ.* **6,** 323–328.

Walker, N. A. (1985). *In* "Inorganic Carbon Uptake by Aquatic Photosynthetic Organisms" (W. J. Lucas and J. A. Berry, eds.), pp. 31–37. Am. Soc. Plant Physiol., Rockville, Maryland.

Walker, N. A., Smith, F. A., and Cathers, I. R. (1980). *J. Membr. Biol.* **57,** 51–58.

Yagawa, Y., Shiraiwa, Y., and Miyachi, S. (1984). *Plant Cell Physiol.* **25,** 775–783.

Yang, S., Tsuzuki, M., and Miyachi, S. (1985). *Plant Cell Physiol.* **26,** 25–34.

Yeoh, H. H., Badger, M. R., and Watson, L. (1981). *Plant Physiol.* **67,** 1151–1155.

Zenvirth, D., and Kaplan, A. (1981). *Planta* **152,** 8–12.

Zenvirth, D., Volokita, M., and Kaplan, A. (1984). *J. Membr. Biol.* **79,** 271–274.

Biochemistry of C_3–C_4 Intermediates

5

GERALD E. EDWARDS
MAURICE S. B. KU

I. Introduction
II. Occurrence: Families, Genera, and Species
III. Properties of Intermediates
A. Leaf Anatomy and Ultrastructure
B. Physiology
C. Biochemistry
IV. Carbon Isotope Composition of Intermediates Relative to Mechanism of Photosynthesis
V. Biochemistry of Photosynthesis in Intermediates in Relation to Developmental and Environmental Factors
A. Leaf Age
B. O_2 Levels, Light Intensity, and Γ Values
C. Growth Conditions
VI. Genetically Based Differences in the Biochemistry of Photosynthesis within an Intermediate Species
VII. Features of Hybrids between Different Photosynthetic Types
A. C_3 and C_4 Species
B. C_3 and Intermediates
C. Intermediates and C_4 Species
VIII. A Theoretical Scheme of Evolution of C_4 Photosynthesis Based on Intermediate Species
IX. Concluding Remarks
References

The Biochemistry of Plants, Vol. 10

I. INTRODUCTION*

In order to define a C_3–C_4 intermediate species, it is necessary to describe briefly C_3 and C_4 plants. The reductive pentose phosphate pathway, or PCR cycle, is the means through which higher plants assimilate CO_2. This cycle is made up of 13 enzymes, which include those of a carboxylation phase (ribulose-1,5-P_2 carboxylase), a reduction phase (3-phosphoglycerate to glyceraldehyde 3-phosphate), and a regeneration phase (glyceraldehyde 3-phosphate to ribulose-1,5-P_2). When CO_2 reacts with ribulose-1,5-P_2, the initial product is a three-carbon compound, 3-phosphoglycerate. Thus, species that fix atmospheric CO_2 directly through ribulose-1,5-P_2 carboxylase are called C_3 plants. Surprisingly, this enzyme has a second catalytic function through ribulose-1,5-P_2 oxygenase. Reaction of ribulose-1,5-P_2 with O_2 results in the synthesis of phosphoglycolate and 3-phosphoglycerate. While reaction of ribulose-1,5-P_2 with CO_2 leads to photosynthesis, reaction with O_2 results in inhibition of photosynthesis. O_2 is a competitive inhibitor with respect to CO_2, and metabolism of phosphoglycolate in the glycolate pathway results in loss of CO_2 through glycine decarboxylation (Fig. 1; also see Andrews and Lorimer, Chapter 3).

When terrestrial plants first evolved, atmospheric O_2 levels are thought to have been lower than in recent geologic time, while the level of CO_2 was higher, perhaps 1 or 2%. Therefore, CO_2 was likely nonlimiting for photosynthesis, and photorespiration minimal, until atmospheric levels of O_2 and CO_2 assumed concentrations near present levels. These are approximately 340 μl/l CO_2 and 21% O_2. This level of O_2 causes about 30% inhibition of photosynthesis at 25°C. The pre-Industrial Age level of CO_2 is estimated to have had a value of 265 μl/l, but since that time it has gradually increased due to the burning of fossil fuels and deforestation (Wigley, 1983).

As atmospheric conditions became more oxidative and favorable for photorespiration, CO_2 became more limiting, and those plants that evolved a means to minimize this problem became more competitive. C_4 plants accomplished this by concentrating CO_2 in the leaf at the site of ribulose-1,5-P_2 carboxylase/oxygenase (Rubisco) (previously reviewed in this series, Edwards and Huber, 1981). Because of the energetic cost of eliminating photorespiration through C_4 photosynthesis, C_4 plants are more competitive only in certain environments, such as high light and warm temperatures. At cool temperatures, C_3 plants may be more competitive. There is currently no evidence that any terrestrial plants have undergone a change in the proper-

---

* Abbreviations: APR, apparent rate of photorespiration; APS, apparent rate of photosynthesis; CE, carboxylation efficiency; intermediate species, C_3–C_4 photosynthetic intermediate species; OAA, oxaloacetate; PCR cycle, photosynthetic carbon reduction cycle; ribulose-1,5-P_2, ribulose-1,5-bisphosphate; Rubisco, ribulose-1,5-bisphosphate carboxylase/oxygenase; Γ, photosynthetic CO_2 compensation point; PEP, phosphoenolpyruvate; PS II, photosystem II; TPR, true rate of photorespiration; TPS, true rate of photosynthesis.

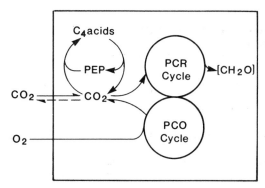

Fig. 1. General scheme showing possible fates of atmospheric CO_2 and photorespired CO_2 in leaves of plants. PCR cycle, photosynthetic carbon reduction cycle; PCO cycle, photosynthetic carbon oxidation cycle; [CH_2O], carbohydrate.

ties of Rubisco that would increase its capacity to react with CO_2 over O_2. C_4 plants have, however, developed a remarkable CO_2-concentrating mechanism through evolution in anatomy and biochemistry. In these species atmospheric CO_2 is fixed through PEP carboxylase in mesophyll cells of the leaf, with OAA as the product. These plants are called C_4 species, because the initial product of CO_2 fixation, OAA, is a C_4 compound. The OAA is converted to malate and aspartate, and these acids are transported to the Kranz cells and metabolized through C_4 acid decarboxylases, where the CO_2 is concentrated prior to fixation in the PCR cycle. The C_4 cycle is complete when the C_3 product of decarboxylation is transported to mesophyll cells and utilized again in the PEP carboxylase reaction. There are currently three known means of decarboxylating C_4 acids in C_4 plants: NADP-malic enzyme, NAD-malic enzyme, and PEP carboxykinase. C_4 species are divided into subgroups on this basis. C_4 species have Kranz type leaf anatomy, which is a term originally used by Haberlandt in 1882 to describe a distinctive inner and outer layer of chloroplast-containing cells surrounding the vascular tissue in some species (see Haberlandt, 1909). The sheath cells of certain grasses were described as having large green "chlorophyll kernels." It is uncertain whether the term Kranz was used in reference to the inner or outer layer of chloroplast-containing cells (in some cases these were called inner and outer sheath). Later, Brown (1975) referred in C_4 plants to the distinctive thick-walled sheath of cells that contain numerous chloroplasts and mitochondria as Kranz cells. In contrast, C_3 plants have mesophyll cells but lack the distinctive Kranz cells (Edwards and Huber, 1981).

It is logical that evolution of C_4 plants from C_3 plants occurred through a number of steps, since the modifications in biochemistry and anatomy associated with C_4 photosynthesis are relatively complex. For example, among the three known C_4 subgroups there are at least 13 enzymes associated with

TABLE I
Anatomical, Biochemical, and Physiological Components of
the Kranz Syndrome in C$_4$ Plants

Anatomical
  Mesophyll cells and Kranz cells
  Kranz cell features:
    Distinctive wreath-like, thick-walled cells surrounding the vascular tissue
    Numerous chloroplasts and mitochondria
    Always internal to the mesophyll cells
Biochemical
  High levels of the complement of the C$_4$-cycle enzymes
  Differential compartmentation of certain photosynthetic enzymes between the mesophyll and
    Kranz cells
  Differential function of chloroplasts and mitochondria between the mesophyll and Kranz cells
  δ^{13}C values of plants = -11 to $-16‰$ (reflects fixation of atmospheric CO$_2$ directly through
    PEP carboxylase)
Physiological
  Photosynthesis saturates at atmospheric [CO$_2$]
  Photorespiration is not apparent by the following criteria:
    CO$_2$ compensation point approaches zero and remains low with increasing [O$_2$]
    No apparent inhibition of photosynthesis by atmospheric levels of O$_2$
    O$_2$ does not reduce quantum efficiency and carboxylation efficiency

the C$_4$ cycle (seven with the NADP-malic enzyme subgroup) (Edwards and
Huber, 1981). Full development of C$_4$ photosynthesis also requires repres-
sion of certain genes encoding for the synthesis of PCR cycle enzymes in the
mesophyll cells (e.g., Rubisco and phosphoribulokinase, which are only
active in the Kranz cells). Undoubtedly, Kranz anatomy is also under multi-
genic control, although the number of genes is unknown. Thus, species
might exist that are in the process of evolving toward C$_4$ photosynthesis, and
that have characteristics intermediate in anatomy, physiology, or biochemistry, between those of C$_3$ and C$_4$ plants. A number of such species have been
identified and are the subject of this review. The intermediates may represent different levels of development toward a reduction in photorespiration.
The intermediate species are of interest in that they may help to elucidate the
mechanism of evolution of C$_4$ plants and mechanism(s) for reduction of
photorespiration.

  C$_4$ plants have a number of essential biochemical, anatomical and physiological features (Table I), and Brown (1977) referred to the combination of all
of these features as the Kranz syndrome. Conversely, plants that lack all of
these characteristics are C$_3$ plants. In general, a C$_3$–C$_4$ intermediate species
can be defined as a species in which (a) one or more of the features of the
Kranz syndrome is "intermediate," that is, the character is at some stage or
level between that of a C$_3$ and a C$_4$ species, or (b) there is a mixture of fully
expressed features of the Kranz syndrome combined with those of species
lacking this syndrome. Some of the intermediates that have thus far been

identified have some obvious intermediate characteristics with respect to physiology, anatomy, and/or biochemistry. If the biochemical and anatomical features of the Kranz syndrome are developed to varying degrees, then a variety of atypical physiological responses in photosynthesis and photorespiration may occur. An important physiological criterion for designating a species intermediate is identification of atypical photorespiratory levels compared to C_3 and C_4 species [e.g., from measuring the CO_2 compensation point (Γ) and O_2 inhibition of photosynthesis]. These features may be intermediate between those of C_3 and C_4 plants; or, for example, Γ values may be closer to those of C_4 plants yet the degree of inhibition of photosynthesis by O_2 may be more like that of C_3 plants.

II. OCCURRENCE: FAMILIES, GENERA, AND SPECIES

Through considerable interest and effort over the last decade, a growing number of naturally occurring plant species with photosynthetic characteristics intermediate to C_3 and C_4 plants have been identified. Most intermediates have in common an anatomy between that of non-Kranz and Kranz, and a partially suppressed photorespiration compared to C_3 plants as indicated by a reduced CO_2 compensation point and/or a reduced sensitivity of net photosynthesis or CE to O_2. So far, 22 species, including both monocots and dicots of seven genera from five families, have been identified as intermediate species (Table II). As shown in Table III, most of the intermediates belong to genera that are known to contain both C_3 and C_4 species (*Mollugo, Alternanthera, Flaveria, Neurachne,* and *Panicum*). However, in the genera *Parthenium* and *Moricandia,* only C_3 and intermediate species are known to exist. Like C_4 species (Moore, 1982), intermediates are found in what are considered to be the most advanced orders, such as Asterales (*Flaveria, Parthenium*), Caryophyllales (*Alternanthera, Mollugo*), Capparales (*Moricandia*), and Poales (*Neurachne, Panicum*). The high frequency of occurrence of intermediates in the genus *Flaveria* is rather unusual. At least ten of the 21 species in the genus have been classified as intermediates. *Flaveria brownii* was previously classified as a C_4 plant based on leaf anatomy and carbon isotope ratio (Powell, 1978). However, a recent study indicates that although *F. brownii* possesses well-developed Kranz cells, it has two types of mesophyll cells and some biochemical characteristics that suggest it is an intermediate species (not complete Kranz syndrome; see Section III,C). It is hypothesized that some of the species in the genus *Flaveria* may be in an active process of evolution from C_3 to C_4 photosynthesis (Powell, 1978). Recently, Fox *et al.* (1986) reported that celery (*Apium graveolens*) has intermediate Γ values and lower O_2 inhibition of photosynthesis than in C_3 plants; it will be of interest to determine how photorespiration may be re-

TABLE II
Occurrence of C_3–C_4 Intermediate Species

Family	Species	Reference
Monocotyledoneae		
Poaceae	*Neurachne minor*	Hattersley *et al.* (1984)
	Panicum decipiens	Morgan and Brown (1979)
	Panicum milioides	Krenzer *et al.* (1975)
		Brown and Brown (1975)
		Kanai and Kashiwagi (1975)
	Panicum schenckii	Morgan and Brown (1979)
Dicotyledoneae		
Aizoaceae	*Mollugo verticillata*	Kennedy and Laetsch (1974)
	Mollugo nudicaulis	Kennedy *et al.* (1980)
Amaranthaceae	*Alternanthera ficoides*	Rajendrudu *et al.* (1986)
	Alternanthera tenella	Rajendrudu *et al.* (1986)
	Flaveria angustifolia	J. Wu and M. S. B. Ku (unpublished)
Asteraceae	*Flaveria anomala*	Apel and Maass (1981)
	Flaveria brownii	Cheng *et al.* (1987)
	Flaveria chloraefolia	Holaday *et al.* (1984)
	Flaveria floridana	Holaday *et al.* (1984)
	Flaveria linearis	Ku *et al.* (1983)
	Flaveria oppositifolia	Brown *et al.* (1986b)
	Flaveria pubescens	Apel and Maass (1981)
	Flaveria ramosissima	Ku *et al.* (1983)
	Flaveria sonorensis	J. Wu and M. S. B. Ku (unpublished)
	Parthenium hysterophorus	Patil and Hegde (1983)
Brassicaceae	*Moricandia arvensis*	Krenzer *et al.* (1975)
		Apel *et al.* (1978)
	Moricandia sinaica	Apel and Ohle (1979)
	Moricandia spinosa	Apel (1980)

duced in this crop species. No doubt, more intermediate species are yet to be identified among higher plants.

With few exceptions, most of the intermediate species occupy a hot and dry or a saline habitat. For example, the intermediate *Moricandia* species are found mainly in the desert regions of Israel and the Sinai (Gomez-Campo, 1980). Similarly, *Neurachne minor,* an intermediate endemic to Australia, occurs in the arid zone (Prendergast and Hattersley, 1985). *Parthenium hysterophorus,* a noxious weed, is another intermediate known to have a wide range of ecological adaptation to extreme environments, such as drought and salt (Hegde and Patil, 1980). Two *Alternanthera* species, recently discovered to be intermediates (Rajendrudu *et al.,* 1986), grow abundantly as weeds in arid regions of India (Pathan and Nimbalkar, 1982). Most of the *Flaveria* intermediates are restricted to, or frequently associated

TABLE III
Families and Genera Known to Contain Both C_3 and C_4 Species[a]

Monocotyledoneae			
Cyperaceae		Poaceae	
Cyperus		*Alloteropsis*	
Scirpus		*Neurachne**	
		*Panicum**	
Dicotyledoneae			
Aizoaceae	Amaranthaceae	Asteraceae	Boraginaceae
*Mollugo**	*Aerva*	*Flaveria**	*Heliotropium*
	*Alternanthera**		
Chenopodiaceae	Convolvulaceae	Euphorbiaceae	Nyctaginaceae
Atriplex	*Evolvulus*	*Chamaesyce*	*Boerhaavia*
Bassia		*Euphorbia*	
Kochia	Nyctaginaceae	Zygophyllaceae	
Suaeda	*Boerhaavia*	*Kallstroemia*	
		Zygophyllum	

[a] Genera known to contain intermediates are designated by an asterisk. For lists and taxonomic distribution of C_4 plants, see Downton (1975), Raghavendra and Das (1978), and Prendergast and Hattersley (1985).

with, saline or gypseous habitats in the coastal regions of Florida and Texas and in the arid, tropical regions of Mexico (Powell, 1978). The general Chihuahuan Desert region is considered the center of origin of *Flaveria* because a number of C_3, C_4, and intermediate members of the genus are concentrated in this area. Unlike other intermediates, the three intermediate *Panicum* species found in North and South America occupy a mesic habitat.

III. PROPERTIES OF INTERMEDIATES

A. Leaf Anatomy and Ultrastructure

Most C_3–C_4 species tend to have a leaf anatomy intermediate to C_3 and C_4 plants. C_4 plants have Kranz cells, consisting of distinctive, wreath-like, thick-walled cells with numerous chloroplasts, which surround the vascular tissue in the leaf. C_3 plants lack Kranz cells, and bundle-sheath cells, when present, contain very few or no chloroplasts. The presence of chloroplast-containing bundle sheath cells is a common feature of all intermediates. However, there is tremendous variation among the intermediate species in the degree of Kranz-cell development, and the leaf anatomy in these species can be divided into three general groups on this basis. These range from a Kranz type in *Neurachne minor* and *Flaveria brownii* to a Kranz-like type in *Panicum milioides* to a poorly developed Kranz anatomy in *Mollugo verticillata* (Table IV). *Neurachne minor* and *F. brownii* possess a typical Kranz

TABLE IV
Leaf Anatomical and Ultrastructural Characteristics of C₃–C₄ Intermediate Species[a]

Species	BSC/vasculature[b]	Package of BSC[c]	BSC organellar feature	Reference
Group I—Kranz anatomy well developed				
Neurachne minor	10–14, Small inner sheath cells, thick-walled, lamella suberized	Over 90%, tightly packed with chloroplasts and mitochondria	Granal chloroplasts	Hattersley *et al.* (1986)
Flaveria brownii	6–8, Thick-walled	30–60%	Granal chloroplasts, prominent mitochondria and peroxisomes centripetally arranged	Holaday *et al.* (1984); Cheng *et al.* (1986)
Group II—Kranz anatomy intermediate to C₃ and C₄ plants				
Panicum milioides *Panicum decipiens* *Panicum schenkii*	6, Large, thick-walled	20–35%	Granal chloroplasts, prominent mitochondria and peroxisomes centripetally arranged	Kanai and Kashiwagi (1975); Brown *et al.* (1983a,b)
Parthenium hysterophorus	5–8, Large, thick-walled	20–35%	Granal chloroplasts, prominent mitochondria and peroxisomes largely arranged in a centripetal position	Moore *et al.* (1987b)
Flaveria anomala *Flaveria ramosissima*	6–8, Thick-walled	20–35%	Granal chloroplasts, mitochondria and peroxisomes centripetally arranged	M. S. B. Ku (unpublished)
Alternanthera ficoides *Alternanthera tenella*	5–6, Thick-walled	20–35[d]	Chloroplasts centripetally located	Rajendrudu *et al.* (1986)

Group III—Kranz anatomy poorly developed

	[b]	[c]		
Flaveria chloraefolia	4–6, Shape and size irregular	15–30%	Granal chloroplasts, prominent mitochondria and peroxisomes centripetally arranged	Holaday *et al.* (1984)
Flaveria floridana *Flaveria pubescens*	10–12, Shape and size irregular	15–30%	Granal chloroplasts, prominent mitochondria and peroxisomes centripetally arranged	Holaday *et al.* (1984), M. S. B. Ku (unpublished)
Flaveria linearis	10–12, Shape and size irregular	10–25%	Granal chloroplasts, mitochondria and peroxisomes evenly distributed throughout the cytosol	Holaday *et al.* (1984)
Moricandia arvensis	8–10	10–25%	Granal chloroplasts, prominent mitochondria and peroxisomes centripetally located	Holaday *et al.* (1981); Winter *et al.* (1982a)
Mollugo verticillata *Mollugo nudicaulis*	5–6	5–10%	Granal chloroplasts and mitochondria evenly distributed in the periphery	Laetsch (1971); Kennedy *et al.* (1980)

[a] It is important to note that bundle sheath cells per vasculature and package of bundle-sheath cells are our estimates from examination of available micrographs in the studies cited. More precise determination of these parameters will require measurements on a large number of micrographs.

[b] Bundle sheath cells (BSC) per vasculature for small veins.

[c] As percent of cell profiles packed with organelles (e.g., chloroplasts, mitochondria, and peroxisomes) of leaf cross sections from the studies cited.

[d] Estimated from hand-drawn illustrations. Additional microscopic examinations will be required to determine specific anatomical features.

anatomy, closely resembling that of their C_4 counterparts. In these two intermediate species, bundle-sheath cells are thick-walled and are tightly packed with numerous chloroplasts and mitochondria. The bundle-sheath cell walls of *N. minor* also contain a suberized lamella, unlike those of other known intermediates. In other intermediate species, bundle-sheath cells are less developed in terms of cell wall thickening, and the proportion of bundle-sheath cells occupied by organelles (e.g., chloroplasts, mitochondria, peroxisomes) seems to be considerably less than in their closely related C_4 counterparts or other C_4 plants. At the opposite end of the spectrum are the intermediate *Mollugo* species, which contain relatively few chloroplasts or mitochondria in their bundle-sheath cells. In most intermediate species, with few exceptions, the bundle-sheath organelles are largely arranged in a centripetal position (C_4 dicots and many NAD-malic enzyme C_4 monocots have centripetally located organelles in Kranz cells). In *F. linearis* and the *Mollugo* intermediates, the organelles in the bundle sheath cells are more evenly distributed along the periphery. Among the *Flaveria* intermediates there is an apparent gradation in leaf anatomy, with *F. brownii* possessing a typical Kranz anatomy, *F. ramosissima* and *F. anomala* possessing a distinct Kranz-like anatomy resembling that of C_4 *Flaveria* species, while other *Flaveria* intermediates show fewer Kranz-like features. However, the variation in leaf anatomy in these *Flaveria* species is not strictly correlated with their photorespiratory and biochemical activities (Section III,B and III,C).

In all intermediates examined, the bundle-sheath chloroplasts have grana and accumulate starch. The granal nature of bundle-sheath chloroplasts would suggest they are capable of PS II activity, unlike agranal bundle sheath chloroplasts in some C_4 plants (see Edwards and Walker, 1983). In addition to numerous chloroplasts, the bundle-sheath cells of most intermediates also have many large, prominent mitochondria located in a centripetal position and in close association with chloroplasts. The presence of these prominent mitochondria in the bundle-sheath cells of intermediates may be important in the suppression of photorespiration. However, in some C_3 species such as *Panicum laxum* and *P. hylaeicum* (Brown *et al.*, 1983a,b) and in *Flaveria pringlei* (Holaday and Chollet, 1984), mitochondria are also prominent in bundle-sheath cells. In contrast to most intermediates, mitochondria are not that apparent in the bundle-sheath cells of *F. linearis* (Holaday *et al.*, 1984). In a recent ultrastructural study of leaves of three intermediate *Panicum* species, some mitochondria in bundle-sheath cells are completely enclosed by chloroplasts (Brown *et al.*, 1983a,b). Whether this unusual ultrastructural feature is related to the low photorespiration in these species remains to be established.

B. Physiology

C_3–C_4 intermediates at least by some measures have levels of apparent photorespiration that are intermediate to those of C_3 and C_4 plants. In considering how photorespiration may be reduced in intermediates, the true rate of photorespiration (TPR) and the apparent rate of photorespiration (APR) need to be defined. TPR is the rate of generation of CO_2 in the leaf in the light. APR is the true rate of release of CO_2 in the light minus the rate of refixation of this CO_2. If all of the photorespired CO_2 were refixed under a given condition or in a certain species, then APR (for example, as measured by the CO_2 compensation point as described below) would be zero. The apparent rate of photosynthesis (APS or net rate of CO_2 uptake) is equal to the true rate of photosynthesis (TPS) minus the true rate of photorespiration. Thus, not including dark respiration, APS = TPS − TPR.

In addition to leaf anatomy, the CO_2 compensation point (Γ) is a primary character for identifying C_3–C_4 intermediate species, and it is one qualitative measure of apparent photorespiration. The term Γ is defined as the ambient CO_2 concentration at which the apparent rate of photosynthesis is just balanced by the apparent rate of photorespiration (and where the true rate of photosynthesis equals the true rate of photorespiration), and therefore Γ is a good measure of photorespiratory loss of CO_2 from the leaf. Typically, at 25°C and 21% O_2, C_3 and C_4 plants have Γ values of about 50 μl/l and near 0 μl/l, respectively. Plants with C_3–C_4 intermediate characteristics exhibit values ranging from 1 to 5 μl/l in *F. brownii* and *N. minor* to 39 μl/l in *M. verticillata* (Table V). Considerable variations in Γ values have been observed in some intermediates, presumably due to differences in methods of measurement (e.g., extrapolation from photosynthetic CO_2 response curves versus steady-state equilibrium), leaf age, conditions of measurement (e.g., light and temperature), and nutrition (Apel *et al.*, 1978; Brown and Morgan, 1980; Winter *et al.*, 1982b). The intermediate Γ values of most C_3–C_4 species indicate that they have lower rates of apparent photorespiration than C_3 plants, yet higher rates than C_4 plants. Exceptions are *N. minor* and some *Flaveria* intermediates (e.g., *F. brownii* and *F. ramosissima*), which could be classified as C_4 species on the basis of their C_4-like Γ values (Table V). Therefore, in classifying intermediates several features of the Kranz syndrome, including other measures of apparent photorespiration, must be considered.

The light-dependent release of CO_2 into CO_2-free air after steady state photosynthesis is reached is another measure of apparent photorespiration that has been used in studying the physiology of intermediates. Although the absolute values vary from species to species, in comparative studies the rates of apparent photorespiration of intermediates in CO_2-free air are invariably lower than those of C_3 counterparts (Brown and Brown, 1975; Brown and Morgan, 1980; Apel, 1980; Apel and Maass, 1981; Holaday *et al.*, 1982). Relative to C_3 species, the intermediates also have lower ratios of apparent photorespiration in CO_2-free air to apparent photosynthesis in normal air (only 30–45% of that for C_3 species).

In C_3 species, Γ increases linearly with increasing [O_2], while in C_4 species Γ remains low as [O_2] increases up to 50%. Compared with C_3 species, Γ for intermediate species is less sensitive to changes in [O_2], especially at low O_2

TABLE V Influence of O_2 on the Photosynthetic CO_2 Compensation Point and Photosynthesis in C_3–C_4 Intermediate Species

Species	CO₂ Compensation point				Percent inhibition by 21% O_2[c]		Reference
	0–2% O_2 (µl/l)	21% O_2 (µl/l)	45–50% O_2 (µl/l)	Apparent break point[a] (% O_2)	Carboxylation efficiency	Photosynthesis	
Alternanthera ficoides		22					Rajendrudu et al. (1986)
Alternanthera tenella		18					Rajendrudu et al. (1986)
Flaveria angustifolia		24					J. Wu and M. S. B. Ku (unpublished)
Flaveria anomala	11	12			46	25	Ku et al. (1983)
Flaveria brownii	10	9	8				Holaday et al. (1984)
	1	1	3				Cheng et al. (1986)
						10[f]	J. Wu and M. S. B. Ku (unpublished)
Flaveria chloraefolia	12	25	46	21[b]			Holaday et al. (1984)
Flaveria floridana	8	8	9	21[b]			Holaday et al. (1984)
Flaveria linearis	15	17	17	21[b]			Holaday et al. (1984)
Flaveria oppositifolia	6	8			42	19	Ku et al. (1983)
		17				20	Brown et al. (1986b)
Flaveria pubescens	11	14			31	21	Ku et al. (1983)
Flaveria ramosissima	5	7			7	22	Ku et al. (1983)
Flaveria sonoransis		30					J. Wu and M. S. B. Ku (unpublished)

Species							Reference
Mollugo nudicaulis		39					Kennedy et al. (1980)
Mollugo verticillata		25–40				12–21	Sayre and Kennedy (1977)
						31	R. H. Brown (unpublished)
Moricandia arvensis	3	28	78	10			Apel et al. (1978)
	4	16	48	15		26	Holaday et al. (1982)
Moricandia sinaica	33	33					Apel and Ohle (1979)
Moricandia spinosa	24			12			Apel (1980)
Neurachne minor	1	5	7		0–10[d]	31[d]	Hattersley et al. (1984); Hattersley et al. (1986)
Panicum decipiens	11	22	75			24	Morgan and Brown (1980)
Panicum milioides	1	17	50	10		21	Keck and Ogren (1976)
	10	16	43			22	Morgan and Brown (1980)
	2	15	40			23	Holaday et al. (1982)
	2	11	40				Hattersley et al. (1986)
					51[e]		Ku and Edwards (1978)
Panicum schenkii	8	14	43			27	Morgan and Brown (1980)
Parthenium hysterophorus	1	22	80				Moore et al. (1987b)

[a] Apparent break point of O_2 concentration in the biphasic response of CO_2 compensation point to changes in O_2, below which little change in CO_2 compensation point occurs. [b] Plotted using the data of Holaday et al. (1984). [c] Calculated as (measurement at 2% O_2 − measurement at 21% O_2) × 100/(measurement at 2% O_2). [d] Estimated from the data of Hattersley et al. (1984, 1986). [e] Calculated from the data of Ku and Edwards (1978). [f] Photosynthesis rates measured at 30°C, 325 μl CO_2/l and 2% or 21% O_2 under a photon flux density of 1600 μmol/m^2 sec.

levels (Table V). The intermediate *F. brownii* exhibits very little increase in
Γ with increasing [O_2] up to 50%. Most of the other intermediate species
examined are characterized by having a biphasic response of Γ to changing
[O_2]: *Panicum* sp. (Keck and Ogren, 1976; Morgan and Brown, 1980; Holaday *et al.*, 1982; Hattersley *et al.*, 1986), *Moricandia* sp. (Apel *et al.*, 1978;
Apel, 1980; Holaday *et al.*, 1982), several *Flaveria* species (Holaday *et al.*,
1985), *Neurachne minor* (although not always observed; Hattersley *et al.*,
1986), and *Parthenium hysterophorus* (Moore *et al.*, 1987b). As [O_2] increases there is a slow rise in Γ until a transition point, after which there is a
more rapid linear rise in Γ with increasing [O_2]. Above the [O_2] where Γ
increases more rapidly, the rate of increase is still always less than that of C_3
plants. The rate of increase in Γ with increasing [O_2] and the [O_2] where there
is an apparent break point varies depending on the intermediate: 10–25% for
Panicum species, 10–15% for *Moricandia* species, 10% for *Parthenium hysterophorus*, and over 50% for *Neurachne minor* (Table V). When comparisons were made with other intermediates, the response of *N. minor* is much
diminished and has a high O_2 transition point. The [O_2] at which there is a
transition in the response and the rate of change in Γ/change in [O_2] above
the transition point must reflect to some extent the efficiency of the biochemical machinery of an intermediate in reducing its apparent photorespiration.
An analysis of the response of Γ to [O_2] should be very useful for interspecific comparison of photorespiratory properties of intermediate species. It
appears that, in intermediates that are capable of operating a limited C_4
pathway of photosynthesis (e.g., *F. floridana* and *N. minor*; see Section
III,C), the transition points are higher and the slopes (change Γ/change [O_2])
lower.

    Another interesting observation with C_3–C_4 intermediates with regard to Γ
is the influence of light intensity. In *Panicum milioides* and *Moricandia
arvensis*, Γ at 21% O_2 increases from intermediate values to values approaching those of C_3 plants as the photosynthetic photon flux density decreases
from about 30% to 5% of full sunlight (Brown and Morgan, 1980; Holaday *et
al.*, 1982). A similar light effect is also observed in the intermediate *Flaveria
linearis* (G. Byrd and R. H. Brown, personal communication). In contrast, Γ
for C_3 and C_4 plants remains essentially unchanged over this range of light
intensities. The increase in Γ at 21% O_2 under low light intensities is not due
to an increase in dark respiratory CO_2 release as light intensity decreases,
since Γ at 2% O_2 for all species remains constant under varying light levels
(Holaday *et al.*, 1982). These results suggest that the proportion of the
photorespired CO_2 that is refixed at low light is less than under high light
intensities. It was suggested that some unknown factor(s) must be involved
in reducing photorespiratory CO_2 loss in these two diverse species, which
varies with a change in light intensity (Holaday *et al.*, 1982). Further research on Γ values in intermediates relative to [O_2] and light intensity may
provide useful information. This includes the effect of light intensity on (a) Γ

in intermediates that possess limited C_4 photosynthesis, such as *F. ramosissima* and *N. minor,* and (b) the response of Γ to varying $[O_2]$ at different light intensities.

Another physiological characteristic of some intermediate species is reduced O_2 inhibition of photosynthesis at atmospheric levels of CO_2 relative to that for C_3 species (Table V). For a number of intermediates the percentage inhibition of photosynthesis by 21% O_2 ranges from 20 to 25%. For *F. brownii,* however, the percentage inhibition of photosynthesis by O_2 is low (8–12%), consistent with its low Γ and insensitivity of Γ to O_2 (Table V). Under similar conditions, photosynthesis by C_3 plants is inhibited by about 30–35% (Ku and Edwards, 1978). There is very little effect of O_2 (21% versus 2%) on photosynthesis in C_4 plants; occasionally 21% O_2 may cause a slight stimulation of photosynthesis in C_4 species. Surprisingly, *N. minor,* which has a C_4-like Γ and low sensitivity of Γ to O_2, has the highest degree of sensitivity of photosynthesis to O_2 (31% inhibition of photosynthesis by 21% O_2; Table V). The percentage inhibition of photosynthesis by O_2 in *N. minor* is comparable to, or even higher than, that in the C_3 *Neurachne* species. The reason for the C_4-like Γ and the C_3-like O_2 inhibition of photosynthesis in *N. minor* is not apparent. The term Γ is a measure of apparent photorespiration that will be influenced by the degree of refixation of the photorespired CO_2, whereas O_2 inhibition of photosynthesis includes both photorespiration and a competitive component (see Section III,C). The O_2 inhibition of photosynthesis in the intermediate *Moricandia arvensis* is also similar to that of C_3 species (30.8% at 30°C and high light over 17 replications; R. H. Brown, personal communication). Therefore, the mechanism of reduction of photorespiration in some species will have to account for a reduced Γ without reduction in O_2 inhibition of photosynthesis. The O_2 sensitivity of photosynthesis in C_3 plants depends on the intercellular $[CO_2]$ (Ku and Edwards, 1977). Therefore, careful control of the intercellular $[CO_2]$ will be necessary for a strict comparison of the O_2 sensitivity of photosynthesis in intermediates versus C_3 species.

The O_2 inhibition of carboxylation efficiency (CE, defined as the initial slope of photosynthesis in response to varying $[CO_2]$ at low levels) has been examined in some intermediates (Table V). In C_3 species O_2 inhibits CE due to the competitive nature of O_2 with respect to CO_2 on Rubisco, but O_2 has no effect on CE in C_4 species because of the CO_2-concentrating mechanism by the C_4 pathway. If the intermediate species possess a CO_2-concentrating mechanism, as do C_4 plants, then an increase in $[O_2]$ from 2 to 21% should have less effect on CE compared to C_3 species. In several intermediates, including *P. milioides, F. anomala,* and *F. linearis,* the degree of O_2 inhibition of CE is equal to that of C_3 species (Monson *et al.,* 1984), suggesting the absence of a CO_2-concentrating mechanism. Only in *F. ramosissima* and *N. minor* is the O_2 inhibition of CE markedly lower than in the C_3 plants, possibly because of a limited CO_2-concentrating mechanism.

C. Biochemistry

1. General Mechanisms by Which Intermediates May Reduce
Apparent Photorespiration

As already noted, there are two means by which O_2 inhibits photosynthesis in C_3 plants. First, O_2 is a competitive inhibitor of Rubisco with respect to CO_2, so that the apparent K_m for CO_2 increases in the presence of O_2. This means, in the presence of O_2, it takes a higher $[CO_2]$ to reach saturation. Second, CO_2 is released during metabolism in the glycolate pathway. Therefore, where APS = TPS − TPR (see Section III,B), O_2 is a competitive inhibitor of TPS with respect to CO_2 and TPR is the rate of release of CO_2 in photorespiration. In C_4 plants, by concentrating CO_2 at the site of Rubisco, both of these means of O_2 inhibition of photosynthesis are largely prevented. Elevated CO_2 will decrease the competition by O_2 and reduce the synthesis of glycolate that leads to photorespiration. However, another means of minimizing losses from photorespiration is the reassimilation of the photorespired CO_2. Although less effective than the CO_2-concentrating mechanism in C_4 plants, the ability to reassimilate photorespired CO_2 may be an early step in the process of evolution of C_4 photosynthesis.

Figure 1 is a general scheme for fixation of atmospheric CO_2, the possible fate of the fixed CO_2, and that of photorespired CO_2. In a C_3 plant, atmospheric CO_2 is fixed only through the PCR cycle, and photorespired CO_2 contributes to O_2 inhibition of photosynthesis. In a C_4 plant, atmospheric CO_2 is fixed solely into C_4 acids, which are then decarboxylated with an increase in the $[CO_2]$ in the leaf. Due to location of the carboxylation phase of the C_4 pathway in the mesophyll cells, and the decarboxylation phase and Rubisco in the Kranz cells, effective donation of carbon from the C_4 to the PCR cycle occurs with a minimal release of photorespired CO_2.

In theory there are several means by which a species could have levels of apparent photorespiration between those of a C_3 and C_4 species, as indicated below.

a. Type I. No C_4 Cycle. If atmospheric CO_2 is fixed only through the PCR cycle, then apparent photorespiration might be reduced by a more efficient reassimilation of photorespired CO_2. In this case, for example at the CO_2 compensation point, the site of release of the photorespired CO_2 in the intermediates would be such that a greater proportion is refixed by Rubisco than in C_3 plants.

b. Type II. With C_4 Cycle. A more complex list of possibilities exists if the intermediate species have a limited, but functional C_4 cycle along with the PCR cycle. There are three possible sources of CO_2 for fixation in the C_4 cycle: CO_2 from the atmosphere, CO_2 from C_4 acid decarboxylation, and

CO_2 from photorespiration (Fig. 1). The extent to which these serve as sources of CO_2 to the C_4 cycle will depend, mainly, on the compartmentation of reactions of the C_4 cycle and PCR cycle in the mesophyll and Kranz-like cells in the leaf if the latter cells are present. The C_4 cycle would be futile if there were continuous carboxylation and decarboxylation of a given molecule of CO_2. Also, apparent photorespiration would not be reduced if either atmospheric CO_2 or photorespired CO_2 were fixed into C_4 acids followed by their decarboxylation in the same cell without an increase in the CO_2 level in the chloroplast. The following are ways in which intermediacy could occur through a C_4 cycle:

(a) If part of the *atmospheric* CO_2 is fixed in the C_4 cycle, the C_4 acids decarboxylated, and the [CO_2] raised to a higher concentration at the site of the Rubisco, O_2 inhibition of photosynthesis will be reduced by a decrease in the competitive inhibition by O_2, and by a reduction in carbon metabolism through the glycolate pathway. If CO_2 is donated from C_4 acids to the PCR cycle with little or no elevation of CO_2 in the leaf, then little reduction in the generation of photorespired CO_2 would occur.

(b) If the *photorespired* CO_2 is fixed in the C_4 pathway, the C_4 acids decarboxylated and the CO_2 then donated to the PCR cycle, apparent photorespiration could be reduced. In order for apparent photorespiration to be reduced by this means, there should be a localized increase in the concentration of photorespired CO_2 around Rubisco (see below).

(c) Another type of intermediate could occur if a species had Kranz anatomy, the C_4 cycle and proper intercellular compartmentation of enzymes (all C_4 plant features), but bundle sheath cells with relatively high conductance. In this case, the cells would be leaky, such that the CO_2 concentration may not be elevated to sufficient levels to totally prevent O_2 inhibition of photosynthesis.

c. Type III. Modified Rubisco. If the properties of Rubisco were modified such that reaction with CO_2 was increased relative to reaction with O_2, photorespiration (true and apparent) would be reduced and the species would have intermediate physiological characteristics.

2. Biochemical Basis for Reduced Apparent Photorespiration among Different Intermediate Species

Among the possibilities discussed above, at the biochemical level there is currently evidence for type I and type II intermediates (Table VI). No intermediate species among higher plants have been identified which have a lower photorespiration through a modification in Rubisco (type III).

In type I intermediates photorespiration is considered to be reduced by refixing photorespired CO_2. In general, these intermediates have low activities of C_4 pathway enzymes, similar to C_3 plants, and they lack a functional C_4 cycle. They possess a Kranz-like anatomy, with mitochondria and chlo-

TABLE VI
List of C_3–C_4 Intermediate Species and Classification
Relative to Mechanism of Reduction of Photorespiration[a]

Species	C_4 cycle not present	Evidence for C_4 cycle	Reference
Alternanthera ficoides	$+^d$		Rajendrudu *et al.* (1986)
Alternanthera tenella	$+^d$		Rajendrudu *et al.* (1986)
Flaveria anomala		+	Monson *et al.* (1987)
Flaveria brownii		+	Bassüner *et al.* (1984)
			Cheng *et al.* (1986)
Flaveria floridana		+	Monson *et al.* (1987)
Flaveria linearis		+	Monson *et al.* (1987)
Flaveria pubescens		+	Bassüner *et al.* (1984)
			Monson *et al.* (1987)
Flaveria ramosissima		+	Rumpho *et al.* (1984)
			Monson *et al.* (1987)
Mollugo verticillata		+	Kennedy and Laetsch (1974)
			Sayre and Kennedy (1977)
Moricandia arvensis	+		Winter *et al.* (1982b)
			Holaday and Chollet (1983)
Neurachne minor		$*^c$	Hattersley and Stone (1986)
Panicum milioides[b]	+		Kanai and Kashiwagi (1975)
			Edwards *et al.* (1982)
Parthenium hysterophorus	+		Hegde and Patil (1981)
			Moore *et al.* (1987b)

[a] It has not been determined whether the intermediate species not listed here (see Table II) have a C_4 cycle. In *M. nudicaulis* only 9–11% of the initial products of CO_2 fixation after a 3-sec pulse is in the C_4 acids malate + aspartate (Kennedy *et al.*, 1980). Further studies on $^{14}CO_2$ assimilation will be required to determine whether a C_4 cycle is functioning in this species.

[b] A functional C_4 cycle was reported for this species (see Rathnam and Chollet, 1980) but no evidence was found for a C_4 cycle under atmospheric conditions in subsequent studies (see Edwards *et al.*, 1982).

[c] Suggested may have C_4 cycle based on enzyme activities but functional C_4 cycle not yet demonstrated.

[d] Based on low activity of C_4 cycle enzymes. Confirmation is needed through pulse-chase/$^{14}CO_2$–$^{12}CO_2$ experiments.

roplasts as prominent in the bundle sheath cells as in the mesophyll cells. A primary site of release of photorespired CO_2 is suggested to be in the bundle-sheath cells, where it is efficiently refixed. As already noted, these species have reduced apparent photorespiration compared to C_3 species of these genera, as shown by lower apparent photorespiration in the light in CO_2 free air and lower Γ.

In type II intermediates there is evidence for significant levels of C_4 pathway enzymes and a functional C_4 pathway. The extent of C_4 pathway activity and the efficiency in donation of carbon from the C_4 pathway to the PCR cycle vary among the species of this type. The plants have Kranz-like anatomy. At least some of the species that are most C_4-like have a greater reduction in photorespiration compared to type I intermediates from analysis of Γ, O_2 inhibition of CE, and O_2 inhibition of photosynthesis. By examining the biochemistry of specific intermediates in some detail, the basis for their reduction in photorespiration and the possible evolution of intermediates can be considered.

 a. Type I. No C_4 Cycle. The best documented cases of reduced apparent photorespiration in intermediates without a functional C_4 cycle are in *Panicum milioides* and *Moricandia arvensis*. In both *P. milioides* (Kanai and Kashiwagi, 1975; Edwards *et al.*, 1982) and *M. arvensis* (Winter *et al.*, 1982b; Holaday and Chollet, 1983), atmospheric CO_2 is fixed via the PCR cycle. It is suggested that a major site of release of photorespired CO_2 is in the mitochondria of bundle sheath cells and that this CO_2 is refixed by bundle-sheath chloroplasts (Brown, 1980; Edwards *et al.*, 1982; Brown and Bouton, 1983; Monson *et al.*, 1984). Therefore the site of release of photorespired CO_2 in the leaf may be such that there is a relative high resistance to its access to the atmosphere relative to its access to chloroplasts. In order for apparent photorespiration to be reduced by this means, a localized increase in $[CO_2]$ should occur in the area where CO_2 is released above what would normally exist in this compartment in the absence of photorespiration. This area needs to be isolated such that the photorespired CO_2 cannot readily escape from the leaf or equilibrate with atmospheric CO_2. Thus the diffusive resistance for CO_2 between the atmosphere and bundle sheath cells would be higher than between the atmosphere and mesophyll cells. For Γ to be reduced by this means compared to a C_3 plant, at least part of the photorespired CO_2 must be confined to a compartment in the leaf where its concentration around Rubisco is greater than that at Γ in a C_3 plant. In theory, whether mitochondria could position themselves within mesophyll cells in such a way as to cause a localized increase in photorespired CO_2 around certain chloroplasts (e.g., if mitochondria were totally surrounded by chloroplasts or located in a centripetal position in the cell) and thereby reduce Γ is not known. The potential for a localized increase in photorespired CO_2 could be limited by a low resistance to diffusion of CO_2 within the cell. Also, the most favorable relative position of chloroplasts and mitochondria for refixing photorespired CO_2 may not provide the best condition for maximum rates of CO_2 fixation under atmospheric conditions. Presently, the species identified of this type possess bundle-sheath cells that contain chloroplasts and mitochondria, which are proposed to account for refixation of photorespired CO_2.

 As illustrated in Fig. 2, in this type of intermediate atmospheric CO_2 is fixed mainly by mesophyll and partly by bundle-sheath chloroplasts through the PCR cycle. Reaction of O_2 with ribulose-1,5-P_2 in the PCR cycle in both

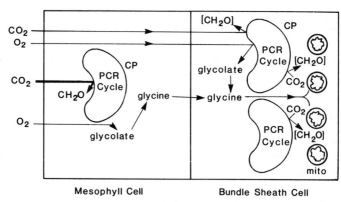

Fig. 2. Proposed mechanism of refixation of photorespired CO_2 in the absence of a C_4 cycle in intermediate species (type I, see text). CP, chloroplast; mito, mitochondria. See legend of Fig. 1 for other abbreviations.

cell types leads to the synthesis of glycolate. Glycolate is metabolized to glycine in peroxisomes and then the glycine is decarboxylated by mitochondria. In bundle-sheath cells the proportion of the photorespired CO_2 that is fixed by the chloroplasts will depend on the diffusive resistance to loss of CO_2 from the compartment and on the energy available (light intensity). Efficient refixation of photorespired CO_2 from mitochondria in bundle-sheath cells and reduction of apparent photorespiration could occur by utilization of glycine generated in the bundle-sheath cells and part of the glycine synthesized in mesophyll cells. In *P. milioides* (and *P. hians,* synonym for *P. milioides*) the activity of Rubisco, glycolate oxidase and catalase are similar in both mesophyll and bundle-sheath cells on a chlorophyll basis (Ku *et al.,* 1976). About 80% of the chlorophyll of the leaf is in mesophyll cells and 20% in bundle-sheath cells (based on measurements on separated cells; G. Edwards, unpublished). Therefore about 80 and 20%, respectively, of the capacity for CO_2 fixation in the PCR cycle, for glycolate synthesis via the oxygenase reaction, and for glycolate conversion to glyoxylate in the peroxisomes is located in mesophyll and bundle-sheath cells, respectively. The distribution of glycine decarboxylase activity between the two cell types is unknown. Knowledge of this distribution and other functions in glycolate metabolism will be required to evaluate this hypothesis in more detail. It should be noted that there is evidence that photorespiration occurs in C_4 plants although this is not apparent in physiological measurements (see Edwards *et al.,* 1985). In C_4 plants glycine decarboxylase and serine hydroxymethyltransferase are located in the mitochondria of bundle-sheath cells (Ohnishi and Kanai, 1983; Gardestrom *et al.,* 1985) and thus the photorespired CO_2 can be efficiently refixed. Therefore, species such as *P. milioides* and *M. arvensis* that may have partially developed this compart-

TABLE VII
Dark/Light Ratios for Evolution of $^{14}CO_2$ during Decarboxylation
of Exogenous [1-^{14}C]Glycine and [1-^{14}C]Glycolate
by Leaf Disks[a]

| Species | Classification | Dark/light ratio of $^{14}CO_2$ evolution | |
		[1-^{14}C]Glycine	[1-^{14}C]Glycolate
Moricandia arvensis	Intermediate	9.1	5.5
Moricandia foetida	C₃	2.5	—
Panicum milioides	Intermediate	11.8	7.6
Panicum laxum	C₃	3.2	2.5
Panicum miliaceum	C₄	Infinity[b]	21.4

[a] From Holbrook *et al.* (1985) with permission from the American Society of Plant Physiologists.
[b] $^{14}CO_2$ evolution in the light.

mentation in photorespiratory metabolism could have some of the biochemical features of C₄ plants.

The best biochemical evidence supporting this hypothesis is a study of the light/dark metabolism of radioactive glycine and glycolate by leaf disks of *M. arvensis* and *P. milioides* versus representative C₃ species (Holbrook *et al.*, 1985) (Table VII). The relatively high dark/light ratios of released radioactive $^{14}CO_2$ from these metabolites in the intermediates suggest they have a relatively greater capacity than the C₃ species to refix photorespired CO_2 in the light. Since the decarboxylation experiments were performed in CO_2 free air, it is uncertain to what extent atmospheric CO_2 would limit refixation of photorespired CO_2 in the various species. If glycine decarboxylation occurs in both mesophyll and Kranz-like cells in the intermediate species, it is unknown whether metabolism of exogenous glycolate and glycine by the two cell types in leaf disks is in the same proportion as that *in vivo*.

Preliminary experiments suggest the true rate of CO_2 fixation through Rubisco in *P. milioides* is not inhibited by O_2 to the same extent as in the C₃ species *P. bisulcatum* (see Holbrook *et al.*, 1985). However, this does not necessarily indicate that O_2 is a less effective inhibitor of the enzyme *in vivo* in this species than in the C₃ species. The state of activation of the enzyme needs also to be considered. Increasing O_2 concentration can increase the state of activation of Rubisco *in vivo* in some species (Schnyder *et al.*, 1984), an effect that could partially offset O_2 inhibition of photosynthesis under some conditions. The reduced O_2 sensitivity in the intermediate could be related to a reduction in photosynthesis rate through a limitation on triose-P (triose phosphate) utilization as proposed by Sharkey (1985). Interestingly, a recent study shows that photosynthesis by a mutant plant of the intermediate *F. linearis* is stimulated rather than inhibited by atmospheric [O_2] (Brown *et al.*, 1986a). It is unlikely that Rubisco in this mutant is altered such that the oxygenase activity has been eliminated, since O_2 still inhibits the carboxylation efficiency measured under low $[CO_2]$.

In both *P. milioides* and *M. arvensis*, the mitochondria of bundle sheath cells are larger than the mitochondria of mesophyll cells (Section III,A). In *P. milioides* and some other species of the *Laxa* group, some of the mitochondria are enclosed by chloroplasts (Brown *et al.*, 1983a,b). Whether this unusual finding has any relevance to the intermediacy in apparent photorespiration is not known. In the intermediates *P. milioides* and *P. schenckii* (synonym *P. spathellosum*), about 25% of the total leaf chloroplasts and 45% of the mitochondria occur in bundle-sheath cells. The bundle-sheath mitochondria are about 1.7 times larger than mesophyll cell mitochondria in these species. If the relative capacity of glycine decarboxylase between mesophyll and bundle-sheath mitochondria is proportional to mitochondrial area (size times number), about 75% of the enzyme potential would reside in the bundle sheath cells in these species. Thus, potentially the bundle-sheath cells could be a major site of release of photorespired CO_2. The C_3 species of the *Laxa* group have photorespiratory characteristics typical of other C_3 plants. Even though *Panicum laxum* has about 11% of the chloroplasts and 17–23% of the mitochondria of the photosynthetic tissue in bundle-sheath cells, this species has photorespiratory characteristics similar to *P. boliviense* (C_3), which has only 2–3% of these organelles in bundle-sheath cells. *Panicum laxum* may have a slightly lower Γ than other C_3 species of this group; however, many measurements of photorespiration under carefully controlled conditions are required to determine if it is any lower in this species than in C_3 species with few or no organelles in the Kranz cells.

In *M. arvensis* there is also a remarkable arrangement of mitochondria in the bundle sheath cells. Arrays of mitochondria are consistently found in the parietal layer of cytoplasm along the inner tangential walls of the bundle sheath cells, and the location of the mitochondria is internal to that of the chloroplasts (Section III,A). The magnitude of Γ depends in part on the relative rates of true photorespiration and the loss of photorespired CO_2 from the leaf. The abundance of mitochondria and their position relative to chloroplasts in bundle-sheath cells of *M. arvensis* may increase the potential for chloroplasts to fix the photorespired CO_2, and thereby reduce the level of Γ. Another potential role of the mitochondria in bundle sheath cells of intermediates is facilitating a C_4 cycle via decarboxylation of malate through NAD-malic enzyme. Although the level of NAD-malic enzyme in most intermediate species is low, this possibility needs to be examined in more detail.

Among the C_4 cycle enzymes that have been measured in leaves of *P. milioides* and *M. arvensis*, only PEP carboxylase shows a significant increase in activity compared to that found in C_3 species (Goldstein *et al.*, 1976; Ku *et al.*, 1976; Morgan *et al.*, 1980; Holaday *et al.*, 1981; Edwards *et al.*, 1982; Winter *et al.*, 1982b). For example, the activities of the C_4 acid decarboxylating enzymes (NADP-malic enzyme, NAD-malic enzyme and PEP carboxykinase) are very low. Also, the activity of pyruvate,Pi dikinase is very low in *P. milioides* (less than 3% of the activity found in maize) and undectable in *M. arvensis*. Low activities of pyruvate,Pi dikinase have been detected in some C_3 species. The activity of this enzyme in *P. milioides* (Edwards *et al.*, 1982) is similar to that found in leaves of wheat (Aoyagi and Bassham, 1983). The activity of PEP carboxylase is generally 1.5- to 2-fold higher in leaves of *P. milioides* and *M. arvensis* than in C_3 species of these genera. However, there is no evidence that PEP carboxylase is contributing to a reduction of apparent photorespiration in these species under atmospheric conditions (Edwards *et al.*, 1982; Winter *et al.*, 1982b; Holaday and Chollet, 1983). In *P. milioides* and *M. arvensis*, pulse-chase experiments with $^{14}CO_2$–$^{12}CO_2$ show a pattern of metabolite labeling typical of C_3 plants, with the majority of the label in the PCR cycle. Only 3–6% of the label appeared in the C_4 acids malate + aspartate, and there was little change in label in the C_4 acids during the chase period. In fact, in some cases there was a continual increase in label of C_4 acids during the chase period. The labeling pattern of aspartate in C-4 versus C-1 + C-2 + C-3 during the pulse-chase period suggests the C_4 acids are synthesized in a secondary carboxylation using labeled 3-phosphoglycerate from Rubisco as the precursor (Edwards *et al.*, 1982). In "middle-aged" leaves of plants of *M. arvensis* grown on nitrate as the N source, only 1.4% of the initial products appeared in C_4 acids after a 10-sec pulse, and during a 5-min chase the label in C_4 acids increased to about 8%, again suggesting a secondary carboxylation (Winter *et al.*, 1982b). The CO_2 for this secondary carboxylation might be derived in part from photorespiration.

Although under normal conditions assimilation of atmospheric CO_2 into C_4 acids does not contribute to intermediate physiological characteristics in *P. milioides* and *M. arvensis*, whether refixation of photorespired CO_2 via the C_4 cycle contributes to reduced photorespiration should be considered. In this case, photorespired CO_2 might be fixed through PEP carboxylase in bundle sheath cells with the PEP being generated from 3-phosphoglycerate (through 3-phosphoglycerate mutase and enolase) or pyruvate (through pyruvate,Pi dikinase). Such CO_2 fixation would not be detected in the primary products of fixation of atmospheric CO_2, but label in C_4 acids might appear during the chase period. The rate of release of CO_2 in photorespiration is estimated to be about 16% of the net rate of photosynthesis (Edwards and Walker, 1983). In both intermediates and C_3 species the activity of PEP carboxylase is more than sufficient to fix photorespired CO_2 (e.g., in *P. milioides* PEP carboxylase activity is about 40% of the rate of photosynthesis). However, the activities of pyruvate,Pi dikinase and of NAD- and NADP-malic enzyme are similar to C_3 plants and only about 5% of the photosynthesis rate, again suggesting the conventional C_4 cycle does not have a role in refixing photorespired CO_2 in these species (Morgan *et al.*,

1980; Edwards *et al.*, 1982). In addition, it is unlikely that a C_4 cycle within bundle-sheath cells would raise the level of CO_2 around Rubisco unless the C_4 acid decarboxylase were located in bundle-sheath chloroplasts.

Earlier it was suggested that PEP carboxylase may contribute to reduction in apparent photorespiration in *P. milioides,* particularly at low $[CO_2]$ (Goldstein *et al.*, 1976; Keck and Ogren, 1976). There is some incorporation of CO_2 into malate and aspartate at subatmospheric $[CO_2]$ in *P. milioides* (Goldstein *et al.*, 1976). However, in *M. arvensis,* pulses with $^{14}CO_2$ concentrations of about 100 $\mu l/l$ resulted in initial incorporation of label into C_4 acids similar to that under higher levels of CO_2 (100 $\mu l/l$ above ambient) (Winter *et al.*, 1982b). Therefore, under these conditions C_4 photosynthesis is not functioning in *M. arvensis.* Whether CO_2 fixation into C_4 acids at Γ in these intermediate species causes any decrease in apparent photorespiration relative to that in C_3 plants is yet to be determined. However, in these species refixation of photorespired CO_2 into the PCR cycle without C_4 photosynthesis may be the major means of reducing apparent photorespiration at Γ. It will be of interest to evaluate the basis for reduction of apparent photorespiration in *Parthenium hysterophorus*, an intermediate whose enzyme composition with respect to the C_4 cycle, including PEP carboxylase, is like that of C_3 plants (Moore *et al.*, 1987b). A recent report indicates that the intermediates *Alternanthera ficoides* and *A. tenella* also have low activities of C_4-cycle enzymes similar to that of C_3 plants (Rajendrudu *et al.*, 1986).

Besides the relatively low activity of PEP carboxylase in *P. milioides* and *M. arvensis* (about 10% of the activity of C_4 plants) there are two other features of PEP carboxylase that indicate these species are not advanced in biochemical evolution toward C_4 photosynthesis. First, the distribution of PEP carboxylase between mesophyll and bundle-sheath cells of *P. milioides* is similar (Ku *et al.*, 1976), indicating no strong preferential location as found in C_4 plants. Second, the PEP carboxylase of *P. milioides* and *M. arvensis* has kinetic properties similar to those of PEP carboxylase from C_3 plants (Holaday and Black, 1981; Holaday *et al.*, 1981). These include a relatively low K_m for PEP, lack of stimulation by glucose-6-P (glucose 6-phosphate) (positive effector with C_4 plant enzyme), and lack of inhibition by malate and aspartate (negative effectors of C_4 enzyme).

b.   Type II. With C_4 Cycle. The genus in which the most evidence exists for intermediate species having a functional C_4 cycle is *Flaveria*. The first indication that C_4 photosynthesis may occur in intermediate *Flaveria* species was the finding that substantial activities of the C_4 cycle enzymes PEP carboxylase, pyruvate,Pi dikinase, NADP-malic enzyme, and NADP-malate dehydrogenase exist in the intermdiates *F. anomala, F. linearis, F. pubescens,* and *F. ramosissima* (Ku *et al.*, 1983). Compared to the C_3 and C_4 *Flaveria* species examined, PEP carboxylase in *F. linearis* and *F. pubescens* had intermediate characteristics in maximum activity on a chlorophyll basis and K_m for PEP. In *F. linearis* the increase in activity of the enzyme by glucose 6-phosphate (a positive effector) at pH 8.0 and limiting PEP was intermediate to that of the C_3 and C_4 species (Nakamoto *et al.*, 1983). In *F. ramosissima*, pulse-chase experiments with $^{14}CO_2$–$^{12}CO_2$ indicated a C_4 cycle is functioning (Rumpho *et al.*, 1984). During a 10-sec pulse about 70% of the label appeared in malate + aspartate, and turnover of the labeled C_4 acids occurred during the 20-min chase period. Evidence for C_4 photosynthesis in the intermediate *F. pubescens* was provided by Bassüner *et al.* (1984). After pulsing with $^{14}CO_2$ for a few seconds, up to 40% of the labeled products were C_4 acids, the percentage of which decreased with longer pulse periods with the labeled CO_2. In a subsequent study of five intermediate *Flaveria* species (*F. anomala, F. floridana, F. linearis, F. pubescens, F. ramosissima*), all showed evidence for C_4 photosynthesis (Monson *et al.*, 1986). The percentage label in C_4 acids as the initial products was lower, and the rate of turnover of labeled C_4 acids during the chase period was slower than in the C_4 species *F. palmeri*. Most recently, evidence was also presented for a CO_2-concentrating mechanism in some C_3–C_4 intermediate and C_4 *Flaveria* species (Moore *et al.*, 1987a).

There are major differences in the degree and relative efficiency of C_4 photosynthesis in the *Flaveria* intermediates, based on the percentage initial labeled products as C_4 acids, the rate of turnover of C_4 acids in the chase period with $^{12}CO_2$, and the quantum yield under various levels of O_2 (Monson *et al.*, 1986). Currently these can be ranked from those having the most efficient C_4 cycle to the least efficient as follows: *F. brownii, F. ramosissima, F. floridana, F. pubescens, F. anomala,* and *F. linearis.*

Flaveria brownii has a typical Kranz anatomy (Section III,A). However, unlike C_4 plants it lacks a strict intercellular compartmentation of certain photosynthetic enzymes. In addition to Kranz cells there are two layers of mesophyll cells, based on size differences, a larger outer layer and a smaller inner layer. There is a gradation of activity (on a chlorophyll basis) of PEP carboxylase from high in the outer layer of mesophyll cells to low activity in the bundle-sheath cells, whereas there is a gradation of activity of ribulose 1,5-bisphosphate (RuBP) carboxylase from bundle-sheath cells to the outer layer of mesophyll cells. Also, this species is capable of fixing part of the atmospheric CO_2 directly via the PCR cycle (Cheng *et al.*, 1986). This may account for the species having slightly more discrimination against $^{13}CO_2$ during photosynthesis as compared to C_4 species (Section IV), lower activity of C_4 enzymes, and about 10% inhibition of photosynthesis by 21% O_2 (Section III). It is therefore classified as an intermediate, and is probably the most C_4-like intermediate currently known.

Among five *Flaveria* intermediates (*anomala, floridana, linearis, pubescens,* and *ramosissima*) there is a range in the percentage of initial products of $^{14}CO_2$ fixation appearing as C_4 acids compared to the C_4 species *F. palmeri* and the C_3 species *F. cronquistii* (Monson *et al.*, 1986). The labeling of C_4

acids in *F. conquistii* was low (5–14% with some tissue-to-tissue variation), and examination of the position of label in C_4 acids indicated CO_2 fixation into C_4 acids occurs to a large extent via a secondary carboxylation (possibly with 3-phosphoglycerate as the precursor). This is consistent with this species having typical C_3 leaf anatomy (Monson *et al.*, 1984). Based on the above criteria, *F. ramosissima* has the most efficient C_4 photosynthesis among these five intermediates. The rate of turnover of label in C_4 acids during the chase period was rather low in *F. anomala* and *F. linearis*, suggesting C_4 acid decarboxylation is limiting in these species or the C_4-acid pool sizes are larger than other intermediates. Thus, there is limited development of a functional C_4 cycle in *F. anomala*, despite it having an advanced Kranz-like anatomy compared to other *Flaveria* intermediates (Section III,A). Other observations in pulse-chase experiments also suggest these *Flaveria* intermediates do not have a completely developed C_4 photosynthesis. The intermediates had 5–15% of the ^{14}C fixed in an initial pulse in citrate + succinate, and most intermediates accumulated substantial label in fumarate during a chase treatment, unlike the C_4 *Flaveria* species (Monson *et al.*, 1986). This suggests some carbon from malate and aspartate may enter the tricarboxylic acid (TCA) cycle.

By analysis of the quantum yield, it was concluded that there was lower efficiency in transfer of CO_2 from the C_4 cycle to the PCR cycle in *F. pubescens* and *F. floridana* than in other *Flaveria* species. In these two intermediate species the quantum yield under atmospheric conditions is 12–15% below that of the C_3 and C_4 species, presumably due to a somewhat futile C_4 cycle, which uses extra energy (Monson *et al.*, 1986). C_3 and C_4 species, including *Flaveria*, have similar quantum yields under atmospheric CO_2. Under atmospheric conditions the quantum yield is reduced below the theoretical maximum for the PCR cycle in C_3 plants, principally because of photorespiration, while it is reduced in C_4 plants due to the additional expense of the C_4 cycle.

Under low O_2 the quantum yield increases in C_3 species due to the elimination of photorespiration, while O_2 has no effect on the quantum yield in C_4 species. In many of the intermediate *Flaveria* species the quantum yield values under low O_2 are in between those of C_3 and C_4 plants. In this case, photorespiration is eliminated, while the C_4 cycle functions and uses energy at some level lower than in C_4 plants, depending on the fraction of atmospheric CO_2 fixed into C_4 acids (Monson *et al.*, 1986).

Neurachne minor is another intermediate species in which a functional C_4 cycle may exist (Hattersley and Stone, 1986). This species has significant activities of PEP carboxylase, pyruvate,Pi dikinase, and NADP-malic enzyme, which are comparable to the highest activities found among the *Flaveria* intermediate species. Also, it clearly has higher activities of C_4 cycle enzymes than the three intermediate grass species in the genus *Panicum*.

There is also some evidence that *Mollugo verticillata,* the first C_3–C_4

intermediate identified (Kennedy and Laetsch, 1974), may have a functional C_4 cycle. In a 5-sec pulse with $^{14}CO_2$, 21% of the initial labeled product was in aspartate + malate (Kennedy and Laetsch, 1974). In a subsequent study among four different populations, from 11 to 40% of the initial products of CO_2 fixation following a 3-sec pulse was in the C_4 acids malate + aspartate (Sayre and Kennedy, 1977). In the Iowa population, 35% of the label was in C_4 acids after a 3 sec pulse, decreasing to a low level within a 10-sec chase period. Yet the rate of photosynthesis was low, and in these studies with *M. verticillata* a rather high percentage of the total labeled products appeared in alanine after a 3-sec pulse (23–54%), which is unusual. Activities of C_4-cycle enzymes, PEP carboxylase, aspartate aminotransferase, and NAD-malic enzyme, are somewhat higher (on a protein basis) in the *M. verticillata* populations than in a C_3 *Nicotiana* species (Sayre *et al.*, 1979). However, on a chlorophyll basis activities of these enzymes were low and similar to C_3 plants. The percentage inhibition of photosynthesis by 21% O_2 in the four populations ranged from 12 to 21%, while three populations had Γ values of 40 μl/l and one population had a Γ value of 25 μl/l (Sayre and Kennedy, 1977). There was no correlation between degree of fixation of CO_2 into C_4 acids and the extent of reduction of apparent photorespiration based on O_2 inhibition of photosynthesis and Γ values. Therefore, in this species refixation of photorespired CO_2 may be a major basis for decreased apparent photorespiration, rather than a C_4 cycle functioning to concentrate atmospheric CO_2 in the leaf.

Kennedy *et al.* (1980) suggested other *Mollugo* species, *M. pentaphylla*, *M. nudicaualis*, and *M. lotoides*, may be C_3–C_4 intermediates. However, *M. lotoides* and *M. pentaphylla* might be more appropriately classified as C_3 species, in that their Γ, inhibition of photosynthesis by 21% O_2, and inhibition of CE by 21% O_2 are similar to C_3 plants, even though these species have chloroplasts in bundle sheath cells. In this respect, they might be similar to *Panicum laxum* (Brown *et al.*, 1983a). *Mollugo nudicaualis* was intermediate based on a reduction in apparent photorespiration (Γ value was 39 μl/l) and the fact that inhibition of photosynthesis by O_2 was lower than in the other two species, and O_2 (21 versus 2%) had no effect on CE. The lack of inhibition of CE by O_2 is surprising in that there was no evidence that a substantial C_4 cycle functions in this species. After 3 sec of $^{14}CO_2$ fixation, only 9–11% of the initial labeled products was C_4 acids malate + aspartate in these three species. As noted in the physiology section, in C_3 plants O_2 inhibition of CE is considered to be due to the competitive effect of O_2 with respect to CO_2 on Rubisco. In C_4 plants, O_2 has no effect on CE, since the high [CO_2] in bundle sheath cells prevents O_2 inhibition of photosynthesis. In intermediate species like *P. milioides*, which lacks a C_4 cycle, O_2 inhibits CE to the same extent as in C_3 plants (Section III,B). Additional studies are needed on *Mollugo* species to determine whether they are type I or II intermediates.

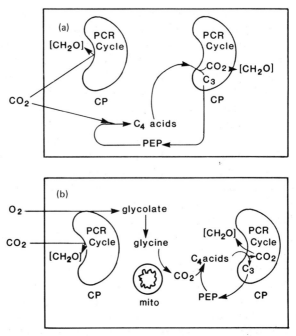

Fig. 3. Proposed mechanisms by which the C_4 cycle may reduce photorespiration in intermediate species. (a) C_4 cycle fixes some atmospheric CO_2 and donates it to the PCA cycle. (b) C_4 cycle fixes some photorespired CO_2 and donates it to the PCA cycle. See legends of Fig. 1 and 2 for abbreviations.

In intermediates that have a functional C_4 cycle, apparent photorespiration may be reduced by fixing part of the atmospheric CO_2 through the C_4 cycle and increasing the $[CO_2]$ at the site of Rubisco, as occurs in C_4 plants [Fig. 3(a)]. Another means by which apparent photorespiration may be reduced in thse species is through an efficient refixation of photorespired CO_2. The C_4 cycle could contribute to the latter process [Fig. 3(B)]. For example, photorespired CO_2 could be refixed by PEP carboxylase in the cytosol, the oxaloacetate converted to malate via NADP-malate dehydrogenase in the chloroplast, and then malate decarboxylated in the chloroplast by NADP-malic enzyme. PEP carboxylase and Rubisco are located in both mesophyll and bundle sheath cells of the *Flaveria* intermediates examined (Bauwe, 1984; Reed and Chollet, 1985). The intercellular compartmentation of NADP-malic enzyme in the *Flaveria* intermediates and in *N. minor* must be determined to consider how the C_4 cycle may function in these species either as a CO_2-concentrating mechanism or in refixation of photorespired CO_2. In both C_4 plants and intermediates having a functional C_4 cycle, it is likely that refixation of photorespired CO_2 contributes to reduction of photorespiration, and this may be most significant at Γ. The ability of *Flaveria* intermediates and *N. minor* to attain very low Γ (Section III,B) might be attributed to refixation of photorespired CO_2 through the C_4 cycle. Presently, the extent that C_4-cycle activity reduces apparent photorespiration in these intermediates by increasing the supply of atmospheric CO_2 to the PCR cycle versus refixation of photorespired CO_2 is unknown. At least in *N. minor* and some *Flaveria* intermediates, the C_3-like carbon isotope values (Section IV) suggest little or no atmospheric CO_2 is fixed through the C_4 cycle and donated to the PCR cycle under normal air.

It is conceivable that CO_2 could be fixed through the C_4 cycle in an intermediate species, and the CO_2 partially donated to the PCR cycle without substantial elevation of CO_2 in the leaf. This is particularly true in an NADP-malic enzyme type intermediate species where the C_4 acid decarboxylase is located in the chloroplast. In this case, reduction in the true rate of photorespiration would be limited and O_2 inhibition of RuBP carboxylase would still occur. Therefore, an evaluation of the pool size of inorganic carbon in leaves during photosynthesis is needed to determine whether the intermediates have a significant capacity for concentrating CO_2 in the leaf through the C_4 cycle. From recent measurements of the labile radioactive pool in leaves during steady-state feeding of $^{14}CO_2$ in the light, the intermediates *F. brownii*, *F. ramosissima*, and *F. floridana* were found to concentrate CO_2 within the range of C_4 *Flaveria* species. However, the intermediate *F. linearis* showed no capacity to concentrate CO_2 in the leaf in the light, and light-dependent accumulation of CO_2 in *F. pubescens* was low (Moore *et al.*, 1987). There is recent evidence that suggests that a high capacity for C_4 photosynthesis requires a high concentration of CO_2 in the leaf, the existence of the inorganic carbon from C_4 acid decarboxylation in the form of CO_2, and little or no carbonic anhydrase in bundle-sheath cells (Furbank and Hatch, 1987). In this respect the fixation of inorganic carbon by Rubisco (CO_2 the active species) in bundle-sheath cells of intermediate species could be limited if these cells have high activity of carbonic anhydrase, which rapidly causes the CO_2 released from C_4 acid decarboxylation to equilibrate to bicarbonate.

Another type of C_3–C_4 intermediate species having a C_4 cycle could exist if the Kranz cells were relatively leaky, and the C_4 cycle not able to sufficiently elevate the CO_2 in the leaf to prevent O_2 inhibition of photosynthesis. This type of intermediacy may occur in *Amaranthus graecizans* (Ku and Edwards, 1980), although presently we have not classified it as an intermediate species. This species is like a C_4 species in having typical Kranz anatomy, localization of PEP carboxylase in mesophyll cells and Rubisco in bundle-sheath cells, and based on its carbon isotope composition ($\delta^{13}C = -14.1‰$) (see Ku and Edwards, 1980). However, the species is atypical in that 21% O_2 inhibited photosynthesis by 9% under atmospheric levels of CO_2, and O_2 caused a similar inhibition of CE at lower CO_2 levels. Also, the quantum yield under 21% O_2 (0.043 mol CO_2/mol quanta) was lower than the quantum yield typically found in other C_4 species (See Monson et al., 1986). Increasing O_2 levels caused a progressive inhibition of photosynthesis, CE, and quantum yield in A. graecizans. The Γ in this species was about 7 $\mu l/l$ under 21% O_2, which is higher than typically found in C_4 species. These observations could be explained by higher conductances for CO_2 and O_2 (more leaky) across the walls of the bundle sheath cells of A. graecizans than in other C_4 species.

c. **Type III Intermediates.** As already indicated, there is no evidence for reduced photorespiration in higher plants through a difference in the kinetic properties of Rubisco. For example, the $K_m(CO_2)$ and CO_2/O_2 specificity factor for purified Rubisco are similar in the intermediate species *Panicum milioides* and *Moricandia arvensis* compared to representative C_3 species *P. laxum* and *M. foetida* (Bauwe and Apel, 1979; Holbrook et al., 1985). The CO_2/O_2 specificity factor ($V_c K_o/V_o K_c$) reflects the relative capacity for carboxylation versus oxygenation of RuBP by Rubisco, where V_c is V_{max} for the carboxylase reaction, K_o is K_m for O_2 in the oxygenase reaction, V_o is V_{max} for the oxygenase reaction, and K_c is the K_m for CO_2 in the carboxylase reaction.

3. *The Biochemistry of Intermediates Relative to Γ, O_2 Inhibition of Photosynthesis, and O_2 Inhibition of Carboxylation Efficiency*

The biochemistry of intermediate species needs to be considered in terms of the consequences on physiological components such as Γ, O_2 inhibition of photosynthesis, and O_2 inhibition of CE. In general, in intermediate species there is a more noticeable reduction in Γ relative to C_3 plants than in the reduction of O_2 inhibition of photosynthesis or in O_2 inhibition of CE. The value of Γ is dependent on the relative rate of CO_2 fixation, true photorespiration, and refixation of photorespired CO_2. Therefore, the reduction in Γ in intermediates is likely reflecting their ability to refix photorespired CO_2.

The following is a simple example of how refixation of photorespired CO_2 could reduce Γ in an intermediate species that lacks the C_4 cycle. In a C_3 plant at Γ, APS = TPS − TPR, and in relative units at Γ, 0 = 100 − 100.

At this same atmospheric concentration of CO_2, in an intermediate species assume, as a simple example, that TPS is the same in the mesophyll cells, that all of the photorespired CO_2 is released and refixed in bundle-sheath cells, and that no atmospheric CO_2 is fixed directly by bundle-sheath cells. Then in the mesophyll cells, APS = TPS − TPR, and in relative units, 100 = 100 − 0.

In the bundle-sheath cells, APS = TPS − TPR, and since in these cells TPS will be equal to TPR, 0 = 100 − 100.

Then overall in the leaf, APS = TPS − TPR, and 100 = 200 − 100.

Thus, the intermediate species in this example will assimilate carbon at an atmospheric concentration equal to Γ for C_3 plants. If the intermediate species had 80% as much Rubisco in mesophyll cells as in C_3 plants, and 20% of the total Rubisco in bundle-sheath cells, then the activities would be reduced 20% (where APS = TPS − TPR and in relative units, 80 = 160 − 80). The efficiency of this process and the benefit to the plant as a means of reducing photorespiration relative to a C_3 plant will depend on a number of factors. These include the distribution of Rubisco between mesophyll and bundle-sheath cells, the amount of atmospheric CO_2 directly fixed by mesophyll cells and bundle-sheath cells, the amount of photorespired CO_2 refixed in bundle-sheath cells, the resistance of the bundle-sheath cell wall to diffusion of CO_2, and the atmospheric $[CO_2]$. The possible carbon gain from refixing photorespired CO_2 is most apparent at a rather low $[CO_2]$ as illustrated above. At atmospheric levels of CO_2, whether refixation of photorespired CO_2 in bundle sheath cells would allow any greater carbon gain than occurs in a C_3 plant is not apparent. Under normal atmospheric conditions, refixation of photorespired CO_2 in the bundle-sheath compartment would represent a carbon gain, but some loss in carbon fixation capacity could occur to the extent that the Rubisco in bundle sheath cells has limited access to atmospheric CO_2 (i.e., the plant cannot have it both ways—a semi-air-tight bundle-sheath compartment to facilitate refixation of photorespired CO_2 would limit direct fixation of atmospheric CO_2 in these cells).

Figure 4 illustrates how refixation of photorespired CO_2 in bundle-sheath cells of an intermediate that lacks a C_4 cycle (type I) could cause a reduction in apparent photorespiration under varying $[CO_2]$. The solid lines are rates of photorespiration and photosynthesis in a C_3 species based on the kinetic properties of Rubisco and considering no photochemical limitation at high $[CO_2]$. $TPS_{21\%O_2}$, which is equal to $APS_{21\%O_2}$ + true $PR_{21\%O_2}$, indicates the true rate of CO_2 fixation through Rubisco. The difference between $TPS_{0\%O_2}$ and $TPS_{21\%O_2}$ represents the competitive component of O_2 inhibition. The difference between $TPS_{21\%O_2}$ and $APS_{21\%O_2}$ represents the photorespiratory component of O_2 inhibition. At Γ a major component of O_2 inhibition of photosynthesis is due to photorespiration, whereas at higher levels of CO_2 the relative contribution by the competitive component increases.

In the example of Fig. 4, in the intermediate species (broken lines) half of the photorespired CO_2 is refixed at the various $[CO_2]$ levels. In order for the Γ value to be lower in the intermediate than in a C_3 plant, the CO_2 generated from photorespiration needs to be concentrated in the leaf (i.e., the bundle-sheath cells), around at least part of the Rubisco, and not in free equilibrium with the CO_2 in the atmosphere. At Γ, $TPS_{21\%O_2}$ equals the uptake of atmospheric CO_2 plus the refixation of photorespired CO_2. Since the intermediate can refix part of the CO_2 from photorespiration in bundle-sheath cells, then Γ, which is the external $[CO_2]$ at which $TPS_{21\%O_2}$ equals the true rate of photorespiration, will be lower than in a C_3 plant. In this case Γ is decreased by about 50%, to one which is truly intermediate to C_3 and C_4 species. Under

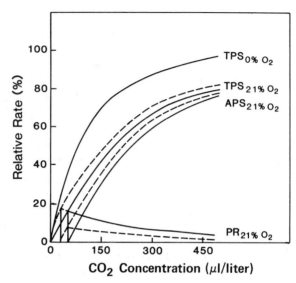

Fig. 4. Model of the rates of photosynthesis and photorespiration in a C_3 species (solid lines) and an intermediate species that refixes half of the photorespired CO_2 in bundle-sheath cells (broken lines) under varying $[CO_2]$. Suggested rates of photorespiration and photosynthesis in the C_3 species at varying $[CO_2]$ are adapted from Monson *et al.* (1984) (copyright 1984, American Institute of Biological Sciences). $TPS_{0\% O_2}$, true rate of photosynthesis at 0% O_2 based on kinetic properties of Rubisco [using $K_m(CO_2)$ of 136 $\mu l/l$ (4 μM CO_2 at 30°C)] and saturating ribulose-1,5-P_2; $TPS_{21\% O_2}$, true rate of photosynthesis at 21% O_2 based on O_2 as a competitive inhibitor with respect to CO_2 [using $K_i(O_2)$ of 200 μM O_2] and saturating ribulose-1,5-P_2; $APS_{21\% O_2}$, predicted response of apparent photosynthesis at 21% O_2 at a given intercellular CO_2 concentration, without any limitation on assimilatory power (regeneration of RuBP) (see Edwards and Walker, 1983); $PR_{21\% O_2}$ (solid line) equals true rate of photorespiration; $PR_{21\% O_2}$ (broken line) equals apparent rate of photorespiration in intermediate species where half of the photorespired CO_2 is refixed in bundle-sheath cells.

atmospheric conditions refixation of photorespired CO_2 does not result in much reduction in O_2 inhibition of photosynthesis. For example, at an intercellular $[CO_2]$ of 275 $\mu l/l$, the percent inhibition of photosynthesis in the C_3 species [where percent inhibition = $(TPS_{0\% O_2} - APS_{21\% O_2}/TPS_{0\% O_2}) \times 100$] is 33. The absolute value of the inhibition will depend on $K_m(CO_2)$, intercellular $[CO_2]$, extent of limitation on regeneration of RuBP, and effect of O_2 on the state of activation of Rubisco. By the same calculation, the percentage inhibition of photosynthesis by 21% O_2 in the intermediate species is 29. Therefore, by refixing photorespired CO_2, intermediate species will reduce Γ to a much greater extent than they will reduce O_2 inhibition of photosynthesis. The actual reduction in O_2 inhibition of photosynthesis in the intermediate species of this type may be even less if there is little localized increase in the concentration of photorespired CO_2 in the leaf when the concentration in the atmosphere is high. However, comparisons of the O_2 inhibition of photosynthesis among species may not be a reliable means for identifying intermediates, in that the intercellular $[CO_2]$ can vary and O_2 might have a differential effect on the state of activation of Rubisco between species (Section III,B). As previously noted, the O_2 inhibition of photosynthesis in *N. minor* and *M. arvensis* is like that of C_3 plants.

In the above example, relative comparisons in rates of photosynthesis bewteen the C_3 and intermediate species were made assuming the same amount of Rubisco capacity in mesophyll cells of the leaf and no direct fixation of atmospheric CO_2 in the bundle-sheath cells. If on a leaf area basis the intermediate species have less Rubisco in mesophyll cells than C_3 species, then with increasing $[CO_2]$ there may be a cross-over point where APS in the intermediate falls below that of the C_3 plant. The Γ could also be reduced in certain intermediates through functioning of a C_4 cycle (type II) analogous to a C_4 plant, in which case CO_2 fixation through PEP carboxylase would influence the true rate of photosynthesis.

A good means of determining whether intermediates significantly reduce the competitive component of O_2 inhibition of photosynthesis is to measure CE at low CO_2 concentrations. CE is dependent on the $K_m(CO_2)$ for carboxylase and on whether a CO_2 concentrating mechanism is functioning. C_4 plants have a high CE under 2 and 21% O_2 because the C_4 cycle increases the $[CO_2]$ around Rubisco. Comparison of the absolute values of CE between C_3 and C_4 plants is best made by expressing photosynthesis rate/Rubisco activity or Rubisco protein (Ku and Edwards, 1978). In *Panicum milioides*, O_2 inhibition of CE is similar to that in C_3 plants, which is consistent with this species lacking a C_4 cycle. However, in *Flaveria ramosissima* and *Neurachne minor*, the O_2 inhibition of CE is lower than in C_3 species, which suggests that the C_4 cycle is increasing $[CO_2]$ in the leaf and limiting O_2 inhibition of photosynthesis (Ku *et al.*, 1983; Monson *et al.*, 1984; Hattersley *et al.*, 1986). It has also been observed that the quantum yield is inhibited less in *F. ramosissima* with increasing $[O_2]$ than in some other *Flaveria* intermediates (Monson *et al.*, 1986). This may be due to it having a more efficient $[CO_2]$ concentrating mechanism. Thus, some differences in physiological characteristics that have been observed between intermediates may be related to the efficiency of the C_4 cycle.

IV. CARBON ISOTOPE COMPOSITION OF INTERMEDIATES RELATIVE TO MECHANISM OF PHOTOSYNTHESIS

Both C_3 and C_4 plants discriminate against $^{13}CO_2$ during photosynthesis, but C_3 plants discriminate against $^{13}CO_2$ to a larger extent, and thus have a more negative $\delta^{13}C$ value relative to C_4 plants (see O'Leary, 1981). The principal basis for discrimination against $^{13}CO_2$ in C_3 plants lies with the

TABLE VIII
$\delta^{13}C$ Values of Various C_3–C_4 Intermediate Species

Species	$\delta^{13}C$ (‰)	Reference
Flaveria augustifolia	−26.8	Smith and Powell (1984)[b]
Flaveria anomala	−26.1	Apel and Maass (1981)
	−28.3 to −30.9	Monson *et al.* (1987)[a]
	−28.5	Smith and Turner (1975)[b]
Flaveria brownii	−15.8	Apel and Maass (1981)
	−17.3	Smith and Powell (1984)
Flaveria chloraefolia	−28.4	Smith and Turner (1975)
Flaveria floridana	−26.3 to −29.9	Monson *et al.* (1986b)
	−27.4	Smith and Turner (1975)
	−30.6	Brown *et al.* (1986b)[c]
Flaveria linearis	−23.8 to −29.3	Monson *et al.* (1987)[d]
	−25.3 to −30.6	Monson *et al.* (1987)
	−27.9	Smith and Turner (1975)
	−30.3 to −33.5	Brown *et al.* (1986b)
Flaveria oppositifolia	−32.2	Brown *et al.* (1986b)
	−27.4	Smith and Turner (1975)
Flaveria pubescens	−26.4	Smith and Turner (1975)
	−27.0	Apel and Maass (1981)
Flaveria ramosissima	−22.5	Smith and Turner (1975)
	−24.4 to −28.5	Monson *et al.* (1987)
Mollugo verticillata	−26.2, −27.4	Smith and Robbins (1975)
	−26.2 to −30.7	Brown and Brown (1975)
Neurachne minor	−21.5 to −27.1	Hattersley and Roksandic (1983)[e]
	−28.4 to −30.5	Hattersley *et al.* (1986)[f]
Moricandia arvensis	−29.3	Apel and Ohle (1979)
Panicum milioides	−25.0	Brown and Brown (1975)
	−25.6 to −27.6	Winter *et al.* (1982b)[g]
	−30.5 to −31.3	Hattersley *et al.* (1986)

[a] Growth chamber (day/night temperature of 27/22°C, 14 h light at 800 μmol quanta m^{-2} sec^{-1}) and greenhouse-grown plants.

[b] Field-collected plants.

[c] Greenhouse-grown plants, maximum day/night temperatures of 32–36/20–22°C during May to July.

[d] Field-collected plants, 10 accessions.

[e] Field-collected plants, nine accessions.

[f] Greenhouse- and growth cabinet-grown plants (day/night temperature of 32/20°C, photon flux density variable 460–1860 μmol m^{-2} sec^{-1}.

[g] Growth chamber-grown plants, day/night temperature of 26/15°C, 12 h light at 600 μmol quanta m^{-2} sec^{-1}, 30 or 70% relative humidity.

enzyme Rubisco. In C_4 plants there is some discrimination due to resistance of diffusion of CO_2 into the leaf, whereas there is little discrimination by PEP carboxylase. Discrimination by Rubisco in bundle-sheath cells will only be apparent to the extent that there is leakage of CO_2 from the bundle-sheath cells. Typically, the $\delta^{13}C$ values in C_3 plants range from −25 to −30‰ and in C_4 plants from −11 to −16‰. Some of the variation within each photosynthetic group may be due to differences in leaf age and growth conditions such as temperature, light, humidity, water supply, $[CO_2]$, salinity, and nutrition. In any case, this difference in isotope composition has become one of the standard methods by which C_4 plants can be distinguished from C_3 plants. If an intermediate species is fixing part of the atmospheric CO_2 directly through the C_4 pathway, which eventually contributes to its growth, one would expect an intermediate $\delta^{13}C$ value between that of a C_3 and C_4 plant. Despite the evidence for considerable C_4 cycle function in some of the intermediates, the $\delta^{13}C$ values of most intermediates, except *F. brownii*, are within the C_3 plant range (Table VIII). For the intermediate species in *Panicum* and *Moricandia,* which assimilate atmospheric CO_2 solely through the PCR cycle, this result is what one would expect. However, for those intermediates capable of fixing CO_2 through a limited C_4 pathway, the $\delta^{13}C$ values are variable, from a C_4-like value in *F. brownii* to a C_3-like value for the others.

Although the reported $\delta^{13}C$ values for *F. brownii* are close to the C_4 plant range, they are always slightly more negative than those of the C_4 *Flaveria* species included in the same studies (Apel and Maass, 1981; Smith and Powell, 1984; Monson *et al.,* 1987). As mentioned earlier, *F. brownii* fixes 65–75% and 25–35% of the atmospheric CO_2 initially via the C_4 and C_3 pathways, respectively (Monson *et al.,* 1986; Cheng *et al.,* 1987). Also, the transfer of carbon from the C_4 to the C_3 pathway in this species is almost as efficient as in other C_4 *Flaveria* species (Cheng *et al.,* 1987). The predominant assimilation of the atmospheric CO_2 through the C_4 cycle plus an efficient transfer of carbon from the C_4 to the PCR cycle in this species are thus consistent with its C_4-like $\delta^{13}C$ value, whereas the minor, direct CO_2 fixation through Rubisco in the PCR cycle or leakiness of the bundle-sheath cells could contribute to its slightly more negative value relative to C_4 plants. Although there is a range in the isotope values reported for some species, relatively positive values have been found among plant material of *F. linearis, F. ramosissima,* and *N. minor* (Table VIII). Most values for the *Flaveria* species, other than that of *F. brownii,* are C_3-like, although there is biochemical evidence for considerable C_4-cycle activity. Thus C_4 photosynthesis may often make a limited contribution to their growth unless there are some unknown complexities in the processes of isotope discrimination. Using the same plant material for biochemical studies and the measurement of leaf carbon isotope ratio, a recent study by Monson *et al.* (1987) revealed that in five *Flaveria* species (*F. anomala, F. floridana, F. linearis, F. pubescens,* and *F. ramosissima*) an increase in the expression of C_4 cycle activity from 14 to 51% of total CO_2 assimilated led to small changes in the carbon isotope ratio ($\delta^{13}C$ values became up to 2.5–3.0‰ less negative). In some species the efficiency of transfer of carbon from the C_4 cycle to the PCR cycle may be limited by the rate of decarboxylation of the C_4 acids or a relatively high leakage of CO_2 from the C_4 cycle (Monson *et al.,* 1987). The most precise assessment of efficiency of transfer of carbon from the C_4 cycle to the PCR cycle might be made by measuring on line discrimination against $^{13}CO_2$ during photosynthesis and the percentage initial fixation of CO_2 into C_4 acids in the same plant material.

V. BIOCHEMISTRY OF PHOTOSYNTHESIS IN INTERMEDIATES IN RELATION TO DEVELOPMENTAL AND ENVIRONMENTAL FACTORS

A. Leaf Age

There is some evidence that leaf age can influence apparent photorespiration in certain intermediate species. In *Moricandia arvensis* there is wide variation in Γ (14 $\mu l/l$ to 50 $\mu l/l$) depending on the age of the leaf, with the "middle-aged" leaves having the lowest Γ values (Apel *et al.*, 1978). In a number of *Flaveria* intermediates mature leaves have an apparent higher capacity for C_4 photosynthesis than young leaves based on incorporation of $^{14}CO_2$ into C_4 acids (Table IX). Whether this is related to a change in the leaf anatomy, distribution of photosynthetic enzymes, and/or activities of photosynthetic enzymes is uncertain.

TABLE IX

Percentage of Label Appearing in C_4 Acids (Malate + Asparate) after an 8-sec Pulse with $^{14}CO_2$ with Various *Flaveria* Species[a]

Species	C_4 Acids (% of total ^{14}C products)	
	First-node leaf[a] (younger)	Third-node leaf (older)
C_3		
F. cronquistii	5	14
C_3–C_4		
F. pubescens	11	41
F. anomala	16	44
F. floridana	25	52
F. ramosissima	42	49
C_4		
F. trinervia	84	83

[a] B. d. Moore, personal communication. First represents the leaf appearing at the first node (youngest) and third represents the leaf appearing at the third node (older). First-node leaves were about 6–10% of full expansion while third-node leaves were about 70–75% of full expansion.

B. O₂ Levels, Light Intensity, and Γ Values

There is a similar biphasic increase in Γ with increasing $[O_2]$, and a similar decrease in Γ with increasing light intensity with *M. arvensis*, *P. milioides*, and some of the *Flaveria* intermediates (Section III,B). Thus, the influence of light intensity and $[O_2]$ on Γ in the intermediate species appears to be independent of a functional C₄ cycle. Peisker and Bauwe (1984) developed a model that simulated the biphasic response of Γ with respect to $[O_2]$ in intermediates based on a limited C₄ cycle. This model requires a relatively high diffusive resistance for CO_2 between mesophyll and bundle sheath cells and a partial C₄ cycle fixing atmospheric CO_2 and functioning in the classical way. The possibility of a similar biphasic response through directly fixing photorespired CO_2 as in Fig. 1 or via a C₄ cycle has not been excluded. In this case, as $[O_2]$ is increased, a point could be reached (i.e., about 10–21% O_2 in *P. milioides* and 10–15% in *M. arvensis*) where the capacity for refixation is exceeded, after which Γ increases more rapidly with increasing $[O_2]$. If a limited CO_2-concentrating mechanism exists, then the transition point may occur at a higher $[O_2]$, which could account for transition points as high as 21% O_2 in certain *Flaveria* intermediates and over 50% in *Neurachne minor* (see Section III,B).

C. Growth Conditions

When the intermediates *N. minor* and *P. milioides* are grown at lower light intensities and measurements are made under the same light levels, Γ values increase by 5 and 10 $\mu l/l$, respectively, compared to higher light regimes (Hattersley *et al.*, 1986). Similarly, plants of *P. milioides*, *F. linearis*, and *M. arvensis* grown under lower light intensities show higher Γ values than plants grown at higher light intensities (measurements under high light; G. Byrd and R. H. Brown, personal communication). There is a shift in the carbon isotope ratio of the intermediate *Flaveria brownii* from more negative values in greenhouse plants grown in the winter ($\delta^{13}C = -21.0 \pm 1.0‰$) to more positive values in plants grown in the summer ($\delta^{13}C = -14.5 \pm 0.5‰$), suggesting an environmental shift in the summer toward C₄ photosynthesis (L. J. Mets, personal communication). Another apparent shift in carbon isotope fractionation caused by differences in environment occurs in the intermediate *F. ramosissima*, being more positive ($\delta^{13}C = -21.1‰$) for plants collected in the field to more negative ($\delta^{13}C = -24.4$ to $-28.5‰$) for plants grown in the growth chamber of greenhouse (see Table V). Whether the shift of $\delta^{13}C$ values in these species is mediated through changes in the physical or biochemical components of carbon isotope fractionation needs further investigation. Since a number of *Flaveria* species are perennials with a relatively long growing season (Powell, 1978), the possibility of acclimation of photosynthetic pathways in intermediates to changing environmental conditions should be considered.

VI. GENETICALLY BASED DIFFERENCES IN THE BIOCHEMISTRY OF PHOTOSYNTHESIS WITHIN AN INTERMEDIATE SPECIES

Among four populations of *Mollugo verticillata* that were collected from the field and grown under the same conditions, there was a wide range in the percentage of C_4 acids (11–40%) formed as initial products of photosynthesis, suggesting there may be genetically based differences in photosynthesis within this species (Sayre and Kennedy, 1977). There are also variations among reports in the literature on whether the leaf anatomy of *Mollugo nudicaulis* is both Kranz and non-Kranz (among various leaves of the same plant; Raghavendra *et al.*, 1978), Kranz (Rathnam and Chollet, 1980), or intermediate (Kennedy *et al.*, 1980). Smith and Robbins (1975) reported that C_3 and C_4 plants may exist within the single taxonomic species *Mollugo cerviana*, based on $\delta^{13}C$ values of two distinct groups of plants. It is quite possible that an ecotypic differentiation with regard to photosynthetic pathway may occur at the intraspecific level. In *Alloteropsis semialata*, a non-Kranz variety with C_3 carbon isotope values and a Kranz variety with C_4 carbon isotope values also have been reported within the same species (Ellis, 1974; Smith and Robbins, 1975; Vogel *et al.*, 1978). Interestingly, Brown (1975) noted from a herbarium specimen a third variety, which exhibits leaf anatomy intermediate between the other two, but which has a C_3 type carbon isotope value. No further study on this variety has been reported, although it could well be another C_3–C_4 intermediate.

The carbon isotope values among different populations of *Flaveria linearis* collected in the Florida Keys varied from -23.8 to $-29.3‰$ (Monson *et al.*, 1987). While this range in isotope values is substantial, they are within the theoretical range for C_3 plants, and the variation could be the result of environmental factors. Further studies of these populations in a common environment will be required to consider whether there may be genetic differences in the populations in the degree of C_4 photosynthesis and the carbon isotope values.

VII. FEATURES OF HYBRIDS BETWEEN DIFFERENT PHOTOSYNTHETIC TYPES

A. C_3 and C_4 Species

In 1968, Björkman and colleagues obtained the first artificial hybrid ever to be produced between a C_3 and C_4 species (see Osmond *et al.*, 1980). In this study, the C_4 species *Atriplex rosea* was used as the female parent and the C_3 species *Atriplex triangularis* as the male parent. The F_1 hybrids were intermediate between the parent plants in a number of anatomical, physiological, and biochemical characteristics. For example, they had a distinct chloroplast-containing bundle sheath surrounded by a radiate mesophyll. The chloroplasts and mitochondria in the bundle sheath cells were larger and more abundant than those of the adjacent mesophyll cells. These features resembled the situation in the C_4 parent *A. rosea*. However, the bundle sheath cells of the F_1 hybrids were thinner and more vacuolated than those of *A. rosea*. Thus, this study showed that the gross leaf anatomy as well as the ultrastructure of the F_1 hybrids is truly intermediate.

The intermediacy of the F_1 hybrids is also expressed in the biochemistry of C_4 photosynthesis. The activities of several key C_4 cycle enzymes, such as PEP carboxylase, pyruvate,Pi dikinase, and alanine and aspartate aminotransferases, are present in the hybrids, although they are lower than in the C_4 parent. Consistent with the enzyme data, the $^{14}CO_2$ labeling pattern of the initial photosynthetic products is also intermediate between those of the parents. Following a 6-sec pulse with $^{14}CO_2$, 44% of the fixed ^{14}C in the leaves of the F_1 hybrids was recovered in the C_4 acids malate and aspartate and 56% in PCR-cycle intermediates. In *A. rosea,* more than 90% of the radioactivity was in the C_4 acids, whereas in *A. triangularis* over 90% was in PCR cycle intermediates. However, the transfer of carbon from the C_4 cycle to the PCR cycle in the F_1 hybrids is greatly impaired as indicated by the pulse-chase experiments.

Despite the intermediate characteristics in leaf anatomy and biochemistry of C_4 photosynthesis, the F_1 hybrids exhibit lower photosynthesis and lower growth rates than either parents. Relative to the C_3 parent, the hybrids have only slightly reduced Γ. The degrees of O_2 inhibition of net photosynthesis and CE are similar to, or even greater than, those observed in the C_3 parent, indicating that there is very little elevation of $[CO_2]$ in the leaf of the hybrids through a C_4 cycle. The $\delta^{13}C$ values of the F_1 hybrids are only slightly lower than those of the C_3 parent, again indicating that an integrated C_4 metabolism does not occur in the hybrids. These physiological features can be best explained in terms of lack of a proper spatial compartmentation for the two sequential CO_2 fixation cycles. Using an *in situ* immunofluorescent labeling technique, Hattersley *et al.* (1977) demonstrated that in the F_1 hybrids Rubisco is present in both mesophyll and bundle-sheath cells, and presumably the C_4 cycle enzymes may not be differentially compartmentalized between the two cell types. Thus, it is conceivable that a futile C_4 cycle (see Section III,C) is operating in these F_1 hybrids.

The Carnegie group also attempted to study the mode of the inheritance of C_4 photosynthesis by examining the recombination of photosynthetic characteristics in the subsequent, segregating generations. Because the chromosomal pairing in the F_1 hybrids is highly irregular and results in progenies with highly variable chromosome numbers, this work met with only partial success. However, some preliminary conclusions regarding the inheritance of C_4 syndrome can be drawn from this work: (1) Kranz anatomy and C_4

biochemistry are not obligately linked in their expression, and (2) C_4 photosynthesis is not simply transmitted through the chloroplast genome and it must involve several nuclear genes. In fact, recent molecular studies with C_4 plants indicate that genes for the key C_4-cycle enzymes, including PEP carboxylase, pyruvate,Pi dikinase, and NADP-malic enzyme, are encoded by the nuclear genome (Sims and Hague, 1981; Hague *et al.*, 1983; Collins and Hague, 1983; Gee *et al.*, 1984). From this genetic work with *Atriplex* hybrids, it is emphasized that "any conclusive statement regarding the mode of inheritance of C_4 photosynthesis and the number of genetic loci involved must await further hybridization experiments and genetic analysis on materials capable of producing fertile and diploid segregating hybrid populations" (Osmond *et al.*, 1980).

Most recently, reciprocal hybrids were obtained from crosses between *A. rosea* and *A. triangularis,* and plants of subsequent generations were produced by selfing (Hinata *et al.*, 1984). Unfortunately, examinations of various photosynthetic characteristics were not complete for all generations. In the F_1 hybrids where the C_4 species (*A. rosea*) served as the female parent, the Γ (12–20 μl/l) is reduced relative to C_3 plants. However, the Γ values for the F_3 plants from either direction of the crosses are within the C_3 plant range. The leaf carbon isotope ratio was measured for the F_1 hybrids and F_2 plants. For the F_1 hybrid *A. triangularis* $\times$ *A. rosea,* the $\delta^{13}C$ value is C_4-like, while for *A. rosea* $\times$ *A. triangularis,* the values are even more negative than those of the C_3 parent. The $\delta^{13}C$ value for another F_1 hybrid, *A. hortensis* f. *lutea* (C_3) $\times$ *A. rosea,* is also similar to that of the C_4 parent *A. rosea.* However, the F_1 hybrid *A. rosea* $\times$ *A. hastata* (C_3) exhibits a $\delta^{13}C$ value intermediate to the two parents. Thus, there is no conclusive evidence for a maternal effect on this parameter. Irrespective of the direction of the crosses, the $\delta^{13}C$ values for all F_2 plants are C_3-like. The photosynthesis rates of most of the F_2 plants are lower than those of the C_3 parent, and none of them approaches the high rates exhibited by the C_4 parent. Similarly, there is no difference between the C_3 parent and F_3 plants in their sensitivity of photosynthesis to O_2. Basically, these results are consistent with those obtained by the Carnegie group. That is, a proper combination of many nuclear genes is very important for the expresion of a functional C_4 system.

The C_3 species *Flaveria pringlei* has been crossed with the C_4-like intermediate species *F. brownii* (male) and some photosynthetic characteristics examined (Holaday *et al.*, 1985). The F_1 hybrid has an intermediate level of apparent photorespiration between the parents based on the Γ, although the Γ was considerably higher than that in naturally occurring C_3–C_4 intermediate *Flaveria* species. The activity of C_4-cycle enzymes, except for NADP-malate dehydrogenase, and the leaf anatomy of the hybrid are more like those of the C_3 parent. Since *F. brownii* is considered to be the most advanced intermediate (in terms of a functional C_4 photosynthesis), it will be of interest to determine the extent a C_4 cycle functions in this hybrid.

B. C_3 and Intermediates

In *Panicum,* three interspecific hybrids have recently been obtained from crosses between two intermediates (females, *P. milioides* and *P. spathellosum*—synonyn *P. schenkii*) and two C_3 species (*P. laxum* and *P. boliviense*) (Brown *et al.,* 1985). Values of O_2 inhibition of photosynthesis, photorespiratory CO_2 loss, and Γ for the hybrids are between those of the parents. All hybrids exhibit leaf anatomical traits intermediate to those of their parents. The amounts of organelles (e.g., chloroplast, mitochondrion, peroxisome) in the vascular bundle-sheath cells of the parents and hybrids are found to be highly correlated to their Γ. It is suggested that reduction in apparent photorespiration in these plants may be largely dependent on the development of organelles in the bundle sheath cells. Progenies were produced from these F_1 hybrids (infertile) using colchicine-induced amphiploids (Bouton *et al.,* 1986). In the F_2 and F_5 populations, O_2 inhibition of photosynthesis and Γ values are typical of C_3 plants. The progenies possess combinations of various traits associated with reduced photorespiration, so that no correlations existed among sensitivity of photosynthesis to O_2, Γ, and leaf anatomical characteristics. The lack of segregates with photorespiratory activity similar to the intermediate parent is thought to be due to the high ploidy level of the hybrids (i.e., a buffering effect).

C. Intermediates and C_4 Species

Several interspecific F_1 hybrids between intermediates and C_4 species have been made in the genus *Flaveria.* Cheng *et al.* (1984) reported that the photorespiratory activity in the two F_1 hybrids *F. linearis* × *F. palmeri* (C_4, male) and *F. pubescens* × *F. palmeri* is further reduced relative to their intermediate parents. These hybrids show a distinct Kranz-like leaf anatomy, low Γ (0–5 μl/l) approaching the value of the C_4 parent, and 10–15% inhibition of photosynthesis by 21% O_2, as compared to 20% in the intermediate parents and over 30% in C_3 plants. The activities of several key C_4-cycle enzymes are more than 50% of those found in the C_4 parent *F. palmeri.* The intercellular compartmentation of C_4-cycle enzymes in these hybrids was also examined using enzymatically isolated mesophyll and bundle-sheath protoplasts (Cheng and Ku, 1985). Although they lack a strict compartmentation of certain photosynthetic enzymes equivalent to C_4 plants, the activities of pyruvate,Pi dikinase and PEP carboxylase in these hybrids are three- to fivefold higher in the mesophyll than in the bundle-sheath cells, whereas Rubisco and NADP-malic enzyme activities are five- to 10-fold higher in the bundle-sheath cells. These results suggest that further reduction of apparent photorespiration and the associated O_2 inhibition of photosynthesis in these hybrids may be due to the differential compartmentation of elevated C_4 cycle enzymes. Such a biochemical compartmentation will result in a more efficient integration of the two carbon fixation cycles, thus providing a higher $[CO_2]$ in the leaf, which will limit the competitive inhibition by O_2. Consistent with these findings, Cheng *et al.* (1987) demonstrated that the reciprocal hybrids between *F. floridana,* an intermediate, and *F. brownii,* a C_4-like intermediate, are capable of fixing a high percentage (over 65%) of atmopsheric CO_2 into C_4 acids, and most importantly, there is an apparent transfer of carbon from the C_4 to the PCR cycle (only slightly lower than that in *F. brownii*). However, the reciprocal hybrids differ in their capacity to fix atmospheric CO_2 into C_4 acids, the fixation of CO_2 into aspartate versus malate, and the rate of turnover of C_4 acids. The biochemical basis for these differences is not known.

Recently, Brown *et al.* (1986b) also produced several similar F_1 hybrids between C_3–C_4 intermediate and C_4 *Flaveria* species. These include *F. linearis* × *F. trinervia* (C_4), *F. trinervia* × *F. linearis, F. trinervia* × *F. floridana,* and *F. trinervia* × *F. oppositifolia* (intermediate). They all show an intermediacy in leaf anatomy, Γ (4–10 μl/l), O_2 sensitivity of photosynthesis (10–17% inhibition by 21% O_2), PEP carboxylase activity, and carbon isotope ratio between the parents. The intermediate δ^{13}C values are consistent with the capacity of these hybrids to partially assimilate atmospheric CO_2 through a functional C_4 cycle (Cheng and Ku, 1985; Cheng *et al.,* 1987). However, the reduction in apparent photorespiration and the associated O_2 inhibition of photosynthesis in these F_1 hybrids did not result in higher net photosynthesis rates per unit leaf area, relative to their C_3–C_4 intermediate parents. There is not strict differential intercellular compartmentation of certain photosynthetic enzymes such as PEP carboxylase and Rubisco, which may limit the efficiency of photosynthesis in the hybrids (Cheng and Ku, 1985). Without compartmentation of appropriate biochemical reactions, the C_4 cycle may be somewhat futile, in which case some CO_2 released from C_4 acid decarboxylation may be refixed by PEP carboxylase instead of by Rubisco, or may even be lost from the leaf. For each turn of the C_4 cycle, two extra ATP are required. However, further studies are needed to determine if under certain environmental conditions (e.g., varying $[CO_2]$ and temperature) the hybrids do have a photosynthetic advantage over the intermediate parent.

VIII. A THEORETICAL SCHEME OF EVOLUTION OF C_4 PHOTOSYNTHESIS BASED ON INTERMEDIATE SPECIES

Since the discovery of C_4 photosynthesis, it has been found that at least 20 genera in 11 families are now known to contain both C_3 and C_4 plants (Section II). These families are very diverse and belong to the more highly evolved orders of angiosperms, such as Asterales, Caryophyllales, Poales, and Cyperales (Moore, 1982). In fact, the great majority of angiosperms evaluated is C_3 plants, and only a relatively few species of the flowering plants have been found to exhibit C_4 metabolism. Since C_4 plants also have C_3 photosynthesis, and no primitive angiosperms are C_4 plants, it appears that C_4 plants are recent in origin. Based on this phylogenetic consideration, it is suggested that C_4 plants have evolved from C_3 plants independently many times in different taxonomic groups (Smith and Robbins, 1975; Björkman, 1976; Brown, 1977; Powell, 1978; Moore, 1982). However, the path of evolution from a C_3 to a C_4 plant at the anatomical, biochemical, and molecular levels remains to be elucidated.

As already discussed, in the past decade several naturally occurring terrestrial plant species with photosynthetic characteristics intermediate to C_3 and C_4 plants have been identified. The number of intermediates currently identified is still relatively low compared to the number of C_4 plants. The reasons for this are unknown. Many more intermediates may exist in certain genera, which will only be revealed by proper and careful analysis. Some species that are currently classified as C_4 plants may be intermediates. As shown by the intermediates *N. minor and F. brownii*, some characters such as Kranz anatomy and Γ can not be used as an exclusive basis for classifying species as C_4 plants. Alternatively, there may be very few intermediate species. In this case, either the properties of intermediates may offer little advantage, or, if intermediates are evolving toward C_4 plants, the process may be less active today than in the past.

A taxonomic survey of the occurrence of intermediates reveals their close relatedness to C_4 plants in distribution (see Section II). Many workers have interpreted these intermediates to be evolutionary "links" between C_3 and C_4 plants (Apel and Maass, 1981; Powell, 1978), but at present there is very little evidence to support this assumption. When considering the intermediates in an evolutionary sense, one must ask whether they represent a stage in the evolution of C_4 photosynthesis, a reversion of C_4 to C_3 photosynthesis, or stabilized derivatives of spontaneous hybridization between closely related C_3 and C_4 species (Holaday and Chollet, 1984; Monson *et al.*, 1984). This question, of course, is a difficult one. To resolve it, intensive taxonomic studies employing cytogenetic, hybridization, biochemical (e.g., isozyme pattern analysis), and molecular (e.g., DNA/RNA homology analysis) approaches will be necessary.

A number of C_4 plants are adapted to shade and temperate conditions. This suggests that C_4 plants, which likely evolved in response to high light, high temperature, or arid conditions, can adapt to other environments without a necessity to revert to C_3 photosynthesis. Provided other adaptations occur (e.g., to stabilize membranes and the photochemistry), C_4 plants may be able to successfully compete with C_3 species in most environments. Several lines of evidence lend some support to the notion that intermediates evolved from C_3 plants rather than being naturally occurring hybrid products of C_3 and C_4 plants. First, although the composite *Parthenium hysterophorus* (Asteraceae) and the crucifer *Moricandia arvensis* are C_3–C_4 intermediates, C_4 plants have not been found in the genera *Parthenium* and *Moricandia* or in the Cruciferae family (Krenzer *et al.*, 1975). Second, in the case of *Flaveria,* Powell (1978) observed no natural hybridization in the field. In the laboratory, the isozyme pattern of 12 *Flaveria* species including C_3, C_4, and many intermediates has been recently analyzed (D. Soltis and M. S. B. Ku, unpublished). This technique has been successfully used in the phylogenetic study of both animals and plants. Twelve enzymes representing a wide spectrum of metabolism were chosen for analysis. These include NADP-malic enzyme, triose-phosphate isomerase, phosphoglucoisomerase, phosphoglucomutase, 6-phosphogluconate dehydrogenase, NAD-malate dehydrogenase, isocitrate dehydrogenase, superoxide dismutase, esterase, and leucine aminopeptidase. Numerous interspecific differences have been observed. These preliminary data suggest that at least some of the intermediate *Flaveria* species are not of hybrid origin. On the other hand, artificial F_1 hybrids included in the investigation clearly show a combined isozyme pattern to their parents. If these intermediates are indeed in the process of evolving toward C_4 plants, they should reveal some important clues to the evolutionary steps leading to the C_4 mechanism.

Table X is a proposed sequence of evolution of C_4 photosynthesis from C_3 plants. As mentioned in Section I, with geologic time our atmopheric environment became more oxidizing and favorable for photorespiration, and CO_2 became more limiting for C_3 photosynthesis, especially under high temperature (i.e., decreased solubility of CO_2; Ku and Edwards, 1977) and drought (i.e., decreased internal [CO_2] due to stomatal closure) conditions. Thus, during evolution of C_4 photosynthesis a mechanism for reducing photorespiratory CO_2 loss, such as an efficient recycling of photorespired CO_2, appears to have evolved before a CO_2 concentrating mechanism through the C_4 cycle (Monson *et al.*, 1984). This is supported by evidence that a common mechanism for suppression of apparent photorespiration may exist in all intermediates. Presumably, this recycling mechanism would have advantages in a stressful environment if CO_2 fixation were severely limited and carbon use efficiency at a premium. The selection for a mechanism to recycle photorespired CO_2 might only occur if some periods exist when the intercellular [CO_2] is relatively low. This mechanism of reducing apparent photorespiration might be achieved initially by development of organelles (e.g., chloroplast, mitochondrion, peroxisome) in the bundle-sheath cells (e.g., *Panicum laxum, Parthenium hysterophorus*). The presence of elevated PEP carboxylase activity in leaves of these plants (e.g., *Moricandia arvensis, Panicum milioides*) may contribute to a reduction in photorespiration at low CO_2 concentrations. For example, under limiting CO_2 some of the photorespired CO_2 may be fixed by PEP carboxylase and malate dehydrogenase to malate in mesophyll cells. The malate could be respired in the

TABLE X

Proposed Evolutionary Scheme from C_3 to Intermediates to C_4 Plants[a]

Type	Biochemistry	Anatomy	Physiology	Suggested representative species
C_3	PCR cycle	BSC with very few or no organelles	High Γ O_2 I of PS O_2 I of CE	*Flaveria cronquistii*
C_3	PCR cycle	BSC with few organelles	High Γ O_2 I of PS O_2 I of CE	*Mollugo pentaphylla*
C_3	PCR cycle	BSC with chloroplasts and mito equivalent to MC	Similar to C_3	*Panicum laxum*
C_3–C_4	PCR cycle	BSC with prominent chloroplast and mito	Reduced Γ Lower O_2 I of PS O_2 I of CE	*Parthenium hysterophorus*
C_3–C_4	Increase in PEP caboxylase activity	BSC with prominent chloroplast and mito	Low Γ O_2 I of CE	*Panicum milioides* *Moricandia arvensis*
C_3–C_4	Increase in activity of C_4-cycle enzymes with limited differential compartmentation between MC and BSC	BSC with prominent chloroplast and mito	Low Γ Lower O_2 I of PS Reduced O_2 I of PS	*Flaveria ramosissima*
C_3–C_4	Increased [CO_2] through C_4-cycle activity, differential compartmentation between MC and BSC imperfect	BSC with thick cell walls, prominent chloroplast and mito	Low Γ Lower O_2 I of CE Lower O_2 I of CE	*Flaveria brownii*
C_4	Differential compartmentation of certain enzymes between MC and BSC	BSC with thick cell walls, prominent chloroplasts and mito	Γ Near zero No O_2 I of PS No O_2 I of CE	*Flaveria trinervia*

[a] Abbreviations: BSC, bundle-sheath cells; MC, mesophyll cells; I, inhibition; PS, photosynthesis; CE, carboxylation efficiency; mito, mitochondria; Γ, CO_2 compensation point.

mitochondria predominantly located in the bundle sheath cells and the CO_2 then refixed in the PCR cycle. NAD-malic enzyme (present in C_3 species) would be required as the first step in respiration of malate and as noted earlier (Section III,C) may be limited. Initially, with pyruvate,Pi dikinase limiting or absent, 3-phosphoglycerate, a product of the PCR cycle, may be used as the precursor for formation of PEP. This would place a restriction on the activity of PEP carboxylase relative to Rubisco. The maximum would be one CO_2 fixed by PEP carboxylase per three CO_2 fixed by Rubisco in the mesophyll cells (the latter generating one PGA as a precursor for PEP). Through this process the CO_2 would need to be elevated in the bundle-sheath cells, where it can be more effectively fixed by Rubisco. Theoretically this could contribute to a lower Γ in these intermediates. Increases in activity of all C_4-cycle enzymes and further development of Kranz anatomy in these intermediates will allow them to synthesize C_4 acids and subsequently donate CO_2 to the PCR cycle (e.g., *Flaveria ramosissima*). Further elaboration and proper intercellular compartmentation of these biochemical steps would eventually lead to the CO_2-concentrating mechanism of C_4 plants (e.g., *Flaveria brownii*). As mentioned in the biochemistry section (Section III,C), many of the C_3–C_4 intermediate *Flaveria* species are capable of assimilating atmospheric CO_2 through a limited C_4 pathway. Thus, they represent true biochemical intermediates between C_3 and C_4 plants. Also apparent is a gradation among the *Flaveria* intermediates in the development of Kranz anatomy and C_4 biochemistry. We believe that the intermediates in the genus *Flaveria* provide the most potential for future evolutionary and genetic studies of the C_4 mechanism. The genus is small (21 species), and independent taxonomic studies have been done recently (Powell, 1978). Interspecific crosses between different photosynthetic species have been quite successful (see Section VIII), and polyploidy does not appear to be a major confusing factor in the cytogenetics of this taxonomic group.

IX. CONCLUDING REMARKS

C_4 plants have no apparent photorespiration under atmospheric conditions, whereas C_3 plants exhibit photorespiration. C_4 plants have Kranz leaf anatomy consisting of distinctive wreath-like, thick-walled cells with numerous chloroplasts, while in C_3 plants, there are no or few organelles in bundle sheath cells. The C_3–C_4 intermediates are species in which apparent photorespiration is detectable but exists at a lower level than that in C_3 plants. These species are of interest in considering means by which photorespiration can be reduced and for studying the evolution of C_4 photosynthesis. All of the intermediates identified thus far have bundle-sheath cells that contain chloroplasts and mitochondria, and the extent to which the anatomy is Kranz-like is species-dependent. However, the occurrence of organelles in bundle-sheath cells (e.g., equivalent to those in mesophyll cells) does not assure the species will be intermediate in physiological and biochemical features (e.g., as in *Panicum laxum* and *Mollugo pentaphylla*, which have C_3 photosynthetic characteristics).

Apparently, more than one strategy exists among the C_3–C_4 intermediate species to reduce their apparent photorespiration. Probably all intermediates reduce apparent photorespiration, at least in part, by recycling photorespired CO_2. In intermediates that lack the C_4 cycle, photorespired CO_2 may be fixed directly by the PCR cycle. The abundance of mitochondria in bundle-sheath cells and their position relative to chloroplasts may make conditions more favorable for refixation of the photorespired CO_2 from the mitochondria. Some intermediates have a C_4 cycle in which part of the atmospheric CO_2 is fixed into the C_4 acids malate and aspartate. In these intermediates, the C_4 cycle may also contribute to the refixation of photorespired CO_2 just as it must do in C_4 species. In intermediates where the C_4 cycle is functioning to fix atmospheric CO_2 and donate it to the PCR cycle, the extent to which this process will reduce apparent photorespiration will depend on whether [CO_2] is elevated in the leaf around Rubisco. In more advanced intermediates that appear closer to developing the complete Kranz syndrome, such as *Flaveria brownii, F. ramosissima,* and *Neurachne minor,* the C_4 cycle may function to increase [CO_2] in the leaf and thus limit apparent photorespiration by reducing O_2 inhibition of photosynthesis through its competitive interactions with respect to CO_2. However, except for *F. brownii,* the carbon isotope values are not C_4-like in these species, which indicates that much of the atmospheric CO_2 is fixed directly by Rubisco (perhaps in mesophyll cells).

While more research is needed on the anatomy and physiology of intermediates, evidence for the mechanism of reduction in apparent photorespiration in these species depends ultimately on biochemical studies. In every genus in which intermediates have been identified there is limited information on photosynthesis at the biochemical level. In general, identification of the intercellular compartmentation of key enzymes, such as malic enzyme and glycine decarboxylase, will be important for elucidating pathways of photosynthetic and photorespiratory metabolism in these species. Research at the biochemical level will be required to determine why increasing light intensity decreases Γ in intermediate species, and why there is often a biphasic response of Γ to increasing [O_2].

By employing techniques in molecular biology, intermediates can be used to study the molecular basis of evolution of C_4 plants from C_3 plants (and to determine if, in some cases, intermediates represent hybrids between C_3 and C_4 species). Within a small genus such as *Flaveria,* which has several closely related C_3, intermediate and C_4 species, studies are needed on chemical and molecular approaches to understanding phylogenetic relationships, on the isozyme patterns of enzymes of the C_4 cycle (e.g., PEP carboxylase; Adams *et al.,* 1986) and other pathways, and to determine whether the genes for C_4 photosynthesis are present in all species and how expression of these genes are regulated. The possibility that genetically based differences in C_3 and C_4

photosynthesis exist within a species (e.g., between different populations) is another area to explore in the future. Similarly, the extent to which leaf age and environmental factors may modulate the reduction of apparent photorespiration in C_3–C_4 intermediates needs to be investigated at the biochemical level.

ACKNOWLEDGMENTS

The authors wish to acknowledge provision of data from R. H. Brown, P. W. Hattersley, L. Mets, and B. d. Moore prior to publication and assistance in translation by C. Critchely. The helpful suggestions by R. H. Brown, S. H. Cheng, M. D. Hatch, P. W. Hattersley, R. K. Monson, and B. d. Moore on the manuscript are much appreciated. The technical assistance of S. Edwards is also kindly acknowledged.

REFERENCES

Adams, C. A., Leung, F., and Sun, S. S. M. (1986). *Planta* **167**, 218–225.
Aoyagi, K., and Bassham, J. A. (1983). *Plant Physiol.* **73**, 853–854.
Apel, P. (1980). *Biochem. Physiol. Pflanz.* **175**, 386–388.
Apel, P., and Maass, I. (1981). *Biochem. Physiol. Pflanz.* **197**, 396–399.
Apel, P., and Ohle, H. (1979). *Biochem. Physiol. Pflanz.* **174**, 68–75.
Apel, P., Ticha, I., and Peisker, M. (1978). *Biochem. Physiol. Pflanz.* **172**, 547–552.
Bassüner, B., Keerberg, O., Bauwe, H., Pyarnik, T., and Keerberg, H. (1984). *Biochem. Physiol. Pflanz.* **179**, 631–634.
Bauwe, H. (1984). *Biochem. Physiol. Pflanz.* **179**, 253–268.
Bauwe, H., and Apel, P. (1979). *Biochem. Physiol. Pflanz.* **174**, 251–254.
Björkman, O. (1976). *In* "CO$_2$ Metabolism and Plant Productivity" (R. H. Burris and C. C. Black, eds.), pp. 287–309. University Park Press, Baltimore, Maryland.
Bouton, J. H., Brown, R. H., Evans, P. T., and Jernsted, J. A. (1986). *Plant Physiol.* **80**, 487–492.
Brown, R. H. (1980). *Plant Physiol.* **65**, 346–349.
Brown, R. H., and Bouton, J. H. (1983). *In* "Current Topics in Plant Biochemistry and Physiology" (D. D. Randall, D. G. Blevins, and R. Larson, eds.), pp. 78–89. Univ. of Missouri Press, Columbia.
Brown, R. H., and Brown, W. V. (1975). *Crop Sci.* **15**, 681–685.
Brown, R. H., and Morgan, J. A. (1980). *Plant Physiol.* **66**, 541–544.
Brown, R. H., Bouton, J. H., Rigsby, L. L., and Rigsby, M. (1983a). *Plant Physiol.* **71**, 425–431.
Brown, R. H., Rigsby, L. L., and Akin, D. E. (1983b). *Plant Physiol.* **71**, 437–439.
Brown, R. H., Bouton, J. H., Evans, P. T., Malter, H. E., and Rigsby, L. L. (1985). *Plant Physiol.* **77**, 653–658.
Brown, R. H., Bouton, J. H., and Evans, P. T. (1986a). *Plant Physiol.* **81**, 212–215.
Brown, R. H., Bassett, C. L., Cameron, R. G., Evans, P. T., Bouton, J. H., Black, C. C., Sternberg, L. O., and DeNiro, M. J. (1986b). *Plant Physiol.* **82**, 211–217.

Brown, W. V. (1975). *Am. J. Bot.* **62**, 395–402.

Brown, W. V. (1977). *Mem. Torrey Bot. Club* **23**, 1–97.

Cheng, S.-H., and Ku, M. S. B. (1985). *Plant Physiol., Suppl.* **77**, 90.

Cheng, S.-H., Ku, M. S. B., and Mets, L. J. (1984). *Plant Physiol., Suppl.* **75**, 58.

Cheng, S.-H., Moore, B. d., and Ku, M. S. B. (1986). *Plant Physiol., Suppl.* **81**, 55.

Cheng, S.-H., Franceschi, V. R., Keefe, D., Mets, L. J., and Ku, M. S. B. (1987). *In* "Progress in Photosynthesis Research" (J. Biggins, ed.), Vol. III, pp. 637–640. Martinus Nijhoff/Dr. W. Junk, The Hague, The Netherlands.

Collins, P. D., and Hague, D. R. (1983). *J. Biol. Chem.* **258**, 4012–4018.

Downton, W. J. S. (1975). *Photosynthetica* **9**, 96–105.

Edwards, G. E., and Huber, S. C. (1981). *In* "The Biochemistry of Plants" (M. D. Hatch and N. K. Boardman, eds.), Vol. 8, pp. 237–281. Academic Press, New York.

Edwards, G. E., and Walker, D. A. (1983). "C₃, C₄: Mechanisms and Cellular and Environmental Regulation of Photosynthesis." Univ. of California Press, Berkeley.

Edwards, G. E., Ku, M. S. B., and Hatch, M. D. (1982). *Plant Cell Physiol.* **23**, 1185–1195.

Edwards, G. E., Ku, M. S. B., and Monson, R. K. (1985). *In* "Phytosynthetic Mechanisms and the Environment" (J. Barber and N. R. Baker, eds.), pp. 287–328. Am. Elsevier, New York.

Ellis, R. P. (1974). *S. Afr. J. Sci.* **70**, 169–173.

Fox, T. C., Kennedy, R. A., and Loescher, W. H. (1986). *Plant Physiol.* **82**, 307–311.

Furbank, R. T., and Hatch, M. D. (1987). *Plant Physiol.* (submitted for publication).

Gardestrom, P., Edwards, G. E., Henricson, D., and Ericson, I. (1985). *Physiol. Plant.* **64**, 29–33.

Gee, S. L., Ruzin, S., and Bassham, J. A. (1984). *Plant Physiol.* **74**, 189–191.

Goldstein, L. D., Ray, T. B., Kestler, D. P., Mayne, B. C., Brown, R. H., and Black, C. C. (1976). *Plant Sci. Lett.* **6**, 85–90.

Gomez-Campo, C. (1980). *In* "Brassica Crops and Wild Allies" (S. Tsunoda, K. Hinata, and C. Gomez-Campo, eds.), pp. 51–63. Jpn. Sci. Soc. Press, Tokyo.

Haberlandt, G. (1909). "Physiologische Pflanzenanatomie," 4th ed. Englemann, Leipzig.

Hague, D. R., Uhler, M., and Collins, P. D. (1983). *Nucleic Acids Res.* **11**, 4853–4865.

Hattersley, P. W., and Roksandic, Z. (1983). *Aust. J. Bot.* **31**, 317–321.

Hattersley, P. W., and Stone, N. E. (1986). *Aust. J. Plant Physiol.* **13**, 399–408.

Hattersley, P. W., Watson, L., and Osmond, C. B. (1977). *Aust. J. Plant Physiol.* **4**, 523–539.

Hattersley, P. W., Watson, L., and Wong, S. C. (1984). *In* "Advances in Photosynthesis Research" (C. Sybesma, ed.), Vol. 3, pp. 403–406. Martinus Nijhoff/Dr. W. Junk, The Hague, The Netherlands.

Hattersley, P. W., Wong, S. C., Perry, S., and Roksandic, Z. (1986). *Plant Cell Environ.* **9**, 217–233.

Hegde, B. A., and Patil, T. M. (1980). *Biovigyanam* **6**, 15–19.

Hegde, B. A., and Patil, T. M. (1981). *Photosynthetica* **15**, 1–4.

Hinata, K., Oguro, H., and Tsunoda, S. (1984). *In* "Genetical Approach in Photosynthesis Research of Higher Plants," Bull. Green Energy Program No. 5, pp. 7–37. Agriculture, Forestry and Fisheries Research Council Secretariat, Ministry of Agriculture, Forestry and Fisheries, Tokyo.

Holaday, A. S., and Black, C. C. (1981). *Plant Physiol.* **67**, 330–334.

Holaday, A. S., and Chollet, R. (1983). *Plant Physiol.* **73**, 740–745.

Holaday, A. S., and Chollet, R. (1984). *Photosynth. Res.* **5**, 307–323.

Holaday, A. S., Shieh, Y.-J., Lee, K. W., and Chollet, R. (1981). *Biochim. Biophys. Acta* **637**, 334–341.

Holaday, A. S., Harrison, A. T., and Chollet, R. (1982). *Plant Sci. Lett.* **27**, 181–189.

Holaday, A. S., Lee, K. W., and Chollet, R. (1984). *Planta* **160**, 25–32.

Holaday, A. S., Talkmitt, S., and Doohan, M. E. (1985). *Plant Sci.* **41**, 31–39.

Holbrook, G. P., Jordan, D. B., and Chollet, R. (1985). *Plant Physiol.* **77**, 578–583.
Kanai, R., and Kashiwagi, M. (1975). *Plant Cell Physiol.* **16**, 669–679.
Keck, R. N., and Ogren, W. L. (1976). *Plant Physiol.* **58**, 552–555.
Kennedy, R. A., and Laetsch, W. M. (1974). *Science* **184**, 1087–1089.
Kennedy, R. A., Eastburn, J. L., and Jensen, K. G. (1980). *Am. J. Bot.* **67**, 1207–1217.
Krenzer, E. G., Moss, D. N., and Crookston, R. K. (1975). *Plant Physiol.* **56**, 194–206.
Ku, M. S. B., and Edwards, G. E. (1977). *Plant Physiol.* **59**, 986–990.
Ku, M. S. B., and Edwards, G. E. (1978). *Plant Cell Physiol.* **19**, 665–675.
Ku, M. S. B., and Edwards, G. E. (1980). *Planta* **147**, 277–282.
Ku, M. S. B., Edwards, G. E., and Kanai, R. (1976). *Plant Cell Physiol.* **17**, 615–620.
Ku, M. S. B., Monson, R. K., Littlejohn, R. O., Nakamoto, H., Fisher, D. B., and Edwards, G. E. (1983). *Plant Physiol.* **71**, 944–948.
Laetsch, W. M. (1971). *In* "Photosynthesis and Photorespiration" (M. D. Hatch, C. B. Osmond, and R. O. Slatyer, eds.), pp. 323–349. Wiley (Interscience), New York.
Monson, R. K., Edwards, G. E., and Ku, M. S. B. (1984). *BioScience* **34**, 563–574.
Monson, R. K., Moore, B. d., Ku, M. S. B., and Edwards, G. E. (1986). *Planta* **168**, 493–502.
Monson, R. K., Teeri, J. A., Ku, M. S. B., Gurevitch, J., Mets, L. J. and S. Dudley (1987). *Planta* (submitted for publication).
Moore, B. d., Ku, M. S. B., and Edwards, G. E. (1987a). *In* "Progress in Photosynthesis Research" (J. Biggins, ed.), Vol. IV, pp. 357–360. Martinus Nijhoff/Dr. W. Junk, The Hague, The Netherlands.
Moore, B. d., Franceschi, V. R., Cheng, S.-H., Wu, J., and Ku, M. S. B. (1987b). *Plant Physiol.* (in press).
Moore, P. D. (1982). *Nature (London)* **295**, 647–648.
Morgan, J. A., and Brown, R. H. (1979). *Plant Physiol.* **64**, 257–262.
Morgan, J. A., and Brown, R. H. (1980). *Plant Physiol.* **65**, 156–159.
Morgan, J. A., Brown, R. H., and Reger, B. J. (1980). *Plant Physiol.* **65**, 156–159.
Nakamoto, H., Ku, M. S. B., and Edwards, G. E. (1983). *Plant Cell Physiol.* **24**, 1387–1393.
Ohnishi, J., and Kanai, R. (1983). *Plant Cell Physiol.* **24**, 1411–1420.
O'Leary, M. H. (1981). *Phytochemistry* **20**, 553–567.
Osmond, C. B., Björkman, O., and Anderson, D. J. (1980). "Physiological Processes: Plant Ecology." Springer-Verlag, Berlin and New York.
Pathan, S. N., and Nimbalkar, J. D. (1982). *Photosynthetica* **16**, 119–122.
Patil, T. M., and Hegde, B. A. (1983). *Photosynthetica* **17**, 64–68.
Peisker, M., and Bauwe, H. (1984). *Photosynthetica* **16**, 119–122.
Powell, A. M. (1978). *Ann. Mo. Bot. Gard.* **65**, 590–636.
Prendergast, H. D. V., and Hattersley, P. W. (1985). *Aust. J. Bot.* **33**, 317–336.
Raghavendra, A. S., and Das, V. S. R. (1978). *Photosynthetica* **12**, 200–208.
Raghavendra, A. S., Rajendrudu, G., and Das, V. S. R. (1978). *Nature (London)* **273**, 143–144.
Rajendrudu, G., Parasad, J. S. R., and Das, V. S. R. (1986). *Plant Physiol.* **80**, 409–414.
Rathnam, C. K. M., and Chollet, R. (1980). *Prog. Phytochem.* **6**, 1–48.
Reed, J. E., and Chollet, R. (1985). *Planta* **165**, 439–445.
Rumpho, M. E., Ku, M. S. B., Cheng, S.-H., and Edwards, G. E. (1984). *Plant Physiol.* **75**, 993–996.
Sayre, R. T., and Kennedy, R. A. (1977). *Planta* **134**, 257–262.
Sayre, R. T., Kennedy, R. A., and Pringnitz, D. J. (1979). *Plant Physiol.* **64**, 293–299.
Schnyder, H., Machler, F., and Nosberger, J. (1984). *J. Exp. Bot.* **35**, 147–156.
Sharkey, T. D. (1985). *Plant Physiol.* **78**, 71–75.
Sims, T. L., and Hague, D. R. (1981). *J. Biol. Chem.* **256**, 8252–8255.
Smith, B. N., and Powell, A. M. (1984). *Naturwissenschaften* **71**, 217–218.
Smith, B. N., and Robbins, M. J. (1975). *Proc. Int. Congr. Photosynth., 3rd, 1974*, pp. 1579–1587.

Smith, B. N., and Turner, B. L. (1975). *Am. J. Bot.* **62,** 541–545.

Vogel, J. C., Fuls, A., and Ellis, R. P. (1978). *S. Afr. J. Sci.* **14,** 211–215.

Wigley, T. M. L. (1983). *Clim. Change* **5,** 315–320.

Winter, K., Holtum, J. A. M., Edwards, G. E., and O'Leary, M. H. (1982a). *J. Exp. Bot.* **33,** 88–91.

Winter, K., Usuda, H., Tsuzuki, M., Schmitt, M., Edwards, G. E., Thomas, R. J., and Evert, R. F. (1982b). *Plant Physiol.* **70,** 616–625.

Control of Photosynthetic Sucrose Formation

6

MARK STITT
STEVE HUBER
PHIL KERR

I. Introduction
II. Biosynthesis of Sucrose in Leaves
   A. Pathway and Intracellular Localization
   B. Why Regulate Sucrose Synthesis?
III. Metabolite Levels during Sucrose Synthesis
   A. Overall and Cytosolic Metabolite Levels
   B. Identification of Nonequilibrium Reactions
   C. Changes of Metabolites during Sucrose Synthesis
IV. Regulatory Properties of the Individual Steps in Sucrose Synthesis
   A. The Phosphate Translocator
   B. The Cytosolic FBPase
   C. Sucrose-Phosphate Synthase
   D. Sucrose Phosphatase
   E. Pyrophosphate : Fructose-6-Phosphate Transferase and Pyrophosphate Metabolism
   F. Supply of Energy
V. Regulation of Fructose 2,6-Bisphosphate Concentration
   A. Presence and Localization
   B. Synthesis and Degradation of $Fru2,6P_2$
   C. Alterations of $Fru2,6P_2$ during Photosynthetic Metabolism
   D. Mechanisms Controlling $Fru2,6P_2$ in Leaves
VI. Coarse Control of SPS
   A. Diurnal Alterations
   B. Effect of Source–Sink Manipulations
   C. Mechanism of Coarse Control of SPS Activity
VII. Coordinate Control of Sucrose Formation by SPS and the Cytosolic Fructose 1,6-Bisphosphatase
   A. Feedforward Control
   B. Feedback Control
VIII. Adaptation of the Regulation of Sucrose Synthesis
   A. Adaptation of Sucrose Synthesis in C_4 and CAM Species
   B. Temperature Dependence of the "Threshold" for Sucrose Synthesis

The Biochemistry of Plants, Vol. 10

327

IX. Limitation of Photosynthesis by Sucrose Synthesis
    A. Limitation by Suboptimal Metabolite Levels
    B. Limitation by Suboptimal Phosphate
    References

I. INTRODUCTION*

Since the last review of sucrose metabolism in this series (Akazawa and Okamoto, 1980), there has been substantial progress in our understanding of the regulation of photosynthetic sucrose formation, which is the topic of this review. In most species, starch and sucrose are the principal end products of photosynthesis. It is clear that the formation of both carbohydrates is highly regulated biochemically. Recent research attention has focused on sucrose synthesis, because it appears that the rate of sucrose formation may indirectly control starch metabolism.

Prior to the last review (Akazawa and Okamoto, 1980), it was thought that sucrose synthesis occured in the cytoplasm of leaf cells and that sucrose-phosphate synthase (sucrose-P synthase, SPS) was the enzyme involved in sucrose production. It has now been well established that sucrose formation occurs exclusively in the cytoplasm (Robinson and Walker, 1979a; Stitt *et al.*, 1980; Kaiser *et al.*, 1982), and progress has been made in elucidating the biochemical mechanisms that control sucrose formation in leaves. It is now recognized that leaf SPS is regulated at two levels: (a) metabolic "fine" control and (b) "coarse" control. The distinction between these two levels of control has been discussed in detail by ap Rees (1980) in relation to the regulation of glycolysis. Briefly, "fine" control of enzyme activity is exerted by metabolic effectors, which instantaneously activate or inhibit catalysis. "Coarse" control refers to slower changes in the extractable maximum activity of an enzyme measured *in vitro* and may be due to protein modification or to protein turnover. Our understanding of the regulation of cytosolic FBPase has also changed markedly. This enzyme, which catalyzes the first irreverible step in the sucrose formation pathway, was first purified and characterized by Zimmermann *et al.* (1978) and since has been considered to be a key control point. However, the key to understanding the regulation of cytosolic FBPase activity was the discovery of a regulator metabolite called

---

* Abbreviations: DHAP, dihydroxyacetone phosphate; FBPase, fructose-1,6-bisphosphatase; Fru1,6P$_2$, fructose 1,6-bisphosphate; Fru2,6P$_2$, fructose 2,6-bisphosphate; Fru6P, fructose 6-phosphate; Fru6P,2-kinase, fructose 6-phosphate,2-kinase; Fru2,6P$_2$ase, fructose 2,6-bisphosphatase; Glc6P, glucose 6-phosphate; hexose-P, the total of glucose 6-phosphate, fructose 6-phosphate, and glucose 1-phosphate; PEP, phosphoenolpyruvate; Pi, inorganic phosphate; PFP, pyrophosphate:fructose-6-phosphate phosphotransferase; PPi, inorganic pyrophosphate; PGA, 3-phosphoglycerate; RuBP, ribulose 1,5-bisphosphate; Rubisco, ribulose-1,5-bisphosphate carboxylase/oxygenase; sucrose-P, sucrose 6-phosphate; triose-P, the sum of dihydroxyacetone phosphate and glyceraldehyde 3-phosphate; UDPGlc, UDP glucose.

Fru2,6P$_2$ in liver (Hers *et al.*, 1982), and its subsequent study in plants. Fru2,6P$_2$ is a potent inhibitor of cytosolic FBPase activity, which also modulates the response of this enzyme to other effectors.

It is becoming apparent that a coordinate regulation of cytosolic FBPase and SPS exists, which provides a framework to understand how sucrose synthesis may be controlled in different situations: when substrate availability is limiting (low photosynthetic rates) and when sucrose accumulates (synthesis exceeds export). The regulation of this pathway also provides a model to explain how the rate of sucrose formation (the principal end product of photosynthesis) may impact on photosynthate partitioning, on the response of photosynthesis to varying environmental conditions, and on the rate of photosynthesis.

Much still remains to be learned, but in this review, we will focus on recent developments that have increased our understanding of the regulation of sucrose formation. Several mini-reviews covering regulation via Fru2,6P$_2$ (Cseke *et al.*, 1984; Stitt, 1985a,b, 1986a, 1987a: Stitt *et al.*, 1987) and regulation of SPS (Huber *et al.*, 1985a,b) have recently been published, which provide a short introduction to developments that will be discussed in more detail in the present review.

II. BIOSYNTHESIS OF SUCROSE IN LEAVES

A. Pathway and Intracellular Localization

1. C$_3$ Plants

There have been conflicting reports over the years as to the intracellular localization of sucrose formation in leaf cells. Initially, it was thought that the chloroplast was the site of sucrose formation. However, it is now well established that the chloroplast envelope is impermeable to sucrose (Heber and Heldt, 1981), and that the enzymes for sucrose biosynthesis are located exclusively in the cytosol (Bird *et al.*, 1974). Consequently, when C$_3$ mesophyll protoplasts that are actively assimilating $^{14}CO_2$ are rapidly fractionated into plastid and extrachloroplast compartments, the labeled sucrose appears exclusively outside the chloroplast (Robinson and Walker, 1979a; Stitt *et al.*, 1980). Understanding the regulation of sucrose formation obviously requires knowledge of the pathway and enzymes involved and their intracellular localization.

Sucrose is synthesized in the cytosol from triose-P that is generated within the chloroplast and released to the cytosol by operation of the phosphate translocator (Fig. 1). The phosphate translocator catalyzes a strict one-to-one exchange of certain phosphate compounds (Heber and Heldt, 1981;

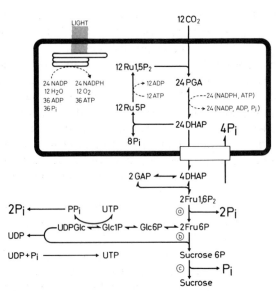

Fig. 1. Simplified scheme showing the intracellular compartmentation of photosynthetic carbon flow to sucrose in the mesophyll cell of a C_3 plant. Cytosolic enzymes that catalyze essentially irreversible reactions are (a) the cytosolic FBPase; (b) sucrose-P synthase; (c) sucrose phosphatase. The numbers show the stoichiometry of fluxes needed to produce one molecule of sucrose, assuming there is no photorespiration. The figures do not include the energy input for regeneration of UTP in the cytosol. The routes of PPi hydrolysis and UTP synthesis are not characterized, but it is assumed one of the Pi deriving from hydrolysis of PPi is used to regenerated UTP from UDP. The other four Pi released in the cytosol can reenter the chloroplast.

Flügge and Heldt, 1984). During sucrose formation, the release of triose-P from the chloroplast is coupled to an influx of the inorganic phosphate (Pi) generated in the cytosol as triose-P is converted to the neutral sucrose molecule. This continual exchange of triose-P and Pi means that the chloroplast and cytosol are metabolically interdependent, and it can be expected that the rate of CO_2 assimilation within the chloroplast will influence the rate of sucrose formation, and vice versa. As discussed in more detail later, there are three enzymes catalyzing effectively irreversible reactions during sucrose synthesis—the cytosolic FBPase, SPS, and sucrose phosphatase (Fig. 1)—and these provide potential points of regulation in the pathway.

Sucrose synthesis also needs an input of energy. For every four molecules of triose-P that are converted to sucrose, one molecule of UTP is expended in the reaction catalyzed by UDP glucose pyrophosphorylase. The products are PPi, released in the same reaction, and UDP, which is released in the reaction catalyzed by SPS. Thus a complete pathway of sucrose synthesis also includes delivery of high-energy phosphate bonds to the cytosol, and the hydrolysis of PPi.

2. Cellular Compartmentation in C_4 Plants

An additional complication is introduced in C_4 plants, where photosynthesis requires cooperation of two different types of cells. The cellular compartmentation of the enzymes involved in primary CO_2 assimilation in C_4 plants has been firmly established for some time (Edwards and Huber, 1981; Edwards and Walker, 1983). Atmospheric CO_2 is initially assimilated by PEPCase in the mesophyll cells to form C_4 dicarboxylic acids, which are then decarboxylated in the bundle-sheath cells. The released CO_2 is assimilated via the reductive pentose phosphate pathway, which occurs only in the bundle-sheath chloroplasts. Strict spatial compartmentation is thought to be essential for the CO_2-concentrating mechanism to function efficiently.

There have been several recent reports that confirm the original suggestion of Downton and Hawker (1973) that there is also a fairly strict intercellular compartmentation of starch and sucrose metabolism in the C_4 plant maize. Starch normally accumulates to a far greater extent in the chloroplasts of the bundle sheath cells compared with the mesophyll, although starch does accumulate in the mesophyll chloroplasts under conditions of prolonged illumination (Downton and Hawker, 1973). The activity of ADPG pyrophosphorylase—a key enzyme of starch synthesis—is almost 20-fold higher in bundle-sheath chloroplasts than in mesophyll chloroplasts (Preiss et al., 1985). In addition, the enzyme from the two chloroplast types exhibits different properties in vitro. The bundle-sheath enzyme is activated over 20-fold by 1 mM PGA (3-phosphoglycerate), much like the spinach leaf enzyme, whereas the mesophyll enzyme is activated less than 3-fold by PGA (Preiss et al., 1985). Thus there appears to be a biochemical limitation of the capacity for starch synthesis in the mesophyll cell.

Conversely, it appears that sucrose synthesis may occur predominantly, or even exclusively, in the mesophyll. Usuda and Edwards (1980) concluded that SPS was highly "enriched" in the mesophyll cell, and this has since been confirmed in maize for both SPS and the cytosolic FBPase (Furbank et al., 1985). In addition a novel technique for rapid separation of the two cell types was used with $^{14}CO_2$ feeding to show that sucrose synthesis occurs in the mesophyll cell. The majority of the Fru2,6P$_2$ (Stitt and Heldt, 1985a) as well as the enzymes involved in synthesizing and degrading Fru2,6P$_2$ (Soll et al., 1985) are also located in the mesophyll of maize leaves. These lines of evidence suggest that sucrose formation occurs largely, if not exclusively, in the mesophyll cell in maize leaves.

The intercellular fluxes of metabolites that would be involved during sucrose formation in maize are shown in Fig. 2, assuming that there is a strict compartmentation of carbohydrate metabolism. For clarity, flow of carbon through the C_4 pathway (which delivers CO_2 to the bundle-sheath cell) and formation of starch in the bundle-sheath chloroplast is not shown. Synthesis of one sucrose molecule requires the net fixation of 12 molecules of CO_2,

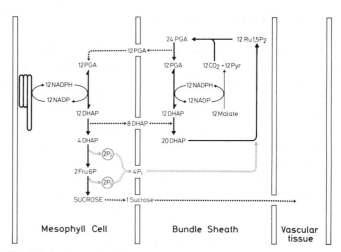

Fig. 2. Simplified scheme showing intercellular compartmentation of sucrose formation in the C_4 plant maize. For clarity, enzymes of the C_4 dicarboxylic acid pathway, the turnover of ATP, and the flow of carbon into starch are not shown. The model assumes that sucrose synthesis occurs exclusively in the mesophyll cells, and that half of the PGA formed in the reductive pentose phosphate pathway must be shuttled to the mesophyll cell for reduction.

which will form 24 molecules of PGA. Maize is an NADP-malic enzyme species that has essentially agranal bundle-sheath chloroplasts. Assuming that the photosystem II deficiency is complete, half of the PGA formed must be shuttled to the mesophyll cell for reduction (the other half could be reduced using the NADP formed by malic enzyme; see Edwards and Huber, 1981). Of the 12 triose-P molecules formed in the mesophyll cell, eight molecules would need to be returned to the bundle sheath (for regeneration of RuBP) and four would be available to form one molecule of sucrose. The net result is that even though sucrose formation may occur exclusively in the mesophyll, intercellular transport of triose-P would still also be necessary. Charge—and phosphate—balance between the two cell types would also require that the Pi released during sucrose synthesis in mesophyll return to the bundle sheath (Fig. 2). The C_4 leaf anatomy (concentric rings of mesophyll and bundle-sheath cells surrounding the vascular bundles) also requires that the sucrose formed in the mesophyll be transported first to the bundle sheath before entry into the phloem.

It is important to note, however, that the strict compartmentation found in maize is probably not a universal characteristic of C_4 plants. Studies conducted in the last decade, using isolated cell types of high purity, established that the distribution of SPS may vary substantially. In *Digitaria pentzi,* more than 90% of the whole leaf SPS activity was found in the bundle-sheath cells (Mbaku *et al.,* 1978), whereas an equal distribution between cell types was observed in *Cyperus rotundus* (Chen *et al.*, 1974) and *Digitaria sanguinalis* (Edwards and Black, 1971). Maize appears to be an example of the other extreme, where SPS is primarily in the mesophyll. It will be important to reevaluate the distribution of sucrose-metabolizing enzymes in various species using current techniques and assay procedures, as maize may represent an extreme situation that is not typical or representative of C_4 plants.

B. Why Regulate Sucrose Synthesis?

1. Adjustment to the Rate of Photosynthesis

The interdependence of chloroplast and cytosolic metabolism implies that coordination of the fluxes and conditions in these compartments is a precondition for rapid photosynthesis. This interaction is defined by the stoichiometry of carbon and phosphate flow during steady-state photosynthesis [Fig. 3(a)]. Chloroplasts convert three CO_2 and one Pi to one molecule of triose-P, and a continuation of photosynthesis depends on the triose-P being removed and more Pi becoming available. This is achieved by exporting triose-P from the chloroplast, in exchange for Pi. However, the rate of this exchange must be coordinated with the rate of CO_2 fixation, because only one-sixth of the triose-P may be removed, representing the net gain of carbon in one turn of the Calvin cycle. The remaining five-sixths must remain in the chloroplast and be reconverted to RuBP to allow more CO_2 to be fixed. Indeed, when photorespiration is occurring, an even smaller proportion of the triose-P may be removed, as some RuBP will be consumed without net fixation of CO_2. The importance of this balance between the removal of triose-P and the supply of Pi is well illustrated by isolated chloroplasts. These must be supplied with Pi to allow photosynthesis, but are inhibited if too much Pi is added, because triose-P is removed too rapidly and the pools of stromal metabolites are depleted so far that the regeneration of RuBP is inhibited (Edwards and Walker, 1983).

In leaves, triose-P is continually removed and Pi is regenerated by sucrose synthesis. The rate at which this occurs needs to be regulated if optimal rates of photosynthesis are to be achieved, analogous to the way in which an optimal Pi concentration must be selected during experiments with isolated chloroplasts. However, in leaves the situation is more complex because the cytosol contains pools of triose-P and PGA as well as Pi, all of which turn over within seconds (Stitt *et al.*, 1980), and also compete for transport on the phosphate translocator (see Section IV,A). Rapid sustained photosynthesis will not be possible unless sucrose synthesis is regulated to allow the rates of CO_2 fixation and the Pi recycling to be balanced at a point where the subcellular concentrations of triose-P and Pi allow all the partial processes of photosynthesis to operate efficiently.

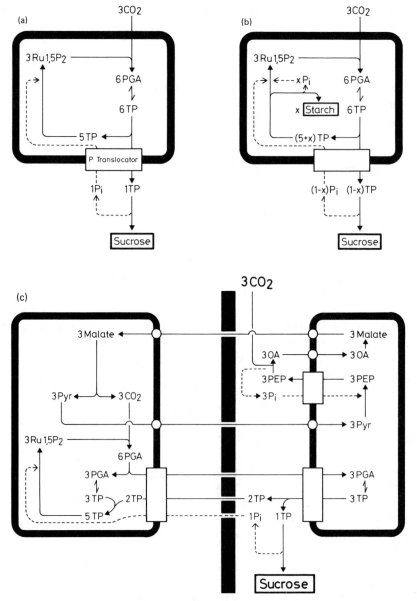

Fig. 3. Scheme for the fluxes of triose-P and Pi during sucrose synthesis. For simplicity the photorespiratory fluxes are omitted. In C_3 plants these would require rates of $Ru1,5P_2$ regeneration and carboxylation, and of PGA reduction which would be about 20–40% of those occurring for reductive carbon fixation. (a) Sucrose as the only product. (b) Starch and sucrose formation. (c) Maize, assuming that only sucrose is formed. The fluxes of Pi and PEP via the phosphate translocator of the mesophyll chloroplast are also shown. The mesophyll and bundle-sheath cells are on the right and left side of the figure, respectively. The remainder of the malate/pyruvate exchange is shown in faint print only. (d) A phosphoenolpyruvate carboxykinase type CAM plant, where malate is converted to extrachloroplastic carbohydrates. Abbreviations in figure: OA, oxaloacetate; PEP, phosphoenolpyruvate; Pyr, pyruvate; TP, triose-P.

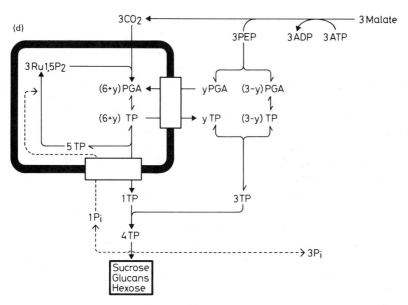

Fig. 3. (*Continued*)

2. *Control of Partitioning*

Although sucrose is the major end product of photosynthesis, its rate of synthesis will need to be adjusted to allow other products to be formed. Some photosynthate is accumulated in the chloroplast as starch and is remobilized during the night for respiration or for synthesis of sucrose, allowing export to be continued during the night (Sharkey and Pate, 1976; Gordon *et al.,* 1980; Fondy and Geiger, 1982). A small proportion of the photosynthate is also used for synthesis of amino acids, following the assimilation of inorganic nitrogen (Robinson and Baysdorfer, 1985).

The key enzyme leading to starch synthesis in ADPGlc pyrophosphorylase (Preiss, 1980), while pyruvate kinase and PEP carboxylase will be crucial enzymes for controlling the synthesis of carbon skeletons for amino acid synthesis (Paul *et al.,* 1978; Baysdorfer and Bassham, 1984). Even when the distribution of photosynthate is modified due to direct regulation of these enzymes, it will still be necessary to adjust the rate of sucrose synthesis in a reciprocal manner, if the total removal of triose-P out of the Calvin cycle is to be kept in balance with the rate of CO_2 fixation [Fig. 3(b)]. However, regulation of sucrose synthesis probably has a more direct impact on the partitioning of photosynthate. For example, it will be discussed later how a restriction of sucrose synthesis actually stimulates the accumulation of starch.

3. Adaptation

a. C₄ and CAM Species. In most C_3 plants, once triose-P is exported from the chloroplasts its major fate is for sucrose synthesis. In contrast, in C_4 and CAM plants there are large fluxes through the cytosolic triose-P pool that are directly involved in the fixation of CO_2. In such plants, sucrose synthesis must be coordinated with these accessory fluxes as well as with the operation of the Calvin cycle [Fig. 3(c) and Fig. 3(d)].

In NADP-malic enzyme species such as maize that have agranal bundle-sheath chloroplasts, intercellular movement of PGA and triose-P is unavoidable because insufficient NADPH is available in the bundle sheath (see above). The intercellular transport involves diffusion along very substantial concentration gradients, which are maintained between the mesophyll and bundle-sheath cells (Leegood, 1985; Stitt and Heldt, 1985a,b). Other C_4 plants have granal chloroplasts in the bundle sheath and could, theoretically, generate enough NADPH to allow all the PGA to be reduced in the bundle sheath. Nevertheless, intercellular movement of triose-P and PGA may occur widely in C_4 plants as a high activity of phosphoglycerate kinase and NADP-glyceraldehyde-3-P dehydrogenase is typical for mesophyll chloroplasts from all types of C_4 plant (Hatch and Osmond, 1976; Edwards and Walker, 1983). Also, representatives from all the C_4 types show a selectively high content of PGA and triose-P compared to C_3 plants, as expected if large amounts of these metabolites are present so that concentration gradients can be built up to drive diffusion between different cells (M. Stitt, unpublished). Figure 3(c) summarizes the fluxes of Pi and triose-P in maize. The synthesis of sucrose will need to be regulated to avoid undermining the concentration gradient of triose-P, which is needed to drive triose-P diffusion back to the Calvin cycle in the bundle sheath. In addition, the fluxes of Pi involved in synthesis and use of triose-P in the mesophyll will have to be coordinated with the additional fluxes of Pi in the mesophyll cell that are required during the conversion of pyruvate to malate.

In CAM plants malate is decarboxylated in the light to provide CO_2, which is refixed in the Calvin cycle. The PEP residue that derives from decarboxylation of malate is converted to starch in the chloroplast, or to extrachloroplastic storage carbohydrates (Osmond, 1978). This means that synthesis of carbohydrate from new photosynthate occurs simultaneously with the flow of carbon from malate to carbohydrate. Similarly, Pi is not only incorporated during photosynthesis, but also during the decarboxylation of malate to CO_2 and PEP. This incorporation and release of Pi during conversion of malate to carbohydrate will need to be coordinated with the turnover of Pi during photosynthesis. Figure 3(d) summarizes the fluxes in a CAM plant, where malate is converted to an extrachloroplast glucan or hexose, showing how a larger proportion of the triose-P pool is removed by reactions in the cytosol than during C_3 photosynthesis. In other CAM plants, starch is accumulated.

The fluxes in these plants will depend on whether the starch is synthesized in the chloroplast from triose-P, or whether the cytosol may be involved. The compartmentation of these processes is not yet known (Edwards and Walker, 1983; Edwards *et al.*, 1985).

b. Different Storage Strategies. Plant leaves vary in the way in which they store that portion of the carbohydrate which is not directly exported. Thus in many plants, such as soybean, sugar beet, pea, peanuts, and to-bacco, most is stored as starch in the chloroplast (Huber, 1981a,b). In others, such as wheat, barley, and oats, relatively little starch is accumulated but large amounts of sucrose (Herold, 1984) are temporarily stored in the vacuole (Kaiser *et al.*, 1983; Gerhardt and Heldt, 1984). Some plants accumulate substantial amounts of both sucrose and starch, such as maize and spinach (Stitt *et al.*, 1983a; Stitt, 1985b). In extreme cases, as in onions and leeks, there is no starch accumulated at all, but only sugar and its derivatives. Often, plants that accumulate sucrose also synthesize large amounts of fructans (Pollock, 1984) or other sugar-based oligosaccarides (Lewis, 1984) in some conditions.

These different strategies for carbohydrate storage will influence the way in which sucrose synthesis is regulated, as the decisive branch point between export and temporary accumulation in the leaf will depend on the form in which carbohydrate is being stored. Figure 4 summarizes the relation between the pathway of sucrose synthesis and the synthesis of these various alternative carbohydrate products. Depending on the plant, storage of photosynthate in the leaf could involve controlling the rate of sucrose synthesis, the transport of sucrose into the vacuole or the phloem, or the polymerization reactions that convert sucrose into other oligosaccharides. For exam-

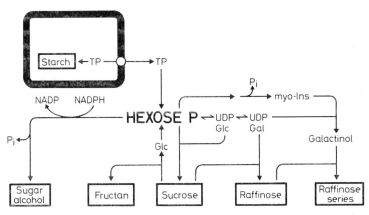

Fig. 4. Relation of alternative carbohydrate end-products of photosynthesis to the pathway for sucrose synthesis.

ple, Pharr and co-workers have found that the activity of galactinol synthase—which catalyzes the first step in the conversion of sucrose to raffinose plus stachyose—is highly correlated with the formation of raffinose saccharides in leaves and seeds of various crop species (Handley et al., 1983), and during leaf development in Cucumis sativus, a stachyose translocator (Pharr and Sox, 1984).

Some plants substitute sugar alcohols partly or entirely for sucrose (Lewis, 1984). Discussion of these plants lies outside the scope of this review, but it may be noted they share a common path for sucrose synthesis as far as Fru6P or Glc6P, which are subsequently reduced and dephosphorylated by unique enzymes. The energetic requirements of this pathway differ from those of sucrose formers. Synthesis of a sucrose molecule from four molecules of triose-P requires an additional input of one UTP, while synthesis of two sugar alcohols from four triose-P molecules requires two NADPH. Thus, sugar alcohols are a more expensive storage form, but would provide a way of utilizing any excess NADPH during photosynthesis and do not require provision of ATP or UTP outside the chloroplast. The possible consequences of this difference between sucrose and sugar alcohols deserves more attention (Loescher et al., 1985).

   c. **Temperature and Osmotic Conditions.** As aerial assimilatory organs with a large surface area to intercept light, leaves are unavoidably exposed to changes in the environmental conditions, including variation in temperature and in water availability. The activity of enzymes and of membrane-associated processes changes as the temperature varies and a shrinkage of cell volume leads to increased ion concentrations, which are thought to be responsible for the inhibition of photosynthesis as the water content decreases (Kaiser, 1982). It is unlikely that all the partial processes involved in photosynthesis respond in the same way to changing temperatures, or ionic conditions, and we might expect that plants will have fine control mechanisms that compensate for the variability in the inherent sensitivity of individual proteins to changes in temperature or ionic concentrations. Further, the major carbohydrate formed from photosynthate may be modified in response to environmental conditions. Low temperature can lead to increased fructan accumulation (Pollock, 1984), while osmotic stress leads to sorbitol accumulation instead of starch in Plantaginaceae (Briens and Larher, 1983).

III. METABOLITE LEVELS DURING SUCROSE SYNTHESIS

   The remainder of this chapter will discuss how the rate of sucrose is coordinated with the rate of photosynthesis, how this is modified to allow starch to be accumulated, and how this control is adapted in C_4 and CAM plants and in response to temperature. Measurements of metabolite levels and of fluxes in a tissue provide a starting point for understanding how a pathway is regulated *in vivo*. The levels of metabolites can be used to calculate the free-energy changes associated with each step in the pathway, pinpointing which enzymes catalyze nonequilibrium and potentially rate-limiting reactions *in vivo*. More detailed studies of the response of metabolite levels to changes in fluxes then reveal which enzymes are playing a dominant role in regulating fluxes in a given set of conditions, and also shows how sensitive the regulatory mechanisms will need to be. Finally, detailed comparison of metabolite levels with the properties of individual enzymes allows the significance of different regulatory mechanisms to be evaluated. This section will deal with the general conclusions that can be drawn from studies of metabolite levels during sucrose synthesis; a detailed comparison of the levels of metabolites and the properties of selected enzymes will be made in later sections.

A. Overall and Cytosolic Metabolite levels

Study of metabolite levels during photosynthetic metabolism is complicated because most of the intermediates are present in both the stroma and the cytosol, and may change independently in these two comaprtments. One way of studying the cytosolic metabolites is to use protoplasts. Several methods are available by which protoplasts can be fractionated and the cytosolic metabolism rapidly quenched for metabolite analysis (Robinson and Walker, 1979a; Wirtz *et al.*, 1980; Hampp *et al.*, 1982; Lilley *et al.*, 1983). Protoplasts provide a useful system for probing the way in which sucrose synthesis responds to rapid changes in the rate of photosynthesis but are, nevertheless, a disturbed system where, for example, the relation between electron transport and carbon metabolism appears to differ from whole leaves (Stitt *et al.*, 1985c). More seriously, the normal route for export of sucrose is disturbed or interupted in protoplast preparations. Although selective export of sucrose has been attained (Huber and Moreland, 1981), most preparations of protoplasts do not export their sucrose (see, e.g., Kaiser and Heber, 1984). Further, the protoplast bounding membrane is impermeable to many chemicals such as Pi, mannose, and weak acids, which penetrate readily into cells in the leaf.

It is therefore important that studies with protoplasts are supplemented by subcellular metabolite analyses of leaf material. Techniques are now available to allow this, although the process is time-consuming. Pioneering studies successfully applied nonaqueous fractionation to leaves 20 years ago (Stocking, 1959; Heber *et al.*, 1963), and these techniques have been used recently to further characterize control points in chloroplast metabolism (Dietz and Heber, 1984; Dietz *et al.*, 1984). Recently, an additional method has been developed to separate not only the chloroplast and cytosol, but also the vacuolar, compartments of leaves (Gerhardt and Heldt, 1984).

Studies using these techniques have shown that some of the metabolites involved in sucrose synthesis are predominantly located in the cytosol, and changes of their overall content provide a reasonable guide to how cytosolic levels are changing. Thus, in protoplasts (Giersch *et al.,* 1980; Stitt *et al.,* 1980, 1983a, 1985c; Hampp *et al.,* 1985) and leaves (Gerhardt and Heldt, 1984; Gerhardt *et al.,* 1987), UDPGlc and Glc6P are largely in the cytosol. However, for other metabolites, the interpretation of overall measurements is more difficult. For example, DHAP is mainly located in the cytosol of protoplasts (Giersch *et al.,* 1980; Stitt *et al.,* 1980, 1983a, 1985c), but recently evidence has been found for a more complex and variable distribution of DHAP in spinach leaves (Gerhardt *et al.,* 1987). Fru1,6P$_2$ and PGA are preferentially located in the stroma. Unexpectedly, Fru6P is not distributed like Glc6P (see above) but is preferentially located in the stroma. This means that the Glc6P : Fru6P ratio varies between the stroma (about 1) and cytosol (3–4) (Dietz and Heber, 1984; Gerhardt and Heldt, 1984). This implies that the phosphoglucose isomerase reaction in the stroma is not at equilibrium, and also provides a way of estimating where the hexose-P are located. When the Glc6P : Fru6P ratio is low, they will be mainly in the stroma, and when this ratio is high, it indicates an increasing pool in the cytosol.

B. Identification of Nonequilibrium Reactions

Although it has been assumed that the cytosolic FBPase, SPS, and sucrose phosphatase catalyze nonequilibrium reactions *in vivo,* there was previously insufficient information about cytosolic metabolite concentrations to allow the free-energy changes during sucrose synthesis to be calculated. In Table I, the available data on the concentration of metabolites in the cytosol of leaves have been used to estimate the mass action ratio of each reaction required during sucrose synthesis. This value is then compared with the standard free-energy change of each reaction, to provide an estimate for the free-energy change during sucrose synthesis *in vivo.*

The values for cytosolic metabolite levels and sucrose are taken from studies using nonaqueous centrifugation (Gerhardt *et al.,* 1987). The estimated concentration of sucrose in the cytosol is a maximum, because this value could include sucrose from the vacuolar system and free space (see Section III,C,5). The Pi concentrations have been estimated from nuclear magnetic resonance (NMR) results as described in Section III,C,3, and cytosolic PPi was measured using nonaqueous fractionation (Weiner *et al.,* 1987). There have been few studies of uridine nucleotides or sucrose-P in plants. Studies with strawberry leaves found 4, 27, and 38 nmol/gfw (gram fresh weight) sucrose-P, UDP, and UTP, respectively, in the light (Isherwood and Selvendran, 1970). The same study found 102 and 57 nmol/gfw ATP and ADP, respectively, and 200, 46, and 20 nmol/gfw Glc6P, Fru6P, and Glc1P, which compare well with those in spinach leaves. A study in

TABLE I

Free-Energy Changes Associated with Synthesis of Sucrose

Estimated cytosolic concentrations (mM)		Enzyme	Estimated molar mass action ratio in the cytosol	Standard $G°$ (kcal/mol)	G in vivo (kcal/mol)
GAP	0.05	Triose-P isomerase	0.1	+1.8	0.4
DHAP	0.3–0.7	Aldolase	4000	−5.5	0.9
FBP	0.1	Fru1,6Pase	0.33	−4.0	−4.7
Fru6P	1.4–3.0	Phosphoglucose isomerase	2.7	−0.4	0.2
Glc6P	4–8	Phosphoglucomutase	0.15	+1.7	0.6
Glc1P	0.6–1.2	UDPGlc pyrophosphorylase	0.35	+0.73	0.1
UDPGlc	1.4–2.3	SPS	0.062	−0.3	−1.9
Sucrose-P	0.2*	Sucrose phosphatase	3.37	−3.3	−2.6
Sucrose	40–50				
PPi	0.3	PFP	2.3	−0.7	−0.2
Pi	10–20	PPase	0.75	−4.6	−3.4
UTP	1.9*				
UDP	1.3*				

[a] The concentrations are estimated from the measured levels of cytosolic intermediates using nonaqueous density gradient fractionation of spinach leaves as in Gerhardt et al. (1987) except for those marked with an asterisk, which are from strawberry leaves (Isherwood and Selvendran, 1970). The latter are assumed to be restricted to the cytosol (volume 20 μl/mg Chl). The range of volumes shown for the concentrations in spinach leaves represents the values for the beginning and end of the photoperiod. The Pi concentration is estimated as described in Stitt et al. (1985c).

tobacco leaves (Meyer and Wagner, 1985) found a similar relation between UTP and UDP but lower amounts, 12 and 5 nmol/gfw, respectively. In Table I, the values for strawberry leaves are included, assuming a value of 1 mg chlorophyll (Chl)/gfw, and that sucrose-P and uridine nucleotides are restricted to the cytosol. Table I also lists the enzymes involved in sucrose synthesis and the standard free-energy change of the reaction they catalyze. For SPS, two very different equilibrium constants have been reported of 3250 and about 2 (Mendicino, 1960; Barber, 1985). The high equilibrium constant is probably in error, due to poor quality of the sucrose-P preparation then available, and the new equilibrium constant of Barber (1985) is used for the calculations in Table I.

The reaction catalyzed by the cytosolic FBPase is clearly removed from equilibrium in vivo, as is the conversion of UDPGlc and Fru6P to sucrose, and is a potential site for regulation. The available measurements of UDP, UTP, and sucrose-P indicate that in vivo this free-energy decrease may be distributed between SPS and sucrose phosphatase, both of which would therefore provide key sites for regulation. The recent evaluation of the equilibrium constant for SPS (Barber, 1985) means this reaction is, in principle, freely reversible. The displacement from equilibrium in vivo therefore depends on a rapid supply of substrates and removal of the products. As phosphoglucose isomerase, phosphoglucomutase, and UDPglucose pyrophosphorylase all catalyze reactions that are close to equilibrium *in vivo*, the Fru6P concentration will depend on the supply of hexose phosphates. The UDPGlc concentration will also depend on the availability of UTP and on PPi being removed. PPi might be removed by PFP or by pyrophosphatase (Section IV,E), and the estimates in Table I suggest that the reaction catalyzed by PFP is probably close to equilibrium, while hydrolysis of PPi to Pi would be irreversible. Continued activity of SPS also requires that its products, sucrose-P and UDP, are continually removed. Sucrose-P may be rapidly removed due to the very high sucrose phosphatase activity in leaves, compared to SPS (Section IV,D). However, any restriction of sucrose phosphatase activity would strongly modify SPS activity. The removal of UDP will be equally important, but it is not known how this occurs.

C. Changes of Metabolites during Sucrose Synthesis

1. Response to Photosynthesis

How do these metabolites fluctuate during a change from respiratory metabolism to photosynthesis, and as the rate of photosynthesis increases? This has been investigated by varying the rate of photosynthesis using different light intensities or CO_2 concentrations, and then quenching leaf material for metabolite analyses. It is important that the leaves are quenched rapidly in the conditions in which they were photosynthesizing, as the pools of photosynthetic intermediates are small and turn over rapidly (Stitt *et al.*, 1980; Sharkey, 1985b). Thus, within 2–3 sec of darkening, metabolites have changed so much that they more closely resemble those found in a leaf in the middle of the night than one that is photosynthesizing.

As the light intensity increases, there are only small changes in the levels of UDPGlc, hexose-P, and PGA in wheat protoplasts (Stitt *et al.*, 1983a) and in spinach (Stitt *et al.*, 1984b) and bean (Badger *et al.*, 1984) leaves. Subcellular metabolite analysis in spinach protoplasts (Stitt *et al.*, 1983a) and leaves (Gerhardt *et al.*, 1987) have confirmed that these metabolites hardly change in the cytosol, or even decrease in the case of UDPGlc. With some plant material, larger changes of these metabolites may be found between the light and dark, including wheat and barley leaves or protoplasts and spinach leaves following a dark pretreatment of over 20 h (Stitt *et al.*, 1985c). In this case, the metabolite levels may be low in the dark due to depletion of starch.

Other metabolites, including triose-P, $Fru1,6P_2$, and RuBP, regularly show larger changes between the dark and light. Thus the triose-P content in spinach (Stitt *et al.*, 1984b), wheat, barley, pea (Stitt *et al.*, 1985c), and bean (Badger *et al.*, 1984) leaves increases severalfold on illumination. Subcellular fractionation confirmed that triose-P increases in the cytosol of spinach leaves after illumination (Gerhardt *et al.*, 1987), and in the cytosol of wheat, barley and pea protoplasts (Stitt *et al.*, 1985c). However, these large changes of triose-P are not essential for regulation of sucrose synthesis. The triose-P content of spinach and bean leaves (Stitt *et al.*, 1984b; Badger *et al.*, 1984) increases markedly at low light intensity, when the rate of photosynthesis is still low, but only increases by another 30–50% as the light is raised until saturating intensities are achieved. Using wheat protoplasts, it was shown that conversion of triose-P to sucrose stopped after darkening, although the cytosolic triose-P level was still over half of that present during photosynthesis (Stitt *et al.*, 1983a).

The relation between metabolite levels and photosynthesis can also be investigated by altering the partial pressure of CO_2. The content of hexose-P, UDPGlc, and triose-P in wheat protoplasts (Stitt *et al.*, 1983a), bean leaves (Badger *et al.*, 1984), and spinach leaves (Stitt *et al.*, 1984b) varies less than twofold as CO_2 is decreased from saturating levels to near or under the compensation point. Measurements of the cytosolic levels in spinach leaves (Gerhardt *et al.*, 1987) and wheat protoplasts (Stitt *et al.*, 1983a) confirmed that the intermediates in the pathway of sucrose synthesis do not change markedly when sucrose synthesis is inhibited by removing CO_2.

The above studies were all carried out at 20–25°C. Evidence is accumulating that the levels of metabolites found during photosynthesis also depend on the temperature. Leegood (1984) reported that the overall metabolite levels remain high at low temperatures in air. Recently it was confirmed that the levels of PGA, RuBP, UDPGlc, Glc6P, Fru6P, triose-P, and Fru1,6P$_2$ all rise as the temperature is decreased, at a range of light intensities or CO_2 concentrations (Stitt, 1987b).

C_4 plants resemble C_3 plants in showing only small fluctuations of their Fru6P, Glc6P, and UDPGlc content between dark, light, and CO_2-free conditions. However, they show a very different behavior for triose-P (Leegood, 1984, 1985; Stitt *et al.*, 1985a,b). The triose-P content rises very markedly in the light, being up to 10-fold above that found in C_3 plants, and there is still four to eight times more triose-P in a maize leaf in CO_2-free air than during rapid photosynthesis in C_3 plants (Stitt and Heldt, 1985b). These high contents of triose-P are due to the high triose-P concentration in the mesophyll (Leegood, 1985; Stitt and Heldt, 1985b), which provides the driving force for diffusion of DHAP back to the bundle sheath. As will be discussed later (Section VIII,A), these high concentrations provide evidence for a modified regulation of the cytosolic FBPase in maize.

Clearly, the pathway of sucrose synthesis can be controlled without requiring large fluctuations in the concentrations of the individual metabolites. Moreover, SPS and the cytosolic FBPase must both be regulated simultaneously, otherwise there would be large fluctuations in the cytosolic hexose pools. The absence of large fluctuations in metabolite pools is important as it decreases the risk that photosynthesis becomes inhibited by depletion of Pi or of triose-P, but it implies that the enzymes in the cytosol must be regulated in a very sensitive and coordinated manner.

2. Photosynthate Partitioning

Metabolites vary, even at a given rate of photosynthesis, depending on the amount of carbohydrate in the leaf. This has been studied in most detail in spinach, where most of the metabolites involved in sucrose synthesis increase two- to threefold as carbohydate accumulates in the leaf. Thus there is a 20–50% increase of hexose-P and there are smaller increases of UDPGlc and triose-P over 4–5 h as sucrose accumulates in spinach leaves during the photoperiod (Stitt *et al.*, 1983b), and similar changes occur within 30–60 min if leaf disks are illuminated, as sucrose can no longer be exported (Stitt *et al.*, 1984c). Hexose-P, UDPGlc, and triose-P also increase when glucose or sucrose is supplied exogenously to spinach leaf disks (Stitt *et al.*, 1983b, 1984c). Recently, nonaqueous fractionation of spinach leaves (Gerhardt *et al.*, 1987; Stitt *et al.*, 1987) has confirmed that these elevated metabolite levels are due to changes in the cytosol, revealing a 30% increase of UDPGlc and a doubling of Glc6P, Fru6P, and triose-P in the cytosol during the day as sucrose accumulated in the leaf and starch synthesis was stimulated.

These results show that a simultaneous restriction of SPS and the cytosolic FBPase leads to an accumulation of their substrates during the day as photosynthate partitioning is shifted in favor of starch. They also show that different factors must control the partitioning of photosynthate and the response of sucrose synthesis to rising rates of photosynthesis. Moderate increases of cytosolic metabolites produce a large stimulation of sucrose synthesis as the rate of photosynthesis increases. In contrast, when partitioning is changing, a restriction of sucrose synthesis is accompanied by a marked accumulation of metabolites in the cytosol.

3. Esterified Phosphate and Phosphate Concentration in the Cytosol

Considerable attention has been paid to the possibility that Pi could be involved in controlling photosynthesis and partitioning (Edwards and Walker, 1983), but direct measurements of the cytosolic or stromal Pi are difficult because most of the Pi is n the vacuole. Even after fractionation, the stromal and cytosolic fractions are still so heavily contaminated with Pi from the vacuole that the cytosolic Pi could not be reliably evaluated (Stitt *et al.*, 1980; R. Gerhardt, unpublished).

Using NMR, it has been estimated that the stromal plus cytosolic Pi averages 20–30 mM in the dark (Foyer *et al.*, 1982b; Waterton *et al.*, 1983; Stitt *et al.*, 1985c), but direct measurements of the cytosolic Pi, or of Pi concentrations during photosynthesis, have not yet been achieved. Such measurements are hampered by technical problems arising from air spaces and paramagnetic ions that lead to broadening of the ^{31}P signal. Also, the leaf material must be packed so tightly that gas exchange and illumination are not possible. Infiltration and use of chelators can improve the peak resolution (Waterton *et al.*, 1983; Stitt *et al.*, 1985c), and measurements using illuminated leaves may become possible using new surface-scanning NMR technology.

Although there are no direct measurements of cytosolic Pi, measurements of phosphorylated intermediates can be used to estimate how Pi may change. This approach assumes that Pi does not move rapidly across the tonoplast or envelope membrane to compensate for changes in the cytosol. The properties of the phosphate translocator preclude a net transfer of phosphate between the chloroplast and cytosol (see Section IV,A), and movement of Pi across the envelope membrane by other means is very slow (Mourioux and Douce, 1981). Kinetics studies (Woodrow *et al.*, 1984) and NMR studies, in which the cytosolic Pi was lowered by mannose (Foyer *et al.*, 1982b), provide evidence that Pi only moves slowly across the tonoplast. In this case, an increase in the amount of phosphate esterified in hexose-P, UDPGlc, FBP, triose-P, PGA, and adenine nucleotides should be accompanied by a reciprocal decrease of the free Pi in the cytosol. Such estimates should, however, be regarded as provisional until it is clearer how the movement of Pi across the tonoplast is regulated in leaves. While rapid movement seems unlikely, slower movement should not be excluded. For example, a redistribution of Pi between the vacuole and the cytosol may occur during the diurnal changes in CAM plants (Black *et al.*, 1985a).

Studies in protoplasts and leaves from several C_3 species (Stitt *et al.*, 1980, 1983a, 1985c; Gerhardt *et al.*, 1987) reveal that esterified phosphate may increase by 5–10 mM as photosynthesis increases. However, the amount of phosphate esterified in the cytosol depends as much on the carbohydrate status of a leaf as on whether it is carrying out photosynthesis or not. Thus, in both the light and the dark, high contents of sucrose and starch are associated with up to 10–15 mM more esterified phosphate in the cytosol, and transitions between dark and light then have little further effect. This contrasts with the stromal pools, where marked changes are seen in the level of esterified phosphate even when that in the cytosol does not change.

Although cytosolic Pi may vary in the short term, evidence is also accumulating for a long-term control over the total amount of phosphate in the cytosol (Rebeille *et al.*, 1983; Foyer and Spencer, 1986). If plants are grown on lower-Pi nutrient, the concentration of Pi in the cytoplasm is maintained relatively constant, while the level in the vacuole decreases. Thus, the vacuole operates as a storage pool for surplus Pi. The importance of controlling the total phosphate in the metabolic compartments will be interlinked with the need to control the concentrations of phosphate-containing cofactors and intermediates in a range where they interact correctly with the kinetic properties of enzymes, and will be discussed again in Section VII.

4. Energy Status in the Cytosol during Photosynthesis

Sucrose synthesis also depends on a supply of energy to the cytosol. Indeed, it has often been supposed that in the light there will be an increase in the cytosolic ATP/ADP ratio, which would suppress respiration and stimulate sucrose synthesis. There is certainly an increase of the overall ATP/ADP ratio in leaves in the light (Heber and Santarius, 1970; Sellami, 1976), but the behavior of the cytosolic pool is very different from that of the whole leaf.

Two different techniques have been developed to fractionate protoplasts and measure metabolites in the chloroplast, cytosol, and mitochondria (Hampp et al., 1982; Lilley et al., 1982). Both reveal that the stromal and cytosolic adenine nucleotides vary in an independent manner. Thus, the stromal ATP/ADP ratio is far lower than that in the cytosol, and increases in the light. In contrast, the cytosolic ATP/ADP quotient does not alter greatly between light and dark apart from transient changes during the transition period (Hampp et al., 1982; Stitt et al., 1982a, 1983a). It would be premature to conclude from these studies that the cytosolic ATP/ADP ratio also remains unchanged during photosynthesis in leaves. Indeed, the extrachloroplast ATP/ADP ratio rises in the light in Elodea and spinach leaves (Heber and Santarius, 1965, 1970). Unfortunately, these results do not provide direct evidence for an increase of the cytosolic ATP/ADP ratio in leaves, because the cytosolic and mitochondrial compartments cannot be separated by these nonaqueous methods. In wheat protoplasts the extrachloroplastic ATP/ADP ratio also increased in the light, but this was due to a change in the mitochondrial ratio, rather than the cytosolic ratio (Stitt et al., 1982a).

The studies with protoplasts demonstrate that sucrose synthesis can be activated and inactivated without this necessitating any alterations of the cytosolic adenine nucleotide levels. However, all these studies used conditions of high CO_2 and need to be extended to more physiological conditions. As will be discussed in Section IV,F, the high CO_2 may have interfered with the control of the cytosolic energy levels by suppressing photorespiration. It therefore remains an open question whether the cytosolic energy status alters during normal photosynthesis, and any such changes would certainly affect the rate of sucrose synthesis (see Sections IV,B and IV,F).

5. Sucrose

Many investigations have searched for relationships between the sucrose content of leaves and the rate of photosynthesis or of sucrose export. While accumulating sucrose has sometimes been correlated with an inhibition of photosynthesis (Neales and Incoll, 1968; Gifford, 1977; Herold, 1980; Herold et al., 1980; Azcon-Bieto, 1983), this relation is often absent (e.g., Austin, 1972; Claussen and Biller, 1977; Stitt et al., 1985c) and is sometimes even reversed (Farrar and Farrar, 1985). Similarly, a clear relation between the sucrose content of leaves and the export rate (Farrar and Farrar, 1985; Fondy and Geiger, 1982) is often absent. In veiw of the complex compartmentation of sucrose in the leaf, it is not surpsing that such simple relations are often absent.

The complex compartmentation of sucrose in the leaf is revealed by its biphasic radioactive labeling kinetics as well as by direct subcellular analysis. Using $^{14}CO_2$ pulse-chase and cell dissection techniques, Outlaw and co-workers (Outlaw and Fisher, 1975; Outlaw et al., 1975) showed sucrose was first labeled in the palisade mesophyll but moved rapidly to the vascular tissue, where the greatest accumulation of newly synthesized sucrose is found. These workers were able to define kinetically distinct pools in these different tissues. However, in addition to this compartmentation between the vascular and mesophyll tissue, there is also a compartmentation between a "transport" pool, which is turned over in 20–50 min, and a "storage" pool, which turns over more slowly. The latter has now been associated with storage of sucrose in the vacuoles of mesophyll cells.

The first evidence for storage of sucrose in the vacuole was obtained by microdissection of centrifuged bean leaves (Fisher and Outlaw, 1979). More recently, nonaqueous fractionation of spinach leaves has shown that almost all of the sucrose accumulation in spinach leaves during the day can be accounted for by temporary storage in the vacuole (Gerhardt and Heldt, 1984; Gerhardt et al., 1987; Stitt et al., 1987). The percentage of the leaf sucrose in the vacuole clearly varies, depending on the tissue as well as the conditions. Thus, in a "sugar storer" like barley, the "storage" pool accounted for about 75–85% of the total (Farrar and Farrar, 1985), while in sugar beet the "storage" pool accounted for 40% of the leaf sucrose (Fondy and Geiger, 1980). Nonaqueous fractionation of spinach leaves showed the vacuolar pool varied between 10 and 75% of the total leaf sucrose, depending on the time of day (Gerhardt et al., 1987).

Sucrose can move quite rapidly between the cytosol and the vacuole. Thus, after a lag of a few minutes, essentially all the sucrose synthesized in barley protoplasts is transferred to the vacuole (Kaiser et al., 1982). In sugar beet, the turnover time of the vacuolar pool was estimated to be about 2 h, sixfold slower than a "cytoplasmic" pool (Fondy and Geiger, 1980). Labeling kinetics suggest that the fluxes of sucrose in and out of the vacuole are also rapid in barley leaves, and may require control as the fluxes did not closely relate to the estimated size of the cytosolic sucrose pool (Farrar and Farrar, 1985). However, it is not kown how this control may be exerted. Studies with isolated barley vacuoles showed that the uptake of sucrose is passive (Kaiser and Heber, 1984), and, in agreement, the estimated concentration of sucrose in the vacuole remains below that in the cytosol of spinach leaves (Gerhardt et al., 1987).

Measurement of the concentration of sucrose in the cytosol remains difficult because neither radioactive kinetics nor subcellular fractionation allows the cytosolic pool to be unambiguously separated from that in the apoplast or the vascular tissue, but they may be used to provide approximate values for the sucrose concentrations we might expect in the cytosol. A maximum estimate of about 50 mM cytosolic sucrose in spinach leaves in the light, falling to 20 mM in the dark, can be estimated, assuming that all the sucrose outside the vacuole is in the cytosol (Gerhardt and Heldt, 1984). A similar value of 20–45 mM (G. Kaiser, personal communication) cytosolic sucrose can be calculated for barley protoplasts, assuming the concentration in the cytosol must be at least as high as that in the vacuoles, since transport over the tonoplast is unenergized (see above). The published ^{14}C kinetics would suggest a "transport" pool of the order of 50–100 mM in barley (Farrar and Farrar, 1985) and sugar beet (Fondy and Geiger, 1980), assuming it is in the cytosol with a volume of about 20 μl.

Two studies on the way in which the cytosolic (including cytosol, vascular, and apoplast) sucrose changes give different results. Using nonaqueous fractionation, Gerhardt *et al.* (1987) found evidence for two- to threefold increase of cytosolic sucrose after illumination, but no further change during the day, as the continued rise in leaf sucrose was all attributable to the vacuole. In contrast, radioactive labeling kinetics (Farrar and Farrar, 1985) suggested a gradual two- to threefold rise of a "transport" pool throughout a 16-h photoperiod in barley. These contrasting results could reflect a difference betweeen spinach and barley, but could just be due to problems in measuring cytosolic sucrose by one or both of these approaches.

Clearly, an accurate evaluation of the sucrose concentration at the site of sucrose synthesis or at the site of mesophyll unloading and phloem loading is not possible at present. This represents a serious gap in our understanding of sucrose synthesis, and hampers critical evaluation of experiments that investigate how sucrose synthesis may be controlled in response to demand. For example, sucrose or glucose can be added exogenously and shown to modify sucrose synthesis and partitioning, but it remains an open question whether the sucrose concentrations occurring *in situ* exert such an effect. On the other hand, correlations between endogenous accumulation of sucrose in leaves and the rate of sucrose synthesis do not provide decisive proof for a direct causal relation, unless it is known whether the cytosolic sucrose concentration changes. It is equally possible that other, as yet unidentified, factors directly control the activity of key enzymes involved in sucrose synthesis.

6. *Ionic Conditions*

Little is known about the ionic conditions in the cytosol during photosynthesis due to the technical difficulty of seperating changes in this compartment from changes in the stroma and vacuole. However, it has recently been shown that the cytosolic Ca^{2+} concentration declines up to 8-fold in the cytosol of giant algae in the light (Miller and Saunders, 1987). Such measurements are not yet available in higher plant cells, but the finding that illumination leads to Ca^{2+} uptake by isolated spinach chloroplasts (Kreimer *et al.,* 1985) suggests that a similar decline of cytosolic Ca^{2+} might occur in leaves. It clearly will be important to establish whether this is the case, and whether such changes could directly modulate enzyme activity or be responsible for the light-dependent alterations of enzyme activities that will be discussed in the later sections.

IV. REGULATORY PROPERTIES OF THE INDIVIDUAL STEPS IN SUCROSE SYNTHESIS

We will now discuss the properties of the individual proteins involved in the synthesis of sucrose. Emphasis will be placed on the enzymes that catalyze nonequilibrium reactions, as these provide the most important sites for regulation. The operation of the phosphate translocator is discussed, as recent evidence suggests that transport may be kinetically limiting *in vivo*. The fate of PPi in the cytosol is also discussed, as this topic has previously received insufficient attention and has acquired new interest through the discovery of PFP and its regulation by $Fru2,6P_2$, and the reevaluation of the equilibrium constant of SPS. Finally, it will be considered how the ATP may be delivered to the cytosol and how the phosphorylation potential is controlled, as this will influence the rate of sucrose synthesis at multiple sites. The way in which these enzymes interact to control the flux to sucrose will be discussed later (Section VII).

A. The Phosphate Translocator

The phosphate translocator catalyzes an exchange of Pi, PGA, and triose-P between the chloroplast stroma and the cytosol (Flügge and Heldt, 1984), and contributes to sucrose synthesis in several ways. It not only catalyzes the export of triose-P to the cytosol in exchange for Pi that has been released during sucrose synthesis, but also allows an exchange of triose-P and PGA whereby ATP and reducing equivalents are transferred to the cytosol. Further, by adjusting the concentrations of Pi, PGA, and triose-P in one subcellular compartment in response to changes in the other, the phosphate translocator allows information to be exchanged about the metabolic status in the chloroplast and the cytosol.

The properties of the phosphate translocator have been reviewed (Heber and Heldt, 1981; Flügge and Heldt, 1984). It catalyzes a strict counterexchange, ensuring that the total sum of the phosphate (esterified and free) in the chloroplast stroma and the cytosol remains constant, despite the large fluxes of the individual substrates between these compartments. The exchange is passive, so the substrates all compete for the available transport capacity, and is also electrically neutral, only the divalent anionic forms being transported. This means that the direction and rate of transport depend on the concentrations of the substrates in the cytosol and stroma.

In the light a pH gradient develops across the envelope membrane (Heber and Heldt, 1981), and this modifies the operation of the phosphate transloca- tor. At physiological pH, Pi and DHAP are predominantly present in the divalent transported form. PGA is present as a mixture of PGA^{2-} and PGA^{3-} ($pK = 7.1$) in the dark, but when the stroma pH increases in the light most of the stromal PGA is present as the nontransported trivalent anion. Studies with liposomes (Flügge et al., 1983) have also revealed an additional direct influence of a pH gradient on the phosphate translocator, leading to PGA being exported preferentially toward the more alkaline compartment. The ionization of PGA and the direct modification of the phosphate translocator reinforce each other, leading to a preferential retention of PGA in the stroma in the light. This ensures that PGA is reduced to triose-P in the stroma, before being exported to the cytosol for sucrose synthesis. The asymmetri- cal transport of PGA and triose-P in the light also favors export of ATP and reducing equivalents to the cytosol via a shuttle in which triose-P is exported to the cytosol and oxidized to PGA before returning to the chloroplast.

The phosphate translocator has a high activity, and in experiments with isolated chloroplasts it is necessary to restrict the translocator activity to allow adequate levels of triose-P and other metabolites to be maintained in the stroma for photosynthesis. This is done by using low Pi concentrations in the medium (Edwards and Walker, 1983), which kinetically limit the translo- cator activity, or by including inhibitors like PPi or pyridoxal P (Edwards and Walker, 1983; Flügge et al., 1980). However, the phosphate translocator may not have a large surplus activity in vivo because the cytosol contains triose-P and PGA as well as Pi, and these will compete for transport back into the chloroplasts. Thus a portion of the carrier activity will be taken up in catalyzing homologous exchange or even in catalyzing uptake of triose-P and release of Pi. Consequently, the net efflux of triose-P and influx of Pi will be considerably lower than the total activity of the translocator.

Indeed, comparison of the stromal and cytosolic PGA/triose-P ratios in spinach leaves suggests that the phosphate translocator exerts a kinetic limitation in vivo. Alkalinization of the stroma should lead to considerably higher PGA/triose-P ratios in the stroma than in the cytosol when leaves are in the light (see above), and this asymmetrical distribution can be observed, provided photosynthetic fluxes are depressed by low CO_2 (Gerhardt et al., 1987), or during transitions between light and dark (Stitt et al., 1980, 1983a). However, this asymmetrical distribution is not found when rapid photosyn- thesis is occurring (Gerhardt et al., 1987). Apparently, the phosphate translocator is no longer able to transport PGA and triose-P rapidly enough to equilibrate them across the envelope membrane when they are being rapidly generated and consumed in the chloroplast and cytosol.

Measurements of the triose-P levels in the cytosol and stroma of spinach leaves reveal the consequences of a limitation by the phosphate translocator *in vivo* (Gerhardt *et al.*, 1987). After 40 min of illumination in air, there is more than twice as much triose-P in the stroma as the cytosol. This gradient, which is presumably required to drive a rapid export of triose-P, disappears when leaves are transferred into CO_2-free conditions to decrease the flux out of the chloroplast. It also decreases later in the day when demand for triose-P in the cytosol decreases because SPS and cytosolic FBPase have been inhibited (see Sections V–VII). These results suggest that the phosphate translocator may exert a colimitation on the rate of sucrose synthesis, as a lowering of the triose-P in the cytosol would decrease the extent to which the cytosolic FBPase is activated (see Section VII). Equally, it will colimit the recycling of Pi to the chloroplast.

More information is needed about how the amount of translocator protein is controlled, and if this is varied in response to different conditions. The phosphate translocator accounts for 15% of the protein in the envelope membrane (Flügge and Heldt, 1984), where it is the major component. There may be an upper limit to the amount of protein that can be incorporated in this membrane.

B. The Cytosolic FBPase

The cytosolic FBPase catalyzes the first irreversible reaction leading to sucrose in the cytosol. Its activity and properties will be decisive in controlling two important metabolic parameters during photosynthesis, namely, (a) the *rate* at which triose-P is removed in the cytosol for sucrose synthesis and (b) the *concentrations* of triose-P that are maintained in the cytosol under given conditions and flux rates. This section will discuss the properties of the cytosolic FBPase; their functional significance will be discussed in Sections VII and VIII.

1. Purification, Structure, and Regulatory Properties in Spinach

The cytosolic FBPase from spinach leaves has been purified to apparent homogeneity (Zimmermann *et al.*, 1978). The enzyme consists of four subunits with a total apparent molecular weight of 130,000, resembling the FBPase from mammals, or chloroplasts. The cytosolic enzyme has a slightly lower molecular weight, and less acidic isoelectric point than the chloroplast enzyme.

The properties of spinach leaf cytosol FBPase are summarized in Table II, and are similar to those of FBPase from mammalian sources (Pontremoli and Horecker, 1970; Marcus *et al.*, 1973; Nimmo and Tipton, 1975a,b; François *et al.*, 1983). Many of the basic characteristics of the leaf enzyme were identified in earlier studies by Latzko and co-workers (Zimmermann *et al.*, 1978; Foyer *et al.*, 1982a). The enzyme possesses a very low K_m for Fru1,6P$_2$

TABLE II
Metabolites and Ions Affecting Activity of Spinach Leaf
Cytosolic Fructose 1,6-Bisphosphatase

Metabolite	Effect	References
FBP	Hyperbolic saturation arise (K_m 2–4 μM)	Zimmermann et al. (1978)
Fru2,6P$_2$	High-affinity inhibitor	Cseke et al. (1982)
	Induces sigmoidal FBP saturation	Herzog et al. (1984)
AMP	Weak inhibitor (competitive to Mg^{2+})	Stitt et al. (1985b)
		Zimmermann et al. (1978)
	Strong synergistic inhibition in presence of Fru2,6P$_2$	Stitt et al. (1985b)
pH	When Mg^{2+} is limiting, shows a marked pH dependence in physiological range	Zimmermann et al. (1978)
	Activity drops as pH decreases	Stitt et al. (1985b)
Mg^{2+}	Necessary for catalysis (probably activator)	Zimmermann et al. (1978)
		Stitt et al. (1985b)
	Mg^{2+} affinity decreases as pH decreases	Stitt et al. (1985b)
	Mg^{2+} affinity markedly increased when AMP and Fru2,6P$_2$ are both present	Stitt et al. (1985b)
K^+	Increases sensitivity to Fru2,6P$_2$	M. Stitt (unpublished)

(2–4 μM) and requires Mg^{2+} for activity. The true substrate is probably the free anion $Fru1,6P_2^{4-}$, with Mg^{2+} acting as an activator (Stitt et al., 1985b), and higher concentrations of Mg^{2+} or $Fru1,6P_2$ are inhibitory (Zimmermann et al., 1978; Habron et al., 1981; Foyer et al., 1982a; Stitt et al., 1985b). The inhibition by $Fru1,6P_2$ occurs at a different site from the catalytic site (Zimmermann et al., 1978). Mg^{2+} can be replaced by Mn^{2+}, which activates at even lower concentrations but also inhibits more strongly (Herzog et al., 1984). The cytosolic FBPase shows no strong pH dependence under optimal conditions, but requires increasingly alkaline conditions as Mg^{2+} becomes limiting (Zimmermann et al., 1978; Stitt et al., 1985b).

The most potent inhibitor of the FBPase is Fru2,6P$_2$. The discovery of this regulatory metabolite (Hers et al., 1982; Uyeda et al., 1982) led to a reevaluation of the regulation of the FBPase in liver (Van Schaftingen and Hers, 1981; François et al., 1983) and subsequently in leaves. Here again, there are remarkable similarities between FBPases from both sources. Fru2,6P$_2$ inhibits the spinach leaf cytosol FBPase at low concentrations (submicromolar) (Cseke et al., 1982; Stitt et al., 1982b), decreasing the affinity for $Fru1,6P_2$ by up to 100-fold and inducing markedly sigmoidal $Fru1,6P_2$ saturation kinetics, with Hill numbers of 2–3 (Herzog et al., 1984; Stitt et al.,

1985b). Fru2,6P$_2$ does not itself markedly alter the pH or Mg^{2+} dependence of spinach leaf cytosolic FBPase (Stitt *et al.*, 1985b) but, as will be discussed below, it strongly modifies the response to several other previously identified effectors.

AMP is a weak inhibitor of the cytosolic FBPase (Zimmermann *et al.*, 1978), inhibiting competitively to Mg^{2+} and noncompetitively to Fru1,6P$_2$ (Stitt *et al.*, 1985b), but only at millimolar concentrations, which are above those found in the cytosol (Stitt *et al.*, 1982b). In the presence of Fru2,6P$_2$, AMP inhibits more strongly (Cseke *et al.*, 1982; Stitt *et al.*, 1982b) at the low concentrations that are representative of those found *in vivo*. Fru2,6P$_2$ and AMP interact synergistically, and when both are present the Mg^{2+} requirement is greatly increased (Stitt *et al.*, 1985b). Since the enzyme becomes more pH-dependent when Mg^{2+} is limiting (see above), there is a marked decrease of activity as the pH is lowered from 7.5 to 7 in the presence of Fru2,6P$_2$, AMP, and 5 mM Mg^{2+} (Stitt *et al.*, 1985b).

Both reaction products inhibit spinach leaf cytosolic FBPase (Habron *et al.*, 1981; Foyer *et al.*, 1982a). Fru6P inhibits noncompetitively to Fru1,6P$_2$, while Pi inhibits competitively (Stitt *et al.*, 1985b). In the presence of Fru2,6P$_2$, Pi still inhibits, acting at concentrations similar to those found *in vivo*, but Fru6P does not inhibit in the presence of Fru2,6P$_2$ (Stitt *et al.*, 1985b). This difference could be explained if Fru2,6P$_2$ were to decrease the affinity for Fru6P more strongly than that for Pi, as has been demonstrated in studies of the back reaction of the mammalian FBPase (Ganson and Fromm, 1982).

Monovalent cations also influence FBPase activity. The mammalian FBPase is stimulated by monovalent cations like K$^+$, NH$_4$, and Na$^+$ (Pontremoli and Horecker, 1970), but is strongly inhibited by low (<1 mM) concentrations of Li$^+$ (François *et al.*, 1983). The spinach leaf enzyme is also inhibited by Li$^+$ (Stitt *et al.*, 1987), which inhibits noncompetitively to Fru1,6P$_2$. Monovalent cations also modify the response of the FBPase to Fru2,6P$_2$. Low concentrations of Li$^+$ decrease the sensitivity of the mammalian (François *et al.*, 1983) and spinach leaf cytosol (Stitt *et al.*, 1986c) FBPase to Fru2,6P$_2$. In the presence of Li$^+$, the Hill number is decreased, and the affinity for Fru1,6P$_2$ increases. In contrast, K$^+$ increases the sensitivity of the spinach leaf FBPase to Fru2,6P$_2$, acting in an antagonistic manner to Li$^+$ (M. Stitt, unpublished). It has been reported that DHAP affects the activity of the cytosolic FBPase (Stitt *et al.*, 1985b). This report is erroneous, the effect being caused by Li$^+$, which is present in some commercial preparations of DHAP. When the Li$^+$ is removed, DHAP has no influence on FBPase activity. The effect of DHAP can be quantitatively accounted for by the Li$^+$ present in the preparations of DHAP.

At first inspection, the properties of the stromal and cytosolic enzyme from leaves are totally different. The stromal enzyme is insensitive to AMP, shows a strong pH and Mg^{2+} dependence, and is activated by thioredoxin via reduction of cysteine residues (Buchanan, 1980). However, the stromal enzyme is inhibited by Fru2,6P$_2$ (Stitt *et al.*, 1982b) and by Fru6P (Gardemann and Heldt, 1986), and the marked similarities between the primary structure of the mammalian FBPase and the spinach stromal FBPase (Harrsch *et al.*, 1986) make it likely that the cytosolic and stromal enzymes from plants also have similar primary sequences. It is intriging that the strong dependence of the stromal enzyme on alkaline pH and high Mg^{2+} is most closely approached by the cytosolic enzyme when Fru2,6P$_2$ and AMP are both present. It can also be generated by partial proteolysis of the mammalian enzyme (Pontremoli and Horecker, 1970). It might be speculated that the difference between the stromal and cytosolic enzyme is due to a modification or loss of regions that are involved in binding of AMP but that maintain a broad pH and Mg^{2+} dependence when AMP is not bound.

2. Alterations of the Properties of the FBPase

Clearly, the cytosolic FBPase responds to a wide range of effectors, including substrate concentration, pH, Mg^{2+}, Fru2,6P$_2$, AMP, and Pi. A change in one of these will lead to change in the response to other effectors. This flexible response may be essential to allow a differentiated response to the varied metabolic and physiological demands to which the cytosolic FBPase is exposed due to its strategic position as the main outlet from the chloroplast. The flexibility of this response is increased because one of the effectors—Fru2,6P$_2$—is itself under a complex control (see Section V). As we discuss later, this interaction between Fru2,6P$_2$ and metabolites leads to a very sensitive control of the cytosolic FBPase, which allows the removal of triose-P for sucrose synthesis to be adjusted to changes in the availability of photosynthate and the demand for sucrose.

In addition, however, the properties of the cytosolic FBPase may vary. This would allow the relation between sucrose synthesis and the prevailing metabolite levels to be altered, and could have considerable significance for adaptive strategies in plants. Two examples of such alterations in properties have already been found. In maize, the cytosolic FBPase has a five- to sixfold higher K_m for Fru1,6P$_2$ (20 μM) than is found in wheat or spinach, and is more sensitive to inhibition by Fru2,6P$_2$ and AMP (Stitt and Heldt, 1985a). These changes are essential to allow the high concentrations of triose-P that are required for intercellular fluxes during photosynthesis in maize (see Section VIII).

The properties of the FBPase also change, depending on the temperature. Similar results are found with the liver FBPase (François *et al.*, 1985) and the spinach leaf cytosolic FBPase (Stitt, 1987b). The K_m values for Fru1,6P$_2$ and Mg^{2+} do not alter greatly as the temperature is varied between 4–40°C, and the Q_{10} of about 2 in V_{max} conditions is typical for soluble enzymes. However, the sensitivity to inhibitors alters in a temperature-dependent way. As the temperature rises, there is a continuous decrease in the sensitivity to Fru2,6P$_2$, and especially to AMP. With the spinach enzyme, the concentrations of Fru2,6P$_2$ and AMP required for a half-maximal inhibition decrease by fourfold and 15-fold, respectively, as the temperature is decreased from 30 to 10°C. These changes are of little physiological significance in the mammalian liver, but have interesting consequences for the temperature response of sucrose synthesis in plants (Section VIII).

Although our knowledge of the cytosolic FBPase has improved during the last 10 years, many crucial gaps remain. First, we do not know whether the alteration of properties in maize is a special case, or whether shifts in the properties of the enzyme occur more widely, and represent a general adaptive strategy. Second, more must be learned about the structure and composition of the enzyme, to elucidate which regions of the protein are important for selected properties. Third, it is not known whether protein modification contributes to controlling cytosolic FBPase activity, as the alterations of liver FBPase activity that result from phosphorylation are subtle and would not have been registered in previous studies in leaves. Finally, we still know nothing of the way in which gene expression and protein turnover control the appearance and total amount of FBPase in leaves in different conditions.

C. Sucrose-Phosphate Synthase

SPS is of primary importance to the biochemical regulation of sucrose formation, as it catalyzes one of the last two reactions in which carbon is removed from the pools of phosphorylated intermediates and converted to carbohydrate. Its regulation is important not only in controlling the rate of sucrose synthesis, but in controlling the partitioning of carbon fixed during photosynthesis between starch and sucrose. Research during the past 5 years has indicated that SPS activity appears to be regulated by two distinct mechanisms: fine control by metabolic activators/inhibitors, and coarse control over the activity of the enzyme *in vivo*. The "coarse" control of SPS activity and its significance to the formation of sucrose will be discussed in Section VI. This section will discuss the current state of knowledge about the biochemical properties of the enzyme. Most of the biochemical information about SPS has been obtained using the enzyme from wheat germ or spinach leaf, but it is becoming increasingly evident that substantial differences exist among species in the properties of SPS, and these differences will be emphasized when present.

1. Physical and Regulatory Properties of SPS

Fundamental information about the physical properties of SPS remain to be determined because the enzyme has not yet been purified to homogeneity. Efforts to purify the enzyme from many sources have been hampered by the lability of the partially purified enzyme. Nevertheless, considerable advances have been made in the understanding of the biochemical properties of SPS since the last review of sucrose metabolism in this series (Akazawa and Okamoto, 1980).

Although conflicting reports of the molecular weight of SPS have been published (cf. Harbron *et al.*, 1981), SPS from most leaf sources has a molecular weight of ~450,000. The enzyme from wheat germ appears to be somewhat smaller in size and has been reported to have a molecular weight of ~380,000 (Salerno and Pontis, 1978). Sedimentation velocity determinations have yielded $S_{20,w}$ values of 13.9 ± 0.5 S for the wheat germ enzyme (Salerno and Pontis, 1978) and 10.4 S for the enzyme from rice scutellum (Nomura and Akazawa, 1974). These data suggest that the tertiary structure of SPS may differ considerably among sources. The basis and significance of these differences in molecular weight are not apparent at present and warrant further investigation. Recent evidence with maize leaf SPS indicates that the molecular weight of SPS is affected by light/dark alterations (Kalt-Torres *et al.*, 1987). The enzyme from illuminated leaves has been estimated by Superose 12B gel filtration to have a molecular weight of 547,000, whereas the enzyme purified from darkened leaves has a molecular weight of 457,000. Further research is needed to determine if these changes in molecular weight in response to altered environmental conditions occurs in other species.

SPS from potato tuber (Slabnik *et al.*, 1968), wheat germ (Salerno and Pontis, 1978), and spinach leaf (Doehlert and Huber, 1985) possess a sulfhydryl group, but pronounced differences are present between these species in the apparent function of the sulfhydryl group. The sulfhydryl group from wheat germ SPS appears to be required for catalysis. In contrast, it appears that the sulfhydryl group is involved in metabolic regulation of the spinach leaf enzyme, rather than catalysis. The alterations of the activity and properties of SPS in response to illumination, however, do not appear to be due to a change in the redox status of the sulfhydryl group. Thus, SPS in spinach leaf does not appear to be light-modulated by a mechanism such as the one that has been suggested to affect phosphoenolpyruvate carboxylase (Slack, 1968) or that allows a light activation of many stromal enzymes via reduction of sulfhydryl groups by thioredoxin (Buchanan, 1980).

Conflicting data on the requirement of metal ions for activity have been reported. The enzyme from potato tuber (Slabnik *et al.*, 1968), wheat germ (Preiss and Greennberg, 1969; Salerno and Pontis, 1978), rice scutellum (Nomura and Akazawa, 1974), and sweet potato roots (Murata, 1972) has been reported to be activated by Mg^{2+}. In contrast, the enzyme from spinach leaf does not appear to be affected by Mg^{2+} (Harbron *et al.*, 1981; Amir and Preiss, 1982) and the enzyme from barley leaf was reported to be slightly inhibited by Mg^{2+} (Murata, 1972). Thus, it appears that Mg^{2+} may affect the enzyme differently in photosynthetic tissues compared to nonphotosynthetic tissues. Monovalent cations such as Na^+ and K^+ do not appear to affect the enzyme (Slabnik *et al.*, 1968; Murata, 1972; Nomura and Akazawa, 1974; Salerno and Pontis, 1978). SPS typically exhibits a broad pH profile with an optimum from 6.5 to 7.5 when assayed under optimum conditions (Leloir and Cardini, 1955; Murata, 1972; Habron *et al.*, 1981; Doehlert and Huber, 1983b).

Contradictory results concerning substrate saturation profiles of SPS have also been reported in the literature and have resulted in much confusion about the metabolic regulation of SPS. Some of this confusion may be due to the properties of SPS varying between species, but it is now apparent that much of the confusion about SPS is due to the use of contaminated enzyme preparations by some investigators (Doehlert and Huber, 1984). Earlier reports of sigmoidal Fru6P substrate profiles with SPS from wheat germ (Preiss and Greenberg, 1969) and spinach leaf (Amir and Preiss, 1982) appear to be an artifact that resulted from contamination of the partially purified SPS preparations with phosphoglucose isomerase. Fru6P saturation profiles in the absence of phosphoglucose isomerase are clearly hyperbolic, and estimates of the K_m(Fru6P) generally range from 2 to 4 mM (Salerno and Pontis, 1977; Harbron et $al.$, 1981; Doehlert and Huber, 1983a). The apparently sigmoidal response to Fru6P results because phosphoglucoseisomerase converts Fru6P into Glc6P, which is an activator of SPS.

A major advancement in the understanding of the biochemical properties of SPS since the last review of sucrose metabolism in this series has been the discovery of the activation of spinach leaf SPS by Glc6P and of the interaction between Glc6P and Pi (Doehlert and Huber, 1983a,b). Glc6P activates the spinach leaf enzyme by increasing the V_{max} and decreasing the K_m(Fru6P). This activation of spinach leaf SPS by Glc6P is antagonized by Pi. When Glc6P is absent, Pi acts as a partial competitive inhibitor with respect to both Fru6P and UDPGlc. Pi inhibition patterns are more complex when Glc6P is present, possibly because of an interaction of Pi at an activation site as well as the catalytic site. The inhibition of SPS by Pi is pH-dependent with inhibition increasing hyperbolically as a function of the dibasic phosphate ion concentration (Dohlert and Huber, 1983b). As a result of the interaction between Glc6P and Pi, SPS activity is a hyperbolic function of the Glc6P/Pi ratio irrespective of the absolute concentration of either effector. At the substrate concentrations estimated to be present in the cytosol, metabolic control of SPS by the Glc6P/Pi ratio will play an important role in the fine control of sucrose formation (Stitt et $al.$, 1987).

It is thought that regulation of SPS by Glc6P and Pi occurs at allosteric sites rather than at the catalytic site. Evidence that supports this postulate has been obtained using sulfhydryl group inhibitors such as p-chlormercuribenzenesulfonic acid and n-ethylmaleimide. These inhibitors have been shown to affect spinach leaf SPS by eliminating Glc6P activation and most of the Pi inhibition without eliminating catalytic activity (Doehlert and Huber, 1985). Although SPS from wheat germ has been reported to be unaffected by Glc6P (Salerno and Pontis, 1978), additional research with SPS from this source is warranted in light of the results obtained with the enzyme from spinach leaf.

UDPGlc saturation profiles for spinach leaf SPS are hyperbolic in the presence of effectors (Pi and Glc6P), whereas in the absence of effectors substrate activation by high concentrations of UDPGlc is observed

(Doehlert and Huber, 1984, 1985). Both hyperbolic (Mendicino, 1960; Salerno and Pontis, 1977) and sigmoidal (Preiss and Greenberg, 1969) UDPGlc saturation profiles have been reported for the wheat germ enzyme. The basis for these conflicting results is not known at present.

Both reaction products inhibit SPS from spinach leaf (Harbron et al., 1981; Doehlert and Huber, 1983b; Amir and Preiss, 1982). UDP inhibits competitively with respect to UDPGlc and noncompetitively with respect to Fru6P (Harbron et al., 1981). Inhibition of wheat germ SPS by UDP is affected by pH with maximum inhibition occurring at ~6.5. The degree of UDP inhibition is also affected by divalent cations (Mg^{2+} and Mn^{2+}) with inhibition decreasing as the cation/UDP ratio increases (Salerno and Pontis, 1978). In spinach, sucrose-P has been reported to be a competitive inhibitor with respect to UDPGlc (Amir and Preiss, 1982), but SPS from wheat germ has been reported to be unaffected by sucrose-P (Mendicino, 1960; Salerno and Pontis, 1977). These results, which indicate a product inhibition, require reevaluation due to a recent report that the reaction catalyzed by SPS may be more reversible than previously thought (Barber, 1985). More attention also needs to be paid to the purity of the sucrose-P preparations used in these studies.

The effect of sucrose on SPS has been shown to vary among species. The enzyme from wheat germ is inhibited about 50% by 50 mM sucrose at high substrate concentrations (Salerno and Pontis, 1978). Sucrose inhibition is noncompetitive with respect to Fru6P. The pattern of inhibition with respect to UDPGlc is more complex, and it appears that SPS exhibits negative cooperativity toward sucrose binding. It has been postulated that forms of wheat germ SPS may exist that differ in sensitivity to sucrose (Salerno and Pontis, 1978). In contrast, SPS from spinach leaf is not significantly affected by sucrose (Amir and Preiss, 1982; Huber et al., 1985c). SPS from maize leaf appears to slightly inhibited by sucrose with about 20% inhibition observed in the presence of 100 mM sucrose (Kalt-Torres et al., 1987), while SPS from tobacco, pea, bean, and peanut leaves are inhibited by sucrose to a greater extent than maize leaf SPS (Huber, 1981b; Huber et al., 1985c). Further research is needed to determine why sucrose inhibition is more pronounced in nonphotosynthetic sources than in leaves, and the degree to which sucrose inhibition in photosynthetic tissues varies among species.

In light of the major importance of Fru2,6P$_2$ to the regulation of the cytosolic FBPase, spinach leaf SPS is not affected significantly by Fru2,6P$_2$ (Doehlert and Huber, 1983b). Conflicting reports of the effect of Fru1,6P$_2$ on spinach leaf SPS have been reported (cf. Harbron et al., 1981; Doehlert and Huber, 1983b). It should be noted that Fru1,6P$_2$ inhibition was observed using preparations where the molecular weight of SPS was observed to be ~275,000 (Harbron et al., 1981) and was not evident in preparations where the molecular weight of SPS was 460,000 (Doehlert and Huber, 1983b).

2. Alterations of the Properties of SPS

As with the cytosolic FBPase, the properties of SPS appear to be altered under certain conditions. For SPS, these differences arise between different species, but also within one species when the enzyme is extracted in different conditions, such as light and dark. The overall significance of these alterations are not fully understood at present but they may play significant roles in the regulation of sucrose formation in plants.

SPS from maize leaf has been recently characterized and found to differ markedly from spinach leaf SPS in several key properties (Kalt-Torres *et al.*, 1987). Activation of maize leaf SPS by Glc6P is observed; however, activation occurs as a result of a decreased K_m(UDPGlc) rather than by a decreased K_m(Fru6P), as is observed with spinach leaf SPS. In addition, maize leaf SPS has a K_m(Fru6P) about 10-fold lower (~0.4 mM) than the K_m(Fru6P) for spinach leaf SPS in the absence of Glc6P (~4 mM). Regulation of SPS by Pi also differs between the species. In maize, Pi alone does not affect activity and thus appears to differ markedly from SPS in other species. Although the mechanism of Glc6P activation of maize leaf SPS is distinctly different from that observed with the spinach leaf enzyme, Pi does interfere with Glc6P activation in maize, as found in spinach. Substrate activation of SPS by high concentrations of UDPGlc is also observed in maize as in spinach. As a result, biphasic UDPGlc saturation profiles are observed with maize leaf SPS and with spinach leaf SPS in the absence of Pi or Glc6P.

Recent evidence has indicated that SPS is barley leaf (Sicher and Kremer, 1984, 1985), *Lolium* (Pollock and Housley, 1985), spinach leaves (Stitt *et al.*, 1987), and maize leaf (Kalt-Torres *et al.*, 1987) is light-activated. Differences in activity persist upon partial purification, and it appears that kinetic properties of the enzyme are affected. In barley, substrate affinities for UDPGlc and Fru6P have been reported to be increased with the light compared to the dark form of the enzyme. It should be noted that sigmoidal Fru6P saturation profiles were observed with both the light and the dark form of the enzyme, and thus these results should be viewed with some caution until preparations free of phosphoglucose isomerase are used. In spinach, the sensitivity to inhibition by Pi decreases in the form of the enzyme obtained from illuminated leaves (Heldt and Stitt, 1987). In maize, the light and dark forms differ in affinity for UDPGlc and in the extent of activation of Glc6P (Sicher and Kremer, 1985; Kalt-Torres *et al.*, 1987). Affinity for Fru6P did not differ significantly between the light and dark form of the enzyme. As mentioned previously, the light form of SPS has a molecular weight about 100,000 greater than the dark form. Thus, it appears that light modulation of SPS occurs in certain species and the modulation involves changes in the physical and kinetic properties of the enzyme.

In soybean leaf, the activity of SPS changes diurnally and is controlled by an endogenous clock mechanism (Kerr *et al.*, 1985). Soybean leaf SPS has been partially purified at times during the diurnal cycle that exhibit high and low activity. Differences in activity persist upon partial purification (Kerr and Huber, 1987), as evidenced by differences in V_{max} between the high- and low-activity forms of the enzyme. In contrast to the results with barley, spinach, and maize, the affinity of SPS for UDPGlc or Fru6P did not differ significantly between the high- and low-activity forms.

Collectively, the results obtained with spinach, barley, maize, and soybean SPS indicate that fundamental properties of SPS are subject to change in response to external and endogenous factors and that the underlying mechanism responsible for these changes may differ among species. Further research with pure enzyme from each of these sources is needed to definitively establish the basis for the observed differences in properties of SPS among species and in response to altered environments.

D. Sucrose Phosphatase

The last enzyme of the sucrose formation pathway is sucrose-P phosphohydrolase, which catalyzes the irreversible hydrolysis of sucrose-P. Thus, sucrose synthesis in higher plants is catalyzed by the sequential action of SPS and sucrose phosphatase. Sucrose phosphatase occurs in a wide range of angiosperms as well as lower plants that synthesize sucrose; however, the enzyme could not be detected in some lower plants that do not contain sucrose and presumably lack SPS activity (Hawker and Smith, 1984). In general, SPS and sucrose phosphatase appear together in tissues, but the activity of sucrose phosphatase exceeds the maximum activity of SPS by about 10-fold (Hawker and Smith, 1984). Although this suggests that the phosphatase may not be limiting for sucrose formation, it is likely that the enzyme is not operating at maximum velocity *in situ*. The potential importance of sucrose phosphatase for the regulation of sucrose synthesis has been increased by the recent reevaluation of the equilibrium constant for SPS (see Section III,B), which shows that SPS will only be able to catalyze a rapid flux in the direction of sucrose synthesis if its products are effectively removed.

Sucrose phosphatase has been partially purified from sugar cane and carrot root tissue (Hawker, 1967; Hawker and Hatch, 1966). The enzyme has a slightly acidic pH optimum, is specific for sucrose-P as substrate, and requires Mg^{2+} for activity. More recently, the enzyme has been purified to homogeneity from pea shoots (Whitaker, 1984). The enzyme consists of two similar subunits (M_r 55 kDa). The kinetic properties of the homogeneous enzyme are very similar to those reported earlier for the partially purified enzyme from sugar-cane stem. However, one apparent difference concerns sensitivity to sucrose. The pea enzyme was not significantly inhibited by physiological concentrations of sucrose (9% inhibition at 100 mM sucrose; 35% inhibition at 500 mM; Whitaker, 1984). In contrast, substantial inhibition by sucrose was reported for the sugar-cane stem and carrot root enzymes (about 70% inhibition at 50 mM sucrose; Hawker, 1967). In a recent survey, Hawker and Smith (1984) reported significant inhibition by sucrose (30–60% by 100 mM sucrose) of sucrose phosphatase from a wide range of plant tissues. While all studies were done with crude extracts, it was judged that there was little interference from nonspecific phosphatases. Because sucrose is a partially competitive inhibitor of sucrose phosphatase (Hawker, 1967), this property could act as an important mechanism to limit the accumulation of sucrose, in some tissues at least. The competitive nature of sucrose inhibition may also explain why inhibition of the pea shoot enzyme (assayed with 1 mM sucrose-P) was low compared to other tissues (usually assayed with 0.06 mM sucrose-P). It will be necessary to carry out assays of sucrose phosphatase at concentrations of sucrose 6-phosphate resembling those found *in vivo*. There is little information available on which to base such estimates, apart from measurements in strawberry leaves (Isherwood and Selvendran, 1970) that would indicate a concentration of about 0.2 mM.

Hawker and Smith (1984) have shown that sucrose phosphatase is associated with isolated vacuoles of red beet, and have postulated that the enzyme might function as a sucrosyl transferase in the transport of sucrose across the tonoplast membrane. Further studies will be required to quantitatively assess the intracellular localisation of this enzyme, and to elucidate the role (if any) in transport of sucrose or sucrose-P across membrane.

E. Pyrophosphate : Fructose-6-Phosphate Transferase and Pyrophosphate Metabolism

During synthesis of most biological polymers, a nucleotide diphosphate derivative acts as a precursor for the polymerization. The synthesis of starch in the chloroplast is a good example of such a reaction sequence. Although the reaction forming a molecule like ADPGlc from ATP and Glc1P is reversible, this reaction is rendered effectively irreversible *in vivo* because the PPi is immediately hydrolyzed by pyrophosphatase.

In contrast, the analogous reaction catalyzed in the cytosol by UDPGlc pyrophosphorylase is reversed during sucrose mobilization via sucrose synthase. This reverse reaction is possible because plants contain significant amounts of PPi (ap Rees *et al.*, 1985; Smyth *et al.*, 1984), but this, in turn, implies that the pyrophosphatases in plants are either regulated or are separated from the PPi by compartmentation. The vast majority of pyrophosphatase in green tissues is in the chloroplast (Bucke, 1970), and a similar plastid location has recently been reported for alkaline pyrophosphatase in soya bean suspension culture cells (Gross and ap Rees, 1986), as required if significant levels of PPi are to be maintained in the cytosol. Recently, it has been shown using nonaqueous fractionation of spinach leaves and membrane filtration of protoplasts that all of the PPi is present in the cytosol, while the alkaline pyrophosphatase is restricted to the chloroplast (Weiner *et al.*, 1987).

The potential importance of PPi is emphasized by two new enzyme activities which have been discovered in the last years. Most plant tissues contain substantial activities of an enzyme, termed PFP, which catalyzes a reversible phosphorylation of fructose 6-phosphate using PPi (ap Rees *et al.*, 1985; Black *et al.*, 1985b). PFP is a cytosolic enzyme (Stitt *et al.*, 1982b; Black *et al.*, 1985b) and is activated by $Fru2,6P_2$. Since the enzyme catalyzes a reversible reaction, it could either utilize PPi as an energy source during glycolysis, or generate PPi by catalyzing the reverse relation in which $Fru1,6P_2$ and Pi are converted to Fru6P and Pi. It has also recently been shown that the tonoplast of storage tissues contains a proton-pumping pyrophophatase, as well as an ATPase (Walker and Leigh, 1981; Rea and Poole, 1985). It is still unknown whether a similar activity is present in leaves.

Turnover of PPi during photosynthetic sucrose synthesis has received little attention. One molecule of PPi will be formed in the reaction catalyzed by UDPGlc pyrophosphorylase for every molecule of sucrose that is synthesized, so PPi must be hydrolyzed at rates of 10–15 μmol/mg Chl h during rapid photosynthesis in spinach plants. It has been tacitly assumed this would occur via pyrophosphatase in the cytosol. Although an earlier study (Klemme and Jacobi, 1974) reported that a "cytosolic" isoenzyme was responsible for a considerable proportion of the activity in spinach leaves, this could not be confirmed in studies with the improved techniques for subcellular fractionation of leaves that are now available (see above). It remains an open question whether a low level of a specific soluble pyrophosphatase is present in the cytosol, or whether a tonoplast pyrophosphatase may be involved. In either case, the levels of PPi that are found mean that any pyrophosphatase present will have to be regulated (Section III,B).

Another way of hydrolyzing PPi would be via a cycle between PFP and the cytosolic FBPase, in which PFP converts PPi and Fru6P to $Fru1,6P_2$ and Pi, and the $Fru1,6P_2$ is then rehydrolysed by the FBPase. The activity of this cycle could depend on the $Fru2,6P_2$ level, because of the differential effects of $Fru2,6P_2$ on the forward and reverse reactions of PFP. Nanomolar concentrations of $Fru2,6P_2$ suffice to activate the glycolytic (PPi-consuming) reaction, but at least 10-fold higher concentrations are needed to activate the gluconeogenetic (PPi-forming) reaction (Cseke *et al.*, 1982; Van Schaftingen *et al.*, 1983). It may be speculated that low $Fru2,6P_2$ favors removal of PPi, but increasing $Fru2,6P_2$ activates the reverse (PPi-generating) reaction and reduces the rate at which PPi is removed. Simultaneous changes of metabolites could amplify this response, because they influence the rate of the reverse reaction, as well as its sensitivity to activation by $Fru2,6P_2$. For example, during a "feedback" inhibition of sucrose synthesis, higher $Fru2,6P_2$ restricts the cytosolic FBPase and leads to an accumulation of

Fru1,6P$_2$, while Pi may decline (Section III). Higher Fru1,6P$_2$ and lower Pi directly increase the rate of the reverse reaction, and also increase its sensitivity to activation by Fru2,6P$_2$ (Van Schaftingen *et al.,* 1983; Cseke *et al.,* 1982).

F. Supply of Energy

Sucrose synthesis needs a continuous input of energy in the form of UTP, and the energy status in the cytosol will affect sucrose synthesis in several ways. UDP will inhibit SPS, as well as the cytosolic FBPase (Foyer *et al.,* 1982a), and adenine nucleotide levels modulate sucrose synthesis via the action of AMP on the cytosolic FBPase. As the cytosolic ATP/ADP ratio decreases, the action of myokinase leads to an increase of AMP (Stitt *et al.,* 1982a). AMP will be an extremely potent inhibitor of the cytosolic FBPase *in vivo* in the presence of Fru2,6P$_2$, acting at concentrations as low as 50–100 μM, by enhancing sensitivity to Fru2,6P$_2$ and increasing the level of substrate needed for activity (Stitt and Heldt, 1985c).

Consequently, rapid sucrose synthesis will depend on the maintenance of a high energetic status in the cytosol. Intuitively, we might expect this would not pose a serious problem, because ATP is generated by photophosphorylation. However, it remains unclear what contribution photophosphorylation makes to the control of the cytosolic phosphorylation potential in the light. As discussed in Section III,C,4, there is still no evidence for a light-dependent increase of ATP in the cytosol. Also, export of ATP from the chloroplast will be a far more complex procedure than, for example, the export of ATP from mitochondria. Mitochondria have an ATP–ADP translocator of high activity whose properties allow mitochondrial electron transport to drive a highly vectorial uptake of ADP and export of ATP (Klingenberg and Heldt, 1982). In contrast, although there is an adenine nucleotide transport system in the chloroplast envelope, its activity is low and its properties favor uptake rather than export of ATP (Heber and Heldt, 1981).

As an alternative, it has been suggested (see Heber and Heldt, 1981) that an exchange of triose-P and PGA allows ATP and NADH to be transferred to the cytosol. These are generated as triose-P is oxidized to PGA in the cytosol [Fig. 5(a)]. Clearly, export of ATP by this route depends on NADH being continually reoxidized in the cytosol. Originally it was thought the NADH might be returned to the chloroplast via reduction of oxaloacetate to malate in the cytosol, followed by uptake of malate into the chloroplast in exchange for more oxaloacetate. This would allow chloroplasts to operate as autonomous exporters of ATP. However, it now seems doubtful whether an exchange of malate and oxaloacetate could occur in this direction during photosynthesis. In the stroma, malate can only be oxidized by an NADP-malate dehydrogenase, and studies in leaves show the stromal NADP(H) is far more reduced than the NAD(H) outside the chloroplast (Santarius and

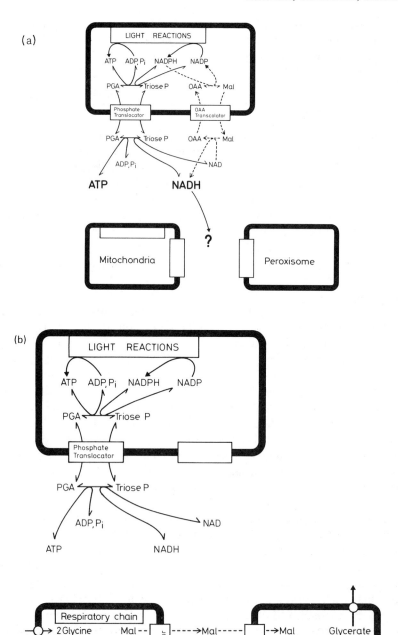

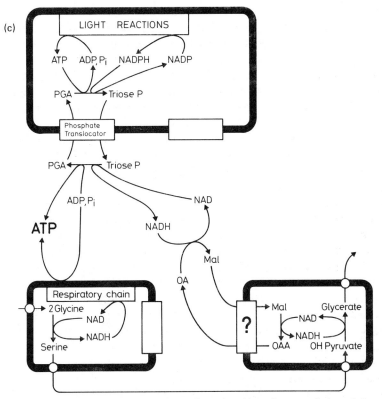

Fig. 5. Relation between ATP transport from the chloroplast, regulation of the cytosolic ATP/ADP ratio, and regulation of mitochondrial NADH oxidation in a leaf carrying out photorespiration. (a) Interaction of PGA/triose-P, and malate/oxaloacetate shuttles, during export of ATP and NADH from the chloroplast. (b) Transport of NADH from the mitochondrial matrix to the peroxisome via an exchange of malate and oxaloacetate. (c) Consequences of oxidation of NADH by mitochondria for supply of ATP to the cytosol from the mitochondria and the chloroplast. For simplicity, the uptake of ATP and ADP into the mitochondrial matrix has been omitted.

Heber, 1965). In this case, any exchange of malate and oxaloacetate across the envelope would draw reducing equivalents out of the chloroplasts. Since the stromal NADP malate dehydrogenase is only activated at high NADPH/ NADP ratios (Scheibe and Jacquot, 1983), the malate/oxaloacetate exchange will only be turned on as the chloroplasts become very reduced, and will allow "surplus" reducing equivalents to be exported from the cytosol. By flooding the cytosol with reducing equivalents, it will even hinder export of ATP via the triose-P/PGA exchange described above. Thus, the available evidence suggests that the chloroplasts may not be able to operate as an automonous system to export ATP to the cytosol. Instead, export of ATP will depend on extrachloroplast metabolism being able to reoxidize the

NADH in the cytosol. Obviously, this would imply that the regulation of the cytosolic phosphorylation potential cannot just depend on the chloroplast, but will also be influenced decisively by reactions in the remainder of the cell.

There is considerable turnover of NAD(H) outside the chloroplast in the light, linked to processes like photorespiration and nitrogen assimilation. For example, in photorespiration NADH is produced during glycine decarboxylation in the mitochondrial matrix, and NADH is oxidized during hydroxypyruvate reduction in the peroxisome. Recently, evidence has accumulated that NADH generated in the mitochondrial matrix may be transferred to the peroxisome via an exchange of malate and oxaloacetate [Fig. 5(b)]. This exchange is made possible by the presence of an active, high-affinity oxaloacetate translocator in mitochondria (Ebbighausen *et al.*, 1987). Theoretically, this exchange could allow an internal redox balance during photorespiration. However, it is also possible that a portion of the NADH generated in the mitochondrial matrix will be oxidized there to produce ATP, and that this shortfall is made up by exporting NADH from the chloroplasts to the peroxisome via the shuttle systems just described [Fig. 5(c)]. Since the rates of photorespiration are very high, the regulation of this NADH exchange between the mitochondria, peroxisomes, and chloroplasts could exert a large influence on the redox potential and phosphorylation potential in the cytosol, at least in C_3 plants. It has recently been shown that oligomycin, which is a fairly specific inhibitor of the mitochondrial ATPase, will inhibit photosynthesis in whole protoplasts at concentrations that have no effect on photosynthesis in intact chloroplasts (Ebbighausen *et al.*, 1987).

For example, if NADH oxidation in the mitochondria were restricted, this would directly decrease the supply of ATP from oxidative phosphorylation and allow an increasing amount of NADH to be transferred from the mitochondria to the peroxisomes. This, in turn, will reduce the ability of chloroplasts to export ATP because less NADH is required from the chloroplasts to support hydroxypyruvate reduction. Conversely, as NADH oxidation in the mitochondria increases, the supply of ATP to the cytosol will be increased due to the higher rates of oxidative phosphorylation, and also because the resulting shortfall of NADH in the peroxisome will draw more NADH out from the chloroplast, favoring an accompanying export of ATP. This relation will be further modified depending on the way in which the phosphorylating and nonphosphorylating "alternative" electron transport is regulated, the contribution of the "external" mitochondrial NADH dehydrogenase, and the use of NADH in biosynthetic and assimilatory reactions, including NO_2 reduction.

We will not be able to give a precise account of how these processes interact until we know more about the control of NADH oxidation in leaf mitochondria. However, it seems plausible that a decreasing ATP/ADP ratio in the cytosol would be one of the signals that increases oxidation of NADH, so more ATP is supplied from the mitochondria and chloroplast, while a rising NADH/NAD$^+$ ratio might lead to overflow into "alternative" pathways or the external NADH dehydrogenase, ensuring that an overreduction in the cytosol does not lead to collapse of the phosphorylation potential. It is of note that the control of the cytosolic FBPase by AMP will not only serve to decrease sucrose synthesis when the cytosolic ATP/ADP ratio decreases, but will also redirect more triose-P toward PGA, allowing more ATP and NADH to be generated in the cytosol. This might provide one way of coordinating sucrose synthesis, and processes like nitrate assimilation, photorespiration, and mitochondrial metabolism.

V. REGULATION OF FRUCTOSE 2,6-BISPHOSPHATE CONCENTRATION

Fru2,6P$_2$ is a regulator metabolite whose discovery in liver has led to a reevaluation of ideas about how glycolysis and gluconeogenesis are controlled (Hers *et al.*, 1982; Uyeda *et al.*, 1982). Its role in plants has recently been reviewed (Cseke *et al.*, 1984; Huber, 1986; Stitt, 1987a). This section will discuss how the concentration of Fru2,6P$_2$ is controlled during photosynthesis. The turnover of Fru2,6P$_2$ can be altered in response to a variety of signals about metabolic conditions in the leaf, allowing these signals to be amplified and integrated as a change in the concentration of Fru2,6P$_2$ (Stitt, 1986a; Stitt and Heldt, 1985c; Stitt *et al.*, 1987). The Fru2,6P$_2$ then interacts with its target enzymes to readjust the fluxes and metabolite pools in the cytosol.

A. Presence and Localization

Fru2,6P$_2$ is extremely acid-labile, and also sensitive to hydrolysis by phosphatases that are present in plant tissues. Techniques used to extract Fru2,6P$_2$ in plants involve either heating in alkali (Huber and Bickett, 1984) or deproteinizing with chloroform and subsequent extraction of Fru2,6P$_2$ by phase partitioning (Stitt *et al.*, 1982b). The exact conditions needed to inactivate the phosphatases without loss of Fru2,6P$_2$, and then to quantitatively extract the Fru2,6P$_2$, vary depending on the plant material (Stitt *et al.*, 1983b; Stitt and Heldt, 1985a; Sicher *et al.*, 1986). It is therefore essential to adjust conditions carefully, and to check that small amounts of Fru2,6P$_2$ can be recovered through the extraction procedure for each tissue that is studied. Fru2,6P$_2$ is normally measured by bioassay, using its ability to activate PFP (Van Schaftingen *et al.*, 1983). Other components in plant extracts can modify the activity of PFP as well as the response to Fru2,6P$_2$, and it is essential that the activation by Fru2,6P$_2$ in the plant extract is calibrated by comparison with an internal standard curve using pure Fru2,6P$_2$ plus acid-treated extract (Stitt *et al.*, 1983b). In extracts that contain high levels of phenols, pretreatment of extracts with polyvinylpyrrolidone or activated charcoal improves the assay (Stitt and Heldt, 1985a), while tissues with high levels of organic acids may require pretreatment with carefully calibrated ion-exchange columns to remove inhibitory acids before adding the extract to PFP (T. Fahrendorf, J. Holtum, and E. Latzko, personal communication).

Fru2,6P$_2$ has been detected in a wide range of leaf material including spinach (Stitt *et al.*, 1983b), pea (Black *et al.*, 1985b), barley (Sicher *et al.*, 1986), soybean (Huber *et al.*, 1985a,b), *Kalanchoe*, pineapple, *Bryophyllum* (T. Fahrendorf, J. Holtum, and E. Latzko, personal communication), and maize (Stitt and Heldt, 1985b). The level normally lies between 80–500 pmol/ mg Chl, which would be equivalent to a total concentration of 4–25 μM, if this Fru2,6P$_2$ were all free in the cytosol (volume 20 μl/mg Chl). These will represent maximal concentrations, as the low concentration of Fru2,6P$_2$ and the high binding affinity for its target proteins means that a significant proportion may be bound on these target proteins *in vivo*. Far higher levels of Fru2,6P$_2$ (2–3 nmol/mg Chl) have been reported in barley leaves at the beginning of the night (Sicher *et al.*, 1986) and in some CAM plants in the dark (T. Fahrendorf, J. Holtum, and E. Latzko, personal communication).

Nonaqueous fractionation of spinach leaf material showed that at least 90% of the Fru2,6P$_2$ was located in the cytosol (Stitt *et al.*, 1983a). In agreement, Fru2,6P$_2$ is absent from isolated chloroplasts (Cseke *et al.*, 1982). However, before concluding that Fru2,6P$_2$ is always restricted to the cytosol, it will be necessary to investigate a wider range of tissues. The subcellular location of the enzymes involved in the synthesis and degradation of Fru2,6P$_2$ (see below) has also not yet been unambiguously established.

In maize leaves most of the Fru2,6P$_2$ is present in the mesophyll cells (Stitt and Heldt, 1985a). The enzymes synthesizing and degrading Fru2,6P$_2$ (Soll *et al.*, 1985) as well as the cytosolic FBPase (Furbank *et al.*, 1985) are also mainly or entirely located in the mesophyll cells. However, it would be premature to conclude that Fru2,6P$_2$ plays no part in bundle sheath metabolism, or that this distribution is typical for C$_4$ plants (see above).

B.   Synthesis and Degradation of Fru2,6P$_2$

1.   The Enzymes

The synthesis and degradation of Fru2,6P$_2$ is catalyzed by specific enzyme activities, termed Fru6P,2-kinase and Fru2,6P$_2$ase, respectively. These activities were first characterized in liver, where they reside at different sites on the same protein. Binding of ligands on this protein, or modification by phosphorylation, allows both the synthetic and degradative reactions to be regulated in a reciprocal manner to generate a large change of the $Fru2,6P_2$ concentration.

Specific Fru6P,2-kinase and $Fru2,6P_2$ase activities have been found in spinach (Cseke and Buchanan, 1983; Cseke *et al.*, 1983) and maize (Soll *et al.*, 1985) leaves, and methods for assaying these activities in plant extracts have been described in detail (Stitt *et al.*, 1985a, 1986b). The spinach Fru6P,2-kinase has been purified 30-fold (Cseke and Buchanan, 1983), and $Fru2,6P_2$ase copurified to this stage (Cseke *et al.*, 1983). Fru6P,2-kinase and $Fru2,6P_2$ase activities also coeluted when spinach leaf extract was fractionated using FPLC (fast protein liquid chromatography) (Stitt *et al.*, 1986b). However, rigorous evidence that these enzymes reside on a single bifunctional protein in plants is still lacking, as the protein has not been purified to homogeneity and the segregation of $Fru2,6P_2$-synthesizing and -degrading activities monitored through the process. Although one recent study suggests Fru6P,2-kinase and the $Fru2,6P_2$ase copurify through an extensive purification procedure (Larondelle *et al.*, 1986), other studies show that plants contain a low-affinity $Fru2,6P_2$ase that does not copurify with Fru6P,2-kinase (MacDonald *et al.*, 1987). More studies are needed to establish the significance of these activities. It should be noted that during both of the above studies several forms of Fru6P,2-kinase were found, which clearly could be shown to result from degradation by proteases during the extraction and purification. It is essential to include protease inhibitors in all studies of these enzymes to avoid the appearance of such artefacts.

2. Metabolite Effectors

The activity of Fru6P,2-kinase and $Fru2,6P_2$ase from liver (Hers *et al.*, 1982) and yeast (François *et al.*, 1985; Yamashoji and Hess, 1985) is modified by metabolic intermediates, which often act in a reciprocal way on the two enzyme activities. An analogous regulation has been found for the spinach leaf enzymes (Table III). Fru6P and Pi both activate Fru6P,2-kinase and inhibit $Fru2,6P_2$ase (Cseke and Buchanan, 1983; Cseke *et al.*, 1983). Fru6P,2-kinase is inhibited by three-carbon compounds including PGA and DHAP at concentrations of under 1 mM (Cseke and Buchanan, 1983; Stitt *et al.*, 1984a; Larondelle *et al.*, 1986), similar to those found in the cytosol. The inhibition is relieved by increasing Fru6P, but shows a complex interaction with Pi. PGA inhibits strongly in the absence of Pi, and the inhibition decreases as the Pi concentration is raised (Stitt *et al.*, 1984a, 1985a).

An interaction of these four metabolites with Fru6P,2-kinase and Fru-$2,6P_2$ase allows a varied response of the $Fru2,6P_2$ concentration to different metabolic conditions (Cseke *et al.*, 1984; Stitt, 1986a,b). Similar regulatory properties are found for Fru6P,2-kinase in maize leaves (Soll *et al.*, 1985), castor bean endosperm (Kruger and Beevers, 1985), corn roots, carrot storage root, potato tuber, maize endosperm, beetroot tuber, onion bulb (Stitt *et al.*, 1985a), and in guard-cell protoplasts from bean leaves (M. Stitt, unpub-

TABLE III
Metabolite Effectors of Spinach Leaf Fructose 6-Phosphate,2-Kinase
and Fructose 2,6-Bisphosphatase[a]

	Activators	Inhibitors	Remarks
Fru6P,2-kinase	Fru6P	PGA	Pi relieves the inhibition by PGA but increases the inhibition by DHAP
	Pi	DHAP	
Fru2,6P$_2$ase	—	Fru6P	Fru6P inhibits noncompetitively, Pi inhibits competitively

[a] From Cseke and Buchanan (1983) and Stitt *et al.* (1984a).

lished). It appears that this pattern of regulation by metabolite effectors may possess a general significance that extends beyond photosynthetic metabolism. It is intriguing that three of these metabolite effectors are substrates for the phosphate translocator and pass between the plastid and cytosol, and that PGA and Pi are involved in regulating starch synthesis in photosynthetic and nonphotosynthetic plant tissues (Preiss, 1980).

3. Protein Modification

In mammalian tissues and yeast the activity of Fru6P,2-kinase and Fru-2,6P$_2$ase is regulated by protein phosphorylation, catalyzed by cAMP-dependent protein kinases. In liver, phosphorylation inhibits Fru6P,2-kinase and activates Fru2,6P$_2$ase (Hers *et al.*, 1982). The resulting decrease of Fru2,6P$_2$ ensures that glucose is released to the blood during glucagon-stimulated glycogenolysis, rather than being respired in the liver. In yeast, phosphorylation acts in the opposite way and activates Fru6P,2-kinase (François *et al.*, 1985; Yamashoji and Hess, 1985). Here, cAMP rises as glucose becomes available, and the resulting phosphorylation of Fru6P,2-kinase increases Fru2,6P$_2$ so that glycolysis is activated and gluconeogenesis is inhibited.

Initially, it appeared that protein modification might not be involved in controlling Fru2,6P$_2$ levels in leaves, as treatment of spinach leaf Fru6P,2-kinase with ATP and liver catalytic subunit did not alter its activity (Cseke and Buchanan, 1983). However, two recent studies suggest protein modification could be involved. First, the activities of Fru6P,2-kinase and Fru-2,6P$_2$ase show diurnal alterations in spinach leaves (Stitt *et al.*, 1986b). Their activity alters in a reciprocal manner; the changes are retained when extracts are desalted, but revert when the extracts stand. The changes do not occur during light–dark transition, but occur gradually during the day as sucrose is accumulating in the leaf (see below). Second, the activity of Fru6P,2-kinase is increased following incubation with ATP and the catalytic subunit (C. Baysdorfer, personal communication). This activation is only clearly seen when more highly purified preparations of Fru6P,2-kinase are used than in earlier studies. However, far more studies are needed to establish whether phosphorylation actually occurs, as well as the significance of the various Fru6P,2kinase and Fru2,6P$_2$ase activities.

It is not known whether the diurnal alterations of Fru6P,2-kinase activity are due to phosphorylation, nor what factors control these changes. Fru6P,2-kinase activity also increases when carrot storage tissue is treated with ethylene (Stitt et al., 1986a) or as guard-cell protoplasts swell in response to light and K$^+$ ions (M. Stitt, unpublished). More work is required to establish whether modification of Fru6P,2-kinase is involved in the response of metabolism to signals like hormones or availability of carbohydrate, and what effect such modification has on carbohydrate metabolism.

C. Alterations of Fru2,6P$_2$ during Photosynthetic Metabolism

The Fru2,6P$_2$ concentration changes in response to several factors during photosynthesis. This makes it important to plan experiments carefully, if a clear response to any one factor is to be obtained. In this section, we will describe how Fru2,6P$_2$ changes in response to different metabolic conditions in leaves. The subsequent section will consider which molecular mechanisms could be responsible for the changes of Fru2,6P$_2$.

1. The Rate of Photosynthesis

As the rate of photosynthesis increases there is a decrease of Fru2,6P$_2$. This can be demonstrated when the light intensity or the CO_2 concentration is varied in short-term experiments, which avoid further complications due to the accumulation of photosynthate in the leaf material (see below). As photosynthesis increases from negligible to maximal rates, Fru2,6P$_2$ decreases continuously over a two- to threefold range in spinach leaf discs (Stitt et al., 1984b). Less detailed studies also find changes of Fru2,6P$_2$ in response to light and CO_2 in maize (Stitt, 1985b), and after illumination of whole attached leaves from spinach (Stitt et al., 1983b), pea (Black et al., 1985b), and soybean (Huber et al., 1985b). In barley, Fru2,6P$_2$ is also lower in the light than in the dark (Sicher et al., 1986), and rises when photosynthesis is inhibited by decreasing the CO_2 partial pressure (R. Sicher, personal communication). The decreased Fru2,6P$_2$ during photosynthesis favors removal of triose-P for sucrose synthesis (Section VII).

The rate of photosynthesis also depends on temperature. C$_3$ photosynthesis shows a complex response to temperature (Berry and Bjorkman, 1984). Often, as the temperature is lowered, photosynthesis becomes light-saturated at lower light intensities. The alterations of the Fru2,6P$_2$ level in spinach leaves at 15°C and 30°C follows the light-saturation curves (Stitt, 1987b). Fru2,6P$_2$ decreased at lower light intensities at 15°C than at 30°C, but the level reached a minimum in each case at the light intensity where photosynthesis became light-saturated.

2. Partitioning of Photosynthate

Marked changes of Fru2,6P$_2$ are associated with changes in the content of soluble carbohydrate and changes in photosynthate partitioning in leaves. Fru2,6P$_2$ increases two- to threefold when 50 mM glucose or sucrose is supplied exogenously to spinach leaf disks (Stitt *et al.*, 1983b) or maize leaf segments (Stitt, 1985b). A similar increase accompanies the gradual accumulation of sucrose during photosynthesis in leaves of whole spinach plants (Stitt *et al.*, 1983a, 1984c), spinach leaf disks (Stitt *et al.*, 1984c), detached spinach leaves (Huber and Bickett, 1984; Stitt *et al.*, 1984c), and maize leaf segments (Stitt, 1985b). In detached material the sucrose accumulates more rapidly, and correspondingly, Fru2,6P$_2$ increases more rapidly (Stitt *et al.*, 1984c). In all cases, an increase of Fru2,6P$_2$ was associated with increased partitioning into starch, as the higher Fru2,6P$_2$ restricts synthesis of sucrose. Gradual alterations of Fru2,6P$_2$ accompany the diurnal alterations of SPS in soybean leaves (Huber *et al.*, 1985b), and are also found during the day in pea leaves (Black *et al.*, 1985b). Another line of evidence for alterations of Fru2,6P$_2$ being associated with changes of partitioning is provided by comparisons of spinach plants grown in short and long-photoperiods. In a short photoperiod more photosynthate is partitioned into starch (Robinson, 1984) and there is a higher Fru2,6P$_2$ level (Robinson and Baysdorfer, 1985).

3. Disease and Stress

There are indications that Fru2,6P$_2$ levels rise when leaves are stressed. Thus, Fru2,6P$_2$ rises in wheat leaves after infection with *Erysiphe* (Cseke *et al.*, 1984). The Fru2,6P$_2$ also rises in wilted spinach leaves (M. Stitt, unpublished), and the increase in the effective concentration will be even larger as wilting is associated with a loss of cell volume. Rapid increases of Fru2,6P$_2$ are found in both the roots and leaves of pea plants in response to flooding (Paz *et al.*, 1986). However, interpretation of such changes is difficult if the rate of photosynthesis is simultaneously decreasing, as this would lead to a rise of Fru2,6P$_2$ anyway (see above). It remains to be established whether Fru2,6P$_2$ alters as a primary or a secondary factor in stressed leaves. Nevertheless, it is noteworthy that an accumulation of starch often occurs after infection by bacteria or virus (Whipps and Lewis, 1981), or after some stresses including cold, and these could provide useful systems for studying what factors are involved in controlling photosynthate partitioning.

4. Specialized Plants

No information is available on Fru2,6P$_2$ levels during photosynthesis in plants like onion or leek, where all storage occurs as extrachloroplast carbohydrate. The role of Fru2,6P$_2$ in interactions between nitrogen and carbon metabolism also still requires attention, especially in legumes where considerable amounts of protein can also be stored in the leaf and where an additional cell layer—the paraveinal mesophyll—appear to have specialized storage functions for starch as well as for proteins.

The short-term alterations of $Fru2,6P_2$ in the C_4 plant maize generally resemble spinach (Stitt, 1985b; Stitt *et al.*, 1985c). However, recent work suggests some differences in the diurnal changes. The $Fru2,6P_2$ concentration remains low throughout the day and increases dramatically in the dark; during the night, $Fru2,6P_2$ concentration remains high and relatively constant (H. Usuda, W. Kalt-Torres, P. S. Kerr, and S. C. Huber, unpublished). There is also a modified behavior of $Fru2,6P_2$ in CAM metabolism (T. Fahrendorf, J. Holtum, and E. Latzko, personal communication). Very high levels of $Fru2,6P_2$ are present in the dark while glycolysis is occuring to provide PEP for dark fixation. In the light, the level is extremely low when stomati are closed during decarboxylation and CAM photosynthesis, but is three- to fourfold higher when the stomata are open and C_3 photosynthesis is occuring using external CO_2. The high levels of $Fru2,6P_2$ during malate synthesis resemble the increase of $Fru2,6P_2$ found in swelling guard-cell protoplasts (Hedrich *et al.*, 1985). The possible significance of the differences between C_3 and CAM metabolites will be discussed in Section VIII.

D. Mechanisms Controlling $Fru2,6P_2$ in Leaves

How far can the known properties of Fru6P,2-kinase and $Fru2,6P_2$ase account for the alterations of $Fru2,6P_2$ in leaves? The discussion will be restricted to the control of $Fru2,6P_2$ levels in spinach leaves, as this is the tissue where the most information is available. In maize, $Fru2,6P_2$ could be controlled in a similar way to spinach (Soll *et al.*, 1985; Stitt, 1985b) but there is far less information available about metabolite levels *in vivo* and the properties of Fru6P,2-kinase. The very high levels of $Fru2,6P_2$ found in the dark in barley leaves and CAM plants cannot be accounted for at present. Four potential regulatory mechanisms in spinach leaves are summarized in Table IV.

1. Response to Photosynthesis

The decrease of $Fru2,6P_2$ with rising rates of photosynthesis is due, at least in part, to regulation of Fru6P,2-kinase by three-carbon effectors like PGA and DHAP. These are transported between the chloroplast and cytosol via the phosphate translocator, and provide information about the metabolic conditions in the chloroplast. In general, an increase in the rate of photosynthesis will be accompanied by an increase in the concentration of these C_3 effectors, which will inhibit Fru6P,2-kinase and lead to a decrease of $Fru2,6P_2$. As the sensitivity of Fru6P,2-kinase to these three-carbon metabolites depends on the Pi concentration (see above), there may be a differential response of $Fru2,6P_2$ to the levels of either PGA or DHAP, depending on the metabolic conditions.

TABLE IV
Potential Mechanisms for Altering Fru2,6P$_2$ Concentration during Photosynthesis

Metabolic condition	Signal	Effect on Fru2,6P$_2$	Influence on metabolism
Increase photosynthesis	DHAP rises	Lowered	Stimulate sucrose synthesis to use DHAP made during photosynthesis and recycle Pi
Increase photosynthesis or decreasing Pi	PGA/Pi ratio rises	Lowered	Same
Low SPS activity (e.g., during diurnal rhythm)	Fru6P rises	Raised	Inhibit FBPase and prevent overaccumulation of hexose-P and sequestration of Pi (more starch made)
Unknown (correlates with accumulation of sucrose)	Fru6P,2-kinase/ Fru2,6P$_2$ase quotient alters	Raised	Decrease sucrose synthesis (more starch made)

When the cytosolic Pi concentration is high, DHAP will inhibit Fru6P,2-kinase while the inhibition by PGA is weakened. In these conditions, it is unlikely that the availability of Pi would restrict the rate of photosynthesis, but depletion of stromal metabolites could lead to photosynthesis being limited by RuBP regeneration. Regulation of Fru6P,2-kinase by DHAP provides a way of adjusting the rate of sucrose synthesis to prevent stromal metabolite pools being depleted too far. DHAP is a particularly suitable signal as it is the starting point for RuBP regeneration, and is linked by equilbrium reactions with many of the stromal metabolites. In contrast, when the Pi concentration in the cytosol is low, Fru6P,2-kinase will only be inhibited weakly by DHAP, but is inhibited strongly by PGA. In these conditions, it is unlikely that photosynthesis will be limited by low levels of metabolites, but it could be limited by Pi availability. Regulation of Fru6P,2-kinase by PGA and Pi provides a way of adjusting Fru2,6P$_2$ to reflect this requirement for Pi. When Pi becomes limiting, there is an accumulation of PGA in the stroma as PGA reduction is restricted (Heldt *et al.*, 1977; Edwards and Walker, 1983). The PGA in the cytosol also rises (Sharkey *et al.*, 1986), and the resulting increase of the PGA/Pi ratio should tend to lower Fru2,6P$_2$ and favor the recyling of Pi during sucrose synthesis.

Metabolite effectors are well suited to adjust the Fru2,6P$_2$ level to rapid changes in the rate of photosynthesis. Leaves contain high activities of Fru6P,2-kinase and Fru2,6P$_2$ase, at least 10-fold above those in other plant tissues (Stitt *et al.*, 1985a), and metabolite effectors provide a way of altering the activity of these enzymes rapidly following a change in the rate of photosynthesis. Recent studies of the changes of Fru2,6P$_2$ during induction and photosynthetic oxcillations suggest it may oscillate in barley leaves with a frequency of 1–2 min (Stitt, 1987b). Equally rapid changes have been seen in pea leaves (Paz et al., 1986).

No evidence was found in spinach for directly light-dependent "coarse" changes of Fru6P,2-kinase or Fru2,6P$_2$ase activity in leaves (Stitt et al., 1986b). However, too few species and conditions have been studied to exclude the possibility that such changes could provide an additional factor linking Fru2,6P$_2$ to photosynthesis.

2. Response to "Demand"

The gradual increase of Fru2,6P$_2$ during the day in spinach leaves as partitioning is changed in favor of starch is due to an interaction between control by metabolites, and "coarse" alterations in the activity of Fru6P,2-kinase and Fru2,6P$_2$ase in the leaf. To illustrate this interaction, Fig. 6 summarizes the alterations of enzyme activities, carbohydrate content, metabolite levels, and Fru2,6P$_2$ in spinach leaves during the day (Stitt et al., 1983a, 1986b; Gerhardt et al., 1987). Sucrose rapidly accumulates in the leaf at the beginning of the photoperiod, but this accumulation plateaus during the course of the day and an increasing amount of photosynthate is later retained as starch in the chloroplast [Fig. 6(f)]. Through this period, Fru2,6P$_2$ rises [Fig. 6(a)]. This increase is partly due to the gradual rise in the Fru6P,2-kinase : Fru2,6P$_2$ase quotient during the day [Fig. 6(b)]. Metabolites will also continue to affect the extent to which the activity of Fru6P,2-kinase and Fru2,6P$_2$ase is utilized in vivo. Thus, as SPS activity declines through the day [Fig. 6(c); see Section VI for fuller discussion], there is an accumulation of Fru6P in the cytosol [Fig. 6(e)] that stimulates Fru6P,2-kinase and inhibits Fru2,6P$_2$ase. The increase of Fru6P thus contributes to the increase of Fru2,6P$_2$ and, by restricting the cytosolic FBPase, will rebalance fluxes into and out of the cytosolic hexose-P pool. The inhibition of the cytosolic FBPase leads to an increase of DHAP in the cytosol [Fig. 6(d)]. This should inhibit Fru6P,2-kinase, but is presumably outweighed by factors that favor a rise of Fru2,6P$_2$. There may also be a decrease in the effectiveness with which DHAP inhibits as the day progresses, if the cytosolic Pi decreases.

While the increase of Fru2,6P$_2$ can be explained in terms of the regulation of SPS and Fru6P,2-kinase/Fru2,6P$_2$ase, the mechanism involved in the "coarse" control of these enzymes and their relation to the accumulation of sucrose remain unexplained. The increase of Fru2,6P$_2$ correlates in varying conditions with the accumulation of sucrose (Stitt et al., 1984c), but we do not know how far it is directly due to the accumulation of sucrose, and how far other, related, processes may be involved.

3. Further Problems

While our understanding of how and why Fru2,6P$_2$ varies during photosynthesis has progressed in the last 3 years, many crucial problems remain. The Fru6P,2-kinase protein still requires rigorous purification and more de-

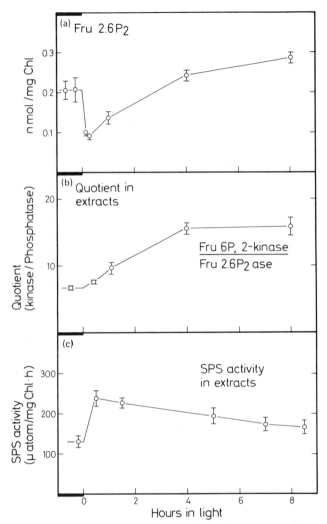

Fig. 6. Feedback control of Fru2,6P$_2$ illustrated during the photoperiod in spinach leaves. (a) Changes of Fru2,6P$_2$. (b) "Coarse" control of Fru6P,2-kinase and Fru2,6P$_2$ase. (c) "Coarse" control of SPS. (d) Alteration of cytosolic DHAP and (e) of cytosolic hexose-P and UDPGlc. (f) Accumulation of starch and sucrose. The distribution of sucrose between the vacuole and the cytosol is also shown. The results are from Stitt *et al.* (1983a, 1986b, 1987) and Gerhardt *et al.* (1987).

tailed structural and functional study. A central problem is a clarification of the possibility that protein modification controls Fru2,6P$_2$ levels and that plants may contain different forms of these enzymes. In addition, information is needed over the control of Fru2,6P$_2$ in a wider range of plant tissues that differ in their physiology.

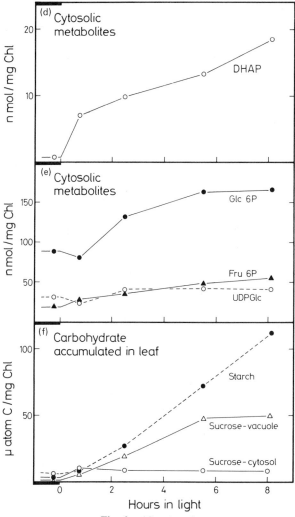

Fig. 6. (*Continued*)

VI. COARSE CONTROL OF SPS

In addition to metabolic fine control of SPS activity, the maximum extractable activity of this enzyme can change over a period of minutes to hours, which is thought to reflect a "coarse" level of control. This coarse control is evidenced by changes in SPS activity in relation to (a) time of day (diurnal fluctuation) and (b) "source–sink" manipulations. The diurnal fluctuations documented to date appear to be of two types. In many species, SPS activity changes rapidly with light/dark transitions, indicating some form of apparent light modulation that may be important in activating sucrose synthesis as photosynthate becomes available. In other species, there is an endogenous rhythm in SPS activity that is independent of light/dark changes, producing gradual changes of SPS that may be important in controlling photosynthate partitioning.

A. Diurnal Alterations

1. Light/Dark Modulation

Rapid changes in SPS activity in response to light/dark transitions were first observed in barley (Sicher and Kremer, 1984). Darkening of illuminated leaves resulted in a rapid decrease in SPS activity that could be fully reversed by subsequent illumination. Similar results have also been obtained with *Lolium temulentum* (Pollock and Housley, 1985). When SPS activity was measured in barley leaves over 1 24-h period (normal light/dark cycle), it was found that SPS activity was relatively constant through the day (i.e., no light activation was observed with the onset of illumination). Rapid dark deactivation occurred during the first 30 min of darkness, but SPS activity then increased throughout the remainder of the night (Sicher and Kremer, 1984). This suggests that the changes of SPS activity in barley leaves that occur during light/dark transitions are not the result of a direct light activation. An alternative explanation is that SPS activity is modulated by some metabolite(s).

Rapid effects of light/dark transition on SPS activity have also been ob-

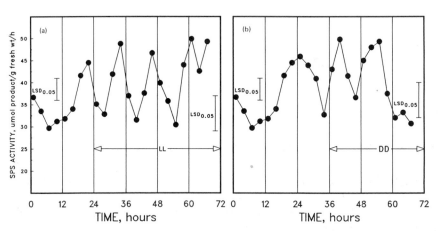

Fig. 7. Sucrose-P synthase activity of the third trifoliate leaf of soybeans in an alternating dark/light cycle (a, 0–24 h; b, 0–36 h), continuous light (a, 24–72 h), and continuous darkness (b, 36–72 h). [Reproduced from Kerr *et al.* (1985), with permission.]

served in maize leaves (Sicher and Kremer, 1985). However, diurnal changes in maize leaf SPS activity are quite different than those reported for barley. Important differences are that in maize, SPS activity reaches a maximum level at midday (coincident with highest light intensity) and was low and constant throughout the night (W. Kalt-Torres, P. S. Kerr, and S. C. Huber, unpublished data). Recently, rapid alterations of SPS activity have also been found in response to illumination and darkening in spinach (Stitt *et al.*, 1987). In spinach, these rapid changes are imposed on a gradual decline of activity through the photoperiod, and a recovery in the dark period [see Fig. 6(c)].

Although the observed diurnal changes in SPS activity are different in barley, spinach, and maize, it is quite possible that enzyme activity is controlled by the same mechanism in these species and the different diurnal patterns could reflect differences in metabolite levels that modulate an underlying mechanism. As will be discussed below, this light/dark modulation of SPS activity may be important in allowing sucrose synthesis to be regulated without a large increase of metabolites, as the rate of photosynthesis rises.

2. Endogenous Rhythms

The first evidence for diurnal fluctuation in SPS activity was obtained in experiments with soybean plants (Rufty *et al.*, 1983). The typical fluctuation observed with mature leaves of vegetative soybean plants exhibits two peaks of SPS activity per 24-h period: one at the beginning of the photoperiod, and one at the beginning of the dark period. Change in SPS activity are not closely aligned with light/dark transitions, and reciprocal transfers of plants between light and dark did not affect activity (Rufty *et al.*, 1983). In a subsequent study (Kerr *et al.*, 1985), it was determined that fluctuation in SPS activity persisted for several cycles in either continuous light or extended darkness (Fig. 7). Thus, the diurnal rhythm in SPS activity in soybean leaves appears to be controlled by an endogenous clock mechanism. The observation that the rhythm persists in extended darkness is significant, because the leaves were essentially depleted of carbohydrate reserves (starch, sucrose) at the end of the normal dark period, that is, before extended darkness began (Kerr *et al.*, 1985). It is important to note that SPS is the only enzyme of the cytosolic sucrose formation pathway that exhibits a pronounced diurnal fluctuation (Rufty *et al.*, 1983).

Gradual diurnal changes in SPS activity have also been documented in mature leaves of tobacco (Huber *et al.*, 1984c), pea (Huber *et al.*, 1985a), cotton (D. Hendrix and S. C. Huber, unpublished), and spinach (Stitt *et al.*, 1987). With pea and cotton, two peaks of SPS activity were observed per 24-h period, whereas with tobacco, only one peak was observed. In spinach, the rapid alteration of SPS activity expected during a dark/light transition was abolished when leaves were not illuminated, but the gradual decline found during the light period still occured in the absence of light (I. Wilke, M. Stitt, and H. Heldt, unpublished). Thus, in all these species, gradual changes in SPS activity occur that are not closely aligned with light/dark transitions or with light intensity. However, it has not been determined whether the fluctuation in SPS activity persists under constant environmental conditions.

The endogenous rhythm in SPS activity has been studied more extensively with soybean than any other species. For example, differences in the amplitude of the diurnal change and timing of peaks have been observed among three soybean cultivars (Huber et al., 1984c). The cultivars tested differed in photoperiod requirements for floral induction. Although the study was limited in scope, it was apparent that the diurnal rhythm in SPS activity was most pronounced in cultivars that were photoperiod-sensitive for floral induction (i.e., cultivars adapted to southern latitudes). Although the mechanistic basis for this relationship is not known, it is worth noting that some correlation is apparent among both soybean and tobacco cultivars (Huber et al., 1984c).

The expression of the endogenous rhythm in SPS activity with a given genotype can be affected by a number of factors. In general, conditions that reduce SPS activity also tend to dampen the diurnal fluctuation: for example, transfer of soybean plants, grown under long days, to a short-day photoperiod (Huber et al., 1984c), nutrient deficiency, and leaf aging (Huber et al., 1985a).

3. Significance of the Diurnal SPS Rhythm

Prior to the discovery that SPS activity fluctuates diurnally, it was realized that the activity of this enzyme in leaf extracts is low (Hawker, 1967), particularly in comparison with other enzymes of the pathway. However, the activity observed is just sufficient to account for the flux of carbon into sucrose; that is, maximum SPS activity may be one of the rate-limiting steps in the pathway. In early studies comparing various species, it was noted that SPS activity was highest in species that partitioned a greater percentage of photosynthate into sucrose and less into starch (Huber, 1981a,b). More recently, an overall positive correlation between assimilate export rate (which usually corresponds closely with sucrose formation rate) and SPS activity has been noted among various species (Huber et al., 1985b). Results such as these have lead to the postulate that SPS activity is regulated and is closely associated with partitioning of photosynthetically fixed carbon between starch and sucrose.

Changes in SPS activity that occur over the course of the photoperiod also appear to be closely aligned with changes in the rate of sucrose formation (Huber et al., 1985a,b). In mature leaves of vegetative soybean plants, export rate usually decreases as the day proceeds, and this corresponds closely with concurrent changes in SPS activity. Thus, the diurnal rhythm in enzyme activity appears to be of physiological significance, and may be one of the biochemical factors that controls the timing of assimilate availability from leaves.

B. Effect of Source–Sink Manipulations

Further evidence for a role of "coarse" control of SPS in controlling export and photosynthate partitioning is provided by experiments showing that the activity of soybean leaf SPS changes in response to alterations in the "demand" for assimilates, that is, source–sink manipulations. The essential observation is that changes in export rate (increase or decrease) are closely aligned with corresponding changes in SPS activity. Export rate from a mature soybean leaf can be increased by removal of other source leaves (partial defoliation). In response to this treatment, formation of starch is dramatically reduced and the rate of sucrose formation and SPS activity are increased (Rufty and Huber, 1983). Conversely, leaf detachment or imposition of a petiole girdle restricts export capacity and SPS activity declines dramatically. Thus, coarse control of SPS activity appears to be an important component of the mechanism that adjusts cellular carbon partitioning in response to changes in "demand." The biochemical mechanism involved is not known at present; it is also unclear what "signal" may regulate the coarse control mechanism. The signal is probably not a metabolic intermediate, because changes in partitioning can occur without a significant change in photosynthetic rate (Rufty and Huber, 1983). Furthermore, soybean leaf SPS activity does not respond in the short-term to changes in metabolite levels or photosynthetic rate. As discussed above, the endogenous rhythm in SPS activity persists in extended darkness when leaves are depleted of endogenous reserves and metabolism has come to a standstill. Also, short-term CO_2 enrichment increases photosynthetic rate (and presumably metabolite levels) but export rate and SPS activity do not increase (Huber *et al.,* 1984a); the additional carbon fixed at elevated CO_2 is partitioned entirely into starch.

If metabolites are not the signal, it is possible that one or more endogenous plant growth regulators (e.g., abscisic acid) may be involved. Abscisic acid (ABA) is formed in leaves, and assuming that the rate of synthesis is constant, leaf ABA concentration will fluctuate with export rate. For example, in a detached leaf, ABA concentration increased dramatically (Setter *et al.,* 1980), and there is some indirect evidence to suggest that the increase in ABA may be linked with the decline in SPS activity (Huber *et al.,* 1984a). However, direct evidence for hormonal control is lacking.

All of the source–sink manipulations discussed above were conducted in the light, that is, under photosynthetic conditions. Similar experiments can be conducted in the dark, where starch mobilization provides the carbon to support sucrose synthesis and export. Defoliation in the dark increases "demand" for assimilates from the remaining source leaves, as evidenced by a reduction in leaf sucrose concentration (Rufty *et al.*, 1984a). Under these conditions, SPS activity is also increased relative to control leaves of untrimmed plants, but the rates of export and starch mobilisation are unaffected (Rufty *et al.*, 1984a). This, along with other evidence, strongly suggests that SPS activity is not limiting the rate of sucrose formation and export at night; presumably other factors, such as the rate of starch mobilization, may be the principal limiting factors. However, it is important to note that SPS activity responds to source–sink manipulations in the dark, even though export rate cannot be increased. Therefore, the "coarse" control mechanism is apparently not light-dependent.

C. Mechanism of Coarse Control of SPS Activity

The biochemical basis for diurnal changes in SPS activity is not known. However, it appears that at least two different mechanisms may exist. In the case of apparent "light modulation" of maize leaf SPS, it is possible that some type of covalent modification may be involved. This is suggested both by the rapidity of the change and by the preliminary evidence that the low activity of the "dark" form was attributed to a decrease in V_{max} and was also associated with decreased substrate affinities (Sicher and Kremer, 1985). In contrast, diurnal changes in soybean leaf SPS activity are not associated with substantial changes in the apparent K_m for either substrate (Kerr and Huber, 1987). Thus, in soybean there is either an endogenous rhythm in synthesis/degradation of SPS protein, or some posttranslational modification mechanism that only affects V_{max}. In all these species, more work is required with the purified enzymes in "high" and "low" activity states. In addition, immunological techniques need to be employed to monitor SPS protein simultaneously with SPS activity.

VII. COORDINATE CONTROL OF SUCROSE FORMATION BY SPS AND THE CYTOSOLIC FRUCTOSE 1,6-BISPHOSPHATASE

In the previous sections, the regulatory properties of individual enzymes of the sucrose formation pathway have been described and the importance of the cytosolic FBPase and SPS to the regulation of the pathway has been emphasized. Although the cytosolic FBPase and SPS are important points for regulation, it is important to realize that regulation is the property of an entire pathway rather than of individual processes. For an individual enzyme to contribute significantly, it is necessary that it possess properties that allow its activity to be controlled, but the impact of the enzyme on the flux depends on its integration into the whole pathway. When flux through a pathway is modified by the activation of one enzyme, this will affect the concentration of the substrates and products of this enzyme. Changes in these metabolites may then potentially affect other enzymes in the pathway, whose activity must be changed if the flux through the entire pathway is to be modified. This analysis (see Kacser and Burns, 1973, 1979) emphasizes that regulation of a given pathway requires an interaction and coordination between enzymes. In this section we will discuss how metabolic "fine" control and "coarse" control of the cytosolic FBPase and SPS may be coordinated to enable sucrose synthesis to respond to alterations in the availability of photosynthate or the demand for sucrose.

A. Feedforward Control

As the rate of photosynthesis is increased by higher light or increased CO_2, changes in the concentration of cytosolic metabolites interact with the regulatory properties of the cytosolic FBPase and SPS to enable sucrose formation to be regulated so that triose-P and Pi concentrations are maintained that allow rapid CO_2 fixation in the chloroplast. For example, in spinach, the concentration of triose-P increases in response to increased rates of photosynthesis, and in turn, the concentration of Fru2,6P$_2$ decreases two- to threefold. As a result, the activity of the cytosolic FBPase rises, and more Fru6P is formed. Since Fru6P and Glc6P are in equilibrium via phosphoglucose isomerase, increased production of Fru6P will result in a greater Glc6P concentration in the cytosol and consequently SPS will be activated. Thus, it is evident that increased rates of photosynthesis will lead to increased rates of sucrose formation as a result of a coordination of the regulatory properties of the cytosolic FBPase and SPS.

The fluctuations of metabolites that are needed to activate and deactivate sucrose synthesis are not large, showing that the regulation of SPS and the cytosolic FBPase must be very sensitive. This sensitive response can be understood in terms of their known properties, as both enzymes are regulated by a number of ligands, which often change simultaneously *in vivo*. The response of the enzymes to these changes has been investigated in simulation experiments, using the levels of metabolites measured in leaves to estimate how the concentrations of their substrates, activators, and inhibitors vary *in vivo*. The activity of the partially purified enzymes were then measured as these changes were simulated (Herzog *et al.,* 1984; Stitt and Heldt, 1985c; Stitt *et al.,* 1987). These experiments suggested that the cytosolic FBPase and SPS will be effectively inactive until a "threshold" concentration of triose-P or hexose-P is attained, respectively. They will then be strongly activated by further small increases of these metabolite pools (Fig. 8). This highly regulated response is partly due to both enzymes having sigmoidal substrate saturation in the conditions expected *in vivo*. Thus, the

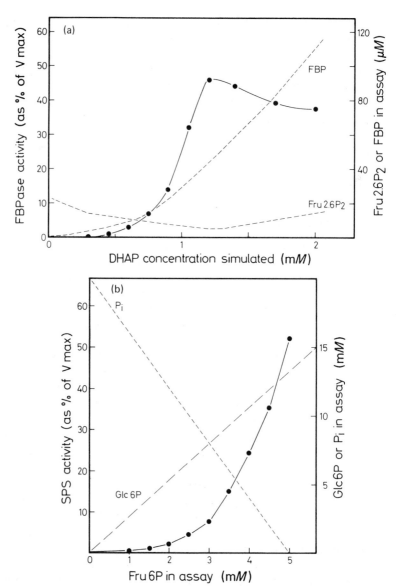

Fig. 8. Simulation of the response of the cytosolic FBPase and SPS to increasing levels of triose-P and hexose-P *in situ*. (a) FBPase. (b) SPS. The simulated enzyme activity is shown as a solid line, and the concentrations of substrate and regulators are shown as dashed lines. In the SPS assay, 2 mM UDPGlc was included. For details, see Stitt *et al.* (1987).

cytosolic FBPase shows a sigmoidal saturation for Fru1,6P$_2$ when Fru2,6P$_2$ is present, and SPS shows a sigmoidal saturation for Fru6P when Pi is present. However, additional amplification is generated because changes of substrate concentration are linked to changes in the concentration of regulators, which, in turn, modify the substrate affinity. Thus, as triose-P increases, the cytosolic FBPase is activated because $Fru1,6P_2$ rises, and also because the decreasing $Fru2,6P_2$ increases the affinity for its substate. In an analogous way, an increase of Fru6P is accompanied by a simultaneous increase of Glc6P, which leads to an increased affinity of SPS for its substrates. In addition, an increase of triose-P and hexose-P levels is likely to be accompanied by a decrease of Pi, which is a competitive inhibitor of both the FBPase and SPS.

This sensitive regulation has important implications in allowing sucrose synthesis to be coordinated with other aspects of cell metabolism. Regulation of the cytosolic FBPase to keep it inactive below a "threshold" concentration of triose-P provides a way of coordinating synthesis of sucrose with the operation of the Calvin cycle. It may be envisaged that the cytosolic FBPase remains inactive until adequate levels of triose-P are present to allow turnover of the Calvin cycle and regeneration of RuBP. Then, as the level of triose-P rises further, the cytosolic FBPase is strongly activated to remove the surplus triose-P for conversion to sucrose. By increasing fluxes to sucrose without requiring a further large increase of triose-P, this strong activation minimizes the risk that the recyling of Pi starts to limit photosynthesis.

The strongly sigmoidal response of SPS to changes in the hexose-P pool also has implications for regulation *in vivo*. On one hand, it will allow significant pools of hexose-P to be maintained for respiratory metabolism in the dark. On the other hand, once the cytosolic FBPase has been activated, SPS will be activated without requiring a large accumulation of hexose-P. This is essential, because considerable amounts of Pi could be sequestered in the hexose-P pool, and also because a large increase of Fru6P could lead, via a rise of $Fru2,6P_2$, to reinhibition of the cytosolic FBPase.

SPS is also clearly subject to "coarse" control during light–dark transitions in many plants, leading to a change in its substrate affinity (Sections IV,D and VI,C). Such "coarse" changes will interact with the "fine" metabolite control and allow SPS to be activated by smaller changes of the hexose-P pool, or even without any changes. This explains why there is no change of hexose-P in the cytosol of spinach protoplasts or leaves during induction of photosynthesis, and UDPGlc even decreases (Stitt *et al.,* 1980; Gerhardt *et al.,* 1987), activation of SPS. This "coarse" activation of SPS may play a crucial role in controlling sucrose synthesis, because, by keeping Fru6P low, it will prevent $Fru2,6P_2$ from rising and inhibiting the cytosolic FBPase. However, the details of interaction will only become clear when the biochemical basis for these changes of SPS activity has been identified.

It is striking that the activation of SPS (and the decline of $Fru2,6P_2$; see Section V) occur in parallel with the rising rate of photosynthesis as, for example, the light intensity or the CO_2 concentration is increased (Stitt, 1987b). This "tuned" response allows the entire pathway of sucrose synthesis to be progressively activated as the supply of photosynthesis increases. In combination with the "threshold" regulation of the cytosolic FBPase and

SPS by metabolites, this provides a highly differentiated and sensitive way of balancing the fluxes in the Calvin cycle and the cytosol, while simultaneously poising the concentrations of metabolites and Pi to allow effective operation of the photosynthetic apparatus in a wide range of conditions.

B. Feedback Control

During periods when sucrose is produced more rapidly than it can be exported, sucrose accumulates within the leaf. It is also evident that as sucrose accumulates within the leaf during the day, the concentration of $Fru2,6P_2$ increases two- to threefold. This increase in $Fru2,6P_2$ would decrease the activity of the cytosolic FBPase and result in decreased rates of sucrose synthesis. The mechanism(s) whereby $Fru2,6P_2$ concentration increases as sucrose accumulates within the leaf is not completely understood; however, it appears that "coarse" control, as well as metabolic "fine" control, is involved. Differences are apparent between species, which may be fundamental, or merely reflect shifts of emphasis.

This interaction between "coarse" and "fine" control has been studied using nonaqueous fractionation of spinach leaves to measure how the cytosolic metabolite levels respond to "coarse" control of SPS (Gerhardt et al., 1987). In spinach, sucrose accumulates in the leaf, but as the day progresses the accumulation of sucrose slows and more starch is synthesized in the chloroplast. The change of partitioning is accompanied by a "coarse" control of SPS (Stitt et al., 1987) and an increase of $Fru2,6P_2$ (Stitt et al., 1983b). This increase of $Fru2,6P_2$ is partly because lower SPS activity leads to twofold increase of Fru6P in the cytosol, and partly because of a simultaneous "coarse" control of the $Fru6P,2$-kinase/$Fru2,6P_2$ase ratio (Stitt et al., 1983b). Thus, it appears that "coarse" control of enzyme activity is part of a feedback mechanism that coordinately controls SPS and the cytosolic FBPase while alterations of the cytosolic Fru6P may operate as a "fine" control that coordinates the response to a "coarse" control at two different sites. Together, these mechanisms restrict the rate of sucrose formation and lead to an accumulation of metabolites in the cytosol. It can be envisaged how the decreased supply of Pi could then lead to a stimulation of starch synthesis (Preiss, 1980). This provides a framework to understand how a leaf alters the partitioning of carbon between sucrose and starch when demand for sucrose is less than its supply [Fig. 9(b)].

In soybean, the activity of SPS changes during the day due to the diurnal rhythm, and these changes in SPS activity are correlated negatively with $Fru2,6P_2$ concentration (Kerr and Huber, 1987). Presumably, the decrease in SPS activity would result in an increase in the Fru6P concentration in the cytosol (see above), which then increases the $Fru2,6P_2$ concentration due to an inhibition of $Fru2,6P_2$ase and an activation of $Fru6P,2$-kinase. As a result, the increase in $Fru2,6P_2$ concentration would inhibit the cytosolic FBPase, and lead to increased starch synthesis.

Reciprocal changes of SPS activity and $Fru2,6P_2$ also occur in barley (Sicher *et al.*, 1986) after darkening at the end of the day. SPS activity decreases sharply and then rises gradually while $Fru2,6P_2$ rises and then declines. The relation between these changes is not known, but they may be involved in restricting further sucrose synthesis until much of the accumulated sucrose has been exported out of the leaf.

It is probable that a similar control mechanism influences sucrose formation under conditions when demand for sucrose is greater than its supply. Source/sink manipulations and growing plants under environmental conditions (altered temperature, mineral nutrition) that increase the growth rate of sinks have been shown to increase the rate of sucrose formation and the activity of SPS (Rufty and Huber, 1983; Rufty *et al.*, 1984b; Kerr *et al.*, 1986), and $Fru2,6P_2$ levels are altered in spinach leaves in response to source–sink manipulations (Stitt *et al.*, 1984c). One can speculate that changes in $Fru6P$ and $Fru2,6P_2$ concentration would result under conditions that increase SPS activity in a manner reciprocal to what occurs when SPS activity is decreased.

The precise mechanisms involved in this "coarse" feedback control, and the factors that trigger them, are unknown. The elucidation of these biochemical and physiological questions is a pressing problem and should yield new insights into how the synthesis and export of photosynthate are coordinated with demand in the remainder of the plant.

In addition to the contribution that "coarse" control makes to feedback control of sucrose formation, feedback control of sucrose formation by sucrose itself may be involved, as earlier proposed by Herold (1980) [Fig. 9(c)]. As mentioned in Section VI, the activity of sucrose phosphatse and SPS is inhibited by sucrose under certain conditions. It appears that significant variation exists among species in the effect of sucrose on SPS and sucorse phosphatase. Inhibition of SPS by 100 mM sucrose varies among species from 10 to 50%. Similar differences among species appear to be present in the effect of sucrose on sucrose phosphatase. Although considerable research remains to done on the effect of sucrose on these enzymes, these results suggest that most species contain one or more enzyme that is sensitive to increased levels of sucrose. In addition, it appears that SPS in leaves is inhibited by sucrose-P. As a result, it appears that changing levels of sucrose could either directly affect SPS or indirectly affect it by changing sucrose-P levels through regulation of sucrose phosphatase. Fluctuations in SPS activity could then affect the cytosolic $Fru6P$ concentration, which would serve to influence the rest of the pathway as described above. Indeed, if sucrose is supplied to spinach or maize leaf segments there is a marked increase of $Fru6P$, which is accompanied by an increase of $Fru2,6P_2$. Con-

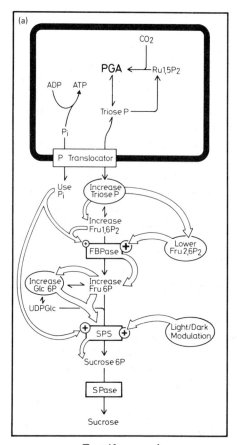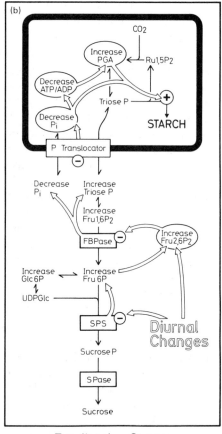

Feedforward   **Feedback − Coarse**

Fig. 9. Scheme for the regulation of sucrose synthesis by a coordinate control of the cytosolic FBPase and SPS. (a) Feedforward control, in response to rising rates of photosynthesis. (b) Feedback control initiated by a "coarse" control of enzymes, as mediated by "fine" control of further enzymes via changes of metabolites. (c) Feedback control initiated by a "fine" feedback inhibition of SPS or sucrose phosphatase (SPase) by sucrose. These three modes are not mutually exclusive and may all interact in a given leaf and condition *in vivo*.

siderable research remains to be done on this aspect of coordinate control; however, it appears that metabolic "fine" control of SPS and/or sucrose phosphatase by sucrose and "coarse" control of enzyme activity function collectively as feedback control mechanisms for surcrose formation.

If feedback control of sucrose synthesis can be exerted both by alteration of cytosolic sucrose and by "coarse" control of SPS and Fru6P,2-kinase, there will be a varied response of the leaf sucrose content to source/sink manipulation, depending on how the "fine" and "coarse" control interact.

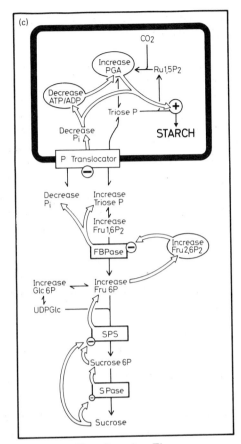

Feedback - Fine

Fig. 9. (*Continued*)

For example, a restriction of sucrose synthesis would be associated with an accumulation of sucrose if "fine" control makes a major contribution, but sucrose could decline if "coarse" control predominates. Alteration of sucrose compartmentation in the leaf could lead to an even more complex picture (Section III,C,5). The variable results found in experiments looking at the relation between sucrose, transport, and photosynthesis may, in part, derive from having operated with an oversimplified model of how a "feedback" control is exerted. We would urge that in the future source/sink manipulation should be accompanied by measurements of multiple parameters, including not just carbohydrate levels, but also the activity of enzymes and levels of metabolites and $Fru2,6P_2$, which all interact to control the rate of sucrose synthesis.

VIII. ADAPTATION OF THE REGULATION OF
SUCROSE SYNTHESIS

The previous sections have described a complex regulatory system by which the rate of sucrose synthesis can be adjusted, allowing concentrations of metabolites, Pi, and adenine nucleotides to be maintained which support an efficient fixation and reduction of CO_2 in the chloroplast. The complexity of this system is probably unavoidable if the balance between CO_2 fixation and sucrose synthesis is to be adjusted in varying environmental and physiological conditions. As discussed already, some features of this regulation may vary, depending on the plant involved. In this section, the adaptation of sucrose synthesis will be discussed in more detail, taking three examples where evidence exists that the regulation of sucrose synthesis is altered as an adaptation to a specialised photosynthetic metabolism, or as a directed response to environmental conditions. The mechanism of the change is different in each case, but they all probably lead to an alteration of the "threshold" level of triose-P at which sucrose synthesis is activated. The resulting adjustment of metabolite levels in the leaf is instrumental in allowing rapid photosynthesis to occur.

A. Adaptation of Sucrose Synthesis in C_4 and CAM Species

1. Modification of the "Threshold" to Allow Intercellular Transport in Maize

Sucrose synthesis in maize occurs in the mesophyll cells using triose-P that has been produced by PGA reduction in the mesophyll chloroplast, and most of the triose-P must return to the Calvin cycle in the bundle sheath to allow more RuBP to be regenerated (see Section II,A,2). Recently, two separate techniques (Leegood, 1985; Stitt and Heldt, 1985c) have confirmed there are large concentration gradients between the different cell types in maize, which allow this intercellular transport to occur by diffusion. The concentration of triose-P in the mesophyll of maize leaves was over 10 mM. This contrasts with C_3 plants where triose-P is exported from the chloroplast and converted to sucrose in the surrounding cytosol and where cytosolic FBPase is regulated so that less than 1 mM triose-P is present in the cytosol. How is the regulation of the FBPase modified to allow this high concentration to be maintained?

So far, no large changes have been found in the way in which the Fru2,6P$_2$ concentration is controlled in maize (Stitt and Heldt, 1985b; Stitt, 1985b). Instead, the properties of the cytosolic FBPase are modified so that it has a lower affinity for its substrate and a higher sensitivity to inhibitors (Stitt and Heldt, 1985a). This shifts the "threshold" for removal of triose-P upward by a factor of 10 (Stitt and Heldt, 1985b) and reconciles synthesis of sucrose in the mesophyll cells with maintenance of the high concentrations of triose-P in the mesophyll that are essential for photosynthesis in this species.

2. Lowered Fru2,6P₂ Levels in CAM

During photosynthesis in C_3 plants, one triose-P may be converted to carbohydrate endproducts per three CO_2 fixed. In contrast, during CAM, a further three PEP molecules must also be converted, via triose-P, to carbohydrate storage products (see Section II,B,3). Thus, per three CO_2 fixed, a total of four triose-P must be converted to carbohydrates, to recycle Pi. How is this large change in the fate of the triose-P pool during CAM metabolism made possible?

It is intriging that CAM plants contain extremely low Fru2,6P₂ in the light when they are photosynthesizing with closed stomati, and are decarboxylating malate (T. Fahrendorf, J. Holtum, and E. Latzko, personal communication). When their stomati are open and atmospheric CO_2 is being fixed, Fru2,6P₂ rises. The low Fru2,6P₂ found during CAM would deinhibit the cytosolic FBPase and stimulate conversion of triose-P to carbohydrates in the cytosol. This explains how PEP can be converted to sucrose or extrachloroplast glucans in species like *Ananas comocus* or *Hoya bella*. Surprisingly, a similar pattern was also found in CAM plants that accumulate starch in the chloroplast, including *Bryophyllum tubiforum* and *Mesembryanthemum crystallinum*. It might be expected that triose-P would be retained in the chloroplast in these starch formers, rather than being converted to hexose-P in the cytosol. However, the compartmentation of the fluxes between malate and starch is not known. There are indications that conversion of starch to malate may involve export from the chloroplast at the hexose stage (see Edwards *et al.*, 1985), and a similar route during malate–starch conversion is equally possible.

B. Temperature Dependence of the "Threshold" for Sucrose Synthesis

A 10°C decrease in the temperature typically leads to a twofold decrease in the activity of enzymes, and the Calvin cycle, which involves an interaction among many enzymes, shows an even higher sensitivity to temperatures (Baldry *et al.*, 1966). Nevertheless, the temperature response curve for photosynthesis of C_3 plants is often very flat in the range from 10–35°C (Berry and Bjorkman, 1984). Many factors are likely to be involved in this temperature response, including reactions in the thylakoids and stroma. Recent evidence suggests sucrose synthesis may also play a role.

The cytosolic FBPase is a temperature-sensitive enzyme (Weenden and Buchanan, 1983), and recently it has been shown that the regulatory properties of this enzyme depend on the temperature at which it is assayed (Section IV,B). Thus, as the temperature decreases the cytosolic FBPase becomes increasingly sensitive to Fru2,6P$_2$ and AMP (Stitt, 1987b). *In vivo*, both these inhibitors act by raising the substrate affinity. This means that increasingly high levels of triose-P will be needed in the cytosol to activate the cytosolic FBPase as the temperature decreases. In agreement, higher levels of triose-P and other intermediates are found in leaves as the temperature decreases (Section III,C). These temperature-dependent alterations of the cytosolic FBPase may allow an automatic adjustment of photosynthesis to changing temperature. As the temperature decreases, the "threshold" concentration of triose-P needed for sucrose synthesis increases, allowing higher levels of metabolites to be maintained in the stroma. These higher metabolite levels may compensate for the decreasing activity of enzymes at low temperatures by increasing the levels of their substrates. Conversely, at higher temperature metabolite levels are lower, but more Pi will be available for rapid photosynthesis.

Further factors could also contribute to reducing sucrose synthesis at low temperatures. The phosphate translocator may become increasingly limiting at low temperatures (see Section IX). Also, SPS may be involved. Although the regulatory properties of the enzyme do not change in a temperature-dependent way (S. C. Huber and M. Stitt, unpublished), it has a relatively high Q_{10} of 3. There are also indications that temperature may affect "coarse" control of SPS in some cases. When soybean plants were transferred from optimal (26°C) to suboptimal (18°C) temperature, there was a decrease of SPS activity and of assimilate export, without substantially reducing the rate of CO_2 assimilation (Rufty *et al.*, 1984b).

IX. LIMITATION OF PHOTOSYNTHESIS BY SUCROSE SYNTHESIS

So far, the flexibility, sensitivity, and adaptability of the regulation of sucrose synthesis have been stressed. However, it is also important to ask whether sucrose synthesis ever directly limits the rate of photosynthesis. How can such a limitation be recognized, in what conditions does it arise, and can it be ascribed to a characterized biochemical mechanism? Generally, it has been concluded that the availability of light and CO_2 and the capacity for electron transport and activity of Rubisco will be the major limitations on the rate of photosynthesis (Farquahar and von Caemmerer, 1982). However, there is some recent evidence that this view may require modification. For example, in saturating light and CO_2, there is a high light-scattering signal from spinach leaves, suggesting the presence of a large thylakoid pH gradient that could not be discharged (Dietz *et al.*, 1984); there is also often a rise in the redox quench (Dietz *et al.*, 1986). Thus, the previous assumption that the maximal rate of photosynthesis is set by the capacity for electron transport may require revision. It is also perplexing that

Rubisco is often partially deactivated even during rapid photosynthesis in saturing light, suggesting that the available Rubisco activity cannot all be utilized.

Such results may be explained by recent evidence that sucrose synthesis can exert a limitation on the rate of photosynthesis in leaves. This evidence has been obtained in experiments in which plant material has been manipulated or exposed to conditions that are not necessarily expected in the field. By using such extreme conditions, it has been possible to identify when carbon metabolism may exert a limitation on photosynthesis, and to show how this limitation can be recognized. Thus, the results that will be discussed do not provide evidence that sucrose synthesis exerts a direct limitation on the rate of photosynthesis in the field, but they do provide a series of tools that will allow this question to be approached.

An imbalance between sucrose synthesis and photosynthesis could lead to two extreme cases where the metabolic conditions limit the rate of photosynthesis. (1) The "threshold" for activating sucrose synthesis is too low, so that stromal metabolites are depleted. Does this reach a point where they restrict the rate of RuBP regeneration to below the rate at which RuBP can be carboxylated, or electron transport can supply ATP and NADPH? (2) Sucrose synthesis does not release Pi from triose-P fast enough and this shortfall cannot be made up by recycling Pi via other processes, so supply of Pi then restricts the rate of photosynthesis that the chloroplasts can achieve.

A. Limitation by Suboptimal Metabolite Levels

Substantial levels of metabolites are maintained in the stroma over a wide range of rates of photosynthesis as light and CO_2 are varied (see Section III,C; also Dietz and Heber, 1984; Dietz et al., 1984), and there is no direct evidence that depletion of metabolites limits the rate of steady-state photosynthesis. However, three considerations suggest that this question requires more attention. First, the crucial experiment has not yet been carried out: this is to develop ways of manipulating plant material to increase the concentrations of metabolites in the stroma and ask whether this increases the rate of photosynthesis. Second, studies are needed in a wider range of conditions that may increase the susceptibility of the Calvin cycle to inhibition by low levels of metabolites, including high temperature (see Section VIII) and the osmotic stress (see Kaiser, 1982). Indeed, a restriction of photosynthesis at low light intensities and high temperatures is associated with a decline of metabolite levels, which is consistent with an overactivation of sucrose synthesis in these conditions (Stitt, 1987b). Third, studies in steady-state conditions may not reveal a colimitation by low metabolite levels because this is masked by regulation that readjusts the balance between CO_2 fixation and sucrose synthesis.

The presence of such colimitations is revealed by subjecting plants to sudden transients, when considerable depletion of metabolites can be seen. One case that has received considerable attention is the requirement for autocatalytic buildup of stromal metabolite pools during the induction period (Edwards and Walker, 1983). Recently, it has also been shown that a temporary inhibition of photosynthesis following a transition from high to low light is associated with a depletion of metabolic pools, which may be due to an overshoot of starch and sucrose synthesis (Prinsley et al., 1986; Prinsley and Leegood, 1986b). Studies of the behavior of Rubisco following a sudden decrease of the light intensity also illustrate how a complex regulatory response may mask a limitation by low metabolites during steady-state photosynthesis (Mott et al., 1984). Immediately after decreasing the light intensity, Rubisco is still highly activated, and RuBP is below the active site concentration, showing how regeneration of RuBP is limiting. However, as Rubisco later deactivates in the low light there is a return to the "normal" picture where RuBP saturates the active sites. It remains an open question how far Rubisco activation has been "down-regulated" to match the available regeneration of RuBP, and how far regeneration of RuBP has been increased by adjusting electron transport or sucrose synthesis.

B. Limitation by Suboptimal Phosphate

1. Evidence that Sucrose Synthesis Can Limit Photosynthesis

Four lines of evidence suggest that short-term availability of Pi may restrict the maximal rate of photosynthesis in saturating light and CO_2. One approach is to manipulate leaf material so that the cytoplasmic Pi is increased, and ask whether his leads to higher rates of photosynthesis. After feeding Pi, there is an increase in the levels of phosphorylated intermediates in the leaf, showing that the Pi is rapidly entering the leaf cells (Dietz and Foyer, 1986). Walker and his associates (Walker and Sivak, 1985; Sivak and Walker, 1985, 1986) have shown that photosynthesis can be transiently increased by feeding Pi to leaves that are photosynthesizing in saturating light and CO_2, while feeding mannose to sequester Pi, reverses this stimulation.

A second approach is based on observations that the net rate of CO_2 assimilation does not always increase in C_3 plants when the O_2 concentration is decreased from 21% to 2% to suppress photorespiration (Jolliffe and Tregunna, 1973; Laisk and Pärnik, 1977; Cornic and Louason, 1980; McVetty and Canvin, 1981; Sharkey, 1986b). This "O_2 insensitivity" is found at moderate temperatures with saturating light and slightly enhanced (500–600 ppm) CO_2 (Sharkey, 1985a; Sharkey et al., 1986) and appears in air at 12°C (Leegood and Furbank, 1986). The lack of sensitivity to O_2 implies that the rate of assimilation is limited by a process for which photorespiration does not compete, that is, that there is enough RuBP regeneration, electron transport capacity and Rubisco activity to support CO_2 fixation and photorespiration simultaneously, without the net rate CO_2 assimilation being affected. Sharkey (1985b) has argued that this could arise if conversion of triose-P to sucrose and starch were limiting photosynthesis, and two recent investigations (Leegood and Furbank, 1986; Sharkey *et al.*, 1986) have confirmed that transferring spinach leaves from 21% to 2% O_2 in O_2-insensitive conditions leads to sequestration of Pi in precursors for starch and sucrose synthesis (e.g., Glc6P, UDPGlc), an accumulation of PGA, and a lowering of the ATP/ADP quotient, as expected if photosynthesis is limited by the recyling of Pi. Feeding Pi to leaves at least partially restores sensitivity to O_2 (Leegood and Furbank, 1986). Conversely, feeding mannose to sequester Pi in the cytosol and induce a Pi limitation (Herold, 1984) leads to appearance of O_2 insensitivity in conditions in which untreated leaves are sensitive to O_2 (Harris *et al.*, 1983).

A third approach involves using a brief interruption of photosynthesis to transiently increase the Pi level in the cytoplasm of the leaf (Stitt, 1986b). When light is decreased, the rate of electron transport changes immediately, and CO_2 fixation slows within 2–3 sec, but sucrose synthesis continues for 30–60 sec at the expense of preexistent pools of phosphorylated intermediates in the cytosol and stroma, only stopping when the cytosolic metabolite pools are depleted to the point where the cytosolic FBPase and SPS are inhibited. As the metabolites are depleted, the free Pi rises by 100–200 nmol/mg Chl, equivalent to 5–10 mM more Pi. If Pi availability restricts the rate of photosynthesis, this transiently enhanced Pi pool should allow a temporary stimulation of photosynthesis above the steady-state rate after returning the leaves back to high light. When leaves in saturating light and CO_2 are subjected to 30 sec in low light and then returned to high light, there is a transient enhancement of photosynthesis (Stitt, 1986b) that is 30–100% above the steady-state rate and that lasts for 20–30 sec. During this period of enhanced photosynthesis, the ATP/ADP ratio rises two- to threefold, the PGA/triose-P ratio decreases two- to threefold, and RuBP rises, showing how electron transport and the turnover of the Calvin cycle have been both stimulated.

A fourth line of evidence comes from the study of photosynthetic oscillations. These appear in conditions of saturating light and CO_2 and can be triggered by increasing the CO_2 or lowering O_2 (Sivak and Walker, 1985, 1986), or by a short period in the dark (Stitt, 1986b). These oscillations are decreased when Pi is supplied to leaves, and increase when mannose is supplied to sequester Pi (Walker and Sivak, 1985; Sivak and Walker, 1986), suggesting that their appearance is linked to a limitation of photosynthesis by Pi. One possible explanation for these oscillations is suggested by the finding that $Fru2,6P_2$ is oscillating out of phase with photosynthesis (Stitt, 1987b). These changes of $Fru2,6P_2$ could be altering the rate at which Pi is returning to the chloroplast, leading to a series of cycles in which the Pi limitation on photosynthesis is being relieved and then reimposed.

It is likely that a deficiency in sucrose synthesis is mainly responsible for the limitation by low Pi, as sucrose is the major end-product of photosynthesis in barley and spinach, which were the plants used in the experiments. Indeed, the capacity of the key enzymes of sucrose synthesis measured in optimal conditions in spinach leaf extracts is not much higher than the maximal rates of photosynthesis that spinach leaves attain in saturating light and CO_2. The activity of SPS and the cytosolic FBPase are 30–40 and 108 μmol hexose/mg Chl h, respectively, at 20°C. In comparison, maximal rates of photosynthesis in spinach at 20°C lie between 220 and 240 μmol CO_2/mg Chl h (Stitt et al., 1987), and sucrose still accounts for about 75% of the photosynthate in these conditions (Stitt et al., 1984c), which would be equivalent to a rate of sucrose synthesis of about 30 μmol hexose/mg Chl h. Thus, the activity of SPS would be almost exhausted, and fluxes equivalent to about 30% of the V_{max} of the cytosolic FBPase would be needed. In the presence of Fru2,6P$_2$ and AMP, it is unlikely that the FBPase would exceed this rate in vivo (Stitt and Heldt, 1985c).

Photosynthesis will probably be limited before the maximal theoretical capacity for sucrose synthesis is exhausted. There is a conflict between the requirement for rapid sucrose synthesis and the requirements for rapid electron transport and CO_2 fixation, which is exacerbated as the rate of photosynthesis increases. Sucrose synthesis is activated by high concentrations of phosphorylated intermediates and inhibited by Pi, but the increasing triose-P/Pi ratio that is needed to achieve faster sucrose synthesis will decrease the rate at which Pi can return to the chloroplast via the phosphate translocator. When the rate of photosynthesis are low, this dilemma can be overcome because "coarse" control of SPS and changes of Fru2,6P$_2$ concentration allow the rate of sucrose synthesis to be stimulated without a large increase of the metabolite concentration having to occur. However, as the rate of photosynthesis approaches the maximal capacity for sucrose synthesis, the ratios of triose-P/Pi and of hexose-P/Pi that are needed to sustain activity of the cytosolic FBPase and SPS will be increased even further, increasing the possibility that recycling of Pi starts to limit the rate of photosynthesis.

This suggests that the maximum rate of sucrose synthesis may depend not only on the enzymatic capacity of the cytosol, but also on the total amount (free plus esterified) of phosphate in the cytosol. If the total phosphate in the cytosol rises, then it will be possible to maintain higher levels of metabolites to activate enzymes, and simultaneously more Pi for the phosphate translocator. These considerations provide a framework to understand how depleting plants of Pi restricts sucrose synthesis (Foyer and Spencer, 1986) and how sucrose synthesis and photosynthesis can be stimulated by feeding Pi to increase the total amount of phosphate in the cytosol (see above).

2. Rate Limitation by Fru2,6P₂ and SPS

If the maximal rates of photosynthesis are already limited by sucrose synthesis, any further increase of $Fru2,6P_2$ or decrease of SPS activity should lead to a direct inhibition of O_2 evolution. This is illustrated by three recent studies of events during the induction of photosynthesis. When barley leaves are illuminated there is a transient doubling of $Fru2,6P_2$, followed by a 20-fold decline over the next 5–10 min (Sicher *et al.,* 1986). This high $Fru2,6P_2$ would be expected to restrict sucrose synthesis during the initial minutes in the light, leading to a restriction of photophosphorylation and PGA reduction. In agreement, there is a marked temporary accumulation of PGA and lowering of the ATP/ADP ratio during the induction of photosynthesis in spinach leaves in some conditions (Prinsley and Leegood, 1986a) as well as in protoplasts (Stitt *et al.,* 1980). Figure 10 summarizes measurements of O_2 evolution, metabolite and $Fru2,6P_2$ levels, and the light activation of stromal and cytosolic enzymes during the induction of photosynthesis in barley leaves (Stitt, 1987b). The induction of photosynthesis in these conditions is clearly biphasic [Fig. 10(a)]. During the first 2 min the chloroplast enzymes activate [Fig. 10(b)] and metabolites rise rapidly [Fig. 10(c)]. High ATP/ADP ratios are maintained, PGA is efficiently reduced, and high rates of O_2 evolution are achieved. However, the rate of sucrose synthesis is still slow, as $Fru2,6P_2$ is present at very high levels [Fig. 10(d)] and SPS has not yet been activated [Fig. 10(e)]. Consequently, Pi cannot be recycled, and once the Calvin cycle pools are filled the O_2 evolution almost stops, the ATP/ADP ratio decreases sharply, and there is a massive accumulation of PGA [Fig. 10(f)]. During the next 5–10 min there is a gradual recovery of the rate of O_2 evolution as the $Fru2,6P_2$ level decreases 20-fold and the activity of SPS increases, allowing synthesis of sucrose to commence and Pi to be recycled. Although the original rates of photosynthesis are achieved following this recovery, the metabolic conditions are very different. In the first burst of photosynthesis, when the Calvin cycle and electron transport are operating independently of sucrose synthesis, there is a high ATP/ADP ratio and high activation of Rubisco. When steady state is reached and the chloroplast and cytosol metabolism are brought into balance, this has not only involved an activation of sucrose synthesis, but also a regulation of chloroplast metabolism that is revealed by the far lower ATP/ADP ratio and the deactivation of Rubisco.

This experiment illustrates an extreme, where the chloroplast processes rapidly become fully active, while the cytosolic reactions needed to recycle Pi are still inactive, and was achieved by using a species where little starch is made and by reilluminating following only a short period in the dark. The

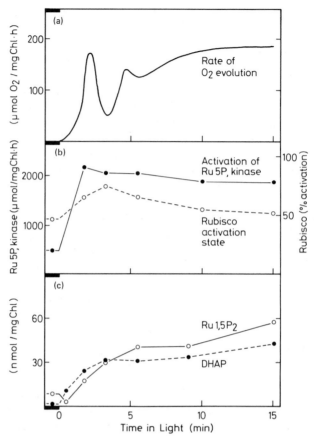

Fig. 10. O_2 evolution, metabolite levels, Fru2,6P$_2$, and activation of Calvin-cycle enzymes and SPS during induction of photosynthesis in barley leaves. Barley was pretreated in low light for 4 h, then 5 min dark, and illuminated in saturating light and CO_2 at 15°C in a leaf disk electrode. (a) Rate of O_2 evolution. (b) Activation of Ru5P,kinase (●) and Rubisco (○). The fully activated Rubisco had an activity of 610 μmol/mg Chl h. (c) Ru1,5P$_2$ (○) and DHAP (●). (d) Fru2,6P$_2$. (e) Activation of SPS, assayed with 2 mM Fru6P, 8 mM Glc6P, 2 mM UDPGlc, pH 7.4. (f) ATP/ADP ratio (○) and PGA (●) content. Note the axis for the PGA is 6.7-fold less amplified than in the figure for Ru1,5P$_2$ and DHAP.

magnitude of the initial burst of O_2 evolution and the subsequent restriction of PGA reduction varies greatly depending on plant material, the length of time that it is left in the dark, and the light intensity used (Prinsley and Leegood, 1986a). These treatments probably all modify the relation between metabolism in the chloroplast and the cytosol. Thus, an extreme limitation by Pi is avoided if CO_2 fixation is activated more slowly, or if the rate of CO_2 fixation is anyway lower, or if sucrose synthesis can be activated more rapidly.

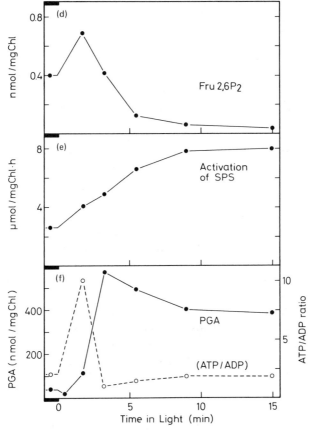

Fig. 10. (*Continued*)

3. Influence of Temperature

A limitation of photosynthesis by phosphate can also be induced by lowering the temperature. At a given light and CO_2 concentration, lowering the temperature leads to O_2 insensitivity (Cornic and Louason, 1980; Leegood and Furbank, 1986; Sharkey *et al.*, 1986), the appearance of a transient enhancement following a short interruption in low light (Stitt, 1986b), and increased sensitivity to stimulation after feeding Pi and inhibition after feeding mannose (M. Stitt, unpublished). Why does low temperature predispose photosynthesis toward being limited by phosphate? Do chloroplasts require higher Pi at low temperatures, or is sucrose synthesis particularly sensitive to temperature, so that Pi decreases at lower temperatures? There is evidence that both could occur.

When isolated chloroplasts photosynthesize at 4°C, optimal rates of photosynthesis require higher concentrations of Pi in the medium than at 20°C

(Machler *et al.*, 1984). The reason for the higher Pi requirement is not known, although a restriction of transport is indicated by the observation that metabolites accumulate in the stroma at low temperatures. Kinetic studies of the temperature response of the phosphate translocator were previously difficult, because the translocator activity above 10°C is too high for measurements using the silicon-oil centrifugation method, but they should now be possible, using a modified approach that has recently been developed by Horowitz and McCarty (1985).

There is also evidence for a selective restriction of sucrose synthesis at low temperatures, as the cytosolic FBPase becomes more sensitive to $Fru2,6P_2$ and AMP, leading to an accumulation of phosphorylated intermediates in the leaf (Sections III, IV, and VIII). As discussed above, maintaining higher metabolite levels may favor Calvin-cycle turnover at moderate rates of photosynthesis. However, it will predispose a leaf toward limitation by Pi as the rate of photosynthesis rises in response to enhanced light or CO_2 availability. This response of sucrose synthesis and Pi availability to changes of temperature may contribute toward the complex response of C_3 photosynthesis to alterations of light and temperature (Berry and Bjorkman, 1984). Often, there is little temperature dependence at low light intensities. Instead, low temperature decreases the light intensity at which photosynthesis becomes saturated. The relatively high rate of photosynthesis in low light intensities and low temperatures may be partly due to the higher metabolite levels in the stroma, which are made possible by an adaptive restriction of sucrose synthesis. Conversely, the lowering of the maximal rate of photosynthesis at low temperature is not due to a ceiling set by electron transport as previously thought, but may involve a capping of the rate of photosynthesis as Pi becomes increasingly limiting. These changes of metabolites and Pi are likely to interact with control of Rubisco activity and with electron transport. For example, the decreased activation of Rubisco at low temperature (Machler and Nosberger, 1980) could be a response to the primary limitation by Pi, Rubisco activation is known to decrease in isolated chloroplasts in limiting Pi (Heldt *et al.*, 1978).

4. Response of Chloroplasts to low Pi

Although low Pi decreases photosynthesis in leaves as well as isolated chloroplasts, the way in which low Pi produces this restriction is not understood. The simplest explanation would be that the decreasing ATP/ADP ratio found in these conditions (Heldt *et al.*, 1977) acts as a signal that Pi is decreasing. Falling Pi would lead to a decrease of the ATP/ADP ratio even if the phosphorylation potential remained constant and would then, via a mass action effect, lead to a restriction of PGA reduction and an accumulation of PGA (Robinson and Walker, 1979b).

The decline of the ATP/ADP ratio might, in addition, reflect an inhibition of the thylakoid ATPase. Even though the apparent concentrations of Pi (2 mM) in Pi-limited chloroplasts (Heldt *et al.*, 1977, 1978) is well above the

TABLE V
Estimated Fluxes and Concentration Gradients of
Triose-P and Pi during Photosynthesis in Isolated
Spinach Chloroplasts in Media Containing Different
Levels of Pi[a]

		Pi in medium at start (mM)			
		1	0.5	0.25	0.1
CO_2 evolution (μmol/mg Chl h)		91	107	106	52
Average net flux	Triose-P export	9	17	14	7
	Pi uptake	14	24	21	9
Triose-P concentration (mM)	Stroma	0.17	0.33	0.40	0.25
	Medium	0.02	0.048	0.041	0.021
Pi concentration (mM)	Stroma	9.6	7.0	4.0	2.2
	Medium	0.96	0.43	0.19	0.08
Apparent concentration gradient (mM)	Triose-P export	−0.14	−0.28	−0.36	−0.23
	Pi uptake	+8.64	+6.57	+3.81	+2.12
$\left(\dfrac{\text{Stromal triose-P}}{\text{Medium triose-P}}\right) \Big/ \left(\dfrac{\text{stromal Pi}}{\text{medium Pi}}\right)$		0.65	0.42	0.46	0.42

[a] The data are from Heldt et al. (1977).

K_m(ATP) (0.6 mM) of the ATP synthase (Afalo and Shavit, 1983; Selman and Selman-Reimer, 1981), some of this Pi may be bound. The finding that Rubisco binds Pi (Parry et al., 1985) means that millimolar concentrations of Pi-binding sites are potentially available in the stroma. Further, evaluation of the concentration gradients of Pi and triose-P across the envelope membrane of isolated chloroplasts (Heldt et al., 1977, 1978) leads to an anomaly that can be resolved if considerable amounts of the stromal Pi were to be bound. The phosphate translocator catalyzes a passive exchange of triose-P and Pi (Section IV,A), but inspection of the published experiments summarized in Table V reveals that the concentration gradient for movement of Pi into the stroma is considerably larger than the gradient for movement of triose-P into the medium. If all of the triose-P and Pi were free, there should be a net uptake of triose-P and release of Pi, which is the opposite of the measured fluxes during photosynthesis. This anomaly would disappear if a considerable proportion of the stroma Pi were to be bound.

Irrespective of the primary site of action, studies with isolated chloroplasts show that Pi affects many aspects of chloroplast metabolism, including electron transport, phosphorylation, Rubisco, and RuBP regeneration. For example, in low Pi the reduction of PGA is restricted, showing that NADPH and ATP generation by electron transport is restricted (Heldt et al., 1977; Edwards and Walker, 1983), and there is a decrease in the activation of Rubisco (Heldt et al., 1977), whose direct cause is still unknown. The resulting changes of metabolites that accompany low Pi also affect several enzymes. For example, Rubisco activity may be lowered directly by low Pi (Bhagwat, 1981; Parry *et al.*, 1985) as well as by the high PGA that accumulates (Badger and Lorimer, 1981). High PGA also inhibits Ru5P,kinase (Gardemann *et al.*, 1983) and may interfere with the light activation of several Calvin-cycle enzymes (Laing *et al.*, 1981). The changed PGA/Pi ratio also stimulates starch synthesis (Preiss, 1980) so that more Pi is recycled within the stroma.

These widespread responses in isolated chloroplasts are matched by the studies in leaves, showing that events in the cytosol can modify electron transport and carbon metabolism in the chloroplast. As discussed above, this is revealed by changes of O_2 evolution, CO_2 fixation, fluorescence, light scattering, the ATP/ADP and PGA/triose-P ratios, pool sizes of RuBP, the activation state of Rubisco, and starch synthesis. Many of these may represent regulatory responses, which adjust chloroplast metabolism to the signals returning from the cytosol in the form of a changed availability of Pi and metabolites. Much remains to be learned about the details of this adaptation and about the extent to which they may modify the rate and efficiency of the reactions in the stroma and thylakoids.

5. *Limitation by Pi in Leaves*

The experiments described above all involved moving plants into conditions where higher rates of photosynthesis were possible, or the temperature was lower, than in the conditions where they grew and do not show that sucrose synthesis exerts a direct limitation on the rate of photosynthesis in the field. Nevertheless, a direct limitation could sometimes occur in plants subjected to large temperature changes during the day, as O_2 insensitivity appeared at 12°C in air (Leegood and Furbank, 1986). It might be noted that a limitation by Pi could also develop, or be exacerbated, if reactions occurring in the chloroplast were to be modified so that higher levels of Pi were required to allow rapid photosynthesis. For example, O_2 insensitivity appears in water-stressed leaves (Sharkey, 1985a). Studies showing that the rate of sucrose synthesis does not always respond to changes in the rate of photosynthesis in field conditions also suggest there may be a limitation in sucrose synthesis. For example, short- or long-term CO_2 enrichment of field-grown soybean plants results in increased rates of photosynthesis but the rate of assimilate export (i.e., sucrose synthesis) remains unchanged (Huber *et al.*, 1984a). Instead, the additional carbon fixed as a result of the CO_2 enrichment (about 30% increase when CO_2 is doubled) is partitioned primarily into leaf starch. Thus, the capacity for sucrose synthesis in soybean leaves may be already saturated under normal conditions (350 ppm CO_2, high light), implying that larger increases in photosynthesis with CO_2 enrichment might be possible if the capacity for sucrose synthesis could be increased.

However, the frequency with which extreme limitation occurs may be less important than two other questions. First, how often may a more subtle colimitation by sucrose synthesis occur? Although we have much to learn, chloroplasts clearly possess regulatory mechanisms that would allow them to adjust to a potential limitation in the cytosol. To what extent are these mechanisms "engaged" during normal photosynthesis, and do they modify the efficiency with which light, CO_2, electron-transport capacity, or Rubisco can be used? For example, chloroplasts and leaves typically have very low ATP/ADP ratios during steady-state photosynthesis, even though high ratios can be generated by isolated thylakoids, or by chloroplast in leaves in special situations like induction (see above). It is also striking that the activation state of Rubisco may be adjusted to optimize pools of RuBP and PGA (Mott *et al.*, 1984). Do such observations reflect a continual modification of thylakoid and stromal processes to maintain a balance with other aspects of carbon metabolism, including sucrose synthesis?

The second question revolves around acclimatization. Our discussion has been restricted to short-term limitation of photosynthesis, when sucrose synthesis can no longer be regulated to maintain levels of metabolites and Pi that allow the available light, CO_2, electron-transport capacity, and Calvin-cycle capacity to be fully utilized. This contrasts with a long-term limitation, in which the development of the overall photosynthetic capacity may be modified. It has been suggested that long-term adaptation of leaves may play an important part in adjusting the assimilation rate to changes in "demand" for photosynthate (Geiger and Fondy, 1985). There is a possibility that some of the developments described in this chapter may allow such adjustments to be studied. For example, following a manipulation to impose a limitation of photosynthesis by Pi, there is a slow readjustment which allows an "escape" from this one-sided limitation. This is seen as a leaf slowly "escapes" from mannose feeding (Sivak and Walker, 1986), or as O_2 sensitivity is regained when a leaf is illuminated in 2% O_2 for 2 h (Sharkey *et al.*, 1986). Recovery from a short-term limitation could occur because a plant adapts, allowing end products to be synthesized faster or chloroplasts to adapt to lower levels of Pi. However, the plant may instead decrease its capacity for electron transport or for regeneration and carboxylation of RuBP, until these processes again colimit photosynthesis (Sharkey *et al.*, 1986). This process would be equivalent to a "stress" and would lead to a plant whose rate of photosynthesis is no longer limited in the short-term by "demand" for sucrose, but nevertheless is limited by sucrose synthesis in the long-term sense.

REFERENCES

Afalo, C., and Shavit, N. (1983). *FEBS Lett.* **154**, 175–179.
Akazawa, T., and Okamoto, K. (1980). *In* "The Biochemistry of Plants" (J. Priess, ed.), Vol. 3, pp. 199–219. Academic Press, New York.
Amir, J., and Preiss, J. (1982). *Plant Physiol.* **69**, 1027–1030.

ap Rees, T. (1980). *In* "The Biochemistry of Plants" (J. Preiss, ed.), Vol. 3, pp. 1–42. Academic Press, New York.

ap Rees, T., Green, J. H., and Wilson, P. M. (1985). *Biochem. J.* **227**, 299–304.

Austin, R. B. (1972). *Photosynthetica* **6**, 123–132.

Azcon-Bieto, J. (1983). *Plant Physiol.* **73**, 681–686.

Badger, M. R., and Lorimer, G. (1981). *Biochemistry* **20**, 2219–2225.

Badger, M. R., Sharkey, T. D., and von Caemmerer, S. (1984). *Planta* **160**, 305–313.

Baldry, C. W., Bucke, C., and Walker, D. A. (1966). *Biochim. Biophys. Acta* **502**, 207–213.

Barber, G. (1985). *Plant Physiol.* **79**, 1127–1128.

Baysdorfer, C., and Bassham, J. A. (1984). *Plant Physiol.* **74**, 374–379.

Berry, J., and Bjorkman, O. (1984). *Annu. Rev. Plant. Physiol.* **31**, 491–543.

Bhagwat, A. S. (1981). *Plant Sci. Lett.* **23**, 197–206.

Bird, I. F., Cornelius, M. T., Keys, M. J., and Whittigham, C. P. (1974). *Phytochemistry* **13**, 59–64.

Black, C. C., Jr., Carnal, N. W., and Kenyon, W. H. (1985a). *In* "Crassulaccan Acid Metabolism" (I. P. Ting and M. Gibbs, eds.), pp. 51–68. Waverly Press, Baltimore, Maryland.

Black, C. C., Jr., Carnal, N. W., and Paz, N. (1985b). *In* "Regulation of Carbon Partitioning in Photosynthetic Tissues" (R. L. Heath and J. Preiss, eds.), pp. 45–62. Waverley Press, Baltimore, Maryland.

Briens, M., and Larher, F. (1983). *Z. Pflanzenphysiol.* **110**, 447–458.

Buchanan, B. B. (1980). *Annu. Rev. Plant Physiol.* **31**, 341–374.

Bucke, C. (1970). *Phytochemistry* **9**, 1303–1309.

Chen, T. M., Dittrich, P., Campbell, W. H., and Black, C. C. (1974). *Arch. Biochem. Biophys.* **163**, 246–262.

Claussen, W., and Biller, E. (1977). *Z. Pflanzenphysiol.* **81**, 189–198.

Cornic, G., and Louason, G. (1980). *Plant, Cell Environ.* **3**, 149–157.

Cseke, C., and Buchanan, B. B. (1983). *FEBS Lett.* **155**, 139–142.

Cseke, C., Weenden, N. F., Buchanan, B. B., and Uyeda, K. (1982). *Proc. Natl. Acad. Sci. U. S. A.* **79**, 4322–4326.

Cseke, C., Stitt, M., Balogh, A., and Buchanan, B. B. (1983). *FEBS Lett.* **162**, 103–106.

Cseke, C., Balogh, A., Wong, J. H., Buchanan, B. B., Stitt, M., Herzog, B., and Heldt, H. W. (1984). *Trends Biochem. Sci.* **9**, 533–535.

Dietz, K. J., and Foyer, C. (1986). *Planta* **167**, 376–381.

Dietz, K. J., and Heber, U. (1984). *Biochim. Biophys. Acta* **767**, 422–443.

Dietz, K. J., Niemanis, S., and Heber, U. (1984). *Biochim. Biophys. Acta* **767**, 444–449.

Dietz, K. J., Schreiber, U., and Heber, U. (1986). *Planta* **166**, 219–226.

Doehlert, D. C., and Huber, S. C. (1983a). *FEBS Lett.* **153**, 293–296.

Doehlert, D. C., and Huber, S. C. (1983b). *Plant Physiol.* **73**, 989–994.

Doehlert, D. C., and Huber, S. C. (1984). *Plant Physiol.* **76**, 250–253.

Doehlert, D. C., and Huber, S. C. (1985). *Biochim. Biophys. Acta* **830**, 267–273.

Downton, W. J., and Hawker, J. S. (1973). *Phytochemistry* **12**, 1551–1556.

Ebbighausen, H., Hatch, M. D., Lilley, R. McC., Kromer, S., Stitt, M., Heldt, H. W. (1987). *In* "Plant Mitochondria: Structural, Functional and Physiological Aspects" (A. L. Moore and R. B. Beechey, eds.). Plenum, New York (in press).

Edwards, G. E., and Black, C. C. (1971). *In* "Photosynthesis and Photorespiration" (M. D. Hatch, C. B. Osmond, and R. O. Slatyer, eds.), pp. 153–168. Wiley (Interscience), New York.

Edwards, G. E., and Huber, S. C. (1981). *In* "The Biochemistry of Plants" (M. D. Hatch and N. K. Boardman, eds.), Vol. 8, pp. 238–282. Academic Press, New York.

Edwards, G. E., and Walker, D. A. (1983). "C_3C_4: Mechanisms, and Cellular and Environmental Control of Photosynthesis." Blackwell, Oxford.

Edwards, G. E., Foster, J. G., and Winter, K. (1985). *In* "Crassulacean Acid Metabolism" (I. P. Ting and M. Gibbs, eds.), pp. 91–111. Waverley Press, Baltimore, Maryland.

Farquahar, G. D., and von Caemmerer, S. (1982). *In* "Encyclopedia of Plant Physiology, New Series" (O. L. Lange, P. S. Nobel, C. B. Osmond, and H. Ziegler, eds.), Vol. 12B, pp. 549–587. Springer-Verlag, Berlin and New York.
Farrar, S. C., and Farrar, J. F. (1985). *New Phytol.* **100**, 271–284.
Fisher, D. B., and Outlaw, W. H., Jr. (1979). *Plant Physiol.* **64**, 481–483.
Flügge, U.-I., and Heldt, H. W. (1984). *Trends Biochem. Sci.* **9**, 550–553.
Flügge, U.-I., Friesl, M., and Heldt, H. W. (1980). *Plant Physiol.* **65**, 574–577.
Flügge, U.-I., Gerber, J., and Heldt, H. W. (1983). *Biochim. Biophys. Acta* **725**, 229–237.
Fondy, B. R., and Geiger, D. R. (1980). *Plant Physiol.* **66**, 528–533.
Fondy, B. R., and Geiger, D. R. (1982). *Plant Physiol.* **70**, 671–676.
Foyer, C., and Spencer, C. (1986). *Planta* **167**, 369–375.
Foyer, C., Walker, D. A., and Latzko, E. (1982a). *Z. Pflanzenphysiol.* **107**, 457–465.
Foyer, C., Walker, D. A., Spencer, D., and Mann, B. (1982b). *Biochem. J.* **202**, 429–434.
François, J., van Schaftingen, E., and Hers, H.-G. (1983). *Eur. J. Biochem.* **134**, 269–273.
François, J., van Schaftingen, E., and Hers, H.-G. (1985). *Eur. J. Biochem.* **145**, 187–193.
Furbank, R. T., Stitt, M., and Foyer, C. H. (1985). *Planta* **164**, 172–178.
Ganson, N. J., and Fromm, H. J. (1982). *Biochem. Biophys. Res. Commun.* **108**, 233–239.
Gardeman, A., Stitt, M., and Heldt, H. W. (1983). *Biochim. Biophys. Acta* **722**, 51–60.
Gardemann, A. G., Schimkat, D., and Heldt, H. W. (1986). *Planta* **168**, 538–545.
Geiger, D. R., and Fondy, B. R. (1985). *Monogr.—Br. Plant Growth Regul. Group* **12**, 177–194.
Gerhardt, R., and Heldt, H. W. (1984). *Plant Physiol.* **75**, 547–554.
Gerhardt, R., Stitt, M., and Heldt, H. W. (1987). *Plant Physiol.* 399–407.
Giersch, C., Heber, U., Kaiser, G., Walker, D. A., and Robinson, S. P. (1980). *Arch. Biochem. Biophys.* **205**, 246–259.
Gifford, R. M. (1977). *Aust. J. Plant Physiol.* **4**, 99–110.
Gordon, A. G., Ryle, G. J. A., and Webb, G. (1980). *J. Exp. Bot.* **31**, 845–850.
Gross, P., and ap Rees, T. (1986). *Planta* **167**, 140–145.
Harbron, S., Foyer, C., and Walker, D. A. (1981). *Arch. Biochem. Biophys.* **212**, 236–246.
Hampp, R., Goller, M., and Ziegler, H. (1982). *Plant Physiol.* **69**, 448–455.
Hampp, R., Goller, M., Füllgraf, H., and Eberk, I. (1985). *Plant Cell Physiol.* **26**, 99–108.
Handley, L. W., Pharr, D. M., and McFeeters, R. F. (1983). *J. Am. Soc. Hortc. Sci.* **108**, 600–605.
Harris, G. C., Cheeseborough, J. K., and Walker, D. A. (1983). *Plant Physiol.* **71**, 102–107.
Harrsch, P. B., Kim, Y., Fox, J. L., and Marcus, F. (1986). *Biochem. Biophys. Res. Commun.* **135**, 374–381.
Hatch, M., and Osmond, C. B. (1976). *In* "Encyclopedia of Plant Physiology," New Series (C. R. Stocking and U. Heber, eds.), Vol. 3, pp. 144–184. Springer-Verlag, Berlin and New York.
Hawker, J. S. (1967). *Biochem. J.* **102**, 401–406.
Hawker, J. S., and Hatch, M. D. (1966). *Biochem. J.* **99**, 102–106.
Hawker, J. S., and Smith, G. M. (1984). *Phytocyemistry* **23**, 245–249.
Heber, U., and Heldt, H. W. (1981). *Annu. Rev. Plant Physiol.* **32**, 139–168.
Heber, U., and Santarius, K. A. (1965). *Biochim. Biophys. Acta* **109**, 390–408.
Heber, U., and Santarius, K. A. (1970). *Z. Naturforsch., B: Anorg. Chem., Org. Chem., Biochem., Biophys. Biol.* **25B**, 718–727.
Heber, U., Pon, M. G., and Heber, M. G. (1963). *Plant Physiol.* **66**, 355–360.
Hedrich, R., Stitt, M., and Raschke, K. (1985). *Plant Physiol.*
Heldt, H. W., and Stitt, M. (1987). *Proc. Int. Conf. Photosynth., 7th,* **III**, 675–684.
Heldt, H. W., Chon, C. J., Maronde, D., Herold, A., Stankovic, Z. S., Walker, D. A., Kraminer, A., Kirk, M. R., and Heber, U. (1977). *Plant Physiol.* **59**, 1146–1155.
Heldt, H. W., Chon, C. J., and Lorimer, G. H. (1978). *FEBS Lett.* **92**, 234–240.
Herold, A. (1980). *New Phytol.* **86**, 131–144.

Herold, A. (1984). *In* "Storage Carbohydrates in Vascular Plants" (D. H. Lewis, ed.), pp. 181–204. Cambridge Univ. Press, London and New York.
Herold, A., McGee, E. E. M., and Lewis, D. M. (1980). *New Phytol.* **85**, 1–13.
Hers, H.-G., Hue, L., and van Schaftingen, E. (1982). *Trends Biochem. Sci.* **7**, 329–331.
Herzog, B., Stitt, M., and Heldt, H. W. (1984). *Plant Physiol.* **75**, 561–565.
Horowitz, K. T., and McCarty, R. E. (1985). *Biochemistry* **24**, 3645–3650.
Huber, S. C. (1981a). *Z. Pflanzenphysiol.* **101**, 49–54.
Huber, S. C. (1981b). *Z. Pflanzenphysiol.* **102**, 443–450.
Huber, S. C. (1986). *Annu. Rev. Plant Physiol.* **37**, 233–246.
Huber, S. C., and Bickett, M. (1984). *Plant Physiol.* **74**, 445–447.
Huber, S. C., and Moreland, D. E. (1981). *Plant Physiol.* **67**, 163–169.
Huber, S. C., Doehlert, D. C., Rufty, T. W., and Kerr, P. S. (1984a). *In* "Advances in Photosynthesis Research" (C. Sybesma, ed.), Vol. 3, pp. 605–608. Martinus Nijhoff/Dr. W. Junk, The Hague, The Netherlands.
Huber, S. C., Rufty, T. W., and Kerr, P. S. (1984b). *Plant Physiol.* **75**, 1080–1084.
Huber, S. C., Rogers, H. H., and Israel, D. W. (1984c). *Physiol. Plant.* **62**, 95–101.
Huber, S. C., Doehlert, D. C., Kerr, P. S., and Kalt-Torres, W. (1985a). *In* "Nitrogen Fixation and CO₂ Metabolism" (P. W. Ludden and J. E. Burris, eds.), pp. 399–407. Am. Elsevier, New York.
Huber, S. C., Kerr, P. S., and Kalt-Torres, W. (1985b). *In* "Regulation of Carbon Partitioning in Photosynthetic Tissue" (R. L. Heath and J. Preiss, eds.), pp. 199–214. Waverley Press, Baltimore, Maryland.
Huber, S. C., Kerr, P. S., and Rufty, T. W. (1985c). *Physiol. Plant.* **64**, 81–87.
Isherwood, F. A., and Selvendran, R. C. (1970). *Phytochemistry* **9**, 2265–2269.
Jolliffe, P. A., and Tregunna, E. B. (1973). *Can. J. Bot.* **51**, 841–853.
Kacser, H., and Burns, J. A. (1973). *Symp. Soc. Exp. Biol.* **27**, 65–104.
Kacser, H., and Burns, J. A. (1979). *Biochem. Soc. Trans.* **7**, 1149–1160.
Kaiser, G., and Heber, U. (1984). *Planta* **161**, 562–568.
Kaiser, G., Martinoia, E., and Wiemken, A. (1982). *Z. Pflanzenphysiol.* **107**, 103–113.
Kaiser, W. M. (1982). *Planta* **154**, 538–545.
Kaiser, W. M., Schwitulla, M., and Wirth, E. (1983). *Planta* **158**, 302–308.
Kalt-Torres, W., Kerr, P. S., and Huber, S. C. (1987). *Physiol. Plant.* (in press).
Kerr, P. S., and Huber, S. C. (1987). *Planta* **170**, 197–204.
Kerr, P. S., Rufty, T. W., and Huber, S. C. (1985). *Plant Physiol.* **77**, 275–280.
Kerr, P. S., Israel, D. W., Huber, S. C., and Rufty, T. W., Jr. (1986). *Can. J. Bot.* **64**, 2020–2028.
Klemme, B., and Jacobi, G. (1974). *Planta* **120**, 147–153.
Klingenberg, M., and Heldt, H. W. (1982). *In* "Metabolic Compartmentation" (H. Sies, ed.), pp. 101–122. Academic Press, London.
Kreimer, G., Melkonian, M., Holtum, J. A. M., and Latzko, E. (1985). *Planta* **166**, 515–522.
Kruger, N. J., and Beevers, H. (1985). *Plant Physiol.* **77**, 358–364.
Laing, W. A. L., Stitt, M., and Heldt, H. W. (1981). *Biochim. Biophys. Acta* 348–359.
Laisk, J. A., and Pärnik, T. (1977). *Photosynthetica* **11**, 251–259.
Larondelle, Y., Mertens, E., Van Schaftingen, E., and Hers, H.-G. (1986). *Eur. J. Biochem.* **161**, 351–357.
Leegood, R. L. (1984). *Photosynth. Res.* **6**, 247–259.
Leegood, R. C. (1985). *Planta* **164**, 163–171.
Leegood, R. C., and Furbank, R. T. (1986). *Planta* **168**, 84–93.
Leloir, L. F., and Cardini, C. E. (1955). *J. Biol. Chem.* **214**, 157–165.
Lewis, D. H. (1984). *In* "Storage Carbohydrates in Vascular Plants" (D. H. Lewis, ed.), pp. 1–52. Cambridge, Univ. Press, London and New York.
Lilley, R. McC., Stitt, M., Mader, G., and Heldt, H. W. (1982). *Plant Physiol.* **70**, 965–970.

Loescher, W. H., Fellmar, J. K., Fox, T. C., Davis, J. M., Redgewell, R. T., and Kennedy, R. A. (1985). *In* "Regulation of Carbon Partitioning in Photosynthetic Tissues" (R. L. Heath and J. Preiss, eds.), pp. 333—357. Waverley Press, Baltimore, Maryland.

MacDonald F., Cseke, C., Chou, Q., and Buchanan, B. B. (1987). *Proc. Natl. Acad. Sci. U. S. A.* (in press).

Machler, F., and Nosberger, J. (1980). *J. Exp. Bot.* **31,** 1485–1491.

Machler, F., Schnyder, H., and Nosberger, J. (1984). *J. Exp. Bot.* **35,** 481–487.

McVetty, P. B. E., and Canvin, D. T. (1981). *Can. J. Bot.* **59,** 721–725.

Marcus, C. J., Geller, A. M., and Byrne, W. L. (1973). *J. Biol. Chem.* **248,** 8567–8573.

Mbaku, S. B., Fritz, G. J., and Bowes, G. (1978). *Plant Physiol.* **62,** 510–515.

Mendicino, J. (1960). *J. Biol. Chem.* **235,** 3347–3352.

Meyer, R., and Wagner, K. G. (1985). *Anal. Biochem.* **148,** 269–276.

Miller, A. J., and Saunders, D. (1987). *Nature* (in press).

Mott, K. A., Jensen, R. G., O'Leary, J. W., and Berry, J. A. (1984). *Plant Physiol.* **76,** 968–971.

Mourioux, G., and Douce, R. (1981). *Plant Physiol.* **67,** 470–473.

Murata, T. (1972). *Agric. Biol. Chem.* **36,** 1877–1884.

Neales, T. F., and Incoll, L. D. (1968). *Bot. Rev.* **34,** 107–125.

Nimmo, H., and Tipton, K. F. (1975a). *Eur. J. Biochem.* **58,** 567–574.

Nimmo, H., and Tipton, K. F. (1975b). *Eur. J. Biochem.* **58,** 875–885.

Nomura, T., and Akazawa, T. (1974). *Plant Cell Physiol.* **15,** 477–483.

Osmond, C. B. (1978). *Annu. Rev. Plant Physiol.* **29,** 379–414.

Outlaw, W. H., Jr., and Fisher, D. B. (1975). *Plant Physiol.* **55,** 599–603.

Outlaw, W. H., Jr., Fisher, D. B., and Christy, A. L. (1975). *Plant Physiol.* **55,** 704–711.

Parry, M. A. J., Schmidt, C. N. G., Cornelius, M. J., Keys, A. T., Millard, B. N., and Gutteridge, S. (1985). *J. Exp. Bot.* **36,** 1396–1404.

Paul, J. S., Kronwell, K. L., and Bassham, J. A. (1978). *Planta* **142,** 49–54.

Paz, N., Wu, X.-P., and Black, C. C., Jr. (1986). *Plant Physiol.* **79,** 1133–1136.

Pharr, D. M., and Sox, H. N. (1984). *Plant Sci. Lett.* **35,** 187–193.

Pollock, C. J. (1984). *In* "Storage Carbohydrates in Vascular Plants" (D. H. Lewis, ed.), pp. 97–114. Cambridge Univ. Press, London and New York.

Pollock, C. J., and Housley, T. L. (1985). *Ann. Bot. (London)* [N.S.] **55,** 593–596.

Pontremoli, S., and Horecker, B. L. (1970). *Curr. Top. Cell. Regul.* **2,** 173–199.

Preiss, J. (1980). *Annu. Rev. Plant Physiol.* **33,** 431–454.

Preiss, J., and Greenberg, E. (1969). *Biochem. Biophys. Res. Commun.* **36,** 289–295.

Preiss, J., Robinson, N., Spilatro, S., and McNamara, K. (1985). *In* "Regulation of Carbon Partitioning in Photosynthetic Tissue" (R. L. Heath and J. Preiss, eds.), pp. 1–26. Waverley Press, Baltimore, Maryland.

Prinsley, R. T., and Leegood, R. C. (1986a). *Biochim. Biophys. Acta* **849,** 244–253.

Prinsley, R. T., and Leegood, R. C. L. (1986b). *Biochim. Biophys. Acta* **849,** 254–263.

Prinsley, R. T., Hunt, S., Smith, A. M., and Leegood, R. C. (1986). *Planta* **167,** 414–420.

Rea, P. A., and Poole, R. J. (1985). *Plant Physiol.* **77,** 46–52.

Rebeille, F., Bligny, R., Martin, J.-B., and Douce, R. (1983). *Arch. Biochem. Biophys.* **225,** 143–148.

Robinson, J. M. (1984). *Plant Physiol.* **75,** 397–409.

Robinson, J. M., and Baysdorfer, C. (1985). *In* "Regulation of Carbon Partitioning in Photosynthetic Tissue" (R. L. Heath and J. Preiss, eds.), pp. 333–357. Waverley Press, Baltimore, Maryland.

Robinson, S. P., and Walker, D. A. (1979a). *FEBS Lett.* **107,** 295–299.

Robinson, S. P., and Walker, D. A. (1979b). *Biochim. Biophys. Acta* **545,** 528–536.

Rufty, T. W., and Huber, S. C. (1983). *Plant Physiol.* **72,** 474–480.

Rufty, T. W., Kerr, P. S., and Huber, S. C. (1983). *Plant Physiol.* **73,** 428–433.

Rufty, T. W., Huber, S. C., and Kerr, P. S. (1984a). *Plant Sci. Lett.* **30,** 7–12.

Rufty, T. W., Huber, S. C., and Kerr, P. S. (1984b). *Plant Sci. Lett.* **34**, 247–252.
Salerno, G., and Pontis, H. (1977). *Arch. Biochem. Biophys.* **180**, 298–302.
Salerno, G., and Pontis, H. (1978). *Planta* **142**, 41–48.
Santarius, K. A., and Heber, U. (1965). *Planta* **102**, 39–54.
Scheibe, R., and Jacquot, J. P. (1983). *Planta* **157**, 548–553.
Sellami, A. (1976). *Biochim. Biophys. Acta* **423**, 524–539.
Selman, B. R., and Selman-Reimer, S. (1981). *J. Biol. Chem.*, 1722–1726.
Setter, T. L., Brun, N. A., and Brenner, M. L. (1980). *Plant Physiol.* **65**, 884–887.
Sharkey, P. J., and Pate, J. S. (1976). *Planta* **128**, 63–72.
Sharkey, T. D. (1985a). *Plant Physiol.* **78**, 71–75.
Sharkey, T. D. (1985b). *Bot. Rev.* **51**, 53–105.
Sharkey, T. D., Stitt, M., Heineke, D., Gerhardt, R., Raschke, K., and Heldt, H. W. (1986). *Plant Physiol.* **81**, 1123–1129.
Sicher, R. C., and Kremer, D. F. (1984). *Plant Physiol.* **76**, 910–912.
Sicher, R. C., and Kremer, D. F. (1985). *Plant Physiol.* **79**, 695–698.
Sicher, R. C., Kremer, D. F., and Harris, W. G. (1986). *Plant Physiol.* **82**, 15–18.
Sivak, M. N., and Walker, D. A. (1985). *Plant, Cell Environ.* **8**, 439–448.
Sivak, M. N., and Walker, D. A. (1986). *New Phytol.* **102**, 499–512.
Slabnick, E., Frydman, R. B., and Cardini, C. E. (1968). *Plant Physiol.* **43**, 1063–1068.
Slack, C. R. (1968). *Biochem. Biophys. Res. Commun.* **30**, 483–488.
Smyth, D. C., Wu, M. X., and Black, C. C., Jr. (1984). *Plant Physiol.* **75**, 862–864.
Soll, J., Wötzel, C., and Buchanan, B. B. (1985). *Plant Physiol.* **77**, 999–1003.
Stitt, M. (1985a). *Monogr.—Br. Plant Growth Regul. Group*, **12**, 35–50.
Stitt, M. (1985b). *In* "Photosynthetic Carbon Partitioning" (R. L. Heath and J. Preiss, eds.), pp. 109–126. Waverley Press, Baltimore, Maryland.
Stitt, M. (1986a). *In* "Phloem Transport" (J. Cronshaw, ed.), pp. 331–347. Alan R. Liss, Inc., New York.
Stitt, M. (1986b). *Plant Physiol.* **81**, 1115–1122.
Stitt, M. (1987a). *Plant Physiol.* (in press).
Stitt, M. (1987b). *Proc. Int. Conf. Photosynth.*, *7th*, **III**, 685–692.
Stitt, M., and Heldt, H. W. (1985a). *Planta* **164**, 179–188.
Stitt, M., and Heldt, H. W. (1985b). *Biochim. Biophys. Acta* **808**, 400–414.
Stitt, M., and Heldt, H. W. (1985c). *Plant Physiol.* **79**, 599–608.
Stitt, M., Wirtz, W., and Heldt, H. W. (1980). *Biochim. Biophys. Acta* **593**, 85–102.
Stitt, M., Lilley, R. McC., and Heldt, H. W. (1982a). *Plant Physiol.* **70**, 971–977.
Stitt, M., Mieskes, G., Söling, H.-D., and Heldt, H. W. (1982b). *FEBS Lett.* **145**, 217–222.
Stitt, M., Wirtz, W., and Heldt, H. W. (1983a). *Plant Physiol.* **72**, 767–774.
Stitt, M., Gerhardt, R., Kürzel, B., and Heldt, H. W. (1983b). *Plant Physiol.* **72**, 1139–1141.
Stitt, M., Cseke, C., and Buchanan, B. B. (1984a). *Eur. J. Biochem.* **143**, 89–93.
Stitt, M., Herzog, B., and Heldt, H. W. (1984b). *Plant Physiol.* **75**, 548–558.
Stitt, M., Kürzel, B., and Heldt, H. W. (1984c). *Plant Physiol.* **75**, 559–560.
Stitt, M., Cseke, C., and Buchanan, B. B. (1985a). *Physiol. Veg.* **23**, 819–827.
Stitt, M., Herzog, B., and Heldt, H. W. (1985b). *Plant Physiol.* **75**, 792–799.
Stitt, M., Wirtz, W., Gerhardt, R., Heldt, H. W., Spencer, C. A., Walker, D. A., and Foyer, C. (1985c). *Planta* **166**, 354–364.
Stitt, M., Cseke, C., and Buchanan, B. B. (1986a). *Plant Physiol.* **80**, 246–248.
Stitt, M., Mieskes, G., Söling, H.-D., Grosse, H., and Heldt, H. W. (1986b). *Z. Naturforsch. C: Biosci.* **41C**, 291–296.
Stitt, M., Gerhardt, R., Wilke, I., and Heldt, H. W. (1987). *Physiol. Plant.* **69**, 377–386.
Stocking, C. R. (1959). *Plant Physiol.* **34**, 56–61.
Usuda, H., and Edwards, G. E. (1980). *Plant Physiol.* **65**, 1017–1022.
Uyeda, K., Furuya, E., and Ricards, C. S. (1982). *Mol. Cell. Biochem.* **48**, 97–120.
Van Schaftingen, E., and Hers, H.-G. (1981). *Proc. Natl. Acad. Sci. U. S. A.* **78**, 2861–2863.

Van Schaftingen, E., Lederer, B., Batrons, R., and Hers, H.-G. (1983). *Eur. J. Biochem.* **129,** 191–195.

Walker, D. A., and Leigh, R. A. (1981). *Planta* **151,** 150–155.

Walker, D. A., and Sivak, M. N. (1985). *In* "Regulation of Carbon Partitioning in Photosynthetic Tissue" (R. L. Heath and J. Preiss, eds.), pp. 93–108. Waverley Press, Baltimore, Maryland.

Waterton, J. C., Bridges, I. C., and Irving, M. P. (1983). *Biochim. Biophys. Acta* **763,** 315–320.

Weenden, N., and Buchanan, B. B. (1983). *Plant Physiol.* **72,** 259–261.

Weiner, H., Stitt, M., Heldt, H. W. (1987). *Biochim. Biophys. Acta* (in press).

Whipps, J., and Lewis, D. H. (1981). *In* "Effects of Disease on the Physiology of the Growing Plant" (P. G. Ayres, ed.), pp. 47–83. Cambridge Univ. Press, London and New York.

Whitaker, D. P. (1984). *Phytochemistry* **23,** 2429–2430.

Wirtz, W., Stitt, M., and Heldt, H. W. (1980). *Plant Physiol.* **66,** 187–193.

Woodrow, I. E., Ellis, J. R., Jellings, A., and Foyer, C. (1984). *Planta* **161,** 525–530.

Yamashoji, S., and Hess, B. (1985). *FEBS Lett.* **178,** 253–256.

Zimmermann, G., Kelly, G. J., and Latzko, E. (1978). *J. Biol. Chem.* **253,** 5952–5956.

Index

A

Activase
  ribulose 1,5-bisphosphate carboxylase,
    activation of, 194–195
Adenine nucleotides
  cytosolic fructose bisphosphatase, inhibi-
    tion of, 352, 363, 367
  cytosolic levels, 363–367
  light, effect of, 346
  response to low P_i, 400–402
  transport, chloroplast, 363
  transport, mitochondria, 363
Adenosine-5'-triphosphate synthase, 49
  amino acid sequences, 103–104
  function, 101
  gene coding of subunits, 103–104
  limitation by P_i, 400–401
  lipids, 84
  localization, 106
  mitochondrial and chloroplast, compari-
    son, 103
  model, 100–101
  polypeptides, 102
  rotational diffusion, 119
  structure, 102
  synthesis, 61

B

Algae
  carbon dioxide concentrating mechanism
    in, 238–247, 254–264
  chromophyte, 11
5-Aminolevulinate,
  cell-free synthesis, 35
  glutamyl-RNA, 35
  light regulation, 36
  synthase, 34–35
  synthesis, 34–35
    C_5 pathway, 34–35
    inhibition by hemin, 36
    porphyrin precursor, 34
Anatomy, leaf, C_3–C_4 intermediates, 281–
  284
Angiosperms, carbon dioxide concentrating
  mechanism in, 254–262

B

Bicarbonate ion, carbondioxide concentrating mechanism, role, 264
Blue-light receptor,
  anthocyanin formation, 31
  gene transcription, 31
  phytochrome action, 33

C

C₃ plants
  carbondioxide assimilation pathway, 276
  leaf anatomy, 281
C₃–C₄ intermediate species
  anatomy, leaf, 281–284
  biochemistry, 290–307
  C₄ pathway enzymes, 291–292, 296–303
  C₄ photosynthesis in, 290–292, 298–304
  carbon dioxide compensation point in,
    285–289
  carbon dioxide concentrating mechanism,
    299, 303, 307
  carbon dioxide release in the light, 285
  carbon isotope composition, 307–310
  carboxylation efficiency, 289, 301, 304,
    307
  definition, 276–279
  developmental factors, 310–312
  ecological distribution, 280
  ecotypic differentiation, 312
  environmental factors, 310–312
  evolution, 316–320
  interspecific hybrids, 312–315
  leaf ultrastructure, 284
  light effect, 288–289, 311
  mitochondria, role, 284, 286
  occurrence, 279–281
  oxygen inhibition of photosynthesis, 285–
    288, 311
  photorespiration, 284–285, 290–307
  photorespiratory CO₂, refixation, 290, 298,
    302–306
  photorespiratory enzymes, 294
  photosynthesis, pathways, 290–307
  quantum yield, 300, 303, 304, 307
  taxonomic distribution, 278–281
C₄ acid decarboxylases, C₃–C₄ intermediates, levels, 296–303
C₄ pathway
  C₃–C₄ intermediates, relation, 290–292,
    298–304
  carbondioxide concentrating mechanism,
    260–262
  enzymes of, in C₃–C₄ intermediates, 291–
    292, 296–303
  ribulose 1,5-bisphosphate carboxylase,
    inhibition 207–208
C₄ plants
  carbon dioxide assimilation pathway, 277
  carbon dioxide concentrating mechanism,
    260–262, 277

  cytosolic metabolites, *see* Metabolites,
    cytosolic levels
  description, 277
  evolution, 316
  fructose, 2,6-bisphosphate, 368, 373
  Kranz cells, 277
  Kranz syndrome, 278–279
  starch synthesis, intercellular distribution,
    331
  sucrose synthesis
    adaptation of, 336, 354, 390
    intercellular distribution, 331–333
CAM, *see* Crassulacean acid metabolism
Carbamylation, ribulose 1,5-bisphosphate
  carboxylase, 153–158, 191–192, 200
Carbon dioxide compensation point,
  C₃–C₄ intermediates, 285–289
  oxygen effect, 285–289
  photosynthetic pathway, effect, 285–289
Carbon dioxide concentrating mechanism
  aquatic phototrophs, in, 219–274
    angiosperms, 262–264
    components of, 224–226
    ecological importance of, 264–269
    freshwater angiosperms and giant algae,
      254–262
    green microalgae, *see* Carbondioxide
      concentrating mechanism, green
      microalgae
    historical perspectives, 221–223
    induction, 247–254
    marine algae and angiosperms, 262–264
    physiological significance, 223–224
  cyanobacteria, in, 226–238
    energetics, 231–235
    inorganic carbon pumping, 226–231
    interconversion of inorganic carbon
      species, 237–238
    leakage of carbon dioxide, 235–237
  general aspects
    C₃–C₄ intermediates, 299, 303, 307
    C₄ plants, 277, 307
    components of, 224–226
    history, 221–223
    role, physiological, 223–224
  green microalgae, 238–247
    carbon dioxide leakage, 244–245
    energetics, 242–243
    inorganic carbon pumping, 239–242
    interconversion of carbon species, 245–
      247
Carbon dioxide enrichment, sucrose synthesis, 402

Carbon isotope ratios, C^{13} in C_3–C_4 intermediates, 307–310
Carbonic anhydrase
  carbon dioxide concentrating mechanism, role in, 237–238, 245–247
  induction, in aquatic phototrophs, 222, 248
  role, in bicarbonate utilization, 237–238, 245–247
Carboxy-3-keto-D-arabinitol 1,5-bisphosphate, ribulose 1,5-bisphosphate carboxylase, 169–175, 182, 189, 200–201
Carboxylation efficiency, C_3-C_4 intermediates, 289, 301, 304, 307
Carboxysomes
  carbon dioxide concentrating mechanism, role in, 230–231
  ribulose 1,5-bisphosphate carboxylase, activity, 208–209
Carotenoids, chloroplast envelope, 40
Chlamydomonas
  5-aminolevulinate synthesis, 34
  chlorophyll synthesis, 36
  chlorophyll *a/b* protein genes, 30, 31
  D1 protein, 12
  D2 protein, 12
  deoxyribonucleic acid replication, 14
  genome, chloroplast, 10
  genome D-loops, 14
  glycerolipid synthesis, 43
  light-harvesting chlorophyll *a/b* proteins, 59
  photosystem 2 assembly, 57
  plastocyanin mRNA, 27
  psbA gene, 12
  psbD gene, 12
  ribosomal protein synthesis, chloroplast, 27
  ribosomal RNA genes, 11
  ribulose, 1,5-bisphosphate carboxylase, small subunit genes, 22
    processing, 47
  synthesis of chlorophyllide *b*, 39
Chlorella, deoxyribonucleic acid, 14
Chlorophyll
  fluorescence, 107, 116, 118
  P680, 93
  synthesis,
    inhibition by hemin, 36
    pathway, 34
    regulation, 36
Chlorophyll *a*, 38
  binding to polypeptides, 89
Chlorophyll *b*,
  synthesis, 39
  thylakoid membrane assembly, role, 59
Chlorophyll RC I, 38, 56, 93
Chlorophyll synthase, 38
Chlorophyllide *a*,
  formation from protochlorophyllide, 37
  phytylation, 38
Chlorophyllide *b*, 39
Chloroplast envelope
  chlorophyll synthesis, 60
  composition, 40
  enzymes, 43
  inner and outer membranes, isolation of, 40
  lipid synthesis, 41–42
  nucleoside triphosphatase, 41
  permeability, 40
  polypeptides, 44
  transport activities, 41
Chloroplast genes, 4–13
  expression, 17–19
    D1 protein, 90–91
    light regulation, 18
    psbA gene sequence, 90
    regulation, 14
  homology, 5
  hydropathy plots, 89
  identification, 4–5
  introns, 11
  library, 5
  light induction, 4
  mutants, *Chlamydromonas,* 39
  products, 10
  promoters, 16–17
  psbB gene, 89
  psbC gene, 89
  psbD gene, 91
  restriction fragments, 4
  sequences, 5, 11, 89
Chloroplast genome, 4–15, 89–91
  algal, 10
  coding capacity, 5
  conservation, 9
  D-loops, 14
  higher plants, 5
  maps, 8
  nucleotide sequence, 5
  replication, 14
  size, 5
Chloroplast proteins,
  precursor, 44
  processing, 44, 46

synthesis, 18–31
transport, 45, 54
Chloroplasts, *see also* Chloroplast envelope;
    Chloroplast genes; Chloroplast genome;
    Chloroplast proteins
cytochromes, 49–53
deoxyribonucleic acid, 6–8
    genetic map, 6
    gene products, 6
    hybridization, 16
    transcription, 16
evolution, 11
genetic map, 4
nuclear genes, role of, 21–31
prokaryotic origin, 9
protein synthesis, 18–31
    posttranscriptional control, 20
respone in low P_i, 400–402
ribonucleic acid, 4–5
    ribosomal genes, 8
    transfer, 5, 13
ribosomal proteins, 6, 12–16
    cDNA library, 28
ribosomal RNA
    genes, 12
    synthesis, 15
RNA polymerase, inhibition, 16
"S" factor, 16
transfer RNA genes, 13
Compartmentation, metabolites
    distribution, 340
    methods, 339
Control, fine and coarse, *see also* Sucrose
        synthesis, regulation of
    definition, 328
    sucrose phosphate synthase, 355, 359,
        376, 377–382
Coupling factor (CF_1), gene transcripts, 19
Crassulacean acid metabolism,
    fructose 2,6-bisphosphate, 368, 373
    intracellular compartmentation, 336, 345,
        391
    scheme, 336
    sucrose synthesis, adaptation, 337, 391
Cyanobacteria, carbondioxide concentrating
        mechanism in, 226–268
Cytochrome b_6, *see* Cytochrome b_{559}
Cytochrome b_6/f complex, 49–52, 84
    function in electron transport, 97
    gene coding, 98
    localization, 50, 106
    model, 99
    polypeptides, 98
    quinone binding, 98

guinone content, 50
structure, 97
Cytochrome b_{559},
    genes, 90
    location, 50
Cytochrome b_{563}, 49–50, 98
    coding, 51
    messenger RNA, 52
    location in thylakoid membrane, 51
    site of synthesis, 51
Cytochrome f
    coding, 51–52
    gene sequence, 99
    molecular weight, 49
    location in thylakoid membranes, 51, 52,
        99
    precursor, 52
    reaction kinetics in electron transport, 115
    redox potential, 98
    site of synthesis, 51–52
    synthesis, 61
Cytosolic metabolites, *see* Metabolites,
    cytosolic levels

D

D1 protein, homology, 12
Diurnal rhythms, *see also* Control, fine and
        course
    comparison of species, 386–387
    cytosolic metabolites, 344
    fructose 2,6-bisphosphate, 372
    fructose 6-P,2-kinase, 370–371
    starch synthesis, 375, 386–387
    sucrose accumulation, 375
    sucrose phosphate synthase, 359–360,
        375–381, 386–387
Divinyl protochlorophyllide, 38
D-loops, *see* Chloroplast genome

E

Ecology, *see* Carbon dixoide concentrating
        mechanism, ecological importance of
Electron transport, photosynthetic, 97–98
    kinetics, 85, 86, 115
    limitation by sucrose synthesis, 392, 394–
        396, 397
    photosystem 1, 93
    Z-scheme, 77
Etioplasts
    galactolipid synthesis, 43

gene expression, light regulation, 18
protein uptake, 46
protochlorophyllide reductase, 37
Euglena
chloroplast genome, 11, 12
ribosomal RNA genes, 11
Evolution
C_3–C_4 intermediates, 316–320
C_4 pathway, 316–320

F

Ferredoxin-NADP oxidoreductase, localization, 106
Ferredoxin mRNA, cDNA, 26
Fluorescence, chlorophyll, 107
Fructans, storage carbohydrate, 337
Fructose bisphosphatase, cytosolic
role, 328–329, 351, 354
free energy change, 341
purification, 351
regulatory properties, 351–354
spinach, 351–354
temperature, effect of, 354, 391–392, 400
Fructose 2,6-bisphosphatase, *see also* Fructose 6-P,2-kinase
activity, 374–375
distribution, 369
properties, 369
Fructose 2,6-bisphosphate,
compartmentation, 368
degradation, *see* Fructose 2,6-bisphosphatase,
distribution, 368
extraction, 367–368
fructose bisphosphatase, inhibition, 352
induction of photosynthesis, 397
photosynthate partitioning, role, 372, 375, 386–389
photosynthesis, response to, 371, 373, 383–385
photosynthetic oscillations, 395–396
pyrophosphate : fructose 6-P-phosphotransfrase, 362–363
stress, response to, 373
synthesis, *see* Fructose 6-P,2-kinase
temperature, response to, 371
Fructose 6-P,2-kinase
distribution, 369
metabolites, response to, 369
protein modification, 370, 375
regulation, 369–371, 373–375

G

Galactinol synthase, activity, 337–338
Galactolipids, *see* Lipids
chloroplast envelope, 40, 42
etioplasts, 43
galactosyl transferase, 43
synthesis, 43
Glutamyl-RNA, 35, 36
Glycolate pathway, 135
Glycolipids, synthesis, 42
Grana
appressed, 105, 106
fatty acid composition, 83
Grana thylakoids
appressed, electron spin resonance, 112
fluidity, 112
Guard cells, fructose 2,6-bisphosphate, effect of, 371, 373

I

Induction
carbon dioxide concentrating mechanism, 247–254
photosynthesis
activation of sucrose phosphate synthase, 397
autocatalysis, 394, 397
changes of fructose 2,6-bisphosphate, 397
metabolite levels, 397
sucrose synthesis, limitation, 397
Inorganic carbon, *see also* Carbon dioxide concentrating mechanism, 219–269
diffusion in solution, 220–221
interconversion of forms, 220–221, 237–238, 245–247
Interspecific hybrids,
C_3 and C_4 species, 312–315
C_4 species and intermediates, 312–315

L

Large subunit, *see* Ribulose 1,5-bisphosphate carboxylase-oxygenase, large subunit
Leaf anatomy,
C_3–C_4 intermediate species, 281–284
C_4 plants, 277, 281–284
LHC, *see* Light-harvesting chlorophyll
Light activation,
Calvin cycle, enzymes, 397–398, 401–402

malate dehydrogenase, NADP⁺, 365
ribulose 1,5-bisphosphate carboxylase activation, 393–400
sucrose phosphate synthase, 359, 377–378, 386

Light-harvesting chlorophyll *a/b* proteins
amino acid sequence, 23
apoproteins, 19
association with chlorophyll, 60
coding, 57–58
encoding by nuclear DNA, 21
genes, 23, 95
   light regulation, 29, 30, 31
   mRNA, 21, 29–30
   promoters of, 29
hydropathy plots, 23
mRNA in bundle-sheath cells, 19
orientation in membrane, 23
pigment composition, 57
polypeptide heterogeneity, 23
polypeptide precursors, 44
posttranslational modifications, 23
posttranslational transport, 48
precursors, 44
reconstitution, 57
sequences, 57
species variations in genes, 23

Light regulation
gene expression, 19

Lipid synthesis, 42, 43
chloroplast envelope, 41, 42

Lipids, thylakoid, *see* Thylakoid membranes, lipids

Lithium ion, cytosolic fructose bisphophatase inhibition, 353

M

Magnesium-2,4-divinyl pheophorbide a₅, 37

Magnesium ions,
cytosolic fructose bisphosphatase, effect on, 352
sucrose phosphatase, effect on, 360
sucrose-P synthase, effect on, 356, 358

Malate dehydrogenase, NADP⁺, light activation, 365

Mannose, phosphate sequestration, 365, 394–395

Metabolite compartmentation
general distribution, 340
methods, 339

Metabolites, cytosolic levels
C₃ plants, 342–344
C₄ plants, 343
carbon dioxide, effect of, 343
during induction, 397
light effect, 342
Pᵢ, 344–346
PPᵢ, 361
sucrose, 346–348

Mitochondria,
adenosine triphosphate transport, 363
C₃–C₄ intermediates, role in, 284–296
control of phosphorylation potential, 352, 363
external NADH dehydrogenase, 367
oxaloacetate transport, 366
photorespiratory metabolism, role, 135
protein import, 45

Mutants, chlorophyll *b*-less, 39

N

NADP : protochlorophyllide oxidoreductase
gene expression, 30, 32
mRNA, 38
synthesis, 38

Nitrogen, supply, carbon dioxide concentrating mechanism, 252, 268–269

NMR, determination of Pᵢ, 340, 344

Nuclear gene expression, chloroplast proteins
light control, 29, 30, 31
mRNA, 30
regulation, 28–31
transcription with isolated nuclei, 30

Nuclear genes, encoding chloroplast proteins, 21–31

O

Oxygen
carbon dioxide concentrating mechanism, effect of, 251
inhibition of photosynthesis
   C₃–C₄ intermediates, 285–288, 301, 304, 307, 311
   C₄ plants, 285–288

Oxygen insensitive photosynthesis, 394

Oxygenase reaction
ribulose 1,5-bisphosphate carboxylase, 133–139

discovery, 133–134
role, hypothesis, 138

P

P680, 57
  kinetics, 85
P700, 56
  binding protein, 93
  electron spin resonance, 93
  light-harvesting complex polypeptides, 94
  light-induced absorption change, 93
  midpoint potential, 93
Partitioning of photosynthate,
  cytosolic metabolites, 337, 347
  diurnal changes, 375
  fructose 2,6-bisphosphate, role of, 372,
    375, 386–389
  sucrose phosphate synthase, role of, 380–
    381, 386–389
PCR cycle, see Photosynthetic carbon re-
    duction cycle
Peroxisomes, photorespiratory metabolism,
    role, 135
pH
  stromal pH, 350, 351
  phosphate translocation, effect on, 350
Phosphate (ortho)
  determination by NMR, 340, 344
  fluxes, during photosynthesis, 333–335
    C_4 plants, 332, 336, 390
    CAM, 337, 391
  fructose 6-P,2-kinase, activation by, 369,
    373–374
  intracellular compartmentation, 344–346
  photosynthesis, limitation, 394–403
  Pyrophosphate : fructose 6-P-phospho-
    transferase inhibition, 362–363
  seqestration by mannose, 345, 394–395
  sucrose-P synthase inhibition, 353
  transport, see Phosphate translocator
  tonoplast, 345, 396
Phosphate translocator
  kinetic limitation by, 351, 396
  properties, 350
  role, 349–351
  temperature, influence of, 400
Phosphoenolpyruvate carboxylase
  C_3–C_4 intermediates, level, 296–299
  genes, 27
  mRNA, 19

Phosphoglucose isomerase, equilibrium,
    stroma, 340
Phospholipids, see Thylakoid membranes,
    lipids
Photorespiration
  adenylate status, cytosol, 363–367
  C_3–C_4 intermediates, in, 284–285
  carbon dioxide concentrating mechanism,
    effect of, 223–224
  carbon dioxide refixation in C_3–C_4 inter-
    mediates, 290, 298, 302–306
  energetics, 135
  enzymes of, in C_3–C_4 intermediates,
    294
  mechanism, 134–139
  oxygenase in, 134–139, 207
  stoichiometry, 135
  suppression, mechanisms, 290–307
Photosynthate
  cytosolic metabolites, 337, 347
  diurnal changes, 375
  partitioning, see Partitioning of photosyn-
    thate
  sucrose phosphate synthase, 380–381,
    386–389
Photosynthesis, see also C_4 plants; Induc-
    tion, photosynthesis
  C_3–C_4 intermediate species, 276–279
  C_4 pathway, 276
  carbon dioxide concentrating mechanism
    energy for, 233–234, 243
    role of, 251–252
  light intensity effect, 109
  oscillations of, 395–396
  oxygen insensitive, 394
  phosphate limitation, 394–403
  quantum yield, 109
  state 1-state 2, 110, 116, 117
  sucrose, effect on, 346–347, 387
Photosynthetic carbon reduction cycle, C_3–
    C_4 intermediates, relation, 276–279
Photosynthetic electron transport, see Elec-
    tron transport
Photosystem 1
  coding, 56
  complexes, 84
  kinetics of electron transport, 93
  light-harvesting complex, 49
    polypeptides, 56, 58
  lipids, 83
  localization, 106
  reaction center, 38
  reaction center apoproteins, 19

418

Index

reaction center complex, 49, 56
    genes, 95
    polypeptides, 94, 95
    synthesis, 60
    thylakoid membrane complexes, 49
Photosystem 2
    antenna size, 106
    bundle-sheath cells, activity, 19
    coding, 56
    complexes, 84
    core proteins, 12
    D2 protein, 90, 91, 93
    herbicide-binding protein, 17–18
        gene, (psbA), 90
    light-harvesting chlorophyll a/b proteins,
        49, 57, 83, 95, 96
        diffusion in thylakoid membrane, 116
        energy distribution, 117–118
        genes, 95
        phosphorylation, 116, 117
        rotational diffusion, 119
        structure, 96–97
    light-harvesting complex, translational
        control, 20
    lipids, 83–84
    localization, 106
    oxygen-evolving complex, 50
        polypeptides, 86–87
    manganese removal, 87
    P680 protein, 18
    quinone binding, 88, 93
    reaction center apoprotein, 19
    reaction center complex, 49, 57
        kinetics, 85–86
        polypeptides, 56, 87–89
        proteins, 13
    reconstitution, 87
    thylakoid membrane complexes, 49
        polypeptides, 86
Photosystems
    antenna size, 110
    assembly, 56
    energy distribution, regulation, 117–118
    model, 58
    stoichiometry, light intensity effect, 110
Phycobiliprotein, genes, 11
Phytochrome, 31–33
    blue light effect, 33
    calcium ions, 33
    gene expression, regulation of, 19, 32
    genes, 32
    location, 33
    mechanism of action, 32

mRNA, 30–31
nuclear gene induction, 30, 32
ribulose-1,5-bisphosphate carboxylase
    gene regulation, 33
    structure, 32
Plastocyanin
    coding, 53
    diffusion
        coefficient, 115
        thylakoid membranes, 112
    electron transport, in, 115
    location, in thylakoid membranes, 53
    precursor, 26, 53
        cDNA, 26
        degradation, 27
        maturation, 27
    processing, 46
    reduction, 50
    structure, 53, 115
    synthesis, 53
    transport, 53, 54
Plastome, mutants, 13–14
Plastoquinol, diffusion in thylakoid mem-
    branes, 112
Plastoquinone-9
    diffusion coefficient, 114
    electron transport, role, 114–115
    lateral diffusion, in thylakoid membranes,
        114
    photosystem 2, 50
    thyalkoid membranes, in, 113–114
Polypeptides, chloroplast envelope, 41
Potassium ion
    fructose bisphosphatase, effect on, 353
    sucrose phosphate synthase, effect on,
        356
Prolamellar body, galactolipid synthesis, 43
Protochlorophyllide
    envelope membranes, 60
    inhibitor of 5-aminolevulinate formation,
        36
    photoconversion, 60
    reduction, 37
    synthesis, 36, 37
Protochlorophyllide oxidoreductase, 37
Proton pump, PP$_i$ dependent, 362
psbA gene
    promoter, 18
    transcription, 17
Pyrophosphatase, role, 362
Pyrophosphate
    role in metabolism, 361–362
    UDPG pyrophosphorylase, 330

Pyrophosphate : fructose 6-P-phosphotrans-
    ferase
    fructose 2,6-bisphosphate, assay, 367
    properties, 361–363
Pyruvate
    cDNA, 27
    precursor protein, 27
Pyruvate,orthophosphate dikinase, C_3-C_4
    intermediates, 297–303

Q

Quantum yield, C_3-C_4 intermediate species,
    300, 303, 304, 307
Quinone
    binding in photosystem 1, 98
    binding in photosystem 2, 88, 93
    electron transport, 98

R

Raffinose, activity, 338
Rhodopseudomonas viridis, reaction center
    structure, 76, 85, 92
Rhodospirullum rubrum, ribulose 1,5-bis-
    phosphate carboxylase, 144, 206
Ribonucleic acid polymerase, genes, 6
Ribosomes, membrane-bound, 60
Ribulose 1,5-bisphosphate
    carboxylase, substrate for, 131–218
    generation of
        limitation by metabolites, 393
        photosynthesis induction, 394, 397
Ribulose 1,5-bisphosphate carboxylase-
    oxygenase, 48, 131–218
    activase, 194–195
    activation state, during induction, 397
    amino acid sequence
        large subunit, 142–145
        small subunit, 145–147, 206
    C_3-C_4 intermediates, relation, 276
    carbamylation, 153–158, 191–192, 200
        mechanism, 153–155
        physical changes, 155–156
    carbondioxide, specificity versus oxygen,
        136–138, 181–188, 206–210
    carboxylase reaction, 133–134
    catalytic bifunctionality, 133–139
        oxygenase role, hypothesis, 138
    catalytic mechanism, 156–189
        affinity labels, 183–185

catalytic site, 183–187
decarboxylation, 173–174, 200
divalent metal ions, role, 175–177
enolization, 161–162, 164–165, 173, 200
hydrolysis, 174–175, 200
oxygenation, 177–181
partial reactions, 173–175, 200
reaction intermediates, 164–169
    2′carboxy-3-keto-D-arabinitol 1,5-
        bisphosphate, 166–169, 169–172,
        173–175, 182, 189, 200–201
    2,3-enediol(ate) of ribulose 1,5-bis-
        phosphate, 136, 164–165, 169–
        172, 181–183, 189
    2′-peroxy-3-keto-D-arabinitol 1,5-
        bisphosphate, 178–181, 182–183
    3-phosphoglycerate, forms of, 169,
        172–173, 189
reaction sequence
    carboxylation, 164
    oxygenation, 178
stereochemistry, 169–173
compartmentation, carboxysomes, 230–
    231
discovery of, 133
effector binding, 156–158, 190–191, 192–
    193, 199–201
    cooperativity, 157–158, 192
evolution, 205–209
improvement, 132–133, 138, 209–211
inhibition,
    2′-carboxy-D-arabinitol-1,5-bisphos-
        phate, by, 169–170, 175, 199–200
    2′-carboxy-D-arabinitol-1-phosphate, by,
        193
    COS, by, 159–160
    H_2O_2, by, 159, 179–180
    2′-peroxy-D-pentitol-1,5-bisphosphate,
        by, 159, 179–180
    time dependent, 159, 193–194
kinetic properties, aquatic phototrophs,
    226, 239, 263, 269
large subunit,
    binding protein, 151–153
    gene, 12, 19
    octamer, catalytic properties, 199–202
light, effect on activity, 394
mRNA, 19–20
mutagenesis
    selective, 211
    site-directed, 187–189, 210
orthophosphate, effect, 400–402
oxygenase reaction, 133–134

discovery of, 133–134
physiological cost, 139
suppression of, 223
photorespiration, role in, 134
photosynthesis, role, 134–139
regulation of activity, 189–195
  carbamylation, 191–192
  kinetics, time dependent, 193–194
  product inhibition, 190–191
  tight-binding inhibitors, 192–193
relative specificity for CO_2 and O_2, 136–
  138, 181–188, 195, 206–210
Rhodospirillum rubrum, from, 144–206
small subunit, 204–205
  degradation, 21, 29
  gene expression, 23
  gene variations, 22
  light regulation, 23, 29
  mRNA, 21
  nuclear DNA coding, 21
  posttranslational transport, 48
  precursors, 44–46, 47, 151
small subunit gene
  mRNA, 29
  promoters, 29–30
structure
  cyrstallography, 147–149, 210
  quaternary, 139–141, 195–196, 205–
    206
substrate binding, 156–158, 160–162, 180
  order of addition, 159–163
subunits
  assembly, 149–153
  hybrids, 202–204
  interactions, 195–205
  reassembly, 196–199, 202–204
  separation, 196–199
sucrose synthesis, effect of, 393–394, 397–
  400
synthesis, 149–153
  eukaryotes, 150–153
  prokaryotes, 149–150
temperature, effect on activity, 400
Ribulose 5-P-kinase, regulation by phosphoglycerate, 402
Rieske iron-sulfur protein, 49–50, 99
  location in membrane, 52
  reaction kinetics in electron transport, 115
  redox potential, 98
  transport, 53
Rubisco, *see* Ribulose 1,5-bisphosphate
  carboxylase-oxygenase
RuBP, *see* Ribulose-1,5-bisphosphate

S

S_{rel}, *see* Ribulose 1,5-bisphosphate carboxylase-oxygenase, relative specificity for
  CO_2 and O_2
Shade plants, photosynthetic antenna size,
  110
Starch synthesis
  diurnal rhythm, 375, 386–389
  location, 331
  pathway, 361
  regulation by P_i, 402
Stress
  fructose 2,6-bisphosphate, effect on, 373
  temperature, *see* Temperature
  water stress, 338, 402
Stroma thylakoids, 105, 107
  fatty acid composition, 83
  fluidity, 112
Sucrose
  accumulation of, 337, 348
  cytosolic concentration, 347–348
  intracellular compartmentation, 346–348
  photosynthesis, effect on, 346–347, 387
  sucrose phosphatase, inhibition of, 360,
    361
  sucrose phosphate synthase, inhibition of,
    358
  transport, tonoplast, 347
Sucrose phosphatase
  distribution, 360
  properties, 360–361
Sucrose phosphate
  levels, 340
  sucrose phosphate synthase, inhibition of,
    358
Sucrose phosphate synthase
  allosteric regulation, 357, 359
  compartmentation, 330, 331
  diurnal rhythm, 359–360, 375–381, 386–
    387
  equilibrium constant, 341–342
  induction of photosynthesis by, 394–403
  kinetic properties, activation, 359
  light activation, 359, 377–379
  limitation of photosynthesis by, 394–403
  molecular weight, 355, 358
  partitioning, role during, 380–381, 386–389
  sink manipulation, 381
  substrate saturation, 357
  sucrose, inhibition by, 358
  temperature, response to, 392
  thiol groups, 356, 357

Sucrose synthesis
  C$_4$ plants, 331, 336, 354, 390
  CAM plants, 337, 391
  free energy changes during, 340
  location of, 328–330
  photosynthesis, limitation by, 392–403
  regulation of, 382, 390, 396
  temperature, adaptation to, 391–392, 400
Sulfolipid, *see* Thylakoid membranes, lipids

T

Taxonomy, C$_3$–C$_4$ intermediates, 279–281
Temperature, effect on
  fructose bisphosphatase, 354, 391, 400
  fructose 2,6-bisphosphate, 371
  light saturation response, 400
  metabolites levels, 343
  photosynthesis, 394
  phosphate translocator, 400
  ribulose 1,5-bisphosphate carboxylase
    activation, 400
  sucrose synthase, 391–392, 400
Thiol groups, sucrose phosphate synthase,
  356, 357
Thylakoid membranes
  assembly, 48–61
  cation binding, 107
  complexes
    light intensity effect, 109
    spatial relationship, 113
    stoichiometry, 109
  conformational changes, 107–109
  cytochrome f, location, 51
  cytochrome b_{563}, location, 51
  diacylgalactosyl diglycerides, 42
  electron microscopy, 108
  fluidity, 110, 111
  fluorescence probes, 110, 111
  lipids, 78–80
    appressed thylakoids, 112
    bilayer structure, 80
    fatty acid composition, 79
    nonappressed thylakoids, 112
    structures, 80–83
  organization, 105–119
  origin, 339

phase changes, 111
photosystem 1, location, 50
photosystem 2, location, 50
plastocyanin diffusion, 112
plastoquinol diffusion, 112
protein transport, 47, 55
proteins
  coding, 84
  genes, 6
  lateral distribution, 105, 109
  membrane insertion, 84
  model for 43-kDa protein, 89
  phosphorylation, 116–117
  rotational diffusion, 119
  salt-induced changes, 107
reconstitution, 111
stacking, 105
structural organization, 83
structure, 105
surface charge density, 107–109
Thylakoid polypeptides synthesis, 6
Transport, inorganic carbon, *see* Carbon
  dioxide concentrating mechanism
Transport, metabolites
  adenine nucleotides, 363
  intracellular in C$_4$ plants, *see* C$_4$ plants
  oxaloacetate, mitochondria, 366
  phosphate, *see* Phosphate translocator
  sucrose, 347
Triose phosphates, in P$_i$ limited photosynthesis, 394–403

U

UDP-galactose : diacylglycerol galactosyl
  transferase, 43
UDP glucose pyrophosphorylase, sucrose
  synthesis, 330, 361

V

Vacuole
  phosphate transport, 345, 397
  sucrose compartmentation, in 346–348

Contents of Other Volumes

Volume 1—The Plant Cell

1. The General Cell
   Eldon H. Newcomb
2. Use of Plant Cell Cultures in Biochemistry
   Paul Ludden and Peter S. Carlson
3. The Primary Cell Walls of Flowering Plants
   Alan Darvill, Michael McNeil, Peter Albersheim, and Deborah P. Delmer
4. The Plasma Membrane
   Robert T. Leonard and Thomas K. Hodges
5. The Cytosol
   Grahame J. Kelly and Erwin Latzko
6. Development, Inheritance, and Evolution of Plastids and Mitochondria
   Jerome A. Schiff
7. Biochemistry of the Chloroplast
   Richard G. Jensen
8. Plant Mitochondria
   J. B. Hanson and D. A. Day
9. Microbodies—Peroxisomes and Glyoxysomes
   N. E. Tolbert
10. The Endoplasmic Reticulum
    Maarten J. Chrispeels
11. Ribosomes
    Eric Davies and Brian A. Larkins
12. The Golgi Apparatus
    Hilton H. Mollenhauer and D. James Morré
13. The Plant Nucleus
    E. G. Jordon, J. N. Timmis, and A. J. Trewavas
14. Protein Bodies
    John N. A. Lott
15. Plant Vacuoles
    Francis Marty, Daniel Branton, and Roger A. Leigh

16.   Cyanobacteria (Blue-Green Algae)
        C. Peter Wolk
Index

Volume 2—Metabolism and Respiration

1.   Assessment of the Contributions of Metabolic Pathways to Plant Respiration
        T. ap Rees
2.   Enzyme Flexibility as a Molecular Basis for Metabolic Control
        Jacques Ricard
3.   Direct Oxidases and Related Enzymes
        V. S. Butt
4.   Electron Transport and Energy Coupling in Plant Mitochondria
        Bayard T. Storey
5.   Nature and Control of Respiratory Pathways in Plants: The Interaction of
     Cyanide-Resistant Respiration with the Cyanide-Sensitive Pathway
        David A. Day, Geoffrey P. Arron, and George G. Laties
6.   Control of the Krebs Cycle
        T. Wiskich
7.   The Regulation of Glycolysis and the Pentose Phosphate Pathway
        John F. Turner and Donella H. Turner
8.   Hydroxylases, Monooxygenases, and Cytochrome P-450
        Charles A. West
9.   One-Carbon Metabolism
        Edwin A. Cossins
10. Respiration and Senescence of Plant Organs
        M. C. J. Rhodes
11. Respiration and Related Metabolic Activity in Wounded and Infected Tissues
        Ikuzo Uritani and Tadashi Asahi
12. Photorespiration
        N. E. Tolbert
13. Effects of Light on "Dark" Respiration
        Douglas Graham
14. Anaerobic Metabolism and the Production of Organic Acids
        David D. Davies
15. Effect of Low Temperature on Respiration
        John K. Raison
16. The Use of Tissue Cultures in Studies of Metabolism
        D. K. Dougall
Index

Volume 3—Carbohydrates: Structure and Function

1.   Integration of Pathways of Synthesis and Degradation of Hexose Phosphates
        T. ap Rees
2.   *myo*-Inositol: Biosynthesis and Metabolism
        Frank A. Loewus and Mary W. Loewus
3.   L-Ascorbic Acid: Metabolism, Biosynthesis, Function
        Frank A. Loewus
4.   Sugar Nucleotide Transformations in Plants
        David Sidney Feingold and Gad Avigad
5.   Branched-Chain Sugars: Occurrence and Biosynthesis
        Hans Grisebach

6. Biosynthesis and Metabolism of Sucrose
    Takashi Akazawa and Kazuo Okamoto
7. Occurrence, Metabolism, and Function of Oligosaccharides
    Otto Kandler and Herbert Hopf
8. Translocation of Sucrose and Oligosaccharides
    Robert T. Giaquinta
9. Structure and Chemistry of the Starch Granule
    W. Banks and D. D. Muir
10. Starch Biosynthesis and Degradation
    Jack Preiss and Carolyn Levi
11. Conformation and Behavior of Polysaccharides in Solution
    David A. Brant
12. Chemistry of Cell Wall Polysaccharides
    Gerald O. Aspinall
13. Structure and Function of Plant Glycoproteins
    Derek T. A. Lamport
14. The Biosynthesis of Cellulose
    J. Ross Colvin
15. Glycolipids
    Alan D. Elbein
16. Biosynthesis of Cell Wall Polysaccharides and Glycoproteins
    Mary C. Ericson and Alan D. Elbein
Index

Volume 4—Lipids: Structure and Function

1. Plant Acyl Lipids: Structure, Distribution, and Analysis
    J. L. Harwood
2. Membrane Lipids: Structure and Function
    John K. Raison
3. Degradation of Acyl Lipids: Hydrolytic and Oxidative Enzymes
    T. Galliard
4. The Role of the Glyoxylate Cycle
    Harry Beevers
5. Lipoxygenases
    T. Galliard and H. W.-S. Chan
6. Biosynthesis of Ethylene
    S. F. Yang and D. O. Adams
7. Biosynthesis of Saturated and Unsaturated Fatty Acids
    P. K. Stumpf
8. The Biosynthesis of Triacylglycerols
    M. I. Gurr
9. Phospholipid Biosynthesis
    J. B. Mudd
10. Phospholipid-Exchange Systems
    Paul Mazliak and J. C. Kader
11. Sulfolipids
    J. L. Harwood
12. Plant Galactolipids
    Roland Douce and Jacques Joyard
13. Biochemistry of Terpenoids
    W. David Loomis and Rodney Croteau

14. Carotenoids
    Sandra L. Spurgeon and John W. Porter
15. Biosynthesis of Sterols
    T. W. Goodwin
16. Sterol Interconversions
    J. B. Mudd
17. Biosynthesis of Acetate-Derived Phenols (Polyketides)
    N. M. Packter
18. Cutin, Suberin, and Waxes
    P. E. Kolattukudy
19. Biosynthesis of Cyclic Fatty Acids
    H. K. Mangold and F. Spener
Index

Volume 5—Amino Acids and Derivatives

1. Biochemistry of Nitrogen Fixation
   M. G. Yates
2. Ultrastructure and Metabolism of the Developing Legume Root Nodule
   J. G. Robertson and K. J. F. Farnden
3. Nitrate and Nitrite Reduction
   Leonard Beevers and Richard H. Hageman
4. Ammonia Assimilation
   B. J. Miflin and P. J. Lea
5. Assimilation of Inorganic Sulfate into Cysteine
   J. W. Anderson
6. Physical and Chemical Properties of Amino Acids
   Peder Olesen Larsen
7. Enzymes of Glutamate Formation: Glutamate Dehydrogenase, Glutamine Synthetase, and Glutamate Synthase
   G. R. Stewart, A. F. Mann, and P. A. Fentem
8. Aminotransferases in Higher Plants
   Curtis V. Givan
9. Synthesis and Interconversion of Glycine and Serine
   A. J. Keys
10. Arginine Synthesis, Proline Synthesis, and Related Processes
    John F. Thompson
11. Synthesis of the Aspartate Family and Branched-Chain Amino Acids
    J. K. Bryan
12. Sulfur Amino Acids in Plants
    John Giovanelli, S. Harvey Mudd, and Anne H. Datko
13. Aromatic Amino Acid Biosynthesis and Its Regulation
    D. G. Gilchrist and T. Kosuge
14. Histidine Biosynthesis
    B. J. Miflin
15. Amino Acid Catabolism
    Mendel Mazelis
16. Transport and Metabolism of Asparagine and Other Nitrogen Compounds within the Plant
    P. J. Lea and B. J. Miflin

17.   Accumulation of Amino Acids and Related Compounds in Relation to
      Environmental Stress
          G. R. Stewart and F. Larher
Index

Volume 6—Proteins and Nucleic Acids

1.   The Nuclear Genome: Structure and Function
          William F. Thompson and Michael G. Murray
2.   Enzymatic Cleavage of DNA: Biological Role and Application to Sequence
      Analysis
          S. M. Flashman and C. S. Levings III
3.   RNA: Structure and Metabolism
          T. A. Dyer and C. J. Leaver
4.   Biosynthesis of Nucleotides
          Cleon W. Ross
5.   DNA and RNA Polymerases
          Tom J. Guilfoyle
6.   Nucleic Acids of Chloroplasts and Mitochondria
          Marvin Edelman
7.   Proteins of the Chloroplast
          Katherine E. Steinback
8.   Plant Proteinases
          C. A. Ryan and M. Walker-Simmons
9.   Proteinase Inhibitors
          C. A. Ryan
10.   Lectins in Higher Plants
          Halina Lis and Nathan Sharon
11.   Seed Storage Proteins: Characterization and Biosynthesis
          Brian A. Larkins
12.   Protein Biosynthesis: Mechanisms and Regulation
          Donald P. Weeks
13.   Tumor Formation in Plants
          M. P. Gordon
14.   Biochemistry of Plant Viruses
          George Bruening
Index

Volume 7—Secondary Plant Products

1.   The Physiological Role(s) of Secondary (Natural) Products
          E. A. Bell
2.   Tissue Culture and the Study of Secondary (Natural) Products
          Donald K. Dougall
3.   Turnover and Degradation of Secondary (Natural) Products
          Wolfgang Barz and Johannes Köster
4.   Secondary Plant Products and Cell and Tissue Differentiation
          Rolf Wiermann
5.   Compartmentation in Natural Product Biosynthesis by Multienzyme
      Complexes
          Helen A. Stafford
6.   Secondary Metabolites and Plant Systematics
          David S. Seigler

7. Stereochemical Aspects of Natural Products Biosynthesis
    Heinz G. Floss
8. Nonprotein Amino Acids
    L. Fowden
9. Amines
    T. A. Smith
10. Coumarins
    Stewart A. Brown
11. Phenolic Acids
    G. G. Gross
12. Enzymology of Alkaloid Metabolism in Plants and Microorganisms
    George R. Waller and Otis C. Dermer
13. Biosynthesis of Plant Quinones
    E. Leistner
14. Flavonoids
    Klaus Hahlbrock
15. Lignins
    Hans Grisebach
16. Cyanogenic Glycosides
    E. E. Conn
17. Glucosinolates
    Peder Olesen Larsen
18. Vegetable Tannins
    Edwin Haslam
19. The Betalains: Structure, Biosynthesis, and Chemical Taxonomy
    Mario Piattelli
20. Phenylalanine Ammonia-Lyase
    Kenneth R. Hanson and Evelyn A. Havir
21. Oxygenases and the Metabolism of Plant Products
    V. S. Butt and C. J. Lamb
22. Transmethylation and Demethylation Reactions in the Metabolism of
    Secondary Plant Products
    Jonathan E. Poulton
23. Glycosylation and Glycosidases
    Wolfgang Hösel
Index

Volume 8—Photosynthesis

1. Thylakoid Membrane and Pigment Organization
    R. G. Hiller and D. J. Goodchild
2. Photosynthetic Accessory Proteins with Bilin Prosthetic Groups
    A. N. Glazer
3. Primary Processes of Photosynthesis
    P. Mathis and G. Paillotin
4. Photosynthetic Electron Transport and Photophosphorylation
    Mordhay Avron
5. Photosynthetic Carbon Reduction Cycle
    S. P. Robinson and D. A. Walker
6. The C_4 Pathway
    G. E. Edwards and S. C. Huber
7. Crassulacean Acid Metabolism
    C. B. Osmond and J. A. M. Holtum

8.   The C_2 Chemo- and Photorespiratory Carbon Oxidation Cycle
      George H. Lorimer and T. John Andrews
9.   Chlorophyll Biosynthesis
      Paul A. Castelfranco and Samuel I. Beale
10.   Development of Photosynthetic Function during Chloroplast
      Biogenesis
      J. W. Bradbeer
11.   Light-Energy-Dependent Processes Other than CO_2
      Assimilation
      J. W. Anderson
Index

Volume 9—Lipids: Structure and Function

1.   Analysis and Structure Determination of Acyl Lipids
      Michael R. Pollard
2. β-Oxidation of Fatty Acids by Specific Organelles
      Helmut Kindl
3.   Oxidative Systems for Modification of Fatty Acids: The Lipoxygenase
      Pathway
      Brady A. Vick and Don C. Zimmerman
4.   Lipases
      Anthony H. C. Huang
5.   The Biosynthesis of Saturated Fatty Acids
      P. K. Stumpf
6.   Biochemistry of Plant Acyl Carrier Proteins
      John B. Ohlrogge
7.   Biosynthesis of Monoenoic and Polyenoic Fatty Acids
      Jan G. Jaworski
8.   Triacylglycerol Biosynthesis
      Sten Stymne and Allan Keith Stobart
9.   Galactolipid Synthesis
      Jacques Joyard and Roland Douce
10.   Sulfolipids
      J. Brian Mudd and Kathryn F. Kleppinger-Sparace
11.   Lipid-Derived Defensive Polymers and Waxes and Their Role in
      Plant–Microbe Interaction
      P. E. Kolattukudy
12.   Lipids of Blue-Green Algae (Cyanobacteria)
      N. Murata and I. Nishida
Index

Volume 11—Biochemistry of Metabolism

1.   Introduction: A History of the Biochemistry of Plant Respiration
      David D. Davies
2.   Control of Metabolism
      H. Kacser
3.   Enzyme Regulation
      Jacques Ricard
4.   The Regulation of Glycolysis and the Pentose Phosphate Pathway
      Les Copeland and John F. Turner

5. Control Involving Adenine and Pyridine Nucleotides
      Philippe Raymond, Xavier Gidrol, Christophe Salon, and Alain Pradet
6. Electron Transfer and Oxidative Phosphorylation in Plant Mitochondria
      Roland Douce, Renaud Brouquisse, and Etienne-Pascal Journet
7. Regulation of Mitochondrial Respiration
      Ian B. Dry, James H. Bryce, and Joseph T. Wiskich
8. Metabolism of Activated Oxygen Species
      Erich F. Elstner
9. Folate Biochemistry and the Metabolism of One-Carbon Units
      Edwin A. Cossins
Index

Volume 12—Physiology of Metabolism

I. Cellular Organization
1. Comparative Biochemistry of Plant and Animal Tubulins
      Peter J. Dawson and Clive W. Lloyd
2. Subcellular Transport of Metabolites in Plant Cells
      Hans Walter Heldt and Ulf Ingo Flügge
3. Compartmentation of Plant Metabolism
      T. ap Rees
4. The Role of Calcium in Metabolic Control
      E. F. Allan and A. J. Trewavas

II. The Metabolism of Organs and Tissues
5. Temperature and Metabolism
      Brian D. Patterson and Douglas Graham
6. Metabolic Responses to Stress
      David Rhodes
7. Plant Responses to Wounding
      Eric Davies
8. Fruit Ripening
      G. A. Tucker and D. Grierson
Index

Volume 13—Methodology

1. Immunochemistry for Enzymology
      Nicholas J. Brewin, David D. Davies, and Richard J. Robins
2. The Use of Mutants for the Study of Plant Metabolism
      P. McCourt and C. R. Somerville
3. The Use of Plant Cell Cultures in Studies of Metabolism
      M. C. J. Rhodes and Richard J. Robbins
4. The Application of Mass Spectrometry to Biochemical and Physiological
    Studies
      John A. Raven
5. NMR in Plant Biochemistry
      Justin K. M. Roberts
6. Electron Spin Resonance
      R. Cammack
Index